ROUGH
MULTIPLE OBJECTIVE
DECISION MAKING

ROUGH MULTIPLE OBJECTIVE DECISION MAKING

JIUPING XU
ZHIMIAO TAO

CRC Press
Taylor & Francis Group
Boca Raton   London   New York

CRC Press is an imprint of the
Taylor & Francis Group an **informa** business

A CHAPMAN & HALL BOOK

MATLAB® is a trademark of The MathWorks, Inc. and is used with permission. The MathWorks does not warrant the accuracy of the text or exercises in this book. This book's use or discussion of MATLAB® software or related products does not constitute endorsement or sponsorship by The MathWorks of a particular pedagogical approach or particular use of the MATLAB® software.

CRC Press
Taylor & Francis Group
6000 Broken Sound Parkway NW, Suite 300
Boca Raton, FL 33487-2742

First issued in paperback 2017

© 2012 by Taylor & Francis Group, LLC
CRC Press is an imprint of Taylor & Francis Group, an Informa business

No claim to original U.S. Government works

ISBN 13: 978-1-138-11271-1 (pbk)
ISBN 13: 978-1-4398-7235-2 (hbk)

Visit the Taylor & Francis Web site at
http://www.taylorandfrancis.com

and the CRC Press Web site at
http://www.crcpress.com

Preface

Since V. Pareto introduced the concept of optimal solution in 1896, multiple objective decision making (also called vector optimization decision making) has been found many important applications in practical decision-making problems, such as in management science, engineering design, transportation, and so on. To find a solution for multiple objective decision-making problems requires the intervention of a decision maker. The main idea is simple: the system generates reasonable alternatives, and the decision maker will make choices. Those choices are used to lead the algorithm to generate more alternatives until the decision maker will reach the solution that pleases him or her most. Helping decision makers to deal with multiple criteria decision and planning problems has been the subject of intensive studies since the 1970s, but many theoretical concepts were defined much earlier. In the 1970s, research focused on the theory of multiple objective mathematical programming and the development of procedures and algorithms for solving such problems. Many ideas originated from the theory of mathematical programming. There are several basic solution approaches such as the weighted-sum approach, the utility function approach, the compromise approach, and the lexicographic ordering approach. Based on the basic solution approach, many researchers presented some comprehensive approaches to multiple objective decision-making problems.

The rough set theory initialized by Z. Pawlak in 1982 has been proved to be an effective mathematical tool in dealing with the vague description of objects. A fundamental assumption is that any object from a universe is perceived through available information, and such information may not be sufficient to characterize the object exactly. A rough set is then defined by a pair of crisp sets, called the lower and upper approximations. Since the day when it was put forward, rough set theory has been applied in many fields such as artificial intelligence, expert systems, civil engineering, medical data analysis, data mining, pattern recognition, and decision theory. This book mainly concentrates on the application of rough set theory and rough approximation techniques to multiple objective decision making and systematically presents a state-of-the-art of multiple objective decision making based rough approximation in both techniques and applications. The application of rough set theory and rough approximation technique to multiple objective decision making includes rough approximation to feasible regions and the assumption that the parameters are rough intervals. For example, when the quantity of an unregulated water source is highly imprecise, it has to be estimated in terms of a decision maker's subjective experiences and objective historical data. An optimal management strategy capable of handling the problem under both normal and special conditions is often preferred. For this type of information, existing methods can neither reflect its dual-layer features nor entirely pass it to the resulting decisions. Two challenges thus emerge: one is to find an effective expression that could reflect dual-layer information (i.e., not only the parameter's most possible value but also its most reliable value), and the other is to use an appropriate method to generate decisions with dual-layer information directly corresponding to the most possible and reliable conditions of the system. Rough intervals can be a suitable concept to express such information. In this book, we apply rough theory to classic multiple objective decision making in two ways. One is using rough sets to approximate the feasible set, and the other is using rough intervals to approximate relative coefficients. For bilevel multiple objective decision making, we approximate some parameters with rough intervals.

In the classical multiple objective decision-making model, all data and information are assumed to be absolutely accurate, and the objectives and constraints are all assumed to be well expressed by mathematical formation. However, it is difficult to clearly describe the objective functions and constraints by mathematical equation in many realistic problems, and thus, the multiple objective decision-making model with certain parameters cannot deal with all real-life problems. The two kinds of uncertainties are randomness and fuzziness. In order to study the randomness and fuzziness of objects, two kinds of theories were produced successively in the history of mathematics, that is, the probability theory and the possibility theory. The probability theory, as a science, originated in the middle of the 17th century with Pascal, Fermat, and Huygens. The real history of the probability theory begins with the work of J. Bernoulli, and De Moivre. After that, S. Poisson, C. Gauss, P. Chebyshev, A. Markov, and A. Lyapunov made important contribution to this theory. The modern period in the development of the probability theory begins with its axiomatization. The first work in this direction was done by S. Bernstein, R. Mises, and E. Borel. In 1933, A. Kolmogorov presented the axiomatic system of probability theory that has become generally accepted and is not only applicable to all the classical branches of probability theory but also provides a firm foundation for the development of new branches that have arisen from questions in the sciences and involve infinite dimensional distribution. Since its introduction in 1965 by L. Zadeh, the fuzzy set theory has been well developed and applied in a wide variety of real problems. The term fuzzy variable was first introduced by S. Kaufmann, and then it was adopted by L. Zadeh and S. Nahmias. The possibility theory was proposed by L. Zadeh, and developed by many researchers such as D. Dubois and H. Prade.

Considering multiple objectives and uncertainties, we can employ the general model to formulate multiple objective uncertain multiple objective decision-making problems by

$$\begin{cases} \max[f_1(\mathbf{x},\boldsymbol{\xi}), f_2(\mathbf{x},\boldsymbol{\xi}), \cdots, f_m(\mathbf{x},\boldsymbol{\xi})] \\ \text{s.t.} \begin{cases} g_r(\mathbf{x},\boldsymbol{\xi}) \leq 0, r = 1, 2, \cdots, p \\ \mathbf{x} \in X \end{cases} \end{cases} \tag{1}$$

where $\mathbf{x} = (x_1, x_2, \cdots, x_n)^T$ is an n-dimensional decision vector, $\boldsymbol{\xi} = (\xi_1, \xi_2, \cdots, \xi_n)$ is a uncertain vector, $f_i(\mathbf{x}, \boldsymbol{\xi})$ are objective functions, $i = 1, 2, \cdots, m, g_r(\mathbf{x}, \boldsymbol{\xi}) \leq 0$ are uncertain constraints, $r = 1, 2, \cdots, p$, and X is a fixed set that is usually determined by a finite number of inequalities and equalities involving functions of $\mathbf{x}$. There exist several methods to deal with the uncertainties of (1). However, the existing methods have significant drawbacks from the viewpoint of information. For random multiple objective decision making, the existing methods may lead to loss of information due to statistical deviation. For fuzzy multiple objective decision making, the existing methods may lead to a loss of information due to subjective deviation. The application of rough theory to (1) can remedy these drawbacks.

This book takes real-life problems as the background and guidance and develops the general framework of rough multiple objective decision making, including the basic theory, model, and algorithm. In addition, the application of the multiple objective decision making model based on rough approximation to the real world are presented. The issues selected are for engineering applications, including construction site layout planning problem, water resource allocation problem, resource-constrained project scheduling problem, and earth-rock work allocation problem.

We shall now give a brief indication of the contents and organization of Rough Multiple Objective Decision making. Chapter 1, on elements of the rough set theory, including materials that will be familiar to most readers: basic concepts and properties of rough sets, rough membership, and rough intervals. These materials are the theoretical basis of our book, which will help readers understand the subsequent chapters better.

Chapter 2 deals with multiple objective rough decision making and its application to construction site layout planning problems. The multiple objective rough decision making referred to here includes two types of models: one in which the feasible set is approximated by the rough set, and the other in which the parameters are assumed as rough intervals. For the latter, expected value rough model using the expected value operator and the chance-constrained rough model, and the dependent-chance rough model using the chance operator are presented. For each model, we deduce the equivalent model of those models in special cases and propose a technique for the rough simulation-based genetic algorithms for general cases. A real application for the Longtan project is illustrated in the last section of Chapter 2.

The main objective in Chapter 3 is the bilevel multiple objective rough decision-making problem and its application to water resource allocation problems. The roughness of this class of model manifests in the rough intervals of parameters. We present three methods similar to the methods in Chapter 2. In some special cases, we obtain the equivalent models and solve them by traditional methods. For general cases, rough simulation-based tabu search algorithms are proposed. The Gan-Fu Plain water resource allocation is discussed in the last part of this chapter.

In order to apply the rough set theory to random multiple objective decision-making problems, we propose random multiple objective rough decision-making models in Chapter 4. This chapter first reviews the literatures on resource-constrained project scheduling problems. The next three sections introduce the random expected value rough model by the expected value operator, the random chance-constrained rough model, and the random dependent-chance rough model by the chance operator. In each section, we deduce the equivalent model for those problems and propose a technique of rough simulation-based multiobjective particle swarm optimization algorithm to deal with those problems. In the last section, the propose models and algorithms have been applied to solve the problems introduced in the first section.

Chapter 5 gives an introduction to random multiple objective rough decision-making. In this chapter, we consider the application of the rough set theory to fuzzy multiple objective decision-making problems called fuzzy multiple objective rough decision-making. The allocation problem literature is reviewed in the first section, and then the fuzzy multiple objective rough decision-making model is proposed. The next three sections introduce the fuzzy expected value rough model by the expected value operator, the fuzzy chance-constrained rough model, and the fuzzy dependent-chance rough model by the chance operator. In each section, we deduce the equivalent model for those problems and propose a simulation-based simulated annealing algorithm to deal with those problems. In the last section, the proposed models and algorithms are applied to solve the real problem introduced in the first section.

As conclusion, Chapter 6 provides the main problems, models, methods, and algorithms to obtain the organic systems and proposes the methodological system for the rough multiple objective decision making.

Additional material including MATLAB® code for some numerical examples will be posted in the Appendix.

This monograph has been supported by the National Natural Science Foundation of China (Grant No. 79760060, 70171021), the National Science Foundation for Distinguished Young Scholars, P. R. China (Grant No. 70425005), and the Key Program of National Natural Science Foundation of China (Grant No. 70831005). We are greatly indebted to a number professors, such as S. Wang and V. Kachitvichyanukul, from whom we have received much help in optimization theory and particle swarm optimization. For discussions and advice, the authors also thank researchers from the Uncertainty Decision-Making Laboratory of Sichuan University, particularly, X. Zhou, L. Yao, Z. Zhang, Y. Tu, Z. Li, Y. Ma, Z. Zeng, and C. Ding, who have done valuable work in this field and have made a number of corrections to this book. Finally, the authors express their deep gratitude to CRC Press, Taylor & Francis Group professional editorial staffs, especially Leong Li-Ming, Amber Donley, Sharma Aastha, Andrew Shih, Jim McGovern and the proofreader.

Sichuan University,   *Jiuping Xu*
December, 2010   *Zhimiao Tao*

List of Tables

x

Acronyms

DM	decision maker
MODM	multiple objective decision making
RMODM	rough multiple objective decision making
RI	rough interval
LA	lower approximation
UA	upper approximation
MORDM	multiple objective rough decision making
GA	genetic algorithm
EVRM	expected value rough model
L-EVRM	linear expected value rough model
NL-EVRM	nonlinear expected value rough model
RWGA	random weight genetic algorithm
CCRM	chance-constrained rough model
L-CCRM	linear chance-constrained rough model
NL-CCRM	nonlinear chance-constrained rough model
AWGA	adaptive weight genetic algorithm
DCRM	dependent-chance rough model
L-DCRM	linear dependent-chance rough model
NL-DCRM	nonlinear dependent-chance rough model
STGA	span tree genetic algorithm
BL-MORDM	bilevel multiple objective rough decision making
TS	tabu search
BL-EVRM	bilevel expected value rough model
BL-LEVRM	bilevel linear expected value rough model
BL-NLEVRM	bilevel nonlinear expected value rough model
ECTS	enhanced continuous tabu search
BL-CCRM	bilevel chance constrained rough model
BL-LCCRM	bilevel linear chance constrained rough model
BL-NLCCRM	bilevel nonlinear chance constrained rough model
PTS	parametric tabu search
BL-DCRM	bilevel dependent-chance rough model
BL-LDCRM	bilevel linear dependent-chance rough model
BL-NLDCRM	bilevel nonlinear dependent-chance rough model
RTS	reactive tabu search

Ra-MORDM	random multiple objective rough decision making
PSO	particle swarm optimization
Ra-EVRM	random expected value rough model
Ra-LEVRM	random linear expected value rough model
Ra-NLEVRM	random nonlinear expected value rough model
APSO	adaptive particle swarm optimization
Ra-CCRM	random chance-constrained rough model
Ra-LCCRM	random linear chance-constrained rough model
Ra-NLCCRM	random nonlinear chance-constrained rough model
Ra-DCRM	random dependent-chance rough model
Ra-LDCRM	random linear dependent-chance rough model
Ra-NLDCRM	random nonlinear dependent-chance rough model
Fu-MORDM	fuzzy multiple objective rough decision making
SA	simulated annealing
Fu-EVRM	fuzzy expected value rough model
Fu-LEVRM	fuzzy linear expected value rough model
Fu-NLEVRM	fuzzy nonlinear expected value rough model
ASA	adaptive simulated annealing
Fu-CCRM	fuzzy chance-constrained rough model
Fu-LCCRM	fuzzy linear chance-constrained rough model
Fu-NLCCRM	fuzzy nonlinear chance-constrained rough model
Fu-DCRM	fuzzy dependent-chance rough model
Fu-LDCRM	fuzzy linear dependent-chance rough model
Fu-NLDCRM	fuzzy nonlinear dependent-chance rough model
PSA	parallel simulated annealing

List of Figures

Contents

Chapter 1

Elements of Rough Set Theory

The rough set theory introduced by Z. Pawlak [240, 241] has often proved to be an excellent mathematical tool for the analysis of a vague description of objects (called actions in decision problems). The adjective vague, referring to the quality of information, means inconsistency or ambiguity that follows from information granulation. The rough sets philosophy is based on the assumption that, with every object of the universe, a certain amount of information (data, knowledge) is associated, expressed by means of some attributes used for object description. Objects having the same description are indiscernible (similar) with respect to the available information. The indiscernibility relation thus generated constitutes a mathematical basis of the rough sets theory; it induces a partition of the universe into blocks of indiscernible objects, called elementary sets, that can be used to build knowledge about a real or abstract world. The use of the indiscernibility relation results in information granulation.

Any subset X of the universe may be expressed in terms of these blocks either precisely (as a union of elementary sets) or only approximately. In the latter case, the subset X may be characterized by two ordinary sets, called lower and upper approximations. A rough set is defined by means of these two approximations, which coincide in the case of an ordinary set. The lower approximation of X is composed of all the elementary sets included in X (whose elements, therefore, certainly belong to X), while the upper approximation of X consists of all the elementary sets that have a nonempty intersection with X (whose elements, therefore, may belong to X). Obviously, the difference between the upper and lower approximation constitutes the boundary region of the rough set whose elements cannot be characterized with certainty as belonging or not to X using the available information. The information about objects from the boundary region is, therefore, inconsistent or ambiguous. The cardinality of the boundary region states, moreover, to what extent it is possible to express X in exact terms on the basis of the available information. For this reason, this cardinality may be used as a measure of vagueness of the information about X.

The rough set theory, dealing with representation and processing of vague information, presents a series of intersections and complements with respect to many other theories and mathematical techniques handling imperfect information, such as probability theory, fuzzy sets theory, discriminant analysis, and metrology. For the multiple objective decision making (MODM) problem, the input information is usually imprecise and insufficient. If we apply rough set to deal with the information, the decision-making process can be more flexible and practical. This chapter introduces the foundations of rough set theory for MODM, including basic concepts and properties of rough sets, rough membership, and rough intervals.

1.1   Basic Concepts and Properties of Rough Sets

In this section we introduce some basic concepts and properties of rough sets, including knowledge and knowledge base.

1.1.1   Knowledge and Knowledge Base

Suppose we are given a set U of objectives we are interested in. We call U a *universe*. In the classic rough set theory [241], the universe is assumed as finite and nonempty, but the universes referred to in this book are general sets, including empty set and infinite set. For a universe U, any subset $X \subseteq U$ will be called a concept or category in U, and any family of concepts in U will be referred to as *abstract knowledge*, or in short, *knowledge*.

Among the various concepts, the most interesting ones are that form a partition of a certain universe U, that is, in the family $\Xi = \{X_1, X_2, \cdots\}$ such that $X_i \subseteq U, X_i \neq \emptyset, X_i \cap X_j = \emptyset$ for $i \neq j$ and $\bigcup X_i = U$. If the family Ξ includes finite elements, it is just the case presented in [241]. A family of partitions over U is called a *knowledge base* over U. Knowledge base represents classification skills.

For mathematical reasons, we shall often use equivalence relation instead of partition. For universe U, a subset $R \subseteq U \times U$ is called a relation over U. If there are $x, y \in U$ such that $(x, y) \in R$, we say x and y have relation R, denoted by xRy. Furthermore, R is called equivalence relation over U if it satisfies (i) $\forall x \in U, xRx$ (reflexivity); (ii) $\forall x, y \in U, xRy \Rightarrow yRx$ (symmetry); (iii) $\forall x, y, z \in U, xRy$ and $yRz \Rightarrow xRz$ (transitivity).

The following two theorems shows the relationship between the partition and equivalence relation over U.

THEOREM 1.1
Let U be a universe, and Ξ a partition over U. Then Ξ determines an equivalence relation over U.

PROOF   Set $\Xi = \{X_1, X_2, \cdots, X_n\}$ is a partition over U, and $xRy \Leftrightarrow x \in X_{i_0}, y \in X_{i_0}, i_0 \in \{1, 2, \cdots, n\}$. It is obvious that xRx for any x, that is R satisfies reflexivity. If xRy, that is, there exists $i_0 \in \{1, 2, \cdots, n\}$ such that $x \in X_{i_0}$ and $y \in X_{i_0}$, then $y \in X_{i_0}$ and $x \in X_{i_0}$, which implies R satisfies symmetry. If xRy and yRz, that is, there exists $i_0 \in \{1, 2, \cdots, n\}$, such that $x \in X_{i_0}, y \in X_{i_0}$ and $z \in X_{i_0}$, then xRz, that is, R satisfies transitivity. Thus R is an equivalence relation over U. The theorem is proved.

THEOREM 1.2

Let U be a universe, and R an equivalence relation over U. Then R determines a partition over U.

PROOF   For $x \in U$, we set $[x]_R = \{y \in U | xRy\}$, where R is a equivalence relation over U. For any $[x]_R$, it follows from xRx that there exist $x \in [x]_R$, that is, $[x]_R \neq \Phi$ and it is clear that $\bigcup [x]_R = U$.

If $[x]_R \cap [y]_R \neq \Phi$, then there exist $z \in [x]_R \cap [y]_R$, that is, xRz, yRz. By symmetry and transitivity of equivalence relation, we have xRy. For any $a \in [x]_R$, that is, xRa, then we have yRa, that is, $a \in [y]_R$, which implies $[x]_R \subseteq [y]_R$. Similarly, for any $b \in [y]_R$, that is, yRb, then we have xRb, that is $b \in [x]_R$, which implies $[y]_R \subseteq [x]_R$. Thus $[x]_R = [y]_R \neq \Phi$ holds. Hence $\{[x]_R | x \in U\}$ is a partition over universe U. The theorem is completed.

From the foregoing two theorems, we know that the two concepts are mutually interchangeable.

If R is a equivalence relation over U, then by U/R we mean the family of all the equivalence classes of R referred to as categories or concepts of R, and $[x]_R$ denotes a category in R containing an element $x \in U$. By a knowledge base we can understand as a relational system, $K = (U, \mathbf{R})$, where $U \neq \Phi$ is a universe and $\mathbf{R}$ is a family of equivalence relations over U. If $\mathbf{P} \subseteq \mathbf{R}$ and $\mathbf{P} \neq \Phi$, then intersection of all equivalence relations belonging to $\mathbf{P}$, denoted by $\bigcap \mathbf{P}$, is also an equivalence relation and will be called an *indiscernibility relationship* over $\mathbf{P}$, denoted by $ind(\mathbf{P})$ and $[x]_{ind\mathbf{P}} = \bigcap_{R \in \mathbf{P}} [x]_R$. Thus $U/ind(\mathbf{P})$ denotes knowledge associated with the family of equivalence relation $\mathbf{P}$, called $\mathbf{P}-$ *basic knowledge* about U in K. For simplicity of notation, we write $U/\mathbf{P}$ instead of $U/ind(\mathbf{P})$, and $\mathbf{P}$ will be also called $\mathbf{P}-$ *basic knowledge* provided it does not cause confusion. Equivalence classes of $ind(\mathbf{P})$ are called *basic concepts of knowledge $ind(\mathbf{P})$*. The *basic concepts* are fundamental building blocks of our knowledge, basic properties of the universe that can be expressed employing knowledge $\mathbf{P}$. Particularly, if $Q \in \mathbf{R}$, then Q will be called Q-*elementary knowledge* about U in K, and equivalence classes of Q are referred to as Q-*elementary concepts* of knowledge $\mathbf{R}$. Thus a basic concept is an intersection of some elementary concepts. Every union of basic $\mathbf{P}$-basic concepts will be called $\mathbf{P}$-concept.

Example 1.1

Given a set of persons as

$$U = \{x_1, x_2, x_3, x_4, x_5, x_6, x_7, x_8, x_9, x_{10}\},$$

assume that these persons are of different ages (*young, middle-aged, old*), figures (*thin, moderate, fat*), and sex (*male, female*). For instance, a person can be young, thin and male, or old, fat and female, etc.

TABLE 1.1   A set of persons

	Age			Figure			Sex	
	Young	Middle-aged	Old	Thin	Moderate	Fat	Male	Female
x_1	Yes			Yes				Yes
x_2		Yes			Yes		Yes	
x_3	Yes					Yes		Yes
x_4		Yes				Yes		Yes
x_5			Yes	Yes				Yes
x_6			Yes		Yes			Yes
x_7	Yes					Yes	Yes	
x_8		Yes				Yes	Yes	
x_9	Yes			Yes			Yes	
x_{10}	Yes			Yes			Yes	

Thus the set of persons U can be classified according to age, figure, and sex, for instance, as shown by Table 1.1.

In other words, by these classifications we define three equivalence relations, R_1, R_2, R_3, having equivalence classes as

$$U/R_1 = \{\{x_1, x_3, x_7, x_9, x_{10}\}, \{x_2, x_4\}, \{x_5, x_6, x_8\}\},$$
$$U/R_2 = \{\{x_1, x_5, x_9, x_{10}\}, \{x_2, x_6\}, \{x_3, x_4, x_7, x_8\}\}$$

and

$$U/R_3 = \{\{x_2, x_7, x_8, x_9, x_{10}\}, \{x_1, x_3, x_4, x_5, x_6\}\},$$

which are elementary concepts in our knowledge base $K = \{U, \{R_1, R_2, R_3\}\}$. Basic concepts are intersection of elementary concepts. For example, set

$$\{x_1, x_3, x_7, x_9, x_{10}\} \cap \{x_3, x_4, x_7, x_8\} = \{x_3, x_7\}, \{x_2, x_4\} \cap \{x_2, x_6\} = \{x_2\}$$

and

$$\{x_5, x_6, x_8\} \cap \{x_3, x_4, x_7, x_8\} = \{x_8\}$$

are (R_1, R_2)-basic concepts, that is, young and fat, middle-aged and moderate, and old and fat, respectively. The sets

$$\{x_1, x_3, x_7, x_9, x_{10}\} \cap \{x_3, x_4, x_7, x_8\} \cap \{x_2, x_7, x_8, x_9, x_{10}\} = \{x_7\},$$
$$\{x_2, x_4\} \cap \{x_2, x_6\} \cap \{x_2, x_7, x_8, x_9, x_{10}\} = \{x_2\}$$

and

$$\{x_5, x_6, x_8\} \cap \{x_3, x_4, x_7, x_8\} \cap \{x_2, x_7, x_8, x_9, x_{10}\} = \{x_8\}$$

are (R_1, R_2, R_3)-basic concepts, young and fat and female, middle-aged and moderate and female, and old and fat and female, respectively. The sets

$$\{x_1, x_3, x_7, x_9, x_{10}\} \cup \{x_2, x_4\}\} = \{x_1, x_2, x_3, x_4, x_7, x_9, x_{10}\},$$
$$\{x_2, x_4\} \cup \{x_5, x_6, x_8\} = \{x_2, x_4, x_5, x_6, x_8\}$$

and

$$\{x_1,x_3,x_7,x_9,x_{10}\} \cup \{x_5,x_6,x_8\}\} = \{x_1,x_2,x_3,x_5,x_6,x_7,x_8,x_9,x_{10}\}$$

are R_1-concepts, young or middle-aged (not old), middle-aged or old (not young), and young or old (not middle-aged).

Note that some concepts are not available in this knowledge base, for instance, the sets

$$\{x_2,x_4\} \cap \{x_1,x_5,x_9,x_{10}\}\} = \emptyset$$

and

$$\{x_1,x_3,x_7,x_9,x_{10}\} \cap \{x_2,x_6\}\} = \emptyset,$$

which means that concepts middle-aged and thin, and young and moderate do not exist on our knowledge base, that is, they are empty concepts.

1.1.2 Approximation and Rough Set

Let U be a universe, and let R be an equivalence relation on U. The pair $A = (U,R)$ will be called an approximation space. If $x,y \in U$ and $(x,y) \in R$, we say that x and y are indistinguishable in A.

Let (U,R) be an approximation space and $X \subseteq U$. We will say that X is R-definable if X is the union of some elementary concepts; otherwise X is R-undefinable. The R-definable sets will be also called R-exact sets and R-undefinable will also be said to be R-inexact or R-rough sets. A rough set can be defined approximately by two exact sets, referred to as the lower and upper approximation of the set.

Suppose we are given an approximation space (U,R), with each subset $X \subseteq U$. We associate the two subsets as

$$\underline{R}X = \bigcup\{Y \in U/R | Y \subseteq X\}$$

and

$$\overline{R}X = \bigcup\{Y \in U/R | Y \cap X \neq \emptyset\},$$

which are called the R-lower approximation and R-upper approximation of X, respectively. The lower approximation and upper approximation can be also presented in an equivalent forms as

$$\underline{R}X = \{x \in U | [x]_R \subseteq X\}$$

and

$$\overline{R}X = \{x \in U | [x]_R \cap X \neq \emptyset\}.$$

The set $bn_R = \overline{R}X - \underline{R}X$ is called the R-boundary of X; $pos_R = \underline{R}X$ is called the R-positive region of X; $neg_R = U - \overline{R}X$ is called R-negative region of X. It is obvious that $\overline{R}X = pos_R \cup bn_R$. The positive region pos_R or the lower approximation of X is the collection of those objects that can be classified with full certainty as members of the set X, using the equivalence relation R. The upper approximation of X consists of objects regarding which R does not allow us to exclude the possibility that those

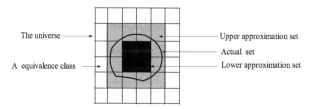

FIGURE 1.1   Rough set.

objects may belong to X. The boundary of X is the undecidable area of the universe, that is, none of the objects belonging to the boundary can be classified with certainty into X or $-X$ as far as R is concerned. The negative region neg_R is the collection of those objects with which it can be determined without ambiguity, employing R, that they do not belong to x, that is, they belong to the complement of X. Figure 1.1 shows the notion of an upper and lower approximation in a two-dimensional approximation space consisting of a rectangle partitioned into elementary squares. It follows from the definition that X is definable $\Leftrightarrow \underline{R}X = \overline{R}X$, and X is R-rough $\Leftrightarrow \underline{R}X \neq \overline{R}X$.

Let us also observe that $\underline{R}X$ is the maximal R-definable set contained in X, whereas $\overline{R}X$ is the minimal R-definable set containing X. The concept of approximation allows us speak precisely about imprecise notations.

Directly from the definition of approximation we can get the following properties of R-lower approximation and R-upper approximation of X.

PROPOSITION 1.1
(Z. Pawlak [241]) Let (U,R) be an approximation space, $X,Y \subseteq U$. Then

1. $\underline{R}X \subseteq X \subseteq \overline{R}X$;
2. $\underline{R}\emptyset = \overline{R}\emptyset = \emptyset, \underline{R}U = \overline{R}U = U$;
3. $\underline{R}(X \cup Y) = \underline{R}X \cup \underline{R}Y$;
4. $\overline{R}(X \cap Y) = \overline{R}X \cap \overline{R}Y$;
5. $X \subseteq Y \Rightarrow \underline{R}X \subseteq \underline{R}Y$;
6. $X \subseteq Y \Rightarrow \overline{R}X \subseteq \overline{R}Y$;
7. $\underline{R}(X \cup Y) \supseteq \underline{R}X \cup \underline{R}Y$;
8. $\overline{R}(X \cap Y) \subseteq \overline{R}X \cap \overline{R}Y$;
9. $\underline{R}(-X) = -\overline{R}X$;
10. $\overline{R}(-X) = -\underline{R}X$;
11. $\underline{R}(\underline{R}X) = \overline{R}(\underline{R}X) = \underline{R}X$;
12. $\overline{R}(\overline{R}X) = \underline{R}(\overline{R}X) = \overline{R}X$.

Let us define two membership functions $\underline{\in}_R, \overline{\in}_R$ (called strong membership and weak membership, respectively), as

$$x\underline{\in}_R X \Leftrightarrow x \in \underline{R}X, x\overline{\in}_R X \Leftrightarrow x \in \overline{R}X.$$

If $x \underline{\in}_R X$, we say that "X surely belongs to X according to R", while $x \overline{\in}_R X$ means that "X possibly belongs to X in according to R."

Directly from the definition of strong and weak membership we can get the following properties of both.

PROPOSITION 1.2

(Z. Pawlak [241]) Let (U,R) be an approximate space, $X \subseteq U$. Then the strong membership and weak membership $\underline{\in}_R, \overline{\in}_R$ has the following properties:

1. $x \underline{\in}_R X \Rightarrow x \in X \Rightarrow x \overline{\in}_R X$;
2. $x \overline{\in}_R (X \cup Y) \Leftrightarrow x \overline{\in}_R X \ or \ x \overline{\in}_R Y$;
3. $x \underline{\in}_R (X \cap Y) \Leftrightarrow x \underline{\in}_R X \ and \ x \underline{\in}_R Y$;
4. $x \underline{\in}_R X \ or \ x \underline{\in}_R Y \Rightarrow x \underline{\in}_R (X \cup Y)$;
5. $x \overline{\in}_R (X \cap Y) \Rightarrow x \overline{\in}_R X \ and \ x \overline{\in}_R Y$;
6. $x \underline{\in}_R (-X) \Leftrightarrow x \overline{\in}_R X \ does \ not \ hold$;
7. $x \overline{\in}_R (-X) \Leftrightarrow x \underline{\in}_R X \ does \ not \ hold$.

In order to know the degree of the upper and lower approximation describing the set X, the concept of the *accuracy* of approximation is proposed by S. Greco et al. [122]:

$$\alpha_R(X) = \frac{|\underline{R}X|}{|\overline{R}X|} \tag{1.1}$$

where $X \neq \emptyset$, $|\cdot|$ expresses the cardinal number of the set X when X is a finite set; otherwise it expresses the Lebesgue measure.

Another ratio defines a *quality* of the approximation of X by means of the attributes from R according to S. Greco et al. [122]:

$$\gamma_R(X) = \frac{|\underline{R}X|}{|X|} \tag{1.2}$$

The quality $\gamma_R(X)$ represents the relative frequency of the objects correctly classified by means of the attributes from R.

REMARK 1.1   For any set A we can represents its frequency of the objects correctly approximated by $(\underline{R}X, \overline{R}X)$ as

$$\beta_R(A) = \frac{|\underline{R}X \cap A|}{|\overline{R}X \cap A|} \tag{1.3}$$

If $\underline{R}X \subseteq A \subseteq \overline{R}X$, namely, A has the upper approximation $\overline{R}X$ and the lower approximation $\underline{R}X$, we have that $\beta_R(A)$ degenerates to the *quality* $\gamma_R(A)$ of the approximation.

As we know, the *quality* $\gamma_R(A)$ of the approximation describes the frequency of A, and when $\gamma_R(A) = 1$, we only have $|A| = |X|$, namely, the set A is well approximated by the lower approximation. If we want to make A be a definable set, it must be that both $\gamma_R(A) = 1$ and $\alpha_R(X) = 1$ hold. So we could make use of the so-called approximation function (which is also called approximation measure) to combine them together. Let $(\underline{R}X, \overline{R}X)$ be a rough set under the equivalence relation R, and A be any set satisfying $\underline{R}X \subseteq A \subseteq \overline{R}X$. Then we define the approximation function as follows, expressing the relative frequency of the objects of A correctly classified into $(\underline{R}X, \overline{R}X)$:

$$\text{Appr}_R(A) = 1 - \eta(1 - \frac{|A|}{|\overline{R}X|}) \tag{1.4}$$

where η is predetermined by the decision maker's preference.

From (1.4) we know that $\frac{|A|}{|\overline{R}X|}$, which keeps accord with $\gamma_R(A)$, describes the relative frequency of the objects correctly classified by R from the view of the upper approximation $\overline{R}X$. Obviously, $\text{Appr}_R(A)$ is a number between 0 and 1 and is increasing along with increase of $|A|$. The extreme case $\text{Appr}_R(A) = 1$ means that $|A| = |\overline{R}X|$, namely, A is completely described by $\overline{R}X$.

LEMMA 1.1
Let $(\underline{R}X, \overline{R}X)$ be a rough set under the equivalence relation R, A a set satisfying $\underline{R}X \subseteq A \subseteq \overline{R}X$. Then

$$Appr_R(A) = \frac{\eta \alpha_R(A) + (1 - \eta)\gamma_R(A)}{\gamma_R(A)} \tag{1.5}$$

PROOF   Since $\underline{R}X \subseteq A \subseteq \overline{R}X$, it means that A has the lower approximation $\underline{X}$ and the upper approximation $\overline{R}X$, and it follows from S. Greco et al. [122] that

$$\alpha_R(A) = \frac{|\underline{R}X|}{|\overline{R}X|}, \quad \gamma_R(A) = \frac{|\underline{R}X|}{|A|}.$$

Thus,

$$\frac{|A|}{|\overline{R}X|} = \frac{\alpha_R(A)}{\gamma_R(A)}.$$

It follows that

$$\text{Appr}_R(A) = 1 - \eta(1 - \frac{|A|}{|\overline{R}X|})$$
$$= 1 - \eta(1 - \frac{\alpha_R(A)}{\gamma_R(A)})$$
$$= \frac{\eta \alpha_R(A) + (1 - \eta)\gamma_R(A)}{\gamma_R(A)}.$$

This completes the proof.

LEMMA 1.2
Let $(\underline{R}X, \overline{R}X)$ be a rough set on the finite universe under the equivalence relation R, A be any set satisfying $\underline{R}X \subseteq A \subseteq \overline{R}X$, and $\eta \in (0,1)$. Then $Appr_R(A) = 1$ holds if and only if $\underline{R}X = A = \overline{R}X$.

PROOF If $\underline{R}X = A = \overline{R}X$ holds, then it is obvious that $\text{Appr}_R(A) = 1$ according to Definition 1.6. Let us prove the necessity of the condition.

Conversely, if $\text{Appr}_R(A) = 1$ holds for any A satisfying $\underline{R}X \subseteq A \subseteq \overline{R}X$, it follows from Lemma 1.1 that

$$\frac{\eta\,\alpha_R(A) + (1-\eta)\,\gamma_R(A)}{\gamma_R(A)} = 1 \Rightarrow \alpha_R(A) = \gamma_R(A) \Rightarrow |\overline{R}X| = |A|$$

for $0 < \eta \le 1$.

Since $A \subseteq \overline{R}X$ and the universe is finite, we have that $A = \overline{R}X$. Because A is any set satisfying $\underline{R}X \subseteq A \subseteq \overline{R}X$, let $A = X$, then we have $X = \overline{R}X$. It follows from the property proposed by Z. Pawlak [239] that $\underline{R}X = X = \overline{R}X$. Thus, we have $\underline{R}X = A = \overline{R}X$.

Lemma 1.1 shows that the approximation function Appr inherits the accuracy and quality of the approximation, and extends it to the relationship between any set A and the rough set $(\underline{R}X, \overline{R}X)$. Lemma 1.2 shows that the approximation function is complete and well describes the property in traditional rough set theory, and describes the property only by one index.

LEMMA 1.3

Let $(\underline{R}X, \overline{R}X)$ be a rough set on the infinite universe under the equivalence relation R, A a set satisfying $\underline{R}X \subseteq A \subseteq \overline{R}X$, and $\eta \in (0,1)$. If $\text{Appr}_R(A) = 1$, then there exists an equivalence relation R^* such that $|\underline{R}X| = |A| = |\overline{R}X|$, where $|\cdot|$ expresses the Lebesgue measure.

PROOF According to Lemma 1.2, we know that $|A| = |\overline{R}X|$ must hold. Let $\underline{R}X = \overline{R}X / \partial\overline{R}X$ under the equivalence relation R^*, where $\partial\overline{R}X$ is composed by all the elements such that $|\partial\overline{R}X| = 0$, namely, the measure of $\partial\overline{R}X$ is 0. Next, we will prove that $\overline{R}X / \partial\overline{R}X \subseteq A$. The argument breaks down into two cases:

1. If $|\overline{R}X| = 0$, then $\overline{R}X / \partial\overline{R}X = \emptyset$. Thus, $|\underline{R}X| = |A| = |\overline{R}X| = 0$.

2. If $|\overline{R}X| \neq 0$, we only need to prove that, for any $x^0 \in \overline{R}X / \partial\overline{R}X$, $x^0 \in A$. In fact, when $x^0 \in \overline{R}X / \partial\overline{R}X$, then $x^0 \in int(\overline{R}X)$ holds, where $int(\overline{R}X)$ is the internal part of $\overline{R}X$. It follows that there exists $r > 0$ such that $N(x^0, r) \subset int(\overline{R}X)$ and $|N(x^0, r)| > 0$. There exist four cases describing the relationship between A and $N(x^0, r)$.

Case 1. $A \cap N(x^0, r) = \Phi$ (see Figure 1.2). Since $N(x^0, r) \subset int(\overline{R}X) \subset \overline{R}X$ and $A \subseteq \overline{R}X$, we have that

$$|\overline{R}X| \ge |N(x^0, r) \cup A| = |N(x^0, r)| + |A|.$$

This conflicts with $|A| = |\overline{R}X|$.

Case 2. $A \cap N(x^0, r) = P$, where the set P includes countable points (see Figure 1.3). Obviously, we have $|P| = 0$, thus $|N(x_0, r)/P| = |N(x_0, r)| > 0$.

FIGURE 1.2   Apartment.

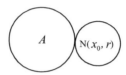

FIGURE 1.3   Tangent.

Then we have

$$|\overline{R}X| \geq |N(x^0,r) \cup A| = |N(x^0,r)/P| + |A|.$$

This also conflicts with $|A| = |\overline{R}X|$.

Case 3. $A \cap N(x^0,r) = P'$, where $P' \subset N(x^0,r)/\{x_0\}$. As Figure 1.4 shows, we can divide it into three parts, namely, $(N(x^0,r)/P) = P' \cup \{x_0\} \cup T$, where P', T, and $\{x_0\}$ do not have common element. Then $|T| > 0$, it follows that

$$|\overline{R}X| \geq |N(x^0,r) \cup A| = |T| + |A|.$$

This also conflicts with $|A| = |\overline{R}X|$.

Case 4. $A \supset (N(x^0,r)/x_0)$ (see Figure 1.5). This means that, for any $x_0 \in int(A)$, $x_0 \notin A$. It follows that $A \cap int(A) = \Phi$, then we have

$$|\overline{R}X| \geq |int(A) \cup A| = |int(A)| + |A|.$$

This also conflicts with $|A| = |\overline{R}X|$. In the foregoing, we can get $\overline{R}X/\partial\overline{R}X \subseteq A$. Thus, there exists the lower approximation $\underline{R}X = \overline{R}X/\partial\overline{R}X$ such that $\underline{R}X \subseteq A \subseteq \overline{R}X$ under the equivalence relation $R*$.

REMARK 1.2   In fact, we can extend approbation function to more

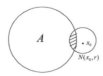

FIGURE 1.4   Intersection.

FIGURE 1.5   Inclusion.

general set. When $\underline{R}X \subseteq A \subseteq \overline{R}X$, we have the following equivalent formula:

$$
\begin{aligned}
\mathrm{Appr}_R(A) &= 1 - \eta\left(1 - \tfrac{|A|}{|\overline{R}X|}\right) \\
&= \tfrac{|A \cap \underline{R}X|}{|\underline{R}X|}\left(1 - \eta\left(1 - \tfrac{|A \cap \overline{R}X|}{|\overline{R}X|}\right)\right) \\
&= \tfrac{|A \cap \underline{R}X|}{|\underline{R}X|} + \eta\left(\tfrac{|A \cap \overline{R}X|}{|\overline{R}X|} - \tfrac{|A \cap \underline{R}X|}{|\underline{R}X|}\right).
\end{aligned}
$$

Furthermore, we get the concept of the approximation function for any set A. Let $(\underline{X}, \overline{X})$ be the rough set generated by X under the equivalence relation R, for any set A. The approximation function of event A by $(\underline{R}X, \overline{R}X)$ is defined as

$$
\mathrm{Appr}_R(A) = \frac{|A \cap \underline{R}X|}{|\underline{R}X|} + \eta\left(\frac{|A \cap \overline{R}X|}{|\overline{R}X|} - \frac{|A \cap \underline{R}X|}{|\underline{R}X|}\right) \tag{1.6}
$$

where η is a given parameter predetermined by the decision maker's preference.

From (1.6) we know that $\mathrm{Appr}_R(A)$ expresses the relationship between the set A and the set $(\underline{R}X, \overline{R}X)$ generated by X, that is, the frequency of A correctly classified into $(\underline{R}X, \overline{R}X)$ according to the similarity relationship R. It has the internal link with the *accuracy* α_R of the approximation and the *quality* γ_R of the approximation to some extent. α_R expresses the degree of the upper and lower approximation describing the set X. $\gamma_R(X)$ represents the relative frequency of the objects correctly classified by means of the attributes from R. Then Appr_R combines both of them together and considers the level at which A has the attributes correctly classified by $(\underline{R}X, \overline{R}X)$ for any A.

LEMMA 1.4
Let $(\underline{R}X, \overline{R}X)$ be a rough set, and A a set. Then the following formula holds:

$$
\mathrm{Appr}_R = \begin{cases}
1, & \text{if } A \supseteq \overline{R}X \\
1 - \eta\left(1 - \tfrac{\alpha_R(A)}{\gamma_R(A)}\right), & \text{if } \underline{R}X \subset A \subset \overline{R}X \\
\tfrac{1 - \eta(1 - \alpha_R(A))}{\gamma_R(A)}, & \text{if } A \subseteq \underline{R}X \\
0, & \text{if } A \cap \overline{R}X = \Phi \\
\tfrac{|A \cap \overline{R}X|}{|\overline{R}X|}\left(\tfrac{\beta_R(A)}{\alpha_R(A)} + \eta\left(1 - \tfrac{\beta_R(A)}{\alpha_R(A)}\right)\right), & \text{otherwise}
\end{cases} \tag{1.7}
$$

PROOF   We prove the five cases one by one.

1. If $A \supseteq \overline{R}X$, we have that $A \cap \underline{R}X = \underline{R}X$ and $A \cap \overline{R}X = \overline{R}X$. Then $\mathrm{Appr}_R = 1$.
2. If $\underline{R}X \subseteq A \subseteq \overline{R}X$, then $A \cap \underline{R}X = \underline{R}X$ and $A \cap \overline{R}X = X$. It follows that $\mathrm{Appr}_R = 1 - \eta(1 - \frac{|A|}{|\overline{R}X|})$.
3. If $A \subset \underline{R}X$, then $A \cap \underline{R}X = A$ and $A \cap \overline{R}X = A$. Thus, $\mathrm{Appr}_R = \frac{1 - \eta(1 - \alpha_R(A))}{\gamma_R(A)}$.
4. If $A \cap \overline{R}X = \emptyset$, then $A \cap \underline{R}X = \emptyset$ and $A \cap \overline{R}X = \emptyset$. So, $\mathrm{Appr}_R = 0$.
5. For the others, we have

$$
\begin{aligned}
\mathrm{Appr}_R(A) &= \frac{|A \cap \underline{R}X|}{|\underline{R}X|} + \eta \left(\frac{|A \cap \overline{R}X|}{|\overline{R}X|} - \frac{|A \cap \underline{R}X|}{|\underline{R}X|} \right) \\
&= \frac{|A \cap \overline{R}X|}{|\overline{R}X|} \left(\frac{|A \cap \underline{R}X|}{|\underline{R}X|} \cdot \frac{|\overline{R}X|}{|A \cap \overline{R}X|} + \eta \left(1 - \frac{|A \cap \underline{R}X|}{|\underline{R}X|} \cdot \frac{|\overline{R}X|}{|A \cap \overline{R}X|} \right) \right) \\
&= \frac{|A \cap \overline{R}X|}{|\overline{R}X|} \left(\frac{\beta_R(A)}{\alpha_R(A)} + \eta \left(1 - \frac{\beta_R(A)}{\alpha_R(A)} \right) \right).
\end{aligned}
$$

This completes the proof.

For other purposes, we can discuss the extreme case as follows.

REMARK 1.3   When $\eta = 1$, $\mathrm{Appr}_R(A) = \frac{|A \cap \overline{R}X|}{|\overline{R}X|}$. It means that the decision maker only considers the level at which A includes the frequency of A correctly classified into $\underline{R}X$ according to the equivalence relation R.

REMARK 1.4   When $\eta = 0$, $\mathrm{Appr}_R(A) = \frac{|A \cap \overline{R}X|}{|\overline{R}X|}$. It means that the decision maker only considers the level at which A includes the frequency of A correctly classified into $\overline{R}X$ according to the similarity relationship R.

1.1.3   Rough Equality and Rough Inclusion of Sets

Three concepts of rough equality of sets will be introduced in this section. Let $K = (U, R)$ be an approximation space , and let $X, Y \subseteq U$. We say that
1. X and Y are roughly lower-equal if and only if $\underline{R}X = \underline{R}Y$, denoted by $X \widetilde{\sim}_R Y$;
2. X and Y are roughly upper-equal if and only if $\overline{R}X = \overline{R}Y$, denoted by $X \backsimeq_R Y$;
3. X and Y are roughly equal if and only if $X \widetilde{\sim}_R Y$ and $X \backsimeq_R Y$, denoted by $X \approx_R Y$.
   It is clear that $X \widetilde{\sim}_R Y$ and $X \backsimeq_R Y, X \approx_R Y$ are equal relations. It is easy to obtain the properties of $\widetilde{\sim}_R, \backsimeq_R, \approx_R$.

LEMMA 1.5
(Z. Pawlak [240]) Let $K = (U, R)$ be an approximation space. Then the following properties hold:
   1. $X \widetilde{\sim}_R Y \Leftrightarrow (X \cap Y) \widetilde{\sim}_R X$ and $(X \cap Y) \widetilde{\sim}_R X$;
   2. $X \backsimeq_R Y \Leftrightarrow (X \cup Y) \backsimeq_R X$ and $(X \cup Y) \backsimeq_R X$;
   3. If $X \backsimeq_R X'$ and $Y \backsimeq_R Y'$, then $X \cup Y \backsimeq X' \cup Y'$;
   4. If $X \widetilde{\sim}_R X'$ and $Y \widetilde{\sim}_R Y'$, then $X \cap X' \widetilde{\sim}_R Y \cap Y'$;
   5. If $X \backsimeq_R Y$, then $X \cup -Y \backsimeq_R U$;

6. If $X \widetilde{\approx}_R Y$, then $X \cap Y \widetilde{\approx}_R \emptyset$;
7. If $X \subseteq Y$ and $Y \simeq_R \emptyset$, then $X \subseteq \emptyset$;
8. If $X \subseteq Y$ and $X \simeq_R U$, then $Y \subseteq U$;
9. $X \simeq_R Y \Leftrightarrow -X \simeq_R -Y$;
10. If $X \widetilde{\approx}_R \emptyset$ or $Y \widetilde{\approx}_R \emptyset$, then $X \cap Y \widetilde{\approx}_R \emptyset$;
11. If $X \subseteq U$ or $Y \subseteq U$, then $X \cup Y \subseteq_R U$.

THEOREM 1.3

Let $K = (U, R)$ be an approximation space, and $X \subseteq U$. Then
   1. $\underline{R}X = \bigcap Y$, where Y satisfies $Y \subseteq U$ and $X \widetilde{\approx}_R Y$;
   2. $\overline{R}X = \bigcup Y$, where Y satisfies $Y \subseteq U$ and $X \subseteq_R Y$.

PROOF   We prove the two conclusion one by one.

1. It follows from $\underline{R}X = \underline{R}Y \subseteq Y$ that $\underline{R}X \subseteq \bigcap Y$. Suppose that $\underline{R}X \neq \bigcap Y$, then there exists a $x_0 \in \underline{R}X$, and $Y_0 \subseteq U$, where $\underline{R}X = \underline{R}Y_0$, such that $x \notin Y_0$. Note that $\underline{R}X = \underline{R}Y \subseteq Y_0$. Then $x \notin \underline{R}X$. A contradiction implies this proof.

2. It follows from $\overline{R}X = \overline{R}Y \supseteq Y$ that $\overline{R}X \supseteq \bigcup Y$. Suppose that $\overline{R}X \neq \bigcup Y$, then there exists a Y_0 with $\overline{R}X = \overline{R}Y_0$ and $y_0 \in Y_0$ such that $y_0 \notin \overline{R}X$. Note that $\overline{R}X \supseteq \bigcup Y \supseteq Y_0$. Then $y_0 \notin Y_0$. A contradiction implies this proof.

Inclusion is a basic concept in Set Theory; a similar concept is also introduced under the scheme of rough set theory. Let $K = (U, R)$ be an approximation space, and let $X, Y \subseteq U$ be the rough inclusion of sets defined as
   1. X are roughly lower-included in Y if and only if $\underline{R}X \subseteq \underline{R}Y$, denoted by $X \subset_{*R} Y$;
   2. X are roughly upper-included in Y if and only if $\overline{R}X \subseteq \overline{R}Y$, denoted by $X \subset_R^* Y$;
   3. X are roughly included in Y if and only if $X \subset_{*R} Y$ and $X \subset_R^* Y$, denoted by $X \subset_{*R}^* Y$.

It is clear that $X \subset_{*R} Y, X \subset_R^* Y$, and $X \subset_{*R}^* Y$ are quasi-orders.

The properties of rough inclusion can be derived directly from the definition.

LEMMA 1.6

(Z. Pawlak [240]) Let $K = (U, R)$ be an approximation space. Then the following properties hold:
   1. If $X \subseteq Y$, then $X \subset_{*R} Y, X \subset_R^* Y$ and $X \subset_{*R}^* Y$;
   2. If $X \subset_{*R} Y$ and $Y \subset_{*R} X$, then $X \widetilde{\approx}_R Y$;
   3. If $X \subset^* R_Y$ and $Y \subset_R^* X$, then $X \simeq RY$;
   4. If $X \subset_{*R}^* Y$ and $Y \subset_{*R}^* X$, then $X \approx_R Y$;
   5. $X \subset_R^* Y \Leftrightarrow X \cup Y \simeq_R Y$;
   6. $X \subset_{*R} Y \Leftrightarrow X \cap Y \widetilde{\approx}_R Y$;
   7. If $X \subseteq Y, X \widetilde{\approx}_R X'$, and $Y \widetilde{\approx}_R Y'$, then $X' \subset_{*R} Y'$;
   8. If $X \subseteq Y, X \simeq_R X'$, and $Y \simeq_R Y'$, then $X' \subset_R^* Y'$;
   9. If $X \subseteq Y, X \approx_R X'$, and $Y \approx_R Y'$, then $X' \subset_{*R}^* Y'$;
   10. If $X' \subset_R^* X$ and $Y' \subset_R^* Y$, then $X' \cup Y' \subset_R^* X \cup Y$;

11. If $X' \subset_{*R} X$ and $Y' \subset_{*R} Y$, then $X' \cap Y' \subset_{*R} X \cap Y$;

12. If $X \subset_{*R} Y$ and $X \approx_R Z$, then $Z \subset_{*R} Y$;

13. If $X \subset_R^* Y$ and $X \leftharpoondown_R Z$, then $Z \subset_R^* Y$;

14. If $X \subset_{*R}^* Y$ and $X \approx_R Z$, then $Z \subset_{*R}^* Y$.

From the foregoing discussion, the rough equality and rough inclusion of sets are dependent on the indiscernibility relation, and differ from the concepts of equality and inclusion. Rough equality and rough inclusion of sets have no absolute sense but are determined by the degree we know it.

1.1.4   Rough Set Based on Similarity

Indiscernibility implies an impossibility to distinguish two objects of U having the same description in terms of the attributes from Q. This relation induces equivalence classes on U, which constitute the basic granules of knowledge. In reality, due to the imprecision of data describing the objects, small differences are often not considered significant for the purpose of discrimination. This situation may be formally modeled by considering similarity or tolerance relations [206, 225, 276, 277].

In general, the similarity relations R do not generate partitions on U; the information regarding similarity may be represented using similarity classes for each object $x \in U$. Precisely, the similarity class of x, denoted by $R(x)$, consists of the set of objects that are similar to x:

$$R(x) = \{y|y \in U, yRx\} \tag{1.8}$$

It is obvious that an object y may be similar to both x and z, while z is not similar to x, that is, yRx and $y \in R(z)$, but $z \notin R(x)$, $x, y, z \in U$. The similarity relation is of course reflexive (each object is similar to itself). R. Slowinski and D. Vanderpooten [276, 277] have proposed a similarity relation that is only reflexive. The abandon of the transitivity requirement is easily justifiable, remembering, for example, Luce's paradox of the cups of tea [204]. As for the symmetry, one should notice that yRx, which means "y is similar to x," is directional; there is a subject y and a referent x, and, in general, this is not equivalent to the proposition "x is similar to y," as maintained by A. Tversky [290]. This is quite evident when the similarity relation is defined in terms of a percentage difference between numerical evaluations of the objects, calculated with respect to the evaluation of the referent object. Therefore, the symmetry of the similarity relation should not be imposed, and then it makes sense to consider the inverse relation of R, denoted by R^{-1}, where $xR^{-1}y$ means "y is similar to x"; $R^{-1}x, x \in U$ is the class of referent objects to which x is similar:

$$R^{-1}(x) = \{y|y \in U, xRy\} \tag{1.9}$$

Given a subset $X \subseteq U$ and a similarity relation R on U, an object $x \in U$ is said to be nonambiguous in each of the two following cases:

1. x belongs to X without ambiguity, that is, $x \in X$ and $R^{-1}(x) \subseteq X$; such objects are also called positive.

2. x does not belong to X without ambiguity (x clearly does not belong to X), that is, $x \in U - X$ and $R_{-1}(x) \subseteq U - X$ (or $R^{-1}(x) \cap X = \emptyset$); such objects are also called negative.

A more general definition of lower and upper approximation may thus be offered (see R. Slowinski and D. Vanderpooten [277]). Let $X \subseteq U$ and R a reflexive binary relation defined on U; the lower approximation of X, denoted by $\underline{R}X$, and the upper approximation of X, denoted by $\overline{R}X$, are defined, respectively, as

$$\underline{R}X = \{x \in U | R^{-1}(x) \subseteq U\}, \overline{R}X = \bigcup_{x \in U} R(x) \qquad (1.10)$$

It may be demonstrated that the key property, $\underline{R}X \subseteq X \subseteq \overline{R}X$, still holds and that

$$\underline{R}X = U - \overline{R}(U - X) \,(\text{ complementarity property})$$

and

$$\overline{R}X = \{x \in U | R^{-1}(x) \cap X \neq \emptyset\}.$$

Moreover, the definitions proposed are the only ones that correctly characterize the set of positive objects (lower approximation) and the set of positive or ambiguous objects (upper approximation) when a similarity relation is reflexive, but not necessarily symmetric nor transitive.

1.1.5   Rough Set Based on Binary Relation

In addition to similarity, rough set is also generated based on reflexive and transitive relations. It was proved that the pair of lower and upper approximation operators induced by a reflexive and transitive relation is exactly a pair of interior and closure operators of a topology [170, 244, 325]. K. Qin et al. [250] investigate the relationship between generalized rough sets induced by reflexive and transitive relations and the topologies on the universe that are not restricted to be finite. It is proved that there exists an one-to-one correspondence between the set of all reflexive and transitive relations and the set of all topologies which satisfy a certain kind of compactness condition.

An extensive research on the algebraic properties of rough sets based on binary relations can be found in the papers [169, 327, 329, 344] . They start from common properties of binary relations to characterize the essential properties of lower and upper approximation operations generated by such relations. In this section, we start from the common properties of classical lower and upper approximation operations listed above. We want to find out what conditions a binary relation should satisfy so that the corresponding relation based lower and upper approximation operations have properties such as normality, extension, and idempotency.

DEFINITION 1.1   *(Y. Yao [326]) Suppose R is a binary relation on a universe U. A pair of approximation operations, $L(R), H(R) : P(U) \to P(U)$*

(P(U) is the power set of U), are defined by

$$L(R)(X) = \{x | \forall y, xRy \Rightarrow y \in X\} = \{x | RN(X) \subseteq X\} \text{ and}$$
$$H(R)(X) = \{x | \exists y, s.t. xRy\} = \{x | RN(X) \cap X \neq \emptyset\}$$

respectively, where $RN(x) = \{y \in U | xRy\}$. They are called the lower approximation operation and the upper approximation operation, respectively.

Example 1.2

Let $U = \{a, b, c, d\}$ and $R = \{(a,a), (a,b), (a,c), (a,d), (b,b), (b,c), (b,d), (c,c), (c,d), (d,d)\}$. Then

$$RN(a) = \{a,b,c,d\}, RN(b) = \{b,c,d\}, RN(c) = \{c,d\}, RN(d) = \{d\},$$
$$LR\{a\} = \emptyset, LR\{b\} = \emptyset, LR\{c\} = \emptyset, LR\{b\} = \emptyset, LR\{d\} = \{d\},$$
$$LR\{a,b\} = \emptyset, LR\{a,c\} = \emptyset, LR\{a,d\} = \{d\}, LR\{b,c\} = \emptyset,$$
$$LR\{b,d\} = \{d\}, LR\{c,d\} = \{d\}, LR\{a,b,c\} = \emptyset, LR\{a,b,d\} = \{d\},$$
$$LR\{a,c,d\} = \{c,d\}, LR\{b,c,d\} = \{b,c,d\}, LR\{a,b,c,d\} = \{a,b,c,d\}$$
$$HR\{a\} = \{a\}, HR\{b\} = \{a,b\}, HR\{c\} = \{a,b,c\}, HR\{d\} = \{a,b,c,d\},$$
$$HR\{a,b\} = \{a,b\}, HR\{a,c\} = \{a,d,c\}, HR\{a,d\} = \{a,b,c,d\},$$
$$HR\{b,c\} = \{a,b,c\}, HR\{b,d\} = \{a,b,d\}, HR\{c,d\} = \{a,b,c,d\},$$
$$HR\{a,b,c\} = \{a,b,c\}, HR\{a,b,d\} = \{a,b,c,d\}, HR\{a,c,d\} = \{a,b,c,d\},$$
$$HR\{b,c,d\} = \{a,b,c,d\}, HR\{a,b,c,d\} = \{a,b,c,d\}.$$

PROPOSITION 1.3

(Y. Yao [326]) Let $X, Y \subseteq U$, and R be a relation on U. Then $L = L(R)$ and $H = H(R)$ satisfy the following properties:
1. $L(U) = U$;
2. $L(X \cap Y) = L(X) \cap L(Y)$;
3. $X \subseteq Y \Rightarrow L(X) \subseteq L(Y)$;
4. $H(\emptyset) = \emptyset$;
5. $H(X \cup Y) = H(X) \cup H(Y)$;
6. $X \subseteq Y \Rightarrow H(X) \subseteq Y$;
7. $L(-X) = -H(X)$.

On the other hand, assume that an operation $L : P(U) \rightarrow P(U)$ satisfies the following properties:
8. $L(U) = U$;
9. $L(X \cup Y) = L(X) \cap L(Y)$.

Then there exists a relation R on U such that $L = L(R)$.

Assume that an operations $H : P(U) \rightarrow P(U)$ satisfies the following properties:
10. $H(\emptyset) = \emptyset$;
12. $H(X \cup Y) = H(X) \cup H(Y)$.

Then there exists a relation R on U such that $H = H(R)$.

R is called serial relation if it satisfies $\forall x \in U, \exists y \in U$, s.t. xRy. For serial relation, the following properties hold.

PROPOSITION 1.4

(Y. Yao [326]) Let R be a serial relation on U, and $X, Y \subseteq U$. Then $L = L(R)$ and $H = H(R)$ satisfy the following properties:

1. $L(U) = U$;
2. $L(\emptyset) = \emptyset$;
3. $L(X \cap Y) = L(X) \cap L(Y)$;
4. $X \subseteq Y \Rightarrow \subseteq Y$;
5. $H(\emptyset) = \emptyset$;
6. $H(U) = U$;
7. $H(X \cup Y) = H(X) \cup H(Y)$;
8. $X \subseteq Y \Rightarrow H(X) \subseteq H(Y)$;
9. $L(-X) = -H(X)$;
10. $L(X) \subseteq H(X)$.

On the other hand, assume that an operation $L : P(U) \rightarrow P(U)$ satisfies the following properties:

11. $L(U) = U$;
12. $L(X \cap Y) = L(X) \cap L(Y)$;
13. $L(\emptyset) = \emptyset$.

Then there exists a serial relation R on U such that $L = L(R)$. Assume that an operation $H : P(U) \rightarrow P(U)$ satisfies the following properties:

15. $H(\emptyset) = \emptyset$;
16. $H(X \cup Y) = H(X) \cup H(Y)$;
17. $H(U) = U$.

Then there exists a serial relation R on U such that $H = H(R)$.

R is called reflexive relation if it satisfies $\forall x \in U$, xRX. For reflexive relation, the following properties hold.

PROPOSITION 1.5

(Y. Yao [326]) Let R be a symmetric relation on U, and $X, Y \subseteq U$. Then $L = L(R)$ and $H = H(R)$ satisfy the following properties:

1. $L(U) = U$;
2. $L(X \cap Y) = L(X) \cap L(Y)$;
3. $L(X) \subseteq X$;
4. $H(\emptyset) = \emptyset$;
5. $H(X \cup Y) = H(X) \cup H(Y)$;
6. $X \subseteq H(X)$;
7. $L(-X) = -H(X)$.

On the other hand, assume that an operation $L : P(U) \rightarrow P(U)$ satisfies the following properties:

8. $L(U) = U$;

9. $(X \cap Y) = L(X) \cap L(Y)$;

10. $L(X) \subseteq X$.

Then there exists one reflexive relation R on U such that $L = L(R)$. Assume that an operation $H : P(U) \to P(U)$ satisfies the following properties:

11. $H(\emptyset) = \emptyset$;

12. $H(X \cup Y) = H(X) \cup H(Y)$;

13. $X \subseteq H(X)$.

Then there exists one reflexive relation R on U such that $H = H(R)$.

DEFINITION 1.2   *(W. Zhu [344]) Let R be a relation on U. If $\forall x, y \in U, xRy \Rightarrow \exists z \in U$ such that xRz and zRy, we say R is a mediate relation on U.*

PROPOSITION 1.6

(W. Zhu [344]) Let R be a mediate relation on U, $X, Y \subseteq U$. Then $L(R)$ and $H(R)$ satisfy the following properties:

1. $L(R)(U) = U$;

2. $L(R)(X \cap Y) = L(R)(X) \cap L(R)(Y)$;

3. $L(R)(L(R)(X) \subseteq L(R)(X)$;

4. $H(R)(\emptyset) = \emptyset$;

5. $H(R)(X \cup Y) = H(R)(X) \cup H(R)(Y)$;

6. $H(R)(H(R)(X)) \subseteq H(R)(X)$;

7. $L(R)(-X) = -H(R)(X)$.

PROPOSITION 1.7

(W. Zhu [344]) Assume that an operation $H : P(U) \to P(U)$ satisfies the following properties:

1. $H(\emptyset) = \emptyset$;

2. $H(X \cup Y) = H(X) \cup H(Y)$;

3. $H(X) \subseteq H(H(X))$.

Then there exists a mediate relation R on U such that $H = H(R)$.

PROPOSITION 1.8

(W. Zhu [344]) Assume that an operation $H : P(U) \to P(U)$ satisfies the following properties:

1. $L(\emptyset) = \emptyset$;

2. $L(X \cap Y) = L(X) \cap L(Y)$;

3. $L(L((X)) \subseteq L(X)$.

Then there exists a mediate relation R on U such that $L = L(R)$.

PROPOSITION 1.9

(Y. Yao [326]) Let R be a mediate and transitive relation on U, $X, Y \subseteq U$. Then $L = L(R)$ and $H = H(R)$ satisfy the following properties:

1. $L(R)(U) = U$;
2. $L(R)(X \cap Y) = L(R)(X) \cap L(R)(Y)$;
3. $L(R)(L(R)(X))$;
4. $H(R) = (\emptyset) = \emptyset$;
5. $H(R)(X \cup Y) = H(R)(X) \cup H(R)(Y)$;
6. $H(R)(H(R)(X)) \subseteq H(R)(X)$;
7. $L(R)(-X) = -H(R)(X)$.

On the other hand, assume that an operation $L : P(U) \to P(U)$ satisfies the following properties:

8. $L(U) = U$;
9. $L(X \cap Y) = L(X) \cap L(Y)$;
10. $L(L(X)) \subseteq L(X)$.

Then there exists one transitive relation R on U such that $L = L(R)$. Assume that an operation $H : P(U) \to P(U)$ satisfies the following properties:

11. $H(\emptyset) = \emptyset$;
12. $H(X \cup Y) = H(X) \cup H(Y)$;
13. $H(H(X)) \subseteq H(X)$.

Then there exists one transitive relation R on U such that $H = H(R)$.

1.2 Rough Membership

In order to measure the degree of membership of x belonging to X, rough membership functions were introduced by Z. Pawlak and A. Skowron.

DEFINITION 1.3 *(Z. Pawlak and A. Skowron [243]) Let (U,R) be an approximate space, $X \subseteq U$, $[x]_R$ be an equivalence class of a element $x \in U$. Then the rough membership function of x belonging to X is*

$$\mu_X^R : U \to [0,1]$$

defined by

$$\mu_X^R(x) = \frac{card([x]_R \cap X)}{card([x]_R)} \tag{1.11}$$

where "card" refers to cardinality of X, calculated by the cardinal number of the set X when X is a finite set; otherwise it expresses the Lebesgue measure.

The value $\mu_X^R(x)$ can be interpreted as the degree that x belongs to X in view of knowledge about x expressed by R or the degree to which the elementary granule $[x]_R$ is included in the set X. This means that the definition reflects a subjective knowledge about elements of the universe in contrast to the classical definition of a set. Observe that the value of the membership function is calculated from the available data, and not subjectively assumed, as it is the case of membership functions of fuzzy sets.

The rough membership function can also be interpreted as the conditional probability that x belongs to X for given R. This interpretation was used by several researchers in the rough set community (see, e.g., [124, 288, 303, 328]). Note also that the ratio on the right-hand side of (1.11) is known as the confidence coefficient in data mining.

Example 1.3

Assume that an equivalent relation of real line $\Re$ is defined by the partition $\{(-\infty, -5), [-5, -4), [-4, -3), [-3, -2), [-2, -1), [-1, 0), [0, 1), [1, 2), [2, 3), [3, 4), [4, 5), [5, +\infty)\}$ and $X = (-3.5, -1.2) \cup (2.1, 4.3)$. Denote $X^\alpha = \{x | \mu_X^R(x) \geq \alpha\}$. It follows from Definition 1.3 that $X^0 = [-4, -1) \cup [2, 5), X^{0.2} = [-4, -2) \cup [2, 5), X^{0.4} = [-4, -2) \cup [2, 4), X^{0.6} = [-3, -2) \cup [2, 4), X^{0.8} = [-3, -2) \cup [2, 4), X^1 = [-3, -2) \cup [3, 4)$.

PROPOSITION 1.10

(Z. Pawlak [242], Y. Yao [325]) Let (U, R) be an approximate space, $X \subseteq U$. Then the rough membership function μ_X^R has the following properties:

1. $\mu_U^R(x) = 1, \forall x \in U$;
2. $\mu_\phi^R(x) = 0, \forall x \in U$;
3. $y \in [x]_R \Rightarrow \mu_X^R(x) = \mu_X^R(y)$;
4. $x \in X \Rightarrow \mu_X^R(x) \neq 0$;
5. $x \notin X \Rightarrow \mu_X^R(x) \neq 1$;
6. $\mu_X^R(x) = 1 \Leftrightarrow x \in \underline{R}(X)$;
7. $\mu_X^R(x) = 0 \Leftrightarrow x \in U - \bar{R}(X)$;
8. $0 < \mu_X^R(x) < 1 \Leftrightarrow x \in BND(X)$;
9. $\mu_{U-X}^R(x) = 1 - \mu_X^R(x)$ *for any* $x \in U$;
10. $\mu_{X \cup Y}^R(x) = \mu_X^R(x) + \mu_Y^R(x) - \mu_{X \cap Y}^R(x)$ *for any* $x \in U$;
11. $X \cap Y = \Phi \Rightarrow \mu_{X \cup Y}^R(x) = \mu_X^R(x) + \mu_Y^R(x)$;
12. $\max\{0, \mu_X^R(x) + \mu_Y^R(x) - 1\} \leq \mu_{X \cap Y}^R(x) \leq \min\{\mu_X^R(x), \mu_Y^R(x)\}$ *for any* $x \in U$;
13. $\max\{\mu_X^R(x), \mu_Y^R(x)\} \leq \mu_{X \cup Y}^R(x) \leq \max\{1, \mu_X^R(x) + \mu_Y^R(x)\}$ *for any* $x \in U$.

REMARK 1.5   For any $n \geq 2$, we have

12′. $\max\{0, \sum\limits_{i=1}^{n} \mu_{X_i}^R(x) - 1\} \leq \mu_{\cap_{i=1}^n X}^R(x) \leq \min\limits_{1 \leq i \leq n}\{\mu_{X_i}^R(x)\}$;

13′. $\max\limits_{1 \leq i \leq n}\{\mu_{X_i}^R(x)\} \leq \mu_{\cup_{i=1}^n X}^R(x) \leq \max\{1, \sum\limits_{i=1}^{n} \mu_{X_i}^R(x)\}$.

Let us use an transportation problem to explain the foregoing discussion. For practical reasons, the problem of the description of transportation quantity may be considered. Common knowledge can define the universe $U = [0, 130]$, that is, trans-

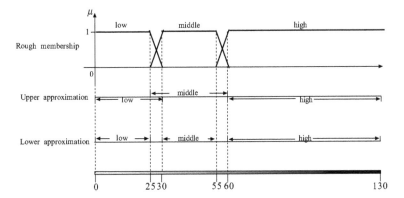

FIGURE 1.6   Rough description for transportation quantity.

portation quantity is less than 130 tons. Based on his knowledge and experience, the DM may divide the universe into three equivalent classes:

1. $[0,25]$: The DM believes that transportation quantity belonging to this interval is "low."

2. $(25,30)$: The DM believes that transportation quantity belonging to this interval is the transition from "low" to "middle."

3. $[30,55)$: The DM believes that transportation quantity belonging to this interval is 'middle."

4. $(55,60)$: The DM believes that transportation quantity belonging to this interval is the transition from "middle" to "high."

5. $[60,130]$: The DM believes that transportation quantity belonging to this interval is 'high."

Then the partition $\{[0,25], (25,30), [30,55], (55,60), [60,130]\}$ by the DM can be regarded as the available information.

For other people, he or she often gives a vague description of the concepts "low," "middle," and "high." For convenience, we assume that a person regards that the low transportation quantity from 0 to 35, the middle transportation quantity from 25 to 60, and the high transportation quantity may be from 50 to 130. According to the partition of the DM, the transportation quantity interval $[0,25]$ is surely low, and the transportation quantity interval $[0,30)$ is possibly low, where any age in $(25,30)$ has rough membership from 1 to 0. Similar analysis may be carried out for the concept "middle" and "high"(see Figure 1.6).

1.3   Rough Intervals

Rough intervals, proposed by M. Rebolledo [251], are used for solving complex problems of existence of the parameters with dual-layer information ([129, 139, 200, 201]). This section will introduce the notation of rough interval and rough interval arithmetic. Some of the content refers to M. Rebolledo [251].

The lower and upper Approximation Interval concepts satisfy the mathematical definition of the rough set's upper and lower approximation adapted to intervals. If a particular qualitative value C must be represented over a variable, the two enveloping intervals, A_* and X^*, can be defined. The implications (1.12) and (1.13) represent the relationship between the variable $x \in U$ (where the universe U can be any complete set of continuous or discrete values), the qualitative value X defined on x, and the intervals X_* and X^*:

$$x \in X_* \Rightarrow x \in X \tag{1.12}$$

and

$$x \notin X^* \Rightarrow x \notin X \tag{1.13}$$

The rough set theory reduces the vagueness of a concept to uncertainty areas at their borders. Within these uncertainty areas, in rough intervals as in rough sets, no definitive conclusion about the problem is possible. Hence, rough membership values must be defined for rough intervals as well. These rough membership values, expressed in rough membership functions, must satisfy all the conditions listed in Proposition 1.10 but also complementarity, monotonicity, and border conditions.

In order to present rough interval arithmetic, a general rough interval can be expressed as

$$RI = ([\underline{RI}_1, \overline{RI}_1], [\underline{RI}_2, \overline{RI}_2])(\underline{RI}_1 \leq \underline{RI}_2 \leq \overline{RI}_1 \leq \overline{RI}_2) \tag{1.14}$$

A rough interval comprises two parts: an upper approximation interval ($[\underline{RI}_2, \overline{RI}_2]$) and a lower approximation interval ($[\underline{RI}_1, \overline{RI}_1]$). For convenience, we denote the upper and lower approximation intervals of RI as RI^* and RI_*, respectively. It means in normal cases that the variable takes values in the lower approximation and in special cases that the variable takes values in the upper approximation, that is to say, there is no value taken by the variable outside the upper approximation interval.

REMARK 1.6   For rough interval RI, if $\underline{RI}_1 = \underline{RI}_2$ and $\overline{RI}_1 = \overline{RI}_2$, that is, no special cases occur, then RI degenerates into a crisp interval. Thus, rough interval is a reasonable generation of the crisp interval.

In order to operate rough interval, M. Rebolledo [251] proposed addition, subtraction, and negation of rough interval arithmetic based on counterpart interval arithmetic [217, 224]. In fact, rough interval arithmetic is just a kind of two-fold interval arithmetic but more complicated. The general rough interval arithmetic is defined as follows.

DEFINITION 1.4 *Let* $RI = ([\underline{RI}_1, \overline{RI}_1], [\underline{RI}_2, \overline{RI}_2])$, *and* $RI' = ([\underline{RI'_1}, \overline{RI'_1}], [\underline{RI'_2}, \overline{RI'_2}])$ *be two rough intervals,* $\circ \in \{+, -, \times, \div\}$ *a binary operation on the set of interval numbers. Then the rough interval arithmetics are defined by*

$$RI \circ RI' = [\underline{RI} \circ \underline{RI'}], [\overline{RI} \circ \overline{RI'}] \tag{1.15}$$

REMARK 1.7 A scalar k can be written as the rough interval form $([k,k],[k,k])$. Then $k \times RI$ can defined as $([k,k],[k,k]) \times RI$, which is called the scalar multiplication of the rough interval RI. It follows from the definition that

$$k \times RI = \begin{cases} ([k\underline{RI}_1, k\overline{RI}_1], [k\underline{RI}_2, k\overline{RI}_2]), \text{ if } k > 0 \\ ([k\overline{RI}_1, k\underline{RI}_1], [k\overline{RI}_2, k\underline{RI}_2]), \text{ otherwise} \end{cases}$$

$k \times RI$ also be written as kRI for simplicity.

REMARK 1.8 As the division $RI \div RI'$, there must be $0 \notin [\underline{RI'_2}, \overline{RI'_2}]$, but not for dividend RI. The division $RI \div RI'$ can also be written as RI/RI'.

Example 1.4
Let $RI = ([4,8],[2,10])$ and $RI' = ([2,5],[1,7])$. Then $RI + RI' = ([6,13],[3,17]), RI - RI' = ([-1,6],[-5,9]), RI \times RI' = ([8,40],[2,70]), RI/RI' = ([4/5,4],[2/7,10]), 2RI = ([8,16],[4,10]), -2RI = ([-16,-8],[-20,-4])$. In order to represent the decision maker's preference, it is necessary to define the order relations between rough intervals.

DEFINITION 1.5 *Let* $RI = ([\underline{RI}_1, \overline{RI}_1], [\underline{RI}_2, \overline{RI}_2])$ *and* $RI' = ([\underline{RI'}, \overline{RI'_1}], [\underline{RI'_2}, \overline{RI'_2}])$. *Then the order relation* $\preceq$ *and* $\prec$ *between* RI *and* RI' *are defined by*

$$RI \preceq RI' \Leftrightarrow \frac{\underline{RI}_1 + \overline{RI}_1}{2} \leq \frac{\underline{RI'_1} + \overline{RI'_1}}{2}, \frac{\underline{RI}_2 + \overline{RI}_2}{2} \leq \frac{\underline{RI'_2} + \overline{RI'_2}}{2}$$

and

$$RI \prec RI' \Leftrightarrow RI \preceq RI' \text{ with } RI \neq RI'$$

respectively.

The order relation $\preceq$ or $\prec$ represents the decision maker's preference for the alternative with upper maximum midpoint both in the normal case and special case for maximization problem. The decision philosophy is to maximize the expected value and minimize the uncertainty. It is clear that the order relation $\preceq$ or $\prec$ are a partial order relation.

REMARK 1.9 Let $a \leq ([\underline{b}_1, \bar{b}_1], [\underline{b}_2, \bar{b}_2])$ denote $([a,a],[a,a]) \preceq ([\underline{b}_1, \bar{b}_1], [\underline{b}_2, \bar{b}_2])$. It follows from Definition 1.5 that $([a,a],[a,a]) \preceq ([\underline{b}_1, \bar{b}_1], [\underline{b}_2, \bar{b}_2]) \Leftrightarrow \frac{a+a}{2} \leq$

$\frac{b_1+\bar{b}_1}{2}, \frac{a+a}{2} \leq \frac{b_2+\bar{b}_2}{2} \Leftrightarrow a \leq \frac{b_1+\bar{b}_1}{2}, a \leq \frac{b_2+\bar{b}_2}{2}$. Thus we have $a \leq ([\underline{b}_1, \bar{b}_1], [\underline{b}_2, \bar{b}_2]) \Leftrightarrow a \leq \min\{\frac{b_1+\bar{b}_1}{2}, \frac{b_2+\bar{b}_2}{2}\}$.

The expected value operator of rough intervals is the base for the expected value model with rough interval parameters.

In order to rank rough intervals, a natural idea is to compare their values. Similar to the the expected value operator in Probability Theory, we define the expected value of a rough interval. More generally, the Expect operator on rough set is introduced first. Note that the rough set in this section is defined by R. Slowinski and D. Vanderpooten [277].

DEFINITION 1.6   *Let X be an event expressed by $\{x|\xi(x) \in B\}$, where ξ is a function from universe U to real line $\Re$, $B \subseteq \Re$, and X is approximated by $(\underline{X}, \overline{X})$ according to the similarity relation R. Then the lower expected value of X is defined by*

$$\underline{E}[\xi] = \int_0^{+\infty} \underline{Appr}\{\xi \geq r\}dr - \int_{-\infty}^0 \underline{Appr}\{\xi \leq r\}dr \qquad (1.16)$$

where $\underline{Appr}(X) = \frac{|X \cap \underline{X}|}{\underline{X}}$.

DEFINITION 1.7   *Let X be an event expressed by $\{x|\xi(x) \in B\}$, where ξ is a function from universe U to real line $\Re$, $B \subseteq \Re$, and X is approximated by $(\underline{X}, \overline{X})$ according to the similarity relation R. Then the upper expected value of X is defined by*

$$\overline{E}[\xi] = \int_0^{+\infty} \overline{Appr}\{\xi \geq r\}dr - \int_{-\infty}^0 \overline{Appr}\{\xi \leq r\}dr \qquad (1.17)$$

where $\overline{Appr}(X) = \frac{|X|}{\overline{X}}$.

DEFINITION 1.8   *Let X be an event expressed by $\{x|\xi(x) \in B\}$, where ξ is a function from universe U to real line $\Re$, $B \subseteq \Re$, and X is approximated by $(\underline{X}, \overline{X})$ according to the similarity relation R. Then the expected value of X is defined by*

$$E[X] = \int_0^{+\infty} Appr\{\xi \geq r\}dr - \int_{-\infty}^0 Appr\{\xi \leq r\}dr \qquad (1.18)$$

The relationship of expected value $E(X)$, lower expected value $\underline{E}(X)$, and upper expected value $\overline{E}(X)$ is stated by the following proposition.

PROPOSITION 1.11
Let X be an event expressed by $\{x|\xi(x) \in B\}$, where ξ is a function from universe U to real line $\Re$, $B \subseteq \Re$, X is approximated by $(\underline{X}, \overline{X})$ according to the

similarity relation R, and η is a given parameter predetermined by the decision maker's preference. Then

$$E(X) = \eta \underline{E}(X) + (1-\eta)\overline{E}(X) \tag{1.19}$$

PROOF Denote the event $\{x|\xi \leq (\geq,<,>)r\}$ by $\{\xi \leq (\geq,<,>)r\}$. It follows from the definition of E and $Appr$ that

$$E[X] = \int_0^{+\infty} Appr\{\xi \geq r\}dr - \int_{-\infty}^0 Appr\{\xi \leq r\}dr$$

$$= \int_0^{+\infty}[\eta \frac{|\{\xi \geq r\} \cap \underline{X}|}{\underline{X}} + (1-\eta)\frac{|\{\xi \leq r\}|}{\overline{X}}]dr - \int_{-\infty}^0 [\eta \frac{|X \cap \underline{X}|}{\underline{X}} + (1-\eta)\frac{|X|}{\overline{X}}]dr$$

$$= \eta \left[\int_0^{+\infty} \frac{|\{\xi \geq r\} \cap \underline{X}|}{\underline{X}} + \int_{-\infty}^0 \frac{|\{\xi \leq r\} \cap \underline{X}|}{\underline{X}} \right] + (1-\eta)\left[\int_0^{+\infty} \frac{|\{\xi \geq r\}|}{\overline{X}} + \int_{-\infty}^0 \frac{|\{\xi \leq r\}|}{\overline{X}} \right]$$

$$= \eta \left[\int_0^{+\infty} \underline{Appr}\{\xi \geq r\} + \int_{-\infty}^0 \underline{Appr}\{\xi \leq r\} \right]$$

$$+ (1-\eta)\left[\int_0^{+\infty} \overline{Appr}\{\xi \geq r\} + \int_{-\infty}^0 \overline{Appr}\{\xi \leq r\} \right]$$

$$= \eta \underline{E}[X] + (1-\eta)\overline{E}[X].$$

The proposition is proved.

THEOREM 1.4
Let $\underline{X} = \{\lambda_1, \lambda_2, \cdots, \lambda_m\}, \overline{X} = \{\lambda_1, \lambda_2, \cdots, \lambda_n\}(m \leq n), \lambda_i \in \Re, \xi(\lambda_i) = \lambda_i, i = 1, 2, \cdots, n$. Then

$$E[\xi] = \sum_{i=1}^n Appr\{\xi = \lambda_i\}\lambda_i \tag{1.20}$$

PROOF Without loss of generality, we assume that $\lambda_1 < \lambda_2 < \cdots < \lambda_k \leq 0 \leq \lambda_{k+1} < \cdots < \lambda_n$. Then

$$E[\xi] = \int_0^{+\infty} Appr\{\xi \geq r\}dr - \int_{-\infty}^0 Appr\{\xi \leq r\}dr$$
$$= (\int_0^{\lambda_{k+1}} Appr\{\xi \geq r\}dr + \int_{\lambda_{k+1}}^{\lambda_{k+2}} Appr\{\xi \geq r\}dr + \cdots + \int_{\lambda_{n-1}}^{\lambda_n} Appr\{\xi \geq r\}dr$$
$$+ \int_{\lambda_n}^{+\infty} Appr\{\xi \geq r\}dr) - (\int_{-\infty}^{\lambda_1} Appr\{\xi \leq r\}dr + \int_{\lambda_1}^{\lambda_2} Appr\{\xi \leq r\}$$
$$+ \cdots + \int_{\lambda_{k-1}}^{\lambda_k} Appr\{\xi \leq r\} + \int_{\lambda_k}^0 Appr\{\xi \leq r\}).$$

For any $\varepsilon > 0$, we have

$$\int_{\lambda_n}^{+\infty} Appr\{\xi \geq r\}dr = \int_{\lambda_n}^{\lambda_n + \varepsilon} Appr\{\xi \geq r\}dr + \int_{\lambda_n + \varepsilon}^{+\infty} Appr\{\xi \geq r\}dr$$
$$= Appr\{\xi = \lambda_n\}\varepsilon + \int_{\lambda_n + \varepsilon}^{+\infty} 0 dr$$
$$= Appr\{\xi = \lambda_n\}\varepsilon.$$

It follows from the extemporariness of ε that

$$\int_{\lambda_n}^{+\infty} Appr\{\xi \geq r\}dr = 0.$$

Similarly, it can be proved prove that

$$\int_{-\infty}^{\lambda_1} Appr\{\xi \leq r\}dr = 0.$$

Therefore, we have

$\int_0^{+\infty} Appr\{\xi \geq r\}dr$

$= \int_0^{\lambda_{k+1}} Appr\{\xi \geq r\}dr + \int_{\lambda_{k+1}}^{\lambda_{k+2}} Appr\{\xi \geq r\}dr + \cdots + \int_{\lambda_{n-1}}^{\lambda_n} Appr\{\xi \geq r\}dr$

$= \lambda_{k+1} \sum_{i=k+1}^{n} Appr\{\xi = \lambda_i\} + (\lambda_{k+2} - \lambda_{k+1}) \sum_{i=k+2}^{n} Appr\{\xi = \lambda_i\} + \cdots + (\lambda_{n-1}$

$-\lambda_n)Appr\{\xi = \lambda_n\}$

$= \lambda_{k+1}Appr\{\xi = \lambda_{k+1}\} + \lambda_{k+2}Appr\{\xi = \lambda_{k+2}\} + \cdots + \lambda_n Appr\{\xi = \lambda_n\}$

and

$\int_{-\infty}^{0} Appr\{\xi \leq r\}dr$

$= \int_{\lambda_1}^{\lambda_2} Appr\{\xi \leq r\} + \int_{\lambda_2}^{\lambda_3} Appr\{\xi \leq r\} + \cdots + \int_{\lambda_{k-1}}^{\lambda_k} Appr\{\xi \leq r\} + \int_{\lambda_k}^{0} Appr\{\xi \leq r\}$

$= (\lambda_2 - \lambda_1)Appr\{\xi = \lambda_1\} + (\lambda_3 - \lambda_2) \sum_{i=1}^{2} Appr\{\xi = \lambda_i\} + \cdots + (\lambda_k - \lambda_{k-1}) \sum_{i=1}^{k-1}$

$Appr\{\xi = \lambda_2\} + (0 - \lambda_{k-1}) \cdot \sum_{i=1}^{k-1} Appr\{\xi = \lambda_i\}$

$= -\lambda_1 Appr\{\xi = \lambda_1\} - \lambda_2 Appr\{\xi = \lambda_2\} - \cdots - \lambda_k Appr\{\xi = \lambda_k\}.$

It follows from

$$E[\xi] = \int_0^{+\infty} Appr\{\xi \geq r\}dr - \int_{-\infty}^{0} Appr\{\xi \leq r\}dr$$

that

$$E[\xi] = \sum_{i=1}^{n} \lambda_i Appr\{\xi = \lambda_i\}.$$

The theorem is proved.

As well known, the rough interval is a special rough set. For the expected value of the rough interval, we have

THEOREM 1.5
Let $A = ([a,b],[c,d])(c \leq a \leq b \leq d)$ be a rough interval. Then the expected value of A is

$$\frac{1}{2}[\eta(a+b) + (1-\eta)(c+d)] \tag{1.21}$$

PROOF   It follows from the Definition of *Appr* that

$$
Appr\{\xi \geq r\} = \begin{cases}
0, & \text{if } d \leq r \\
(1-\eta)\frac{d-r}{(d-c)}, & \text{if } b \leq r \leq d \\
(1-\eta)\frac{d-r}{d-c} + \eta\frac{b-r}{b-a}, & \text{if } a \leq r \leq b \\
(1-\eta)\frac{d-r}{d-c} + \eta, & \text{if } c \leq r \leq a \\
1, & \text{if } r \leq c
\end{cases}
$$

and

$$
Appr\{\xi \leq r\} = \begin{cases}
0, & \text{if } r \leq c \\
(1-\eta)\frac{r-c}{(d-c)}, & \text{if } c \leq r \leq a \\
(1-\eta)\frac{r-c}{d-c} + \eta\frac{r-a}{b-a}, & \text{if } a \leq r \leq b \\
(1-\eta)(\frac{r-c}{d-c} + \eta), & \text{if } b \leq r \leq d \\
1, & \text{if } d \leq r
\end{cases}
$$

There are five cases when we compute the expected value of ξ. Let us discuss every case in turn.

Case 1. $0 \leq c \leq a \leq b \leq d$.

$$
E[A] = \int_0^{+\infty} Appr\{\lambda \in \Lambda | \xi \geq r\} dr - \int_{-\infty}^0 Appr\{\lambda \in \Lambda | \xi \leq r\} dr
$$

$$
= \int_0^c 1 dr + \int_c^a \left[(1-\eta)\frac{d-r}{d-c} + \eta\right] dr + \int_a^b \frac{1}{2}(\frac{d-r}{d-c} + \frac{b-r}{b-a}) dr + \int_b^d \frac{d-r}{2(d-c)} dr
$$

$$
= \frac{1}{2}[\eta(a+b) + (1-\eta)(c+d)].
$$

Case 2. $c \leq d \leq a \leq b \leq d$.

$$
E[\xi] = \int_0^{+\infty} Appr\{\lambda \in \Lambda | \xi \geq r\} dr - \int_{-\infty}^0 Appr\{\lambda \in \Lambda | \xi \leq r\} dr
$$

$$
= \int_0^a \left[(1-\eta)\frac{d-r}{d-c} + \eta\right] dr + \int_a^b \left[(1-\eta)\frac{d-r}{d-c} + \eta\frac{b-r}{b-a}\right] dr
$$

$$
+ \int_b^d (1-\eta)\frac{d-r}{(d-c)} dr - \int_c^0 (1-\eta)\frac{r-c}{(d-c)} dr
$$

$$
= \frac{1}{2}[\eta(a+b) + (1-\eta)(c+d)].
$$

Case 3. $c \leq a \leq 0 \leq b \leq d$.

$$
E[\xi] = \int_0^{+\infty} Appr\{\lambda \in \Lambda | \xi \geq r\} dr - \int_{-\infty}^0 Appr\{\lambda \in \Lambda | \xi \leq r\} dr
$$

$$
= \int_0^b ((1-\eta)\frac{d-r}{d-c} + \eta\frac{b-r}{b-a}) dr + \int_b^d (1-\eta))\frac{d-r}{(d-c)} dr
$$

$$
- \int_c^a (1-\eta)\frac{r-c}{(d-c)} dr - \int_a^0 \left[(1-\eta)\frac{r-c}{d-c} + \eta\frac{r-a}{b-a}\right] dr
$$

$$
= \frac{1}{2}[\eta(a+b) + (1-\eta)(c+d)].
$$

Case 4. $c \leq a \leq b \leq 0 \leq d.$

$$E[\xi] = \int_0^{+\infty} Appr\{\lambda \in \Lambda | \xi \geq r\} dr - \int_{-\infty}^0 Appr\{\lambda \in \Lambda | \xi \leq r\} dr$$

$$= \int_0^d (1-\eta)\frac{d-r}{(d-c)} dr - \int_c^a (1-\eta))\frac{r-c}{(d-c)} dr$$

$$- \int_a^b \left[(1-\eta)\frac{r-c}{d-c} + \eta\frac{r-a}{b-a} \right] dr - \int_b^0 [(1-\eta)\frac{r-c}{d-c} + \eta] dr$$

$$= \tfrac{1}{2}[\eta(a+b) + (1-\eta)(c+d)].$$

Case 5. $c \leq a \leq b \leq 0 \leq d.$

$$E[\xi] = \int_0^{+\infty} Appr\{\lambda \in \Lambda | \xi \geq r\} dr - \int_{-\infty}^0 Appr\{\lambda \in \Lambda | \xi \leq r\} dr$$

$$= -\int_c^a (1-\eta)\frac{r-c}{(d-c)} dr - \int_a^b \left[(1-\eta)\frac{r-c}{d-c} + \eta\frac{r-a}{b-a} \right] dr$$

$$- \int_b^d \left[(1-\eta)\frac{r-c}{d-c} + \eta\frac{r-a}{b-a} \right] + dr \int_D^0 1 dr$$

$$= \tfrac{1}{2}[\eta(a+b) + (1-\eta)(c+d)].$$

So we always have

$$E[A] = \frac{1}{2}[\eta(a+b) + (1-\eta)(c+d)].$$

This theorem is proved.

REMARK 1.10   If $\eta = 0.5$, the expected value $E[([a,b],[c,d])] = \frac{1}{4}(a+b+c+d)$.

THEOREM 1.6
Let $\xi = ([x_1, x_2], [x_3, x_4])$, and a and b be any real numbers. Then

$$E[a\xi + b\eta] = aE[\xi] + bE[\eta] \tag{1.22}$$

PROOF   Since the trust measure *Appr* has the same property as the probability measure *Pr*, we may prove the linearity of the rough expected value operator by a similar way as for the stochastic expected value operator.

Chapter 2

Multiple Objective Rough Decision Making

Multiple objective decision making (MODM)–also called vector programming–is a recent development in mathematical decision making and emerged from an attempt to tackle the problems raised by the current developments in science, engineering, industry, economy, etc. Due to the complexity of these problems, several objectives had to be incorporated in the optimization process. Basically, the problem consists of optimizing several objective functions (some of which must be maximized, some minimized) provided the variables satisfy the linear and nonlinear constraints. Nowadays, mathematical programming problems under uncertain environments are of increasing focus. In this chapter, two kinds of multiple objective rough decision making (MORDM) models are discussed:

1. The objective function is a deterministic function with constraints based on rough approximations.

2. The parameters in the objectives and constraints are rough intervals.

For the second type of model, we present three methods to deal with it, that is, the expected value rough model (EVRM), the chance-constrained rough model (CCRM) and the dependent-chance rough model (DCRM).

Finally, an application of the construction site layout planning problem is detailed to show the effectiveness of the above three models. Readers can refer to the following content to know the details.

2.1 Site Layout Planning Problem

Efficient layout planning of a construction site is fundamental to any successful project undertaking. Since I. Yeh [330], P. Zouein, and I. Tommelein [348] identified the construction site layout problem, a large number of research outputs have emerged in this field. A review of the facility layout problem can be found in D. Russell and K. Gau [212].

The site layout problem can be formulated as a quadratic assignment problem (QAP) aimed at the optimal assignment of L facilities to L predetermined locations [79]. Most previous research papers only under a objective function [100, 101]. While, in fact, the objective of site layout is to position temporary facilities both geographically and at the correct time so that construction work can proceed satisfactorily with minimal costs, improved safety, and a better working environment. In addition, a large portion of the previous research on construction site layout planning heavily concentrated on static problems [182, 330]. In static construction site layout problems, the facilities serviced in the different construction phases in accordance with the requirements of the construction work during the entire duration of a construction project are the same [349]. The models for static layout planning ignore the possible reuse of site space to accommodate different resources at different times, to relocate resources, and vary the space needs of resources over time [227]. There is an increasing number of studies focusing on solving dynamic construction site layout planning problems. A. Baykasoglu and N. Gindy [24] made use of a simulated annealing algorithm to solve the dynamic layout problem. A. Baykasoglu et al. [23] made the first attempt to show how an ant colony optimization algorithm can be applied to a dynamic construction site layout planning problem with budget constraints. X. Ning et al. [227] used a continuous dynamic searching scheme to guide the max-min ant system algorithm, which is one of the ant colony optimization algorithms, to solve the dynamic construction site layout planning problem.

In practice, there are two types of uncertainties in layout planning problems [268]: the first one is due to internal disturbances, such as equipment breakdown, variable task times, queuing delays, rejects, and rework, etc., and the second one is caused by external forces, such as uncertainties in the level of demand, product prices, product mix, etc. In these two kinds of uncertainties, objective information can be dealt with by randomness and subjective information can be dealt with by fuzzyness. Current research on uncertainty in construction site layout planning problem is concentrated on fuzzy uncertainty [94, 157]. For instance, F. Karray et al. [157] used a fuzzy rule-based system to calculate the facility closeness relationship. A. Soltani and T. Fernando [278] studied the fuzzy-based multiple objective path planning of construction sites, and so on.

Although random variables, fuzzy variables, and interval numbers are capable of handling some kinds of uncertainty, they are not useful for solving complex problems on the existence of parameters with dual-layer information. To deal with this type of parameter, the rough interval was proposed by M. Rebolledo [251] to depict the most possible and reliable conditions. Recently, Several research works have been published in this field, such as H. Lu et al. [200, 202]. In this book, the variables mentioned above are highly imprecise. They have to be estimated in terms of a decision maker's subjective experiences and objective historical data. Obviously, the strategy capable of handling the problem under both normal and special conditions is preferred. Based on such considerations, decision makers identified this kind of variable ranges from a units to b units in most cases, and in some special periods, the range may vary from c units to d units, where $c \leq a \leq b \leq d$. Hence rough interval $([a,b],[c,d])$ can be used to deal with this kind of uncertain parameter. How to model and solve the dynamic construction site layout planning problem with rough interval parameters is a new area of research interest. To the best of the authors' knowledge, so far, there is little research in this area. Realistically, the construction site layout planning of large-scale construction projects has to be economical and safe and should meet other objectives. Moreover, in practice, most construction site layout planning is dynamic in nature, especially in large scale construction projects where activities within the construction site are generally highly dynamic and complex. The dynamic construction site layout planning problem of large-scale construction projects cannot be settled satisfactorily by static, single objectives and certain models. In this book, we propose an MORDM model for the dynamic construction site layout planning problem.

As pointed out by M. Mawdesley and S. Al-Jibouri [209], the trend in research for facility or site layout problems is concentrated in four areas:

1. Developing more suitable models
2. Extending existing models to include a time element (dynamic layout)
3. Adding uncertainty
4. Or adding multiple criteria for evaluation, which are also special cases for specific types of problems

Our study on MORDM model for the dynamic construction site layout planning problem of large-scale construction projects explores these four areas.

Two types of methods are employed in attacking the construction site layout problem: heuristics and exact . Exact algorithms are developed to obtain, in theory, optimal solutions, but because of the combinatorial nature of the problem, they are only applicable for small-scale problems. Heuristic methods are usually used to larger problems [209]. For large-scale construction projects, heuristic methods have a distinct advantage. Literature reviews show that artificial intelligence (AI), evolutionary algorithm (EA) or evolutionary programming (EP), swarm intelligence (SI), and computer-aided design (CAD) were developed to solve construction site layout planning problems. In the application of AI, the merit is that a search strategy and inference mechanism can be specified. Moreover, experts' knowledge can be elaborated, and thus the heuristic judgment and experience of experts are transformed into a knowledge base, and the decision-making process for designing site layout can be simulated with a computer (X. Ning et al. [227]). Among the models using EA in solving construction site layout planning problems, genetic algorithm (GA) is mostly used ([128, 182, 183]). The merit of GA is its strong evolutionary process to find an optimal solution using crossover, selection, and mutation of the parents' generation. However, the randomly generated initial generation at the beginning of the algorithm's hall affects the solution quality because of the bad gene inherited from the parent generation. Moreover, the search capability is reduced as GA does not rely on gradient or derivative information [96]. On the other hand, most of the GA approaches adopt a sequence-based representation to encode candidate solutions to the layout problem, where each element in the sequence represents a facility, and the number in the element represents the location to place the facility. Reproduction of sequence-represented solutions based on the operators (e.g., crossover and mutation) may lead to infeasible solutions in which several elements have the same value of numbers, that is, overlay of multiple facilities at one location. Some modified GA methods have been proposed to avoid such infeasibility (H. Li and P. Love [182]; M. Mawdesley and S. Al-Jibouri [209]). As another evolutionary algorithm, particle swarm optimization (PSO) was first proposed by J. Kennedy and R.

Eberhart in 1995 [162], and has become one of the most important swarm intelligence paradigms. PSO has a superior search performance for a lot of hard optimization problems with faster and more stable convergence rates compared with other population-based stochastic optimization methods. H. Zhang and J. Wang [343] used PSO to solve the single objective static construction site layout planning problem. J. Xu and Z. Li [308] consider a dynamic construction site layout planning problem that can be expressed by Figure 2.1.

X. Ning et al. [227] presents a new method using a continuous dynamic searching scheme to guide the maxmin ant system algorithm, which is one of the ant colony optimization algorithms proposed to solve the dynamic construction site layout planning problem under the two congruent objective functions of minimizing safety concerns and reducing construction costs. The model they considered is formulated by

$$
\begin{cases}
\min f_1 = \sum_{i=1}^{n} \sum_{j=1}^{n} \sum_{l=1}^{m} \sum_{k=1}^{m} S_{ij} d_{kl} x_{ik} x_{jl} \\
\min f_2 = \sum_{i=1}^{n} \sum_{j=1}^{n} \sum_{l=1}^{m} \sum_{k=1}^{m} C_{ij} d_{kl} x_{ik} x_{jl} \\
\text{s.t.} \begin{cases}
\sum_{i=1}^{n} x_{ik} = 1 \\
\sum_{l=1}^{m} x_{jl} = 1 \\
x \in \{0,1\}
\end{cases}
\end{cases}
\tag{2.1}
$$

where the first objective function f_1 can be represented by minimizing the representative score of safety/environment concerns associated with the construction site layout. The second objective function f_2 can be represented by minimizing the total handling cost of interaction flows between the facilities associated with the construction site layout. Let S_{ij} in f_1 and C_{ij} in f_2 be the closeness relationship values for safety/environment concerns and the total handling cost of interaction flows (MF, IF, PF, EF, SE, and UP) between facilities i and j, respectively and d_{kl} be the distance between facilities k and l. x_{ik} means when facility i is assigned to location k, and x_{jl} means that when the facility j is assigned to location l. n is the number of the facilities, and m is the number of the locations. The constraint of x_{ik} will be a binary variable that takes the value 1 if facility i is assigned to location k, and 0 otherwise. The constraint of x_{jl} will be a binary variable that takes value 1 if facility j is assigned to location l, and 0 otherwise. The two constraints guarantee that one facility can be assigned to one location, and one location can accommodate one facility. The location of each facility is then assigned and located in accordance with these constraints. More details about formulas (2.1) can be seen in [227].

This chapter mainly concentrates on discussing the dynamic construction site layout planning problem with rough interval parameters. Before the real application, the general MODM with rough interval parameters is introduced. By *Appr* measure, we present the expected value model, chance-constrained model, and depended chance model to transform uncertain models to crisp ones. In addition, another MORDM, MODM based rough approximation for feasible regions, is considered.

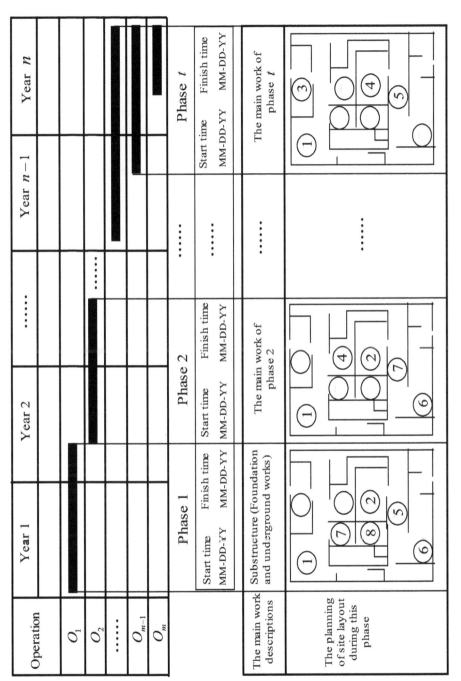

FIGURE 2.1 A dynamic construction site layout planning problem.

2.2 MODM Based on Rough Approximation for Feasible Region

In this section, two classes of rough multiple objective decision making models are presented. In the first one, the feasible region is approximated by the lower approximation (LA) and the upper approximation (UA). The second one adopts a rough membership function. The former is a specific case of the latter.

2.2.1 MODM-Based LA and UA for Feasible Region

Let $U \subseteq \Re^n$ be a nonempty universal set, and $E \subseteq U \times U$ be an equivalence relation defined on U. Let M be a rough subset of U with respect to E. A rough decision making (RDM) model is formulated as follows, and the discussion of this section is based on this model:

$$\begin{cases} \min[f_1(\mathbf{x}), f_2(\mathbf{x}), \cdots, f_m(\mathbf{x})] \\ \text{s.t. } \mathbf{x} \widetilde{\in} X = \{\mathbf{x} | g_r(\mathbf{x}) \leq 0, r = 1, 2, \cdots, p\} \end{cases} \qquad (2.2)$$

where $f_i(\mathbf{x})$ and $g_r(\mathbf{x})$ are a real valued functions, represent objectives and constraints, respectively: $i = 1, 2, \cdots, m; r = 1, 2, \cdots, p$; and $\widetilde{\in}$ refers to $\mathbf{x} \in \underline{E}X$ (surely belonging)), and $\mathbf{x} \overline{\in} X$ refers to $\mathbf{x} \in \overline{E}X$ (possibly belonging).

In order to explain the background of this model, a practical problem proposed by E. Youness [333] is used, although it is single objective programming problem. However, the factor is the approximation of the feasible region, having nothing to do with the objectives. Thus, the example in [333] is appropriate to explain our problem without loss of generality.

Example 2.1

Let f be the effecting of the drug D that was developed to treat livers cancer, that is, $f : X \to R$, where X is the set of all cells of the liver, and $f(x) = y \in \Re$ means the effecting of D on the cell x is y. Suppose

1. Using the drug for 1 year implies killing 30% of cancer cells.

2. Using the drug for 2 years implies killing 50% of cancer cells and 0.1% of normal cells.

3. Using the drug for 3 years implies killing 70% of cancer cells and 0.3% of normal cells.

4. Using the drug for 4 years implies to killing 100% of cancer cells and 0.5% of normal cells.

Define an equivalence relation E on X as $xEy \Leftrightarrow x, y$ are killed in the same year, or x, y are not killed at the end of treatment. Let $A \subseteq X$ be the set of all cancer cells. The defined equivalence relation E makes a partition to X as

$E_1 = \{x \in A | x$ was killed in the first year$\}$.
$E_2 = \{x \in A, y \in X - A | x, y$ were killed in the second year$\}$.
$E_3 = \{x \in A, y \in X - A | x, y$ were killed in the third year$\}$.
$E_4 = \{x \in A, y \in X - A | x, y$ were killed in the forth year$\}$.
$E_5 = \{y \in X - A | y$ were not killed in the end of treatment$\}$.

It is clear that the set A is rough set with respect to the defined equivalence relation, where

$$\underline{E}A = E_1, \text{ and } \overline{E}A = E_1 \cup E_2 \cup E_3 \cup E_4.$$

The problem of finding a cell (or cells) $x \widetilde{\in} A$, on which the effecting of the drug D is as maximum as possible, is formulated as

$$\begin{cases} \max f(x,t) \\ \text{s.t. } x \widetilde{\in} A \end{cases} \tag{2.3}$$

where $f : A \times T \to \Re$ is the drug effecting on a cell (or cells) $x \widetilde{\in} A$ at certain year $t \in T$. For model (2.2), x^* is said to be a surely feasible solution if $x \underline{\in} A$ and x^* is said to be a possibly feasible solution of problem if $x \widetilde{\in} A$.

Obviously, a surely feasible solution must be possibly feasible solution since possibly feasible region is lager than the surely feasible region. The more general model can be formulated as

$$\begin{cases} \max f(\mathbf{x}) \\ \text{s.t. } \mathbf{x} \widetilde{\in} X \end{cases} \tag{2.4}$$

For convenience, we call model (2.4) as rough decision making (RDM) model.

DEFINITION 2.1 $\mathbf{x}^* \widetilde{\in} X$ *is said to be a local efficient solution of RDM if and only if it satisfies* $f_i(\mathbf{x}) \geq f_i(\mathbf{x}^*)$ *for* $i = 1, 2, \cdots, n$ *with strict inequality holding for at least one i for each* $\mathbf{x} \in N_\varepsilon(\mathbf{x}^*) = \{\mathbf{x} | d(\mathbf{x}, \mathbf{x}^*) \leq \varepsilon\}$, *where d is the Euclid distance.*

If the solution $\mathbf{x}^*$ is in the lower approximation of X, then $\mathbf{x}^*$ is called surely local optimal solution with respect to E. If $N_\varepsilon(\mathbf{x}^*) \cap X \neq \emptyset$, then $\mathbf{x}^*$ is called possibly local optimal solution with respect to E.

DEFINITION 2.2 $\mathbf{x}^* \widetilde{\in} X$ *is said to be a global efficient solution of RDM if and only if it satisfies* $f_i(\mathbf{x}) \geq f_i(\mathbf{x}^*)$ *for* $i = 1, 2, \cdots, n$ *with strict inequality holding for at least one i for each* $\mathbf{x} \widetilde{\in} X$.

If $x \underline{\in} X$, then it is called surely global optimal solution with respect to E, and if $x \overline{\in} X$, then it is called possibly global optimal solution with respect to E.

It is well known that convexity plays a important role in mathematical programming problem. A set $X \subseteq \Re^n$ is called a convex set if $\lambda \mathbf{x} + (1 - \lambda)\mathbf{y} \in X$ for all $\mathbf{x}, \mathbf{y} \in X$, and $0 \leq \lambda \leq 1$. For a rough set, the convexity can be defined as

DEFINITION 2.3   *(E. Youness [333]) Let U be a nonempty set in $\Re^n$ and $E \subseteq U \times U$ be an equivalence relation on U. A rough set $M \subseteq U$ with respect to E is said to be a convex rough set if its upper approximation $\overline{E}X$ is convex.*

Based on this definition, a convex set X is an exact E-definable set for which $\underline{E}M = \overline{E}M = M$.

Example 2.2
Let U be a universal set defined as

$$U = \{x = (x_1, x_2) \in \Re^2 | -4 \le x_1 \le 4, -4 \le x_2 \le 4\}$$

and $g(x) = x_1 - x_2$. Define an equivalence relation on U as $xEy \Leftrightarrow g(x) < 0, g(y) < 0$ or $g(x) > 0, g(y) > 0$ or $g(x) = g(y) = 0$. Then we have the following classes:

$$E_1 = \{x \in U | g(x) = 0\},$$
$$E_2 = \{x \in U | g(x) > 0\},$$
$$E_3 = \{x \in U | g(x) < 0\}.$$

A subset $X_1 \subseteq U$, which is defined as

$$X_1 = \{x \in U | x_2 \ge x_1^2 + \frac{1}{2}\} \cup \{x \in U | x_1 = x_2\},$$

is a convex rough set (see Figure 2.2). In fact, $\underline{E}X_1 = E_1, \overline{E}X_1 = E_1 \cup E_3$, which is a convex set. The set

$$X_2 = \{x \in U | x_2 \ge x_1^2 + \frac{1}{2}\} \cup \{x \in U | (x_1 - 1)^2 + (x_2 + 1)^2 \le 1\}$$

is a convex rough set since $\overline{E}X_2 = E_2 \cup E_3$ (see Figure 2.2).

It is known that the intersection of two exact convex sets is convex but their union is not necessarily convex. In the case of rough sets, the following theorem shows that the intersection and union of two convex rough sets are convex when the upper approximation of their union equals the upper approximation of their intersection.

THEOREM 2.1
(E. Youness [333]) Let U be a nonempty set in $\Re^n$, and $E \subseteq U \times U$ be an equivalence relation on U. If X_1 and X_2 are two convex rough sets in U with respect to E and $\overline{E}(X_1 \cup X_2) = \overline{E}(X_1 \cap X_2)$, then $X_1 \cup X_2$ and $X_1 \cap X_2$ are convex rough sets with respect to E.

PROOF   Since $\overline{E}(X_1 \cup X_2) = \overline{E}X_1 \cup \overline{E}X_2$ and $\overline{E}(X_1 \cap X_2) \subseteq \overline{E}X_1 \cap \overline{E}X_2$, then from supposition, we get $\overline{E}(X_1 \cup X_2) = \overline{E}(X_1 \cap X_2) = \overline{E}X_1 = \overline{E}X_2$. That is $X_1 \cup X_2$ and $X_1 \cap X_2$ are convex rough sets with respect to E.

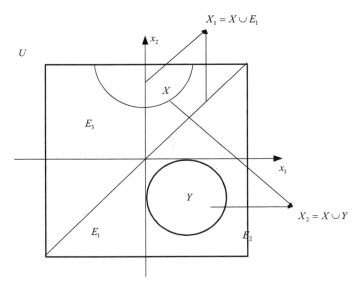

FIGURE 2.2   Convex rough set and nonconvex rough set.

DEFINITION 2.4   *A function $f : U \to \Re$ is said to be a convex function if and only if*

$$f(\lambda \mathbf{x} + (1 - \lambda)\mathbf{y}) \leq \lambda f(\mathbf{x}) + (1 - \lambda)f(\mathbf{y}) \tag{2.5}$$

for each $\mathbf{x}, \mathbf{y} \in U$ and $0 \leq \lambda \leq 1$.

The decision-making problem $\min_{\mathbf{x} \in X}[f_1(\mathbf{x}), f_2(\mathbf{x}), \cdots, f_m(\mathbf{x})]$ is called convex multiple objective decision making if X is a convex set and $f_i(\mathbf{x})$ are convex functions. In an ordinary convex multiple objective convex programming, every local efficient solution is a global efficient solution, and the following proposition shows that the same result is valid in rough programming when both the objectives and the rough set X are convex.

PROPOSITION 2.1
Let X be a convex rough set, and let $f_i(\mathbf{x})$ be convex functions on universe for $i = 1, 2, \cdots, m$. Then, any local efficient solution of RDM is a global efficient solution.

PROOF   Suppose $\mathbf{x}^*$ is a local efficient solution of (2.2). In order to verify that x^* is also a global efficient solution of (2.2), the proof process will be divided into three steps.

First, we claim that $\mathbf{x}^*$ is the local optimal solution of the following auxiliary

$n-1$ programming problems

$$(P_i(\varepsilon)) \quad \begin{cases} \min f_i(\mathbf{x}) \\ \text{s.t.} \begin{cases} f_j(\mathbf{x}) \le \varepsilon_j, j \ne i \\ \mathbf{x} \in X^\alpha \end{cases} \end{cases}$$

where $\varepsilon = (\varepsilon_1, \cdots, \varepsilon_{i-1}, \varepsilon_{i+1}, \cdots, \varepsilon_m)$ are predetermined by DM. Suppose that there exists i such that x^* is not a local solution of $(P_i(\varepsilon))$. Let $\varepsilon = (f_1(\mathbf{x}^*), \cdots, f_{i-1}(\mathbf{x}^*), f_{i+1}(\mathbf{x}^*), \cdots, f_m(\mathbf{x}^*))$. Then there exists $x \in \overline{EX} \cap N(\mathbf{x}^*, \delta)$ such that $f_i(\mathbf{x}) < f_i(\mathbf{x}^*)$ and $f_j(\mathbf{x}) \le f_j(\mathbf{x}^*)$ for $j \ne i$, which contradicts that $\mathbf{x}^*$ is a local efficient solution of (2.2). Then the claim is proved.

After that, we claim that $\mathbf{x}^*$ is the global optimal solution of $(P_i(\varepsilon))$ for all $i = 1, 2, \cdots, m$. Suppose there exists i such that $\mathbf{x}^*$ is not a global optimal solution of $(P_i(\varepsilon))$. Then there exists $\bar{\mathbf{x}} \in X^\alpha$ such that $f_i(\bar{\mathbf{x}}) \le f_i(\mathbf{x}^*)$. Since X is absolutely convex, then $\overline{EX}$ is a convex set, and hence, $\lambda \bar{\mathbf{x}} + (1-\lambda)\mathbf{x}^* \in \overline{EX}$. It follows from the convexity of f_i that

$$\begin{aligned} f_i(\lambda \bar{\mathbf{x}} + (1-\lambda)\mathbf{x}^*) &\le \lambda f_i(\bar{\mathbf{x}}) + (1-\lambda)f_i(\mathbf{x}^*) \\ &\le \lambda f_i(\mathbf{x}^*) + (1-\lambda)f_i(\mathbf{x}^*) \\ &= f_i(\mathbf{x}^*). \end{aligned}$$

Let $\lambda = \frac{\delta_i}{2\|\mathbf{x}^* - \bar{\mathbf{x}}\|}$. Then $\|(\lambda \bar{\mathbf{x}} + (1-\lambda)\mathbf{x}^* - \mathbf{x}^*)\|\frac{\delta_i}{2} \le \delta_i$, that is, $\lambda \bar{\mathbf{x}} + (1-\lambda)\mathbf{x}^* \in N_{\delta_i}(\mathbf{x}^*)$, implying that $\mathbf{x}^*$ is the global optimal solution of $(P_i(\varepsilon))$ for all $i = 1, 2, \cdots, m$. Then the claim is proved.

Finally, we claim that $\mathbf{x}^*$ is an efficient solution of (2.2). Suppose $\mathbf{x}^*$ is not a global efficient solution of (2.2). Then there exists $\bar{\mathbf{x}} \in \overline{EX}$ such that $f_i(\bar{\mathbf{x}}) \le f_i(\mathbf{x}^*), i = 1, 2, \cdots, m$ with strict inequality holding for at least one i. Without loss of generality, we assume that $f_i(\bar{\mathbf{x}}) \le f_i(\mathbf{x}^*)$. Let $\varepsilon = (f_2(\mathbf{x}^*), f_3(x^*), \cdots, f_m(\mathbf{x}^*))$, then $\mathbf{x}^*$ is not a global optimal solution of

$$(P_1(\varepsilon)) \quad \begin{cases} \min f_1(\mathbf{x}) \\ \text{s.t.} \begin{cases} f_i(\mathbf{x}) \le \varepsilon_i, i = 2, \cdots, m \\ \mathbf{x} \in \overline{EX} \end{cases} \end{cases}$$

a contradiction. Therefore $\mathbf{x}^*$ s a global efficient solution of (2.2). This proof is complete.

PROPOSITION 2.2

Let S be a set of all global rough efficient solutions of RDM. If $\bar{x} \in S$ implies $[\bar{x}] = S$, then S is E-definable.

PROOF   Let $\bar{x} \in S$ and $[\bar{x}] = S$, then the lower approximation $\underline{E}S$ of S is $\bar{x}$. Also, the upper approximation $\overline{E}S$ of S is $\bar{x}$, and hence, $\underline{E}S = \overline{E}S = S$, that is, S is E-definable.

PROPOSITION 2.3

If RDM is convex, then the set of rough efficient solutions is either a convex exact set or a convex rough set.

PROOF Let S be the set of all global rough efficient solution for problem (2.2). If $x^* \in S$ and $[x^*] = S$, then, from Proposition 2.2, S is exactly defined by E. Since $[x^*] \subset \overline{E}M$, then, from the convexity of rough set X, we get

$$\lambda y_1 + (1-\lambda)y_2 \in \overline{E}X, \forall y_1, y_2 \in [x^*], \lambda \in [0,1]$$

and

$$f(\lambda y_1 + (1-\lambda)y_2) \leq f(x), \forall x \tilde{\in} X, \lambda \in [0,1],$$

that is, S is a convex set. On the other hand, if $[x^*]$, for each $x^* \in S$, contains other points not in S, then $\overline{E}S = \bigcup_{x^* \in S}[x^*]$ and $\underline{E}S = \emptyset$. Since $\overline{E}S \subset \overline{E}X$, then, for any two points $y_1, y_2 \in S, 0 \leq \lambda \leq 1$, we get $\lambda y_1 + (1-\lambda)y_2 \in \overline{E}X$. Thus, from the onvexity of f_i we get $\lambda y_1 + (1-\lambda)y_2 \in S$, which implies that $\lambda y_1 + (1-\lambda)y_2 \in \overline{E}S$. Therefore, S is a convex rough set.

2.2.1.1 Linear Weighted-Sum Approach

The rough decision-making problem (2.2) can be divined into two subproblems:

$$\begin{cases} \min[f_1(\mathbf{x}), f_2(\mathbf{x}), \cdots, f_m(\mathbf{x})] \\ \text{s.t. } \mathbf{x} \underline{\in} X = \{\mathbf{x} | g_r(\mathbf{x}) \leq 0, r = 1, 2, \cdots, p\} \end{cases} \tag{2.6}$$

and

$$\begin{cases} \min[f_1(\mathbf{x}), f_2(\mathbf{x}), \cdots, f_m(\mathbf{x})] \\ \text{s.t. } \mathbf{x} \overline{\in} X = \{\mathbf{x} | g_r(\mathbf{x}) \leq 0, r = 1, 2, \cdots, p\} \end{cases} \tag{2.7}$$

which are two crisp multiple objective programming problems. In principle, multiple objective optimization problems are very different from single objective optimization problems. In the single objective case, one attempts to obtain the best solution, which is absolutely superior to all other alternatives. In the case of multiple objectives, there does not necessarily exist a solution that is best with respect to all objectives because of incommensurability and conflict among objectives. A solution may be best in one objective but worst in other objectives. Therefore, there usually exists a set of solutions for the multiobjective case, which cannot simply be compared with each other. There are many approaches to solving the multiple objectives programming problem. In this section, the linear weighted-sum approach is used.

For a general multiple objective decision-making problem

$$\begin{cases} \min[f_1(\mathbf{x}), f_2(\mathbf{x}), \cdots, f_m(\mathbf{x})] \\ \text{s.t. } \mathbf{x} \in X \end{cases} \tag{2.8}$$

assign a group of nonnegative coefficients $(w_1, w_2, \cdots, w_m)$ to each objective's function, where w_i can be interpreted as the relative emphasis or worth of that objective compared to other objectives. In other words, the weight can be interpreted as representing our preference over objectives. The bigger w_i is, the more important the objectives f_i is; The smaller w_i is, the less important the objectives f_i is. Then combine them into a single objective function $\sum_{i=1}^{m} w_i f_i$. We called these nonnegative coefficient *weight*. Sometimes, we require that $\sum_{i=1}^{m} w_i = 1$ to normalize it. The vector comprised by weight coefficients $w = (w_1, \cdots, w_m)$ is called it weigh vector. This approach is called linear weighted-sum approach for its characteristic. The procedure of linear weighted-sum approach is summarized as follows:

Step 1. Assign weight coefficients. Assign weight coefficients $w_1, \cdots, w_m$ to objectives according the degree of importance of the objective in problem. It is required that $w_i \geq 0$ and $\sum_{i=1}^{m} w_i = 1$.

Step 2. Solve the single objective programming problem, in which the objective is the weighted-sum objective function. Report the optimal solution as the result. The Linear weighted-sum approach can be represented as

$$
\begin{cases}
\min \sum_{i=1}^{m} w_i f_i(x) \\
\text{s.t. } \mathbf{x} \in X
\end{cases}
\tag{2.9}
$$

The following theorem guarantees that the optimal solution of problem (2.9) is an efficient solution to the problem (2.8).

THEOREM 2.2
Assume that $w_i > 0, i = 1, 2, \cdots, m$. If $\mathbf{x}^* \in X$ is an optimal solution (2.9), then it is an efficient solution of problem (2.8).

PROOF Suppose that $\mathbf{x}^*$ is not an efficient solution to problem (2.8). Then it follows from the definition of an efficient solution that there exists $\mathbf{x} \in X$ such that $f_i(\mathbf{x}) \geq f_i(\mathbf{x}^*)$ for $i = 1, 2, \cdots, n$ with strict inequality holding for at least one i. Observe that $w_i > 0$, then

$$w_i f_i(\mathbf{x}) \leq w_i f_i(\mathbf{x}^*)$$

for $i = 1, 2, \cdots, n$ with strict inequality holding for at least one i. Sum the above inequalities, and we have

$$\min \sum_{i=1}^{m} w_i f_i(\mathbf{x}) < \min \sum_{i=1}^{m} w_i f_i(\mathbf{x}^*),$$

which contradicts that $\mathbf{x}^* \in X$ is a optimal solution to (2.9). Thus, $\mathbf{x}^*$ is an efficient solution to problem (2.8). This proof is complete.

By changing $\mathbf{w}$, we can obtain a set composed of the efficient solutions of the problem (2.8) by solving the problem (2.9).

2.2.1.2 Numerical Example

We use the following example to explain the the multiple objective rough programming and the linear weight-sum method.

Example 2.3
Consider the problem

$$\begin{cases} \min f_1(\mathbf{x}) = x_1 + x_2 \\ \min f_2(\mathbf{x}) = 6x_1^2 + \sqrt{x_2} \\ \text{s.t. } \mathbf{x} \tilde{\in} X. \end{cases} \tag{2.10}$$

Let U be a universal set defined as

$$U = \{\mathbf{x} = (x_1, x_2) \in \Re^2 | x_1^2 + x_2^2 \leq 9\}$$

and K be a polytope generated by the following closed half planes:

$$g_1(\mathbf{x}) = x_1 + x_2 - 2 \leq 0, g_2(x) = x_2 - x_1 - 2 \leq 0,$$

$$g_3(\mathbf{x}) = x_2 - x_1 + 2 \geq 0, g_4(\mathbf{x}) = x_2 + x_1 + 2 \geq 0.$$

Define an equivalence relation as, for each $\mathbf{x}, \mathbf{y} \in U$,

$$\mathbf{x} E \mathbf{y} \Leftrightarrow \begin{cases} \mathbf{x} \text{ is an interior point of } K \text{ implies } \mathbf{y} \text{ is an interior.} \\ \mathbf{x} \text{ is a boundary point of } K \text{ implies } \mathbf{y} \text{ is a boundary.} \\ \mathbf{x} \text{ is an exterior point of } K \text{ implies } \mathbf{y} \text{ is an exterior.} \end{cases}$$

It is clear that E deduces the following classes on U as (see Figure 2.3):

$E_1 = \{\mathbf{x} \in U | g_i(\mathbf{x}) < 0, i = 1, 2, g_i(\mathbf{x}) > 0, i = 3, 4\}$,
$E_2 = \{\mathbf{x} \in U | g_i(\mathbf{x}) > 0, \text{ for at least one } i = 1, 2, g_i(x) < 0, \text{ for at least one } i = 3, 4\}$,
$E_3 = \{\mathbf{x} \in U | g_i(\mathbf{x}) = 0, \text{ for at least one } i = 1, 2, 3, 4\}$.

A subset X of U is defined as

$$X = \{\mathbf{x} \in U | x_1^2 + x_2^2 \leq 4\}$$

with respect to E, with

$$\underline{E}X = E_1 \cup E_3, \overline{E}X = E_1 \cup E_2 \cup E_3.$$

Problem (2.10) can be divided into two crisp multiple objective programming problems. The multiple objective decision-making problems is based on lower approximation

$$\begin{cases} \min f_1(\mathbf{x}) = x_1 + x_2 \\ \min f_2(\mathbf{x}) = 6x_1^2 + \sqrt{x_2} \\ \text{s.t. } \begin{cases} x_1 + x_2 - 2 \leq 0 \\ x_2 - x_1 - 2 \leq 0 \\ x_2 - x_1 + 2 \geq 0 \\ x_2 + x_1 + 2 \geq 0 \end{cases} \end{cases} \tag{2.11}$$

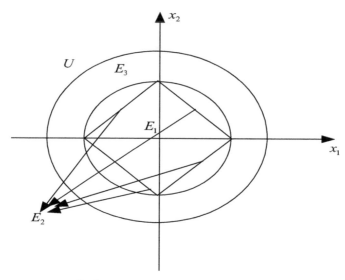

FIGURE 2.3   Rough feasible region.

The multiple objective decision-making problem is based on the upper approximation

$$
\begin{cases}
\min f_1(\mathbf{x}) = x_1 + x_2 \\
\min f_2(\mathbf{x}) = 6x_1^2 + \sqrt{x_2} \\
\text{s.t. } x_1^2 + x_2^2 \le 9
\end{cases}
\tag{2.12}
$$

For problem (2.11), if we think the two objectives have equal importance, the weights are set $w_1 = w_2 = 0.5$. By solving the following single objective programming problems:

$$
\begin{cases}
\min\ \{0.5(x_1 + x_2) + 0.5(6x_1^2 + \sqrt{x_2})\} \\
\text{s.t. }
\begin{cases}
x_1 + x_2 - 2 \le 0 \\
x_2 - x_1 - 2 \le 0 \\
x_2 - x_1 + 2 \ge 0 \\
x_2 + x_1 + 2 \ge 0
\end{cases}
\end{cases}
\tag{2.13}
$$

we obtain a pair of efficient solution $\mathbf{x}^* = (-0.0833, 0.25)$ and the corresponding objective values $(f_1(\mathbf{x}^*), f_2(\mathbf{x}^*)) = (0.167, -0.458)$. If we think the first objective is more important than the second one, we assign a larger weight to the first one, for example $w_1 = 0.7, w_2 = 0.3$. By solving the following single objective programming problems:

$$
\begin{cases}
\min\ \{0.7(x_1 + x_2) + 0.3(6x_1^2 + \sqrt{x_2})\} \\
\text{s.t. }
\begin{cases}
x_1 + x_2 - 2 \le 0 \\
x_2 - x_1 - 2 \le 0 \\
x_2 - x_1 + 2 \ge 0 \\
x_2 + x_1 + 2 \ge 0
\end{cases}
\end{cases}
\tag{2.14}
$$

TABLE 2.1 Efficient solutions with different weights

w_1	w_2	x_1^*	x_2^*	$f_1(x^*)$	$f_2(x^*)$
0.1	0.9	0.001	2.000	1.999	-1.414
0.2	0.8	-0.121	1.988	1.976	-1.409
0.3	0.7	-0.036	1.361	1.325	-1.159
0.4	0.6	-0.056	0.562	0.507	-0.771
0.5	0.5	-0.083	0.250	0.167	-0.458
0.6	0.4	-0.125	0.111	-0.014	-0.240
0.7	0.3	-0.194	0.046	-0.149	0.013
0.8	0.2	0.333	-0.016	-0.318	0.542
0.9	0.1	-0.750	-0.003	-0.747	3.319

we obtain a pair of efficient solution $\mathbf{x}^* = (-0.194, 0.046)$ and the corresponding objective values $(f_1(\mathbf{x}^*), f_2(\mathbf{x}^*)) = (-0.149, 0.013)$. On the contrary, if we think the first objective is less important than the second one, we assign a smaller weight to the first one, for example $w_1 = 0.3, w_2 = 0.7$. By solving the following single objective programming problems:

$$
\begin{cases}
\min \ \{0.3(x_1 + x_2) + 0.7(6x_1^2 + \sqrt{x_2})\} \\
\text{s.t.} \ \begin{cases}
x_1 + x_2 - 2 \le 0 \\
x_2 - x_1 - 2 \le 0 \\
x_2 - x_1 + 2 \ge 0 \\
x_2 + x_1 + 2 \ge 0
\end{cases}
\end{cases}
\tag{2.15}
$$

we obtain a pair of efficient solution $\mathbf{x}^* = (-0.036, 1.361)$ and the corresponding objective values $(f_1(\mathbf{x}^*), f_2(\mathbf{x}^*)) = (1.325, -1.159)$.

By changing the weights, we can get different efficient solutions, which are prepared to be selected by decision maker (see Table 2.1).

2.2.2 MODM with Rough Membership

The general multiple objective decision-making model can be represented as follows;

$$
\begin{cases}
\min[f_1(\mathbf{x}), f_2(\mathbf{x}), \cdots, f_m(\mathbf{x})] \\
\text{s.t. } \mathbf{x} \in X = \{\mathbf{x} \in \mathfrak{R}^n | g_r(\mathbf{x}) \le 0, r = 1, 2, \cdots, p\}
\end{cases}
\tag{2.16}
$$

where $\mathbf{x}$ is n-dimensional decision vector, and f_i and g_r are n-dimensional real-valued functions denoting objectives and constraints, respectively: $i = 1, 2, \cdots, m; r = 1, 2, \cdots, p$.

In practical cases, the DM will give a large region U as estimation for X, i.e. $X \subseteq U$. If a equivalence relation $R \subseteq U \times U$ is defined on the universe U, then (U, R) is an approximate space.

As the roughness is taken into account in the universe, a type of rough model for

(2.16) is formulated as follows:

$$\begin{cases} \min[f_1(\mathbf{x}), f_2(\mathbf{x}), \cdots, f_m(\mathbf{x})] \\ \text{s.t.} \begin{cases} \mu_X^R(\mathbf{x}) \geq \alpha, \alpha \in (0, 1] \\ X = \{\mathbf{x} \in \Re^n | g_j(\mathbf{x}) \leq 0, j = 1, 2, \cdots, p\} \end{cases} \end{cases} \quad (2.17)$$

where $\mu_X^R(\mathbf{x})$ is the rough membership function of x belonging to X, and $\alpha (\in (0, 1])$ is predetermined rough membership by DM, representing the DM's required accuracy.

REMARK 2.1 If X is a definable set in universe U, then the rough model (2.17) degenerates into the crisp model (2.16). In fact, for any $x \in U$, the rough membership of $\mathbf{x}$ to X only has two values, that is,

$$\mu_X^R(\mathbf{x}) = \begin{cases} 1, \text{ if } \quad \mathbf{x} \in X \\ 0, \text{ otherwise} \end{cases}$$

Thus $\mu_X^R(x) \geq \alpha$ is equivalent to $x \in X$ for any $\alpha \in (0, 1]$.

2.2.2.1 Feasible Solutions

In this section, the feasibility of solutions of (2.17) is discussed. For convenience, the set of x satisfying $\mu_X^R(\mathbf{x}) \geq \alpha$ is denoted to be X^α. For problem (2.17), the rough membership α, given by the DM, affect the feasible region notably. It is natural that the smaller the values of rough membership, that is, the lower the accuracy the DM requires, the larger the feasible region. The result is shown by the following proposition.

PROPOSITION 2.4
If $\alpha_1, \alpha_2 \in [0, 1]$ with $\alpha_1 \leq \alpha_2$, then $X^{\alpha_1} \supseteq X^{\alpha_2}$.

PROOF Let $x \in X^{\alpha_2}$, that is, $\mu_X^R(\mathbf{x}) \geq \alpha_2$. It follows from $\alpha_1 \leq \alpha_2$ that $\mu_X^R(\mathbf{x}) \geq \alpha_1$, that is, $\mu_X^R(x) \geq \alpha_1$. That is, $\mathbf{x} \in X^{\alpha_1}$. This proof is complete.

Although the DM may give the rough membership function from 0 to 1 continuously, the feasible region will no longer narrow down as the rough memberships reach some threshold. This threshold is meaningful since it is not necessary for the DM the increase the required rough membership if he or she wants to narrow down the feasible region when the threshold has been reached.

DEFINITION 2.5 $\bar{\alpha} (\in [0, 1])$ *is said to be the* supremum *of rough membership if and only if $X^\alpha = X^{\bar{\alpha}}$ holds for any $\alpha \geq \bar{\alpha}$.*

DEFINITION 2.6 $\underline{\alpha} (\in [0, 1])$ *is said to be the* infimum *of rough membership if and only if $X^\alpha = X^{\underline{\alpha}}$ holds for any $\alpha \leq \underline{\alpha}$.*

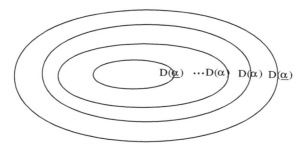

FIGURE 2.4 The procedure for the feasible region narrows down.

The next propositions provide methods to obtain the *supremum* and *infimum* of rough membership for (2.17).

PROPOSITION 2.5

Let $\bar{\alpha} = \sup_{x \in U} \mu_X^R(x)$. Then $\bar{\alpha}$ is the supremums of rough membership.

PROOF   Let $\alpha \geq \bar{\alpha}$. It follows from the monotonicity of $X^{\cdot}$ that $X^\alpha \subseteq X^{\bar{\alpha}}$. If there exists $x^0 \in X^\alpha$ but $x^0 \notin X^{\bar{\alpha}}$, then $\alpha \leq \mu_X^R(x^0) < \bar{\alpha}$, which contradicts the fact that $\alpha \geq \bar{\alpha}$. Then $X^\alpha \supseteq X^{\bar{\alpha}}$. Furthermore, we have $X^\alpha = X^{\bar{\alpha}}$, that is, $\bar{\alpha}$ is the *supremum* of rough membership for the constraint in (2.17). Then the proof is complete.

PROPOSITION 2.6

Let $\underline{\alpha} = \inf_{x \in U} \mu_X^R(x)$. Then $\underline{\alpha}$ is the infimum of rough membership.

PROOF   Let $\alpha \leq \underline{\alpha}$. It follows from the monotonicity of X^α that $X^\alpha \supseteq X^{\underline{\alpha}}$. If there exists $x^0 \in X^{\underline{\alpha}}$ but $x^0 \notin X^\alpha$, then $\underline{\alpha} \leq \mu_X^R(x^0) < \alpha$, which contradicts the fact that $\underline{\alpha} \leq \alpha$. Then $X^{\underline{\alpha}} \supseteq X^\alpha$. Furthermore, we have $X^\alpha = X^{\underline{\alpha}}$, that is, $\underline{\alpha}$ is the *infimum* of rough membership for the constraint in (2.17). Then the proof is complete.

The varying procedure of the feasible region are shown by Figure 2.4.

If the DM approximates the constraints one by one, then he or she can employ model (2.18):

$$\begin{cases} \min[f_1(\mathbf{x}), f_2(\mathbf{x}), \cdots, f_m(\mathbf{x})] \\ \text{s.t.} \begin{cases} \mu_{X_r}^R(\mathbf{x}) \geq \alpha_r \\ X_r = \{\mathbf{x} \in U | g_r(\mathbf{x}) \leq 0\} \\ r = 1, 2, \cdots, p \end{cases} \end{cases} \qquad (2.18)$$

where $\alpha_1, \alpha_2, \cdots, \alpha_p$ are predetermined rough memberships by DM, representing the DM's required accuracy for the rth constraint.

REMARK 2.2 If the feasible region is defined by only one constraint, then there are no differences between (2.17) and (2.18).

The relationship of model (2.17) and (2.18) can be shown by the following proposition.

PROPOSITION 2.7
Assume that $\alpha_1 = \alpha_2 = \cdots \alpha_n = \alpha$, and denote D_1 and D_2 to be the feasible region of (2.17) and (2.18). Then $D_2 \subseteq D_1$.

PROOF For any $x \in D_2$, that is, $\mu_X^R(\mathbf{x}) \geq \alpha$. By virtue of the relationship between X and $X_j, j = 1, 2, \cdots, p$, $\mu_X^R(\mathbf{x}) \geq \alpha$ is equivalent to $\mu_{X_1 \cap X_2 \cap \cdots \cap X_p}^R(\mathbf{x}) \geq \alpha$. It follows from Lemma 1.10 that $\mu_{X_1 \cap X_2 \cap \cdots \cap X_p}^R(\mathbf{x}) \leq \min\{\mu_{X_1}^R(\mathbf{x}), \mu_{X_2}^R(\mathbf{x}), \cdots, \mu_{X_p}^R(\mathbf{x})\}$. Thus, $\mu_{X_j}^R(\mathbf{x}) \geq \alpha, j = 1, 2, \cdots, p$, that is, $\mathbf{x} \in D_1$. Hence $D_2 \subseteq D_1$. This proof is complete.

REMARK 2.3 The strict inclusion relationship $D_2 \subset D_1$ may hold. For examples, let U be a universal set defined as

$$U = \{\mathbf{x} = (x_1, x_2) \in \mathfrak{R}^2 | 0 \leq x_1 \leq 3, 0 \leq x_2 \leq 2\}.$$

Give a partition of U as follows,

$$E_1 = \{\mathbf{x} \in U | 0 \leq x_1 < 1, 0 \leq x_2 \leq 2\},$$

$$E_2 = \{\mathbf{x} \in U | 1 \leq x_1 < 2, 0 \leq x_2 \leq 2\},$$

$$E_3 = \{\mathbf{x} \in U | 2 \leq x_1 < 3, 0 \leq x_2 \leq 2\},$$

then an equivalence relation of U can be defined as $xRy \Leftrightarrow x, y \in E_1$ or $x, y \in E_2$ or $x, y \in E_3$. A subset $X_1 \subset U$ which is defined as

$$X_1 = \{\mathbf{x} \in U | x_1 - x_2 \leq 0\}.$$

A subset $X_2 \subset U$, which is defined as

$$X_2 = \{\mathbf{x} \in U | x_1 - x_2 - 1 \leq 0\}.$$

Assume that rough membership $\alpha = 0.2$. Denote $D_1 = \{\mathbf{x} \in U | \mu_{X_1}^R(x) \geq \alpha, \mu_{X_2}^R(\mathbf{x}) \geq \alpha\}, D_2 = \{\mathbf{x} \in U | \mu_{X_1 \cap X_2}^R \geq \alpha\}$, then $D_1 = E_2$ but $D_2 = \Phi$ (see Figure 2.5).

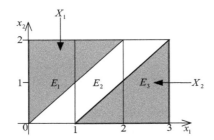

FIGURE 2.5 Comparison of D_1 and D_2.

The result of Proposition 2.7 tells us that if the DM approximates the constraints one by one but not in whole, then the feasible region will be larger, that is, it is possible to obtain the more feasible solutions with the same rough membership.

REMARK 2.4 If there is only one decision variable $x \in \Re$, and the equivalence relationship defined on R is an identical relation, that is, $xRy \Leftrightarrow x = y$ for any $x, y \in \Re$, then models (2.17) and (2.18) degenerate into crisp model (2.16). In other words, the special case of models (2.17) and model (2.18) are a degeneration of model (2.16) in the approximation space.

2.2.2.2 Efficient Solutions

If the rough membership α is given, then (2.17) is a crisp multiple objective programming model. In the theory of multiple objective programming, efficient (Pareto-optimal, nondominated, noninferior) solution and weakly efficient solution are important concepts. We will define the three kinds of solutions for (2.17).

DEFINITION 2.7 $\mathbf{x}^*$ *is said to be an α-efficient solution if and only if $\mathbf{x}^* \in X^\alpha$ and there exists no other $\mathbf{x} \in X^\alpha$ such that $f_i(\mathbf{x}) \le f_i(\mathbf{x}^*)$ for all $i = 1, 2, \cdots, m$ with strict inequality for at least one i.*

If the condition in the definition of the α-efficient solution is relaxed, the α-weakly efficient solution can be defined as

DEFINITION 2.8 $\mathbf{x}^*$ *is said to be an α-weakly efficient solution if and only if $\mathbf{x}^* \in X^\alpha$ and there exists no other $\mathbf{x} \in X^\alpha$ such that $f_i(\mathbf{x}) < f_i(\mathbf{x}^*)$ for all $i = 1, 2, \cdots, m$.*

If all the objectives reach the optimal values simultaneously, the α-absolutely optimal solution can be defined as

DEFINITION 2.9 $\mathbf{x}^*$ *is said to be an* α-*absolutely optimal solution if and only if* $\mathbf{x}^* \in X^\alpha$ *and* $f_i(\mathbf{x}^*) \leq f_i(\mathbf{x})$ *for all* $x \in X^\alpha$ *and* $i = 1, 2, \cdots, m$.

For convenience of notation, we shall denote the set of all α-efficient solution by E^α, the α-weakly efficient solutions set by E_w^α, and the α-absolutely optimal solutions set by O_a^α.

For a fixed $\alpha \in (0, 1]$, (2.17) is a crisp multiple objective programming problem. The relationship of E^α, E^α, and O_a^α can be shown by the following proposition.

PROPOSITION 2.8

Assume that the rough membership $\alpha \in (0, 1]$ *is given. Then*

1. $E_w^\alpha \subseteq E^\alpha$,
2. $E^\alpha = O_a^\alpha$ *if* $O_a^\alpha \neq \emptyset$.

PROOF   The proof of the first result is direct according to Definition 2.7 and Definition 2.8. Now we focus on the proof of the second result. Suppose that $O_a^\alpha \not\subseteq E^\alpha$. Then there exists $\mathbf{x}^* \in E_s^\alpha$ but $\mathbf{x}^* \notin E^\alpha$. It follows from the definition of the α-efficient solution that there exists $\mathbf{x}' \in X^\alpha$ such that $f_i(x') \leq f_i(\mathbf{x}^*)$ for all $i = 1, 2, \cdots, m$ with strict inequality for at least one i, which contradicts the fact that x^* is an α-strong solution. Thus, $O_a^\alpha \subseteq E^\alpha$. Conversely, suppose that $E^\alpha \not\subseteq O_a^\alpha$. Then there exists $\mathbf{x}^* \in E^\alpha$ but $\mathbf{x}^* \notin O_a^\alpha$. Since $O_a^\alpha \neq \emptyset$, there exist $\mathbf{x}^0 \in O_a^\alpha$. It follows from the definition of the α-absolutely optimal solution that $f_i(\mathbf{x}^0) \leq f_i(\mathbf{x})$ for all $\mathbf{x} \in X^\alpha$ and $i = 1, 2, \cdots, m$. Furthermore, we have $f_i(\mathbf{x}^0) \leq f_i(\mathbf{x}^*)$ for all $i = 1, 2, \cdots, m$. It is noted that $\mathbf{x}^* \notin E_s^\alpha$. Thus, one strict inequality holds at least for $f_i(\mathbf{x}^0) \leq f_i(\mathbf{x})$ $i = 1, 2, \cdots, m$, which contradicts the fact that $\mathbf{x}^* \in E^\alpha$. Thus, $E^\alpha \subseteq O_a^\alpha$. Therefore $E^\alpha = O_a^\alpha$. This proof is complete.

DEFINITION 2.10   *A* δ-*neighborhood of a point* $x_0 \in U$ *is the set of all points* $x \in U$ *such that* $||x - x_0|| \leq \delta$, *where* $|| \quad ||$ *is any norm of interest.*

For convenience, we denote a δ-neighborhood of a point $x_0 \in U$ by $N(x_0, \delta)$. Euclidean distance, that is, $d(\mathbf{x}, \mathbf{y}) = \sqrt{\sum_{i=1}^{n} (x_i - y_i)^2}$ for $\mathbf{x} = (x_1, x_2, \cdots, x_n), \mathbf{y} = (y_1, y_2, \cdots, y_n)$, is selected as the norm in this book.

DEFINITION 2.11 $\mathbf{x}^*$ *is said to be an* α-*local efficient solution if and only if there exists* $\delta > 0$ *such that* $\mathbf{x}^*$ *is efficient in* $X^\alpha \cap N(\mathbf{x}^*, \delta)$, *that is, there exists no other* $\mathbf{x} \in X^\alpha \cap N(\mathbf{x}^*, \delta)$ *such that* $f_i(\mathbf{x}) \leq f_i(\mathbf{x}^*)$ *for all* $i = 1, 2, \cdots, m$ *with strict inequality for at least one* i.

As well known, convexity plays an important role in mathematical programming.

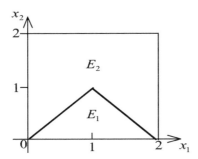

FIGURE 2.6 (U,R) is not a convex approximation space.

The convexity of the crisp set is defined as

DEFINITION 2.12   *A set X is said to be a convex set if and only if*

$$\lambda \mathbf{x} + (1-\lambda)\mathbf{y} \in X \qquad (2.19)$$

holds for any $\mathbf{x},\mathbf{y} \in X$ *and* $0 \leq \lambda \leq 1$.

If roughness is taken into account in the decision-making problem, a new kind of convexity of approximation space is introduced.

DEFINITION 2.13   *An approximation space* (U,R) *is said to be a convex approximation space if and only if the equivalent class* $[x]_R$ *is a convex set for any* $x \in U$.

By the definition, it is easy to verify that the approximation space (U,R) in Remark 2.3 is a convex approximation space. As shown in the definition, the convexity of the approximation space depends on the equivalence relation R.

Example 2.4
Let U be a universal set defined as

$$U = \{\mathbf{x} = (x_1,x_2) \in \Re^2 | 0 \leq x_1 \leq 2, 0 \leq x_2 \leq 2\}$$

and the equivalence relation R is defined by the following partition on U:

$$E_1 = \{\mathbf{x} \in U | x_1 - x_2 \leq 0, x_1 + x_2 - 2 \leq 0\} \text{ and } E_2 = U \backslash E_1.$$

It is obvious that E_1 is a convex set but E_2 is not (see Figure 2.6). Therefore, (U,R) is not a convex approximation space.

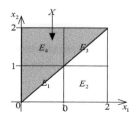

FIGURE 2.7 α-Convexity of X on approximation space (U,R).

In the remainder of this section, the approximation space (U,R) refers to the convex approximation space. The approximation to a fixed set is composed of elementary $[x]_R$. It is known that the intersection of two crisp convex sets is convex, but their union is not necessarily convex. The number of elementary sets forming the approximation to X depends on the approximation, that is, rough membership.

DEFINITION 2.14   *A set $X \subseteq U$ is said to α-convex if and only if X^α is a convex set.*

Example 2.5
Let U be a universal set defined as

$$U = \{x = (x_1,x_2) \in \Re^2 | 0 \le x_1 \le 2, 0 \le x_2 \le 2\}.$$

The equivalence relation R is defined by the following partition on U:

$$E_1 = \{x \in U | 0 \le x_1 \le 1, 0 \le x_2 < 1\},$$

$$E_2 = \{x \in U | 1 < x_1 \le 2, 0 \le x_2 \le 1\},$$

$$E_3 = \{x \in U | 1 \le x_1 \le 2, 1 < x_2 \le 2\},$$

$$E_4 = \{x \in U | 0 \le x_1 < 2, 1 \le x_2 \le 2\}.$$

A subset $X \subseteq U$ is defined as

$$X = \{x \in U | x_1 - x_2 \le 0\}.$$

If the rough membership $0 \le \alpha \le 0.5$, $X^\alpha = E_4$ is a convex set, and if $0.5 < \alpha \le 1$, $X^\alpha = E_i \cup E_2 \cup E_3$ is not a convex set (see Figure 2.7).

DEFINITION 2.15   *A set $X \subseteq U$ is said to absolute convex if and only if X is α-convex for any $0 \le \alpha \le 1$.*

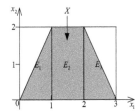

FIGURE 2.8   Absolutely convex set.

Example 2.6
Let (U, R) be an approximate space defined in Remark 2.3. Assume that

$$X = \{x \in U \mid -2x_1 + x_2 \leq 0, 2x_1 + x_2 - 6 \leq 0\}.$$

If the rough membership $0 \leq \alpha \leq 0.5$, $X^\alpha = U$ is a convex set. And if $0.5 < \alpha \leq 1$, $X^\alpha = E_2$ also is a convex set (see Figure 2.8). Then X is an absolutely convex set.

DEFINITION 2.16   *A function $f : U \mapsto \Re$ is said to be a convex function if and only if*

$$f(\lambda \mathbf{x} + (1-\lambda)\mathbf{y}) \leq \lambda f(\mathbf{x}) + (1-\lambda)f(\mathbf{y})$$

for any $\mathbf{x}, \mathbf{y} \in U$ and $0 \leq \lambda \leq 1$. As strict inequality holds, f is said to be a strictly convex function.

PROPOSITION 2.9
Let X be an absolutely convex set and $f_i(\mathbf{x})$ are convex functions on universe U, $i = 1, 2, \cdots, m$. Then, any α-local efficient solution is also an α-efficient solution.

PROOF   Assume that $\mathbf{x}^*$ is an α-local efficient solution of (2.17). In order to verify that $\mathbf{x}^*$ is also an α-efficient solution of (2.17), the proof process will be divided into three steps.

First, we claim that $\mathbf{x}^*$ is the local optimal solution of the following auxiliary $n - 1$ programming problems:

$$(P_i(\varepsilon)) \quad \begin{cases} \min f_i(\mathbf{x}) \\ \text{s.t.} \begin{cases} f_j(\mathbf{x}) \leq \varepsilon_j, j \neq i \\ \mathbf{x} \in X^\alpha \end{cases} \end{cases}$$

where $\varepsilon = (\varepsilon_1, \cdots, \varepsilon_{i-1}, \varepsilon_{i+1}, \cdots, \varepsilon_m)$ are predetermined by DM. Suppose that there exists i such that $\mathbf{x}^*$ is not a local solution of $(P_i(\varepsilon))$. Let $\varepsilon = (f_1(\mathbf{x}^*), \cdots, f_{i-1}(\mathbf{x}^*), f_{i+1}(\mathbf{x}^*), \cdots, f_m(\mathbf{x}^*))$. Then, for all $\delta > 0$, there exists $\mathbf{x} \in X^\alpha \cap N(\mathbf{x}^*, \delta)$

such that $f_i(\mathbf{x}) < f_i(\mathbf{x}^*)$ and $f_j(\mathbf{x}) \leq f_j(\mathbf{x}^*)$ for $j \neq i$, which contradicts that x^* is an α-local efficient solution of (2.17). Then the claim is proved.

After that, we claim that $\mathbf{x}^*$ is the global optimal solution of $(P_i(\varepsilon))$ for all $i = 1, 2, \cdots, m$. Suppose there exists i such that $\mathbf{x}^*$ is not a global optimal solution of $(P_i(\varepsilon))$. Then there exists $\bar{x} \in X^\alpha$ such that $f_i(\bar{x}) \leq f_i(\mathbf{x}^*)$. Since X is absolutely convex, then X^α is a convex set, and hence, $\lambda\bar{x} + (1 - \lambda)\mathbf{x}^* \in X^\alpha$. It follows from the convexity of f_i that

$$f_i(\lambda\bar{x} + (1-\lambda)x^*) \leq \lambda f_i(\bar{x}) + (1-\lambda)f_i(\mathbf{x}^*)$$
$$\leq \lambda f_i(\mathbf{x}^*) + (1-\lambda)f_i(\mathbf{x}^*)$$
$$= f_i(\mathbf{x}^*).$$

Let $\lambda = \frac{\delta_i}{2\|\mathbf{x}^* - \bar{x}\|}$. Then $\|(\lambda\bar{x} + (1-\lambda)\mathbf{x}^* - \mathbf{x}^*)\|\frac{\delta_i}{2} \leq \delta_i$, that is, $\lambda\bar{x} + (1-\lambda)\mathbf{x}^* \in N_{\delta_i}(\mathbf{x}^*)$, implying that $\mathbf{x}^*$ is the global optimal solution of $(P_i(\varepsilon))$ for all $i = 1, 2, \cdots, m$. Then the claim is proved.

Finally, we claim that $\mathbf{x}^*$ is an α-efficient solution of (2.17). Suppose $\mathbf{x}^*$ is not a α-efficient solution of (2.17). Then there exists $\bar{x} \in X^\alpha$ such that $f_i(\bar{x}) \leq f_i(\mathbf{x}^*), i = 1, 2, \cdots, m$ with strict inequality holding for at least one i. Without loss of generality, we assume that $f_i(\bar{x}) \leq f_i(\mathbf{x}^*)$. Let $\varepsilon = (f_2(\mathbf{x}^*), f_3(\mathbf{x}^*), \cdots, f_m(\mathbf{x}^*))$. Then $\mathbf{x}^*$ is not a global optimal solution of

$$(P_1(\varepsilon)) \quad \begin{cases} \min f_1(\mathbf{x}) \\ \text{s.t.} \begin{cases} f_i(\mathbf{x}) \leq \varepsilon_i, i = 2, \cdots, m \\ \mathbf{x} \in X^\alpha \end{cases} \end{cases}$$

a contradiction. Therefore, $\mathbf{x}^*$ s an α-efficient solution of (2.17). This proof is complete.

REMARK 2.5 If X is not an absolute convex set, a local efficient solution may not a global efficient solution. For example, let U be a universal set defined as

$$U = \{x \in \Re | 0 \leq x \leq 4\}.$$

The equivalence relation R is defined by the following partition: $\{[0, 1), [1, 2), [2, 3), [3, 4), [4, 5]\}$. A subset $X \subseteq U$ is defined as

$$X = \{x \in U | 0.8 \leq x \leq 1.8 \quad or \quad 3.8 \leq x \leq 4.6\}.$$

Then $X^\alpha = [1, 2) \cup [4, 5)$ for $\alpha \geq 0.5$. Consider the following programming problem:

$$\begin{cases} \min[f_1(x), f_2(x)] \\ \text{s.t. } x \in X^\alpha, 0.5 \leq \alpha \leq 1 \end{cases}$$

where $f_1(x)$ and $f_2(x)$ are defined as

$$f_1(x) = (x - 2.5)^2 \text{ and } f_2(x) = \begin{cases} -x + 9/4, & \text{if } x \in [0, 2) \\ 0.25, & \text{if } x \in [2, 3) \\ x + 11/4, & \text{if } x \in [3, 5] \end{cases}$$

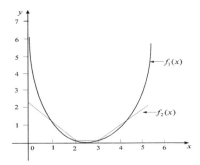

FIGURE 2.9 α-convexity of X on approximation space (U,R).

It is easy to know that $x = 4$ is a local efficient solution not included in the global efficient solution set $[2,3]$ (see Figure 2.9).

As shown above, the α-efficient solution is an α-weakly efficient solution, and it is not true for the inverse. However, the two concepts are identical in some special cases.

PROPOSITION 2.10

Let X be an absolutely convex set and $f_i(\mathbf{x})$ a strictly convex function, $i = 1, 2, \cdots, m$. Then, any α-weakly efficient solution is also an α-efficient solution for any $0 \leq \alpha \leq 1$.

PROOF   Suppose there exists an α_0-weakly efficient solution $\mathbf{x}^*$ is not an α_0-efficient solution for some $\alpha_0 \in (0,1]$. Then there exists some $\mathbf{x} \in X^{\alpha_0}$ such that $f_i(\mathbf{x}) \leq f_i(\mathbf{x}^*)$ for all $i = 1, 2, \cdots, m$ with strict inequality for at least one i. Since X is an absolutely convex set, X^{α_0} is a convex set. Thus $\lambda \mathbf{x} + (1-\lambda)\mathbf{x}^* \in X^{\alpha_0}, 0 \leq \lambda \leq 1$. Since $f_i(\mathbf{x})$ are strictly convex functions. Then we have

$$f_i(\lambda \mathbf{x} + (1-\lambda)\mathbf{x}^*) < \lambda f_i(\mathbf{x}) + (1-\lambda)f_i(\mathbf{x}^*)) \leq \lambda f_i(\mathbf{x}^*) + (1-\lambda)f_i(\mathbf{x}^*)) = f_i(\mathbf{x}^*)$$

for $i = 1, 2, \cdots, m$. It is noted that $\lambda \mathbf{x} + (1-\lambda)\mathbf{x}^* \in X^{\alpha_0}$, which contradicts that $\mathbf{x}^*$ is weakly α_0-efficient solution. Thus $\mathbf{x}^*$ is a efficient solution. This proof is complete.

From Proposition 2.4, we know that the feasible region narrows down as the rough membership increases. However, the result is not suitable for the sets of α-efficient (α-weakly efficient) solutions due to their complexity. Next, we discuss the relationship among $E_w^{\alpha}, E^{\alpha}, O_a^{\alpha}$ with different rough memberships.

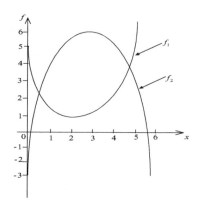

FIGURE 2.10   Illustration of Remark 2.6.

PROPOSITION 2.11
If $E^{\alpha_1}, E^{\alpha_2} \subseteq X^{\alpha_2}$ with $\alpha_1 < \alpha_2 (\in [0,1])$, then $E^{\alpha_1} \subseteq E^{\alpha_2}$.

PROOF   Suppose $E^{\alpha_1} \not\subseteq E^{\alpha_2}$. Then there exists $\mathbf{x}^* \in E^{\alpha_1} (\subseteq X^{\alpha_2})$, but $\mathbf{x}^* \notin E^{\alpha_2} (\subseteq X^{\alpha_2})$, which implies that there exists $\mathbf{x}_0 \in X^{\alpha_2}$ such that $f_i(\mathbf{x}_0) \leq f_i(\mathbf{x}^*)$ for all $i = 1, 2, \cdots, n$ with strict inequality for at least one i. It follows from $\alpha_1 < \alpha_2$ and Proposition 2.4 that $X^{\alpha_2} \subseteq X^{\alpha_1}$. Thus $\mathbf{x}_0 \in X^{\alpha_1}$, which implies that $\mathbf{x}^* \notin E^{\alpha_1}$, a contradiction. Hence $E^{\alpha_1} \subseteq E^{\alpha_2}$. The proposition is proved.

REMARK 2.6   If the condition $E^{\alpha_1}, E^{\alpha_2} \subseteq X^{\alpha_2}$ is not satisfied, the conclusion may be not true. For example, let U be a universal set defined as

$$U = \{x \in \Re | 0 \leq x \leq 6\}.$$

The equivalence relation R is defined by the following partition: $\{[0,1), [1,2), [2,3), [3,4), [4,5), [5,6)\}$. A subset $X \subseteq U$ is defined as

$$X = \{x \in U | 0.2 \leq x \leq 3.4\}.$$

Consider the problem
$$\begin{cases} \min[f_1(x), f_2(x)] \\ \text{s.t. } x \in X^{\alpha}, 0 \leq \alpha \leq 1 \end{cases}$$

where $f_1(x) = (x-2)^2 + 1$ and $f_2(x) = -(x-3)^2 + 6$. It is easy to know that $X^{0.2} = [0,4), X^{0.5} = [0,3]$ and $E^{0.2} = [0,2] \cup [3,4) \not\subseteq X^{0.5}, E^{0.5} = [0,2]$. Then $E^{0.2} \not\subseteq E^{0.5}$ (see Figure 2.10).

REMARK 2.7   The strict conclusion relation may hold if the conditions of Proposition 2.11 are satisfied. For example, let U be a universal set defined

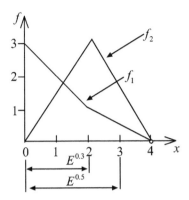

FIGURE 2.11 Illustration of Remark 2.7.

as

$$U = \{x \in \Re | 0 \le x < 4\}.$$

The equivalence relation R is defined by the following partition: $\{[0,1),[1,2),[2,3],(3,4)]\}$. A subset $X \subseteq U$ is defined as

$$X = \{x \in U | 0.2 \le x \le 3.4\}.$$

Consider the problem

$$\begin{cases} \min[f_1(x), f_2(x)] \\ \text{s.t. } x \in X^{\alpha}, 0 \le \alpha \le 1 \end{cases}$$

where $f_1(x)$ and $f_2(x)$ are defined as

$$f_1(x) = \begin{cases} -x+3, & \text{if } x \in [0,2) \\ \frac{1}{2}x+2, & \text{if } x \in [2,4) \end{cases} \quad \text{and} \quad f_2(x) = \begin{cases} \frac{3}{2}x, & \text{if } x \in [0,2) \\ -\frac{3}{2}x+6, & \text{if } x \in [2,4] \end{cases}$$

It is easy to know that $X^{0.3} = [0,4), X^{0.5} = [0,3], E^{0.3} = [0,2] \subseteq X^{0.5}$, and $E^{0.5} = [0,3] \subset X^{0.5}$. Then, $E^{0.3} \subsetneq E^{0.5}$ (see Figure 2.11).

PROPOSITION 2.12
If $E_w^{\alpha_1}, E_w^{\alpha_2} \subseteq X^{\alpha_2}$ with $\alpha_1 < \alpha_2(\in [0,1])$, then $E_w^{\alpha_1} \subseteq E_w^{\alpha_2}$.

PROOF Suppose $E_w^{\alpha_1} \not\subseteq E_w^{\alpha_2}$. Then there exists $\mathbf{x}^* \in E_w^{\alpha_1}(\subseteq X^{\alpha_2})$, but $\mathbf{x}^* \notin E_w^{\alpha_2}(\subseteq X^{\alpha_2}$, which implies that there exists $\mathbf{x}^0 \in X^{\alpha_2}$ such $f_i(x_0) < f_i(x^*)$ for all $i = 1,2,\cdots,n$. It follows from $\alpha_1 < \alpha_2$ and Proposition 2.4 that $X^{\alpha_2} \subseteq X^{\alpha_1}$. Thus $\mathbf{x}^0 \in X^{\alpha_1}$, which implies that $x^* \notin E_w^{\alpha_1}$, a contradiction. Hence $E_w^{\alpha_1} \subseteq E_w^{\alpha_2}$. The theorem is proved.

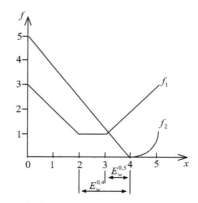

FIGURE 2.12   Illustration of Remark 2.8.

REMARK 2.8   If the condition $E_w^{\alpha_1}, E_w^{\alpha_2} \subseteq X^{\alpha_2}$ is not satisfied, the conclusion may be not true. For example, let U be a universal set defined as

$$U = \{x \in \Re | 0 \le x \le 5\}.$$

The equivalence relation R is defined by the following partition: $\{[0,1), [1,2), [2, 3), [3,4), [4,5]\}$. A subset $X \subseteq U$ is defined as

$$X = \{x \in U | 2.6 \le x \le 4.8\}.$$

Consider the problem

$$\begin{cases} \min[f_1(x), f_2(x)] \\ \text{s.t. } x \in X^\alpha, 0 \le \alpha \le 1 \end{cases}$$

where $f_1(x)$ and $f_2(x)$ are defined as

$$f_1(x) = \begin{cases} -x+3, & \text{if } x \in [0,2) \\ 1, & \text{if } x \in [2,3) \\ x-2, & \text{if } x \in (3,5] \end{cases} \text{ and } f_2(x) = \begin{cases} -\frac{9}{4}x+5, & \text{if } x \in [0,4) \\ (x-4)^2, & \text{if } x \in [4,5] \end{cases}$$

It is easy to know that $X^{0.4} = [2,5], X^{0.5} = [3,5]$, and $E_w^{0.4} = [2,4] \not\subseteq X^{0.4}, E_w^{0.5} = [3,4]$. Then $E_w^{0.4} \not\subseteq E^{0.5}$ (see Figure 2.12).

REMARK 2.9   The strict conclusion relation may hold if the conditions of Proposition 2.12 are satisfied. For example, let U be a universal set defined as

$$U = \{x \in \Re | 0 \le x < 3\}.$$

The equivalence relation R is defined by the following partition: $\{[0,1), [1,2), [2, 3], (3,4)]\}$. A subset $X \subseteq U$ is defined as

$$X = \{x \in U | 0.2 \le x \le 3.4\}.$$

Consider the problem

$$\begin{cases} \min[f_1(x), f_2(x)] \\ \text{s.t. } x \in X^\alpha, 0 \le \alpha \le 1 \end{cases}$$

where $f_1(x)$ and $f_2(x)$ are defined as

$$f_1(x) = \begin{cases} -x+3, & \text{if } x \in [0,2) \\ 1, & \text{if } x \in [2,3) \\ -x+4, & \text{if } x \in (3,4) \end{cases} \text{ and } f_2(x) = \begin{cases} \frac{3}{2}x, & \text{if } x \in [0,2) \\ -\frac{3}{2}x+6, & \text{if } x \in [2,4] \end{cases}$$

It is easy to know that $X^{0.3} = [0,4), X^{0.5} = [0,3], E_w^{0.3} = [0,2] \subseteq X^{0.5}$, and $E_w^{0.5} = [0,3] \subseteq X^{0.5}$. Then $E_w^{0.3} \nsubseteq E^{0.5}$.

2.2.2.3   Compromise Solutions

For many real multiple objective programming problems, the set of efficient solutions may be very large. In order to select an *optimal* solution from the set of efficient solutions, the preference structure of DM is required. For multiple objective programming under a rough environment, we defined the α-compromise solution as follows:

DEFINITION 2.17 $\mathbf{x}^*$ *is said to be an* α-*compromise solution of (2.17) if and only if* $\mathbf{x}^* \in E^\alpha$ *and* $[f_1(\mathbf{x}^*), f_2(\mathbf{x}^*), \cdots, f_m(\mathbf{x}^*)] \preceq [f_1(\mathbf{x}), f_2(\mathbf{x}), \cdots, f_m(\mathbf{x})]$ *for any* $\mathbf{x} \in X^\alpha$, *where* $\preceq$ *is the "less than" relation in the objective functions space based on the DM's explicit or implicit selection criteria.*

The above definition implies that x^* as an α-compromise solution must satisfy two conditions. First, x^* should be an α-efficient solution. Second, x^* is selected based on the DM's preference.

If α is fixed, (2.17) is a crisp multiple objective programming problem. A lot of researchers describ the act of finding a compromise solution to a multiple objective programming problem, such as L. Zadeh and L. Desoer [338], M. Zeleny [339], and A. Wierzbicki [302]. Based on the former researchers' work, S. Gass and P. Roy [111] summarized three basic approaches to obtain comprise solution:

1. A priori articulation of preferences: the DM states a set of criteria weights and the corresponding parametric system is solved.
2. Progressive articulation of preferences: the DM is an interactive participant in the search for an acceptable compromise efficient solution.
3. A posteriori articulation of preferences: the finding of the totality or subset of efficient solutions (usually including the set of ideal solutions), with the DM then choosing one solution based on some explicit or implicit algorithm.

In what follows, we use the interactive fuzzy programming method [260] to obtain an α-compromise solution of (2.17).

Because of the imprecise nature of the DM's judgment for each objective function of (2.17), the fuzzy goals such as "make $f_i(x)$ approximately larger than a certain value" are introduced. Then (2.17) is converted into

$$
\begin{cases}
\max \ [\delta_1(f_1(\mathbf{x})), \delta_2(f_2(\mathbf{x})), \cdots, \delta_m(f_m(\mathbf{x}))] \\
\text{s.t. } \mathbf{x} \in X^\alpha
\end{cases} \tag{2.20}
$$

where the fuzzy goal is characterized by the membership function

$$
\delta_i(f_i(\mathbf{x})) = \begin{cases}
0, & f_i(\mathbf{x}) > f_i^1 \\
\frac{f_i^1 - f_i(\mathbf{x})}{f_i^1 - f_i^0}, & f_i^0 \le f_i(\mathbf{x}) \le f_i^1 \\
1, & f_i(\mathbf{x}) < f_i^0
\end{cases}
$$

where f_i^1 and f_i^0, respectively, denote the maximal and minimal values of the objective functions $f_i(\mathbf{x})$ as follows:

$$
f_i^0 = \min_{\mathbf{x} \in X^\alpha} f_i(\mathbf{x}), \quad f_i^1 = \max_{\mathbf{x} \in X^\alpha} f_i(\mathbf{x}), \quad i = 1, 2, \cdots, m.
$$

The symbol δ will be used to represent the fuzzy membership instead of the generally used μ, with the intention of explicitly marking the difference with rough membership symbol μ, which has been employed.

For each objective function $\delta_i(f_i(\mathbf{x}))$, assume that we can specify the so-called reference membership function value $\bar{\delta}_i$, which reflects the membership function value of $\delta_i(f_i(\mathbf{x}))$. The corresponding optimal solution, which is nearest to the requirements in the minimax sense or better than that if the reference membership function value is attainable, is obtained by solving the following minimax problem:

$$
\min_{\mathbf{x} \in X^\alpha} \max_{i=1,2,\cdots,m} \{\bar{\delta}_i - \delta_i(f_i(\mathbf{x}))\} \tag{2.21}
$$

By introducing the auxiliary variable λ, problem (2.21) is equivalent to

$$
\begin{cases}
\min \lambda \\
\text{s.t. } \begin{cases}
\bar{\delta}_i - \delta_i(f_i(\mathbf{x})) \le \lambda, \ i = 1, 2, \cdots, m \\
0 \le \lambda \le 1 \\
\mathbf{x} \in X^\alpha
\end{cases}
\end{cases} \tag{2.22}
$$

The relationship between the optimal solution of problem (2.22) and the efficient solution of problem (2.17) can be characterized by the following theorem.

THEOREM 2.3
1. If $\mathbf{x}^* \in X$ is a unique optimal solution to problem (2.22) for some $\bar{\delta}_i, i = 1, 2, \cdots, m$, then $\mathbf{x}^*$ is an α-efficient solution to problem (2.17). 2. If $\mathbf{x}^*$ is an α-efficient solution to problem (2.17) with $0 < \delta_i(f_i(\mathbf{x}^*)) < 1$ holding for all i, then $\mathbf{x}^*$ is an optimal solution to problem (2.22).

PROOF   1.   If a unique optimal solution $\mathbf{x}^*$ to problem (2.22) is not the α-efficient solution to problem (2.17), then there exists $\bar{\mathbf{x}} \in X^\alpha$ such that $f_k(\bar{\mathbf{x}}) < f_k(\mathbf{x}^*)$ for some $k \in \{1, 2, \cdots, m\}$.

Then

$$\frac{f_k^1 - f_k(\bar{\mathbf{x}})}{f_k^1 - f_k^0} > \frac{f_k^1 - f_k(\mathbf{x}^*)}{f_k^1 - f_k^0}$$

$$\Leftrightarrow \delta(f_k(\bar{\mathbf{x}})) > \delta(f_k(\mathbf{x}^*))$$

$$\Leftrightarrow \bar{\delta}_k - \delta(f_k(\bar{\mathbf{x}})) < \bar{\delta}_k - \delta(f_k(\mathbf{x}^*)).$$

This means that there exists a $\bar{\lambda}$ satisfying $\bar{\lambda} < \lambda^*$. It follows that x^* is not the optimal solution to problem (2.22), which contradicts the assumption that $\mathbf{x}^*$ is a unique optimal solution to problem (2.22).

2. If $\mathbf{x}^*$ is not an optimal solution to problem (2.22), then there exists $\bar{\mathbf{x}} \in X^\alpha$ such that $\bar{\delta}_i - \delta(f_i(\bar{\mathbf{x}})) < \bar{\delta}_i - \delta(f_i(\mathbf{x}^*)), i = 1, 2, \cdots, m$. Since $0 < \delta(f_i(\mathbf{x}^*)) < 1$, then

$$\delta(f_i(\mathbf{x})) > \delta(f_i(\mathbf{x}^*))$$

$$\Leftrightarrow \frac{f_i^1 - f_i(\bar{\mathbf{x}})}{f_i^1 - f_i^0} > \frac{f_i^1 - f_i(\bar{\mathbf{x}})}{f_i^1 - f_i^0}$$

$$\Leftrightarrow f_i(\bar{\mathbf{x}}) < f_i(\mathbf{x}^*).$$

This shows that x^* is not an efficient solution to problem (2.17), which contradicts the assumption that x^* is a efficient solution to problem (2.17). The proof is complete.

If x^*, an optimal solution to problem (2.22), is not unique, then the α-efficiency test for x^* can be performed by solving the following problem:

$$\begin{cases} \max \sum_{i=1}^m \varepsilon_i \\ \text{s.t.} \begin{cases} \delta_i(F_i(x^*)) + \varepsilon_i = \bar{\delta}_i \\ \varepsilon_i \geq 0, i = 1, 2, \cdots, m \\ x \in X^\alpha \end{cases} \end{cases} \qquad (2.23)$$

From Theorem 2.3, we know that the unique optimal solution x^* of problem (2.22) is an α-efficient solution to problem (2.17). Then the interactive fuzzy satisfying method can be constructed to obtain a compromise solution of problem (2.17).

Based the above discussion, the solution procedure are summaried as follows:

Step 1. The DM is required to present reference membership values $\bar{\delta}_i, i = 1, 2, \cdots, m$.

Step 2. The optimal solution to problem (2.22), which is also an α-efficient solution to problem (2.17), is considered as an α-comprise solution to problem (2.17).

Step 3. If the obtained $\delta_i(f_i(x^*))$ are satisfying, the process stops and x^* is selected as a comprise solution to problem (2.17). Otherwise, the DM should update his or her reference membership values $\bar{\delta}_i$ and return to Step 2.

The optimal solution will improve as the rough membership α decreases for fixed reference membership values $\bar{\delta}_i$. However, the efficient solutions set of (2.17) would not necessarily be larger as rough membership α decreases. Then the compromise solution would not would not necessarily improve as α decreases. For special cases, that is, $E^{\alpha_1} \supseteq E^{\alpha_2}$ with $\alpha_1 < \alpha_2$, the compromise solution will improve in the first situation than in the second situation.

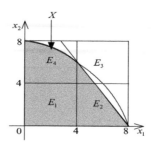

FIGURE 2.13 Rough feasible region of Example 2.7.

Example 2.7
Consider the problem

$$\begin{cases} \min f_1(\mathbf{x}) = 3x_1^2 - 6\sqrt{x_2} \\ \min f_2(\mathbf{x}) = -x_1 + 3x_2 \\ \text{s.t.} \begin{cases} x_1^2 + x_2^2 \leq 64 \\ \sqrt{3}x_1 + x_2 \leq 8\sqrt{3} \\ x_1, x_2 \geq 0 \end{cases} \end{cases} \qquad (2.24)$$

The universe $U = \{(x_1, x_2) \in \Re^2 | 0 \leq x_1 \leq 8, 0 \leq x_2 \leq 8\}$. The equivalence relation R is defined by the following partition on U:

$$E_1 = \{x \in U | 0 \leq x_1 \leq 4, 0 \leq x_2 < 4\},$$

$$E_2 = \{x \in U | 4 < x_1 \leq 8, 0 \leq x_2 \leq 4\},$$

$$E_3 = \{x \in U | 4 \leq x_1 \leq 8, 4 < x_2 \leq 8\},$$

$$E_4 = \{x \in U | 0 \leq x_1 < 4, 4 \leq x_2 \leq 8\}.$$

Let $X = \{(x_1, x_2 \in U | x_1^2 + x_2^2 \leq 64, \sqrt{3}x_1 + x_2 \leq 8\sqrt{3}, x_1, x_2 \geq 0\}$ (see Figure 2.13). The different approximation of X with different rough membership can be calculated as follows:

$$X^\alpha = \begin{cases} E_1, & \text{if } \sqrt{3}/2 < \alpha \leq 1 \\ E_1 \cup E_4, & \text{if } 1 - \sqrt{3}/6 < \alpha \leq \sqrt{3}/2 \\ E_1 \cup E_2 \cup E_4, & \text{if } 32\sqrt{3}/3 - 16 < \alpha \leq 1 - \sqrt{3}/6 \\ E_1 \cup E_2 \cup E_3 \cup E_4, & \text{if } 0 \leq \alpha < 32\sqrt{3}/3 - 16 \end{cases}$$

As $\alpha \in (\sqrt{3}/2, 1]$, $f_i^1, f_i^0 (i = 1, 2)$ can be calculated as

$$f_1^0 = -12, f_1^1 = 48, f_2^0 = -4, f_2^1 = 12.$$

TABLE 2.2 Interactive process as $\sqrt{3}/2 < \alpha \leq 1$

$\bar{\delta}_1$	$\bar{\delta}_2$	f_1	f_2	$\delta(f_1)$	$\delta(f_2)$	x_1	x_2	λ
1.00	1.00	0.571	−0.649	0.209	0.791	0.784	0.045	0.209
0.95	1.00	2.353	−0.974	0.761	0.844	1.049	0.025	0.189
1.00	0.95	−0.968	−0.257	0.816	0.766	0.542	0.095	0.183
1.00	0.90	−2.223	0.208	0.837	0.737	0.377	0.195	0.163
1.00	0.85	−3.253	0.734	0.854	0.704	0.286	0.340	0.146
1.00	0.80	−4.130	1.300	0.869	0.669	0.233	0.511	0.131

TABLE 2.3 Interactive process as $1 - \sqrt{3}/6 < \alpha \leq \sqrt{3}/2$

$\bar{\delta}_1$	$\bar{\delta}_2$	f_1	f_2	$\delta(f_1)$	$\delta(f_2)$	x_1	x_2	λ
1.00	1.00	−4.337	1.454	0.805	0.805	0.223	0.559	0.195
0.95	1.00	−3.040	0.613	0.785	0.835	0.302	0.305	0.165
1.00	0.95	−5.444	2.379	0.822	0.722	0.180	0.853	0.178
0.90	1.00	−1.442	−0.101	0.761	0.861	0.473	0.124	0.139
0.85	1.00	0.538	−0.642	0.730	0.880	0.780	0.046	0.120
0.80	1.00	2.853	−1.051	0.695	0.895	1.117	0.022	0.105

By model (2.22), we have

$$
\begin{cases}
\min \lambda \\
\text{s.t.} \begin{cases}
\bar{\delta}_1 - \frac{48 - 3x_1^2 + 6\sqrt{x_2}}{60} \leq \lambda \\
\bar{\delta}_2 - \frac{12 + x_1 - 3x_2}{16} \leq \lambda \\
0 \leq x_1 \leq 4 \\
0 \leq x_2 < 4 \\
0 \leq \lambda \leq 1
\end{cases}
\end{cases}
\tag{2.25}
$$

The compromise solutions are listed in Table 2.2. For the initial reference membership value 1, each membership function value and the solution x as well as the objective functions $f_i(\mathbf{x}), i = 1, 2$ are obtained (see the first row in Table 2.2). If the DM wishes to increase $f_2(\mathbf{x})$ by sacrificing $f_1(x)$, then the reference membership value needs to be updated, such as $(0.95, 1)$, or else should be updated, such as $(1, 0.95)$. The results are listed in the second and third rows, respectively. Suppose that the DM is satisfied with the solution when the reference membership value is $(1, 0.80)$. Then the interaction process is terminated and the satisfactory solution and the compromise solution is $x^* = (0.233, 0.511)$. In addition, we have $(f_1^*, f_2^*) = (-4.130, 1.300)$.

Similarly, the results of interactive process as $1 - \sqrt{3}/6 < \alpha \leq 3/2$ and $2\sqrt{3}/3 - 16 < \alpha \leq 1 - \sqrt{3}/6$ are listed in Table 2.3, Table 2.4 and Table 2.5, respectively.

TABLE 2.4   Interactive process as $32\sqrt{3}/3 - 16 < \alpha \le 1 - \sqrt{3}/6$

$\bar{\delta}_1$	$\bar{\delta}_2$	f_1	f_2	$\delta(f_1)$	$\delta(f_2)$	x_1	x_2	λ
1.00	1.00	18.823	−2.518	0.829	0.829	2.530	0.004	0.171
0.95	1.00	26.313	−2.971	0.793	0.843	2.980	0.003	0.157
1.00	0.95	11.739	−2.000	0.862	0.812	2.020	0.007	0.138
1.00	0.90	5.339	−1.378	0.893	0.791	1.20	0.014	0.107
1.00	0.85	0.196	−0.566	0.917	0.768	0.725	0.053	0.082
1.00	0.80	−2.932	0.554	0.9933	0.792	0.310	0.288	0.067

TABLE 2.5   Interactive process as $0 \le \alpha < 32\sqrt{3}/3 - 16$

$\bar{\delta}_1$	$\bar{\delta}_2$	f_1	f_2	$\delta(f_1)$	$\delta(f_2)$	x_1	x_2	λ
1.00	1.00	18.823	−2.518	0.829	0.829	2.530	0.004	0.171
0.95	1.00	26.313	−2.971	0.793	0.843	2.980	0.003	0.157
1.00	0.95	11.739	−2.000	0.862	0.812	2.020	0.007	0.138
1.00	0.90	5.339	−1.378	0.893	0.791	1.20	0.014	0.107
1.00	0.85	0.196	−0.566	0.917	0.768	0.725	0.053	0.082
1.00	0.80	−2.932	0.554	0.993	0.792	0.310	0.288	0.067

2.3   EVRM

Rough intervals (RIs) can be a suitable concept to express dual-layer information [201, 251]. For example, in the problem of water resource management, when the quantity of an unregulated water source is highly imprecise, it has to be estimated in terms of a decision maker's subjective experiences and objective historical data. An optimal management strategy capable of handling the problem under both normal and special conditions is often preferred. Based on the investigation, decision makers identified that the water quantity ranges from 12 to $18 \times 10^6 m^3$ per year under most cases; however, in several special years, the range may change to 10 to $20 \times 10^6 m^3$. For this type of information, the existing methods cannot reflect its dual-layer feature. Two challenges thus emerge: one is to find an effective expression means that could reflect dual-layer information (i.e., not only the parameter's most possible value but also its most reliable value), and the other is to use an appropriate method to generate decisions with dual-layer information directly corresponding to the most possible and reliable conditions of the system. Then the dual-layer information can be expressed by the rough interval $([12 \times 10^6, 18 \times 10^6], [10 \times 10^6, 20 \times 10^6])$.

A rough interval consists of an upper approximation interval (UAI) and a lower approximation interval (LAI). A variable defined as an RI is assumed to be inside its UAI and cannot get the qualitative value outside of this interval. The second element of an RI, LAI, can also be defined on this basis. RIs fulfill all the rough set's properties and core concepts, such as operation methods of rough sets [251].

Through embedding UAI and LAI, RIs are capable of reflecting complex information where the most reliable and possible variation ranges can be provided. The solution method can also provide decision supports through the embedded dual-layer information corresponding to the management strategies under common and special conditions, respectively.

If rough intervals are taken account into the multiple objective decision-making model, we have the following model:

$$\begin{cases} \max[f_1(\mathbf{x},\boldsymbol{\xi}),f_2(\mathbf{x},\boldsymbol{\xi}),\cdots,f_m(\mathbf{x},\boldsymbol{\xi})] \\ \text{s.t. } g_r(\mathbf{x},\boldsymbol{\xi}) \leq 0, r = 1,2,\cdots,p \end{cases} \tag{2.26}$$

where $\boldsymbol{\xi} = (\xi_1,\xi_2,\cdots,\xi_n)$ is a vector comprised by rough intervals $\xi_1,\xi_2,\cdots,\xi_n$.

It is necessary for us to know that the model (2.26) is a conceptual model rather than a mathematical model because we cannot maximize an uncertain quantity. There does not exist a natural ordership in an uncertain world. So, we need to transform the uncertain multiple objective model into an approximate certain models to describe the uncertain model.

From Section 2.3 to Section 2.5, three kinds of models, that is, Expected value rough model(EVRM), chance-constrained rough model (CCRM), and dependent chance rough model (DCRM) are used to deal with the multiple objective model with rough interval parameters.

2.3.1   General Model for EVRM

The general expected value rough model is formulated as follows:

$$\begin{cases} \max[E[f_1(\mathbf{x},\boldsymbol{\xi})],E[f_2(\mathbf{x},\boldsymbol{\xi})],\cdots,E[f_m(\mathbf{x},\boldsymbol{\xi})]] \\ \text{s.t. } \begin{cases} E[g_r(\mathbf{x},\boldsymbol{\xi})] \leq 0, r = 1,2,\cdots,p \\ x_j \geq 0, j = 1,2,\cdots,n \end{cases} \end{cases} \tag{2.27}$$

where $\boldsymbol{\xi} = (\xi_1,\xi_2,\cdots,\xi_n)$ is a vector comprised by rough intervals $\xi_1,\xi_2,\cdots,\xi_n$.

DEFINITION 2.18   *A feasible point* $\mathbf{x}^*$ *is said to be an efficient solution to Problem (2.27) if and only if there exists no other feasible solution* $\mathbf{x}$ *such that* $E[f_i(\mathbf{x},\boldsymbol{\xi})] \geq E[f_i(\mathbf{x}^*,\boldsymbol{\xi})]$ *for* $i = 1,2,\cdots,n$ *with strict inequality holding for at least one i.*

Clearly, problem (2.27) is a multiple objective problem with crisp parameters. Then we can convert it into a single objective problem by the traditional method of weighted sum:

$$\begin{cases} \max \sum_{i=1}^{m} \omega_i E[f_i(\mathbf{x},\boldsymbol{\xi})] \\ \text{s.t. } \begin{cases} E[g_j(\mathbf{x},\boldsymbol{\xi})] \leq 0, r = 1,2,\cdots,p \\ \omega_1 + \omega_2 + \cdots + \omega_m = 1 \end{cases} \end{cases} \tag{2.28}$$

As shown by Theorem 2.2, the efficient solution to problem (2.27) is the optimal solution to problem (2.4) and the optimal solution to problem (2.4) is the efficient solution to problem (2.27).

THEOREM 2.4
Let $\boldsymbol{\xi} = (\xi_1, \xi_2, \cdots, \xi_n)$ be a rough interval vector, and let f_i and g_r be convex continuous functions with respect to $\mathbf{x}$, $i = 1, 2, \cdots, m; r = 1, 2, \cdots, p$. Then the expected value programming problem is a convex programming.

PROOF Let $\mathbf{x}^1$ and $\mathbf{x}^2$ be two feasible solutions. Because $g_j(\mathbf{x}, \boldsymbol{\xi})$ is a convex continuous function with respect to x, then

$$g_r(\rho\mathbf{x}^1 + (1-\rho)\mathbf{x}^2, \xi) \leq \rho g_r(\mathbf{x}^1, \boldsymbol{\xi}) + (1-\rho)g_j(\mathbf{x}^2, \boldsymbol{\xi}),$$

where $0 \leq \rho \leq 1$, $j = 1, 2, \cdots, p$. We can have

$$E[g_r(\rho\mathbf{x}^1 + (1-\rho)\mathbf{x}^2, \xi)] \leq E[\rho g_r(\mathbf{x}^1, \boldsymbol{\xi}) + (1-\rho)g_r(\mathbf{x}^2, \boldsymbol{\xi})].$$

By the linearity of the expected operator, we have

$$E[g_r(x^1, \boldsymbol{\xi}(\lambda)) + (1-\rho)g_r(\mathbf{x}^2, \boldsymbol{\xi}(\lambda))] = \rho E[g_r(\mathbf{x}^1, \boldsymbol{\xi}(\lambda))] + (1-\rho)E[g_r(\mathbf{x}^2, \boldsymbol{\xi}(\lambda))].$$

Then $E[g_r(\rho\mathbf{x}^1 + (1-\rho)\mathbf{x}^2, \boldsymbol{\xi})] \leq \rho E[g_r(\mathbf{x}^1, \boldsymbol{\xi})] + (1-\rho)E[g_r(\mathbf{x}^2, \boldsymbol{\xi})] \leq 0$. This means that $\rho \mathbf{x}^1 + (1-\rho)\mathbf{x}^2$ is also a feasible solution. Then X is a convex feasible set.

For every i, $f_i(\mathbf{x}, \boldsymbol{\xi})$ is a convex continuous function with respect to $\mathbf{x}$, and it follows that

$$f_i(\rho\mathbf{x}^1 + (1-\rho)\mathbf{x}^2, \xi) \leq \rho f_i(\mathbf{x}^1, \boldsymbol{\xi}) + (1-\rho)f_i(\mathbf{x}^2, \boldsymbol{\xi}).$$

Hence

$$E[f_i(\rho\mathbf{x}^1 + (1-\rho)\mathbf{x}^2, \boldsymbol{\xi})] \leq \rho E[f_i(x^1, \boldsymbol{\xi})] + (1-\rho)E[f_i(\mathbf{x}^2, \boldsymbol{\xi})]$$

and

$$\sum_{i=1}^{m} \omega_i E[f_i(\rho\mathbf{x}^1 + (1-\rho)\mathbf{x}^2, \boldsymbol{\xi})] \leq \rho \sum_{i=1}^{m} \omega_i E[f_j(\mathbf{x}^1, \boldsymbol{\xi})] + (1-\rho) \sum_{i=1}^{r} \omega_i E[f_j(\mathbf{x}^2, \boldsymbol{\xi})].$$

This means that function $\sum_{i=1}^{m} \omega_i E[f_i(\mathbf{x}, \boldsymbol{\xi})]$ is convex. Above all, we can conclude that the expected value programming problem (2.4) is a convex programming.

We can also obtain the rough expected value goal programming as follows:

$$
\begin{cases}
\min \left(\sum\limits_{i=1}^{m} P_i(u_i d_i^+ + v_i d_i^-) + \sum\limits_{r=1}^{m} P_r(u_r d_r^+ + v_r d_r^-) \right) \\
\text{s.t.}
\begin{cases}
E[f_i(\mathbf{x}, \boldsymbol{\xi})] + d_i^- - d_i^+ = q_i, i = 1, 2, \cdots, m \\
E[g_j(\mathbf{x}, \boldsymbol{\xi})] + d_j^- - d_j^+ = 0, j = 1, 2, \cdots, p \\
d_i^-, d_i^+, d_j^-, d_j^+ \geq 0 \\
u_i, v_i, u_j, v_j = 0 \text{ or } 1
\end{cases}
\end{cases}
\tag{2.29}
$$

where P_i, P_j are the priority coefficients that express the importance of goals.

2.3.2 L-EVRM and Minimax Point Method

In this subsection we concentrate on the multiple objective linear decision-making problem with rough interval coefficients

$$
\begin{cases}
\max \left[\tilde{\mathbf{c}}_1^T \mathbf{x}, \tilde{\mathbf{c}}_2^T \mathbf{x}, \cdots, \tilde{\mathbf{c}}_m^T \mathbf{x} \right] \\
\text{s.t.}
\begin{cases}
\tilde{\mathbf{a}}_r^T \mathbf{x} \leq \tilde{b}_r, r = 1, 2, \cdots, p \\
\mathbf{x} \geq 0
\end{cases}
\end{cases}
\tag{2.30}
$$

where $\tilde{\mathbf{c}}_i = (\tilde{c}_{i1}, \tilde{c}_{i2}, \cdots, \tilde{c}_{in})^T$, $\tilde{\mathbf{a}}_r = (\tilde{a}_{r1}, \tilde{a}_{r2}, \cdots, \tilde{a}_{rn})^T$ are rough interval vectors, and $\tilde{b}_r$ are rough intervals, $i = 1, 2, \cdots, m; r = 1, 2, \cdots, p$. By the expected value operator we can get the following model:

$$
\begin{cases}
\max \left[E[\tilde{\mathbf{c}}_1^T \mathbf{x}], E[\tilde{\mathbf{c}}_2^T \mathbf{x}], \cdots, E[\tilde{\mathbf{c}}_m^T \mathbf{x}] \right] \\
\text{s.t.}
\begin{cases}
E[\tilde{\mathbf{a}}_r^T \mathbf{x}] \leq E[\tilde{b}_r], r = 1, 2, \cdots, p \\
\mathbf{x} \geq 0
\end{cases}
\end{cases}
\tag{2.31}
$$

which is called the linear expected value rough model (L-EVRM).

2.3.2.1 Crisp Equivalent Model

One way of solving an expected value multiple objective decision-making problem is to convert the objectives and constraints of problem (2.31) into their respective crisp equivalents, and then solve them with traditional multi-objective decision making methods. However, this process is usually a hard work and only successful for some special cases. Next, we will consider a special case and present the result in this subsection.

Considering problem (2.31), if these rough vectors and rough variables have special forms, then we have the following theorem.

THEOREM 2.5
Let $\tilde{c}_{ij} = ([c_{ij1}, c_{ij2}], [c_{ij3}, c_{ij4}])(c_{ij3} \leq c_{ij1} \leq c_{ij2} \leq c_{ij4})$, $\tilde{a}_{rj} = ([a_{rj1}, a_{rj2}], [a_{rj3}, a_{rj4}])(a_{rj3} \leq a_{rj1} \leq a_{rj2} \leq a_{rj4})$, *and* $\tilde{b}_r = ([b_{r1}, b_{r2}], [b_{r3}, b_{r4}])(b_{r3} \leq b_{r1} \leq b_{r2} \leq$

b_{r4}) *for* $i = 1, 2, \cdots, m; r = 1, 2, \cdots, p; j = 1, 2, \cdots, n$. *Then problem (2.31) is equivalent to the conventional multiple objective decision-making problem*

$$
\begin{cases}
\max[\frac{1}{2} \sum\limits_{j=1}^{n} \eta((c_{1j1} + c_{1j2}) + (1 - \eta)(c_{1j3} + c_{1j4}))x_j, \frac{1}{2} \sum\limits_{j=1}^{n} (\eta(c_{2j1} + c_{2j2}) \\
\quad + (1 - \eta)(c_{2j3} + c_{2j4}))x_j, \cdots, \frac{1}{2} \sum\limits_{j=1}^{n} (\eta(c_{mj1} + c_{mj2}) + (1 - \eta)(c_{mj3} + c_{mj4}))x_j] \\
\text{s.t.} \begin{cases} \sum\limits_{j=1}^{n} (a_{rj1} + a_{rj2} + a_{rj3} + a_{rj4})x_j \leq b_{r1} + b_{r2} + b_{r3} + b_{r4}, r = 1, 2, \cdots, p \\ x_j \geq 0, j = 1, 2, \cdots, n \end{cases}
\end{cases}
$$

(2.32)

where $0 \leq \eta \leq 1$ *is predetermined by the decision maker.*

PROOF   It follows from the nonnegativity of $x_j (j = 1, 2, \cdots, n)$ and linearity of the rough expected value operator that

$$
E[\tilde{\mathbf{c}}_i^T \mathbf{x}] = E[\sum_{j=1}^{n} \tilde{c}_{ij} x_j] = \sum_{j=1}^{n} E[\tilde{c}_{ij}] x_j = \sum_{j=1}^{n} \frac{1}{2}(\eta(c_{ij1} + c_{ij2}) + (1 - \eta)(c_{ij3} + c_{ij4}))x_j
$$
$$
= \frac{1}{2} \sum_{j=1}^{n} (\eta(c_{ij1} + c_{ij2}) + (1 - \eta)(c_{ij3} + c_{ij4}))x_j
$$

for $i = 1, 2, \cdots, m$,

$$
E[\tilde{\mathbf{a}}_r^T \mathbf{x}] = E[\sum_{j=1}^{n} \tilde{a}_{rj} x_j] = \sum_{j=1}^{n} E[\tilde{a}_{rj}] x_j = \sum_{j=1}^{n} \frac{1}{2}(\eta(a_{rj1} + a_{rj2}) + (1 - \eta)(a_{rj3} + a_{rj4}))x_j
$$
$$
= \frac{1}{2} \sum_{j=1}^{n} (\eta(a_{rj1} + a_{rj2}) + (1 - \eta)(a_{rj3} + a_{rj4}))x_j
$$

and

$$
E[\tilde{b}_r] = \frac{1}{2}(\eta(b_{r1} + b_{r2}) + (1 - \eta)(b_{r3} + b_{r4}))
$$

for $r = 1, 2, \cdots, p$. Thus the theorem is proved.

REMARK 2.10   If $\eta = 1/2$, the equivalent model can be rewritten as

$$
\begin{cases}
\max[\frac{1}{4} \sum\limits_{j=1}^{n} (c_{1j1} + c_{1j2} + c_{1j3} + c_{1j4})x_j, \frac{1}{4} \sum\limits_{j=1}^{n} (c_{2j1} + c_{2j2} + c_{2j3} + c_{2j4})x_j, \cdots, \\
\quad \frac{1}{4} \sum\limits_{j=1}^{n} (c_{mj1} + c_{mj2} + c_{mj3} + c_{mj4})x_j] \\
\text{s.t.} \begin{cases} \sum\limits_{j=1}^{n} (a_{rj1} + a_{rj2} + a_{rj3} + a_{rj4})x_j \leq b_{r1} + b_{r2} + b_{r3} + b_{r4}, r = 1, 2, \cdots, p \\ x_j \geq 0, j = 1, 2, \cdots, n \end{cases}
\end{cases}
$$

(2.33)

REMARK 2.11   If the rough interval vectors (rough intervals) degenerate to interval numbers, that is, $\tilde{c}_{ij} = (c_{ij1}, c_{ij2}), \tilde{a}_{rj} = (a_{rj1}, a_{rj2}), \tilde{b}_r = (b_{r1}, c_{r2})$ for

$i = 1, 2, \cdots, m, r = 1, 2, \cdots, mj = 1, 2, \cdots, n$, then the result of Theorem 2.5 can be rewritten as

$$
\begin{cases}
\max[\frac{1}{2} \sum_{j=1}^{n} (c_{1j1} + c_{1j2})x_j, \frac{1}{2} \sum_{j=1}^{n} (c_{2j2} + c_{2j2})x_j, \cdots, \frac{1}{2} \sum_{j=1}^{n} (c_{mj1} + c_{mj2})x_j] \\
\text{s.t.} \begin{cases} \sum_{j=1}^{n} (a_{rj1} + a_{rj2})x_j \leq b_{r1} + b_{r2}, r = 1, 2, \cdots, p \\ x_j \geq 0, j = 1, 2, \cdots, n \end{cases}
\end{cases}
\tag{2.34}
$$

2.3.2.2 Minimax Point Method

In this section we use the minimax point method to deal with the crisp multiple objective problem (2.32). To maximize the objectives, the minimax point method first constructs an evaluation function by seeking the minimal objective value after computing all objective functions, that is, $u(\mathbf{H}(x)) = \min_{1 \leq i \leq m} H_i(x)$, where $\mathbf{H}(x) = (H_1(x), H_2(x), \cdots, H_m(x))^T$. Then the objective function of problem (2.32) comes down to solving the maximization problem as follows:

$$
\max_{x \in X'} u(\mathbf{H}(x)) = \max_{x \in X'} \min_{1 \leq i \leq m} H_i(x)
\tag{2.35}
$$

Sometimes, decision makers need to consider the relative importance of various goals, and then the weight can be combined into the evaluation function as follows:

$$
\max_{x \in X'} u(\mathbf{H}(x)) = \max_{x \in X'} \min_{1 \leq i \leq m} \{\omega_i H_i(x)\}
\tag{2.36}
$$

where the weight $\sum_{i=1}^{m} \omega_i = 1 (\omega_i > 0)$ and is predetermined by decision makers.

THEOREM 2.6
Let x^* be the optimal solution to the problem (2.36). Then , x^* also is the weak efficient solution to the problem (2.32).

PROOF Assume that $\mathbf{x}^* \in X'$ is the optimal solution to problem (2.36). If there exists an $\mathbf{x}$ such that $H_i(\mathbf{x}) \geq H_i(\mathbf{x}^*)(i = 1, 2, \cdots, m)$, we have

$$
\min_{1 \leq i \leq m} \{\omega_i H_i(\mathbf{x}^*)\} \leq \omega_i\, H_i(\mathbf{x}^*) \leq \omega_i H_i(\mathbf{x}), \ 0 < \omega_i < 1.
$$

Denote $\delta = \min_{1 \leq i \leq m} \{\omega_i H_i(\mathbf{x})\}$, then $\delta \geq \min_{1 \leq i \leq m} \{\omega_i H_i(\mathbf{x}^*)\}$. This means that $\mathbf{x}^*$ is not the optimal solution to problem (2.36). This conflicts with the condition. Thus, there does not exist $\mathbf{x} \in X'$ such that $H_i(\mathbf{x}) \geq H_i(\mathbf{x}^*)$, namely, $\mathbf{x}^*$ is a weak efficient solution to problem (2.32).

By introducing an auxiliary variable, the minimax problem (2.36) can be converted into a single objective problem. Let

$$
\lambda = \min_{1 \leq i \leq m} \{\omega_i H_i(\mathbf{x})\}.
$$

Then problem (2.36) is converted into

$$
\begin{cases}
\max \lambda \\
\text{s.t.} \begin{cases} \omega_i \mathbf{H}_i(\mathbf{x}) \geq \lambda, i = 1, 2, \cdots, m \\ \mathbf{x} \in X' \end{cases}
\end{cases}
\tag{2.37}
$$

THEOREM 2.7

Problem (2.36) is equivalent to problem (2.37).

PROOF   Assume that $\mathbf{x}^* \in X'$ is the optimal solution to problem (2.36), and let $\lambda^* = \min_{1 \leq i \leq m} \{\omega_i \, \mathbf{H}_i(\mathbf{x}^*)\}$. Then it is apparent that $\mathbf{H}_i(\mathbf{x}^*) \geq \lambda^*$. This means that $(\mathbf{x}^*, \lambda^*)$ is a feasible solution to problem (2.37). Assume that $(\mathbf{x}, \lambda)$ is any feasible solution to problem (2.37). Since $\mathbf{x}^*$ is the optimal solution to problem (2.36), we have

$$
\lambda^* = \min_{1 \leq i \leq m} \{\omega_i \mathbf{H}_i(\mathbf{x}^*)\} \geq \min_{1 \leq i \leq m} \{\omega_i \mathbf{H}_i(\mathbf{x})\} \geq \lambda,
$$

namely, $(\mathbf{x}^*, \lambda^*)$ is the optimal solution to problem (2.37).

Conversely, assume that $(\boldsymbol{x}^*, \lambda^*)$ is an optimal solution to problem (2.37). Then, $\omega_i H_i(\boldsymbol{x}^*) \geq \lambda^*$ holds for any i, which means $\min_{1 \leq i \leq m} \{\omega_i H_i(\boldsymbol{x}^*)\} \geq \lambda^*$. It follows that, for any feasible $\mathbf{x} \in X'$,

$$
\min_{1 \leq i \leq m} \{\omega_i \mathbf{H}_i(\mathbf{x})\} = \lambda \leq \lambda^* \leq \min_{1 \leq i \leq m} \{\omega_i \mathbf{H}_i(\mathbf{x}^*)\}
$$

holds, namely, $\mathbf{x}^*$ is the optimal solution to problem (2.36).

The minimax point method can be summarized as follows:

Step 1. Compute the weight for each objective function by solving the two problems, $\max_{\boldsymbol{x} \in X'} H_i(\boldsymbol{x})$ and $\omega_i = H_i(\mathbf{x}^*) / \sum_{i=1}^{m} H_i(\mathbf{x}^*)$.

Step 2. Construct the auxiliary problem as follows:

$$
\begin{cases}
\max \lambda \\
\text{s.t.} \begin{cases} \omega_i \mathbf{H}_i(\mathbf{x}) \geq \lambda, i = 1, 2, \cdots, m \\ \mathbf{x} \in X' \end{cases}
\end{cases}
$$

Step 3. Solve the above problem to obtain the optimal solution.

2.3.2.3 Numerical Example

Example 2.8

Consider the problem

$$
\begin{cases}
\max \left[E[\sum\limits_{j=1}^{6} \xi_j x_j], E[\sum\limits_{j=7}^{12} \xi_j x_j] \right] \\
\text{s.t.} \begin{cases}
E[\xi_1 x_1 + 3\xi_2 x_2 + 5\xi_3 x_3 + 7\xi_4 x_4 + 9\xi_1 x_5 + \xi_1 x_1] \le E[110\xi_6] \\
E[2\xi_7 x_1 + 4\xi_8 x_8 + 6\xi_9 x_9 + 8\xi_{10} x_{10} + 10\xi_{11} x_{11}] \le E[120\xi_{12}] \\
x_1 + x_3 + x_5 + x_7 + x_9 - x_{11} \ge 150 \\
x_2 + x_4 + x_6 + x_8 + x_1 0 + x_{12} \le 400 \\
x_j \ge 0, j = 1, 2, \cdots, 12
\end{cases}
\end{cases} \tag{2.38}
$$

where $\xi_j, j = 1, 2, \cdots, 12$ are rough intervals characterized as

$$
\begin{aligned}
&\xi_1 = ([8, 10], [7, 11]), \quad \xi_2 = ([9, 11], [8, 12]), \quad \xi_3 = ([10, 12], [9, 13]), \\
&\xi_4 = ([11, 13], [10, 14]), \quad \xi_5 = ([12, 14], [11, 15]), \quad \xi_6 = ([13, 15], [12, 16]), \\
&\xi_7 = ([14, 16], [13, 17]), \quad \xi_8 = ([15, 17], [14, 18]), \quad \xi_9 = ([16, 18], [15, 19]), \\
&\xi_{10} = ([17, 19], [16, 20]), \quad \xi_{12} = ([18, 20], [17, 21]), \quad \xi_{12} = ([19, 21], [18, 2]).
\end{aligned}
$$

Set $\eta = 1/2$, by model (2.34), we know that problem (2.38) is equivalent to

$$
\begin{cases}
\max F_1(\mathbf{x}) = 9x_1 + 10x_2 + 11x_3 + 12x_4 + 13x_5 + 14x_6, \\
\max F_2(\mathbf{x}) = 15x_7 + 16x_8 + 17x_9 + 18x_{10} + 19x_{11} + 20x_{12} \\
\text{s.t.} \begin{cases}
9x_1 + 30x_2 + 55x_3 + 84x_4 + 117x_5 \le 1540 \\
30x_7 + 64x_8 + 102x_9 + 144x_{10} + 190x_{11} \le 2400 \\
x_1 + x_3 + x_5 + x_7 + x_9 - x_{11} \ge 150 \\
x_2 + x_4 + x_6 + x_8 + x_{10} + x_{12} \le 400 \\
x_j \ge 0, j = 1, 2, \cdots, 12
\end{cases}
\end{cases} \tag{2.39}
$$

According to the minimax point method, we first compute the weight by solving the two single objective models by

$$
w_1 = F_1^* / (F_1^* + F_2^*) = 7140 / (7140 + 9200) = 0.437
$$

and

$$
w_2 = F_2^* / (F_1^* + F_2^*) = 9200 / (7140 + 9200) = 0.563.
$$

Then, according to (2.37), we construct the following model:

$$
\begin{cases}
\max \lambda \\
\text{s.t.} \begin{cases}
0.437(9x_1 + 10x_2 + 11x_3 + 12x_4 + 13x_5 + 14x_6) + \\
0.563(15x_7 + 16x_8 + 17x_9 + 18x_{10} + 19x_{11} + 20x_{12}) \ge \lambda \\
9x_1 + 30x_2 + 55x_3 + 84x_4 + 117x_5 \le 1540 \\
30x_7 + 64x_8 + 102x_9 + 144x_{10} + 190x_{11} \le 2400 \\
x_1 + x_3 + x_5 + x_7 + x_9 - x_{11} \ge 150 \\
x_2 + x_4 + x_6 + x_8 + x_{10} + x_{12} \le 400 \\
x_j \ge 0, j = 1, 2, \cdots, 12
\end{cases}
\end{cases} \tag{2.40}
$$

After solving the model (2.40, we can get an efficient solution as follows:
$\mathbf{x}^* = (171.11, 0, 0, 0, 0, 198.92, 80, 0, 0, 0, 0, 201.08)$ and $(f_1^*, f_2^*) = (1540, 9200)$.

2.3.3   NL-EVRM and Rough Simulation-Based RWGA

In the case of some nonlinear expected value rough model (NL-EVRM) problems, it is usually difficult to convert them into crisp ones and obtain their expected values. For example, let us consider the problem $\max_{\mathbf{x} \in X} E[\sqrt{(x_1 - \xi_1)^2 + (x_2 - \xi_2)^2}]$, where ξ_1, ξ_2 are rough intervals. As we know, it is almost impossible to convert it into a crisp one. Thus, an intelligent algorithm should be provided to solve it. The technique of rough simulation-based genetic algorithms (GA) is a useful and efficient tool when dealing with them. Let us consider the following multiple objective problem:

$$\begin{cases} \max[E[f_1(\mathbf{x}, \boldsymbol{\xi})], E[f_2(\mathbf{x}, \boldsymbol{\xi})], \cdots, E[f_m(\mathbf{x}, \boldsymbol{\xi})]] \\ \text{s.t.} \begin{cases} E[g_j(\mathbf{x}, \boldsymbol{\xi})] \le 0, j = 1, 2, \cdots, p \\ \mathbf{x} \in X \end{cases} \end{cases}$$

where $f_i(\mathbf{x}, \boldsymbol{\xi})$ or $g_j(\mathbf{x}, \boldsymbol{\xi})$ or both of them are nonlinear with respect to $\mathbf{x}$ and $\boldsymbol{\xi}$; $i = 1, 2, \cdots, m$, $j = 1, 2, \cdots, p$. $\boldsymbol{\xi} = (\xi_1, \xi_2, \cdots, \xi_n)$ is a rough interval vector. Because of the existence of the nonlinear functions, we cannot usually convert it into the crisp one. In this section, we will apply rough simulation-based random weigh genetic algorithm (RWGA) to solve it.

2.3.3.1   Rough Simulation 1 for Expected Value

Let $\boldsymbol{\xi}$ be an n-dimensional rough interval vector, and $f : \mathbf{R}^n \to \mathbf{R}$ a measurable function. Then $f(\boldsymbol{\xi})$ is a rough interval. In order to calculate the expected value $E[f(\boldsymbol{\xi})]$, we sample $\underline{\lambda}_1, \underline{\lambda}_2, \cdots, \underline{\lambda}_N$ from LAI and $\overline{\lambda}_1, \overline{\lambda}_2, \cdots, \overline{\lambda}_N$ from UAI randomly. Then the value $E[f(\boldsymbol{\xi})]$ can be estimated by

$$\frac{1}{N} \sum_{k=1}^{N} \frac{\eta(f(\underline{\lambda}_k) + (1 - \eta)f(\overline{\lambda}_k)}{N} \tag{2.41}$$

provided that N is sufficiently large.

Then the rough simulation 1 for the expected value can be summarized as follows:

Step 1. Set $\underline{L} = \overline{L} = 0$.
Step 2. Generate $\underline{\lambda}_1, \underline{\lambda}_2, \cdots, \underline{\lambda}_N$ from LAI randomly.
Step 3. Generate $\overline{\lambda}_1, \overline{\lambda}_2, \cdots, \overline{\lambda}_N$ from UAI randomly.
Step 4. $\underline{L} \to \underline{L} + f(\xi(\underline{\lambda})), \overline{L} \to \overline{L} + f(\xi(\overline{\lambda}))$.
Step 5. Return $L/2N$.

Example 2.9
Assume that the rough variable $\xi = ([-1, 1], [-2, 2])$. We employ the rough simulation to compute the expected value of $(1 + \xi)/(1 + \xi^2)$. Let $\eta = 0.5$, a

run of the rough simulation with 2000 cycles obtains $E[(1+\xi)/(1+\xi^2)] = 0.67$. The Matlab® file is presented in A.1.

2.3.3.2 RWGA

Since the 1960s, there has been increasing interest in imitating living beings to develop powerful algorithms for difficult optimization problems. A term now in common use refer to such techniques evolutional computation. The best as known algorithms in this class include GAs, developed by J. Holland [135]; evolution strategies, developed by I. Rechenberg and M. Eigen [252], H. Schwefel [267]; and genetic programming developed by J. Koza [171]. There are also many hybrid versions that incorporate various features of the foregoing paradigms. State-of-the-art overviews of the field of evolutionary computation have been given by T. Bäck et al. [13], and Z. Michalewicz [214].

Genetic algorithms, as powerful and broadly applicable stochastic search and optimization techniques, are perhaps the most widely known types of evolutionary computation methods today. In the past few years, the genetic algorithm community has turned much of its attention to optimization problems in industrial engineering, resulting in a fresh body of research and application [39, 66, 178]. A bibliography on genetic algorithms has been prepared by J. Alander [6].

In general, a genetic algorithm has five basic components as summarized by Z. Michalewicz [214]:

1. A genetic representation of solutions to the problem.

2. A way to create an initial population of solutions.

3. A evaluation function rating solutions in terms of their fitness.

4. Genetic operators that alter the genetic composition of children during reproduction.

5. Values for the parameters of genetic algorithms.

H. Ishibuchi et al. [148] proposed a weight-sum based fitness assignment method, called RWGA to obtain a variable search direction toward the Pareto frontier. The weighted-sum approach can be viewed as an extension of methods used in the conventional approach to the multiple objective optimizations to GA. It assigns weights to each objective function and combines the weighted objectives into a single objective function. Typically, there are two types of search behavior in the objective space: fixed direction search and multiple direction search, as demonstrated in Figures 2.14 and 2.15, respectively. The random-weight approach gives the genetic algorithms a tendency to demonstrate a variable search direction, therefore, and to sample the area uniformly over the entire frontier.

This section attempts to apply rough simulation to convert the uncertain multiobjective problem into deterministic one and make use of random-weight genetic algorithm to solve the multiple objective problem. For the following model:

$$\begin{cases} \max[E[f_1(\mathbf{x},\boldsymbol{\xi})], E[f_2(\mathbf{x},\boldsymbol{\xi})], \cdots, E[f_m(\mathbf{x},\boldsymbol{\xi})]] \\ \text{s.t.} \begin{cases} E[g_r(\mathbf{x},\boldsymbol{\xi})] \leq 0, r = 1, 2, \cdots, p \\ x_j \geq 0, j = 1, 2, \cdots, n \end{cases} \end{cases}$$

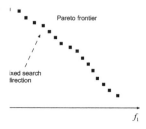

FIGURE 2.14   Search in a fixed direction in criterion space.

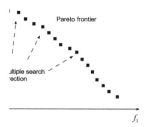

FIGURE 2.15   Search in multiple directions in criterion space.

we can first simulate its expected value by random rough simulation and apply genetic algorithm to solve the multiple objective programming problem. It can be summarized as follows:

Representation. A vector $\mathbf{x}$ is chosen as a chromosome to represent a solution to the optimization problem.

Handling the objective and constraint function. To obtain a determined multiple objective programming problem, we can apply the technique of rough simulation to deal with them.

Initializing process. Suppose that the decision maker is able to predetermine a region which contains the feasible set. Generate a random vector x from this region until a feasible one is accepted as a chromosome. Repeat the above process $N_{popsize}$ times, then we have initial feasible chromosomes $\mathbf{x}^1, \mathbf{x}^2, \cdots, \mathbf{x}^{N_{popsize}}$.

Evaluation function. Decision maker's aim is to obtain the maximum expected value of every goal. Suppose $E[f(\mathbf{x}, \boldsymbol{\xi})] = \sum_{i=1}^{m} E[f_i(\mathbf{x}, \boldsymbol{\xi})]$, where the weight coefficient w_i expresses the importance of $E[f_i(\mathbf{x}, \boldsymbol{\xi})]$ to the decision maker. Then the evaluation function could be given as follows:

$$eval(\mathbf{x}) = \sum_{i=1}^{m} E[f_i(\mathbf{x}, \boldsymbol{\xi})],$$

where the random weight is generated as

$$w_i = \frac{r_i}{\sum_{i=1}^m r_i}, i = 1, 2, \cdots, m,$$

where r_i are nonnegative random numbers.

Selection process. We can apply the roulette wheel method to develop the selection process. Each time a single chromosome for a new population is selected in the following way: Compute the total probability q,

$$q = \sum_{j=1}^{N_{popsize}} eval(\mathbf{x}^j).$$

Then compute the probability of the ith chromosome $q_i, q_i = \frac{eval(\mathbf{x}^i)}{q}$. Generate a random number r in $[0,1]$ and select the ith chromosome $\mathbf{x}^i$ such that $q_{i-1} < r \leq q_i, 1 \leq i \leq N_{popsize}$. Repeat the above process $N_{pop-size}$ times and we obtain $N_{pop-size}$ copies of chromosomes. The selection probability can be computed by the following function:

$$p_i = \frac{eval(\mathbf{x}^i) - eval(x)_{min}}{\sum_{j=1}^{popsize} eval(\mathbf{x}^i) - eval(\mathbf{x})_{min}},$$

where $eval(\mathbf{x})_{min}$ is the minimum fitness value of current population.

Crossover operation. Generate two random numbers λ_1, λ_2 from the open interval $(0, 1)$ satisfying $\lambda_1 + \lambda_2 = 1$, and the chromosome $\mathbf{x}^i$ is selected as a parent provided that $\lambda_i < P_{\lambda_i}$, where parameter P_{λ_i} is the probability of the crossover operation. Repeat this process $N_{popsize}$ times and $P_{\lambda_i} \cdot N_{popsize}$ chromosomes are expected to be selected to undergo the crossover operation. The crossover operator on x^1 and x^2 will produce two children y^1 and y^2 as follows:

$$\mathbf{y}^1 = \lambda_1 \mathbf{x}^1 + \lambda_2 \mathbf{x}^2, \quad \mathbf{y}^2 = \lambda_1 \mathbf{x}^2 + \lambda_2 \mathbf{x}^1.$$

If both children are feasible, then we replace the parents with them, or else we keep the feasible one if it exists. Repeat the above operation until two feasible children are obtained or a given number of cycles is finished.

Mutation operation. Similar to the crossover process, the chromosome x^i is selected as a parent to undergo the mutation operation provided that random number $m < P_m$, where parameter P_m is the probability of the mutation operation. $P_{\lambda_i} \cdot N_{popsize}$ are expected to be selected after repeating the process $N_{popsize}$ times. Suppose that x^1 is chosen as a parent. Choose a mutation direction $\mathbf{d} \in \mathbf{R}^n$ randomly. Replace x with $\mathbf{x} + M \cdot \mathbf{d}$ if $\mathbf{x} + M \cdot \mathbf{d}$ is feasible; otherwise we set M as a random between 0 and M until it is feasible or a given number of cycles are finished. Here, M is a sufficiently large positive number.

We illustrate the rough simulation-based random weight genetic algorithm procedure as follows:

Step 1. Initialize $N_{popsize}$ chromosomes whose feasibility may be checked by fuzzy random simulation.

Step 2. Update the chromosomes by crossover and mutation operations and rough simulation is used to check the feasibility of offspring. Compute the fitness of each chromosome based on weighted-sum objective.

Step 3. Select the chromosomes by spinning the roulette wheel.

Step 4. Make the crossover operation.

Step 5. Make the mutation operation for the chromosomes generated by crossover operation.

Step 6. Repeat the second to fourth steps for a given number of cycles.

Step 7. Report the best chromosome as the optimal solution.

Above all, we combine the rough simulations and RWGA to obtain the rough simulation-based RWGA. The flowchart of this algorithm is shown by Figure 2.16.

2.3.3.3 Numerical Example

Example 2.10

Consider the problem

$$
\begin{cases}
\max f_1(\mathbf{x}, \boldsymbol{\xi}) = 4\xi_1 x_1 + 3.5\xi_2\sqrt{x_2} + 5\sqrt{\xi_3 x_3} + 1.5\xi_4 x_4 \\
\max f_2(\mathbf{x}, \boldsymbol{\xi}) = 7x_1 + 4\xi_2 x_2 + 3\xi_3 x_3 + 2\xi_4 x_4 \\
\text{s.t.} \begin{cases}
x_1 + x_2 + x_3 + x_4 \le 60 \\
10x_1 + 3\xi_2 x_2 + 3\xi_3 x_3 + 3\xi_4 x_4 \ge 120 \\
x_1, x_2, x_3, x_4 \ge 0
\end{cases}
\end{cases} \tag{2.42}
$$

where $\xi_i (i = 1, 2, 3, 4)$ are all rough intervals defined by

$$
\xi_1 = ([2,4],[1,5]), \quad \xi_2 = ([3,5],[2,6]),
$$
$$
\xi_3 = ([4,6],[3,7]), \quad \xi_4 = ([5,7],[4,8]).
$$

Set $\eta = 1/2$. By the expected value operator, we have the following expected model of problem (2.42):

$$
\begin{cases}
\max H_1(\mathbf{x}, \boldsymbol{\xi}) = E[5\xi_1 x_1 + 7\xi_2 x_2 + 5\sqrt{\xi_3 x_3} + \xi_4^2 x_4] \\
\max H_1(\mathbf{x}, \boldsymbol{\xi}) = E[6x_1 + 4x_2 + 2x_3 + x_4] \\
\text{s.t.} \begin{cases}
x_1 + x_2 + x_3 + x_4 \le 60 \\
E[2\xi_1 x_1 + 3\xi_2 x_2 + \xi_3 x_3 + \xi_4 x_4] \ge 120 \\
x_1, x_2, x_3, x_4, x_5 \ge 0
\end{cases}
\end{cases} \tag{2.43}
$$

Since there exist nonlinear objective function and constraint, we cannot transform it into its crisp equivalent model. In order to solve it, we use the rough simulation based RWGA to deal with it.

Let the probability P_c of the crossover process be 0.6, and the probability P_m of the mutation process be 0.3. Perform the rough simulation-based GA with 5000 cycles and we obtain the optimal solutions under different weights as shown in Table 2.6.

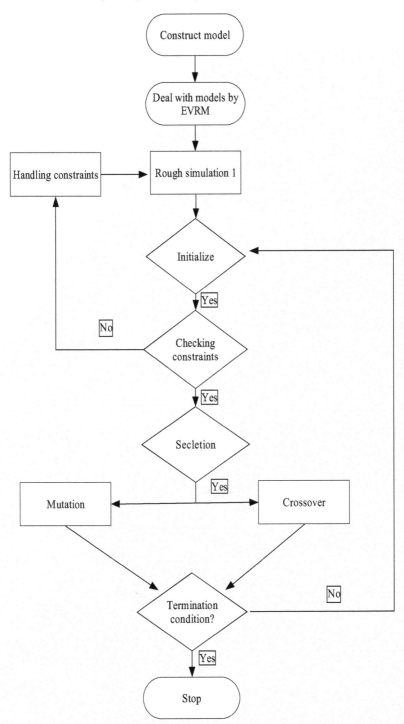

FIGURE 2.16 Flowchart of rough simulation-based RWGA.

TABLE 2.6　Efficient solutions with different weights

w_1	w_2	x_1^*	x_2^*	x_3^*	x_4^*	$f_1(x^*)$	$f_2(x^*)$
0.1	0.9	0	59.68	0.32	0	114.46	959.68
0.2	0.8	0	58.71	1.29	0	119.99	958.71
0.3	0.7	0	57.06	2.94	0	124.92	957.06
0.4	0.6	0	54.78	5.22	0	129.17	954.78
0.5	0.5	31.91	5.44	1.95	20.70	617.44	546.71
0.6	0.4	57.94	1.36	0.70	0	720.94	437.88
0.7	0.3	58.84	0.74	0.43	0	725.35	430.05
0.8	0.2	59.17	0.52	0.31	0	726.34	427.14
0.9	0.1	59.34	0.40	0.26	0	726.65	25.67

2.4　CCRM

In the theory of stochastic programming, chance-constrained programming (CCP) was proposed by A. Charnes and W. Cooper [52, 53] as a means of dealing with randomness by specifying a confidence level at which stochastic constraints hold. Motivated by the idea, a similar technique is produced to deal with multiple objective decision-making problem with rough interval parameters. We call it chance-constrained rough model (CCRM). For the rough problem, the chance operator is actually the $\underline{Appr}, \overline{Appr}$, or $Appr$ measure.

2.4.1　General Model for CCRM

For the following multiple objective model with rough interval parameters,

$$\begin{cases} \max[f_1(\mathbf{x},\boldsymbol{\xi}), f_2(\mathbf{x},\boldsymbol{\xi}), \cdots, f_m(\mathbf{x},\boldsymbol{\xi})] \\ \text{s.t. } g_r(\mathbf{x},\boldsymbol{\xi}) \leq 0, r = 1, 2, \cdots, p \end{cases} \tag{2.44}$$

where $\boldsymbol{\xi} = (\xi_1, \xi_2, \cdots, \xi_n)$ is a vector comprised by rough intervals $\xi_1, \xi_2, \cdots, \xi_n$.
　　The general CCRM is formulated as

$$\begin{cases} \max[\bar{f}_1, \bar{f}_2, \cdots, \bar{f}_m] \\ \text{s.t. } \begin{cases} Ch\{f_i(\mathbf{x},\boldsymbol{\xi}) \geq \bar{f}_i\} \geq \beta_i, i = 1, 2, \cdots, m \\ Ch\{\mathbf{x},\boldsymbol{\xi}) \leq 0\} \geq \alpha_r, r = 1, 2, \cdots, p \end{cases} \end{cases} \tag{2.45}$$

where α_i, β_r are predetermined confidence levels, $i = 1, 2, \cdots, m; r = 1, 2, \cdots, p$.
　　We adopt $\underline{Appr}, \overline{Appr}$, or $Appr$ measure to measure the rough event, and then we have three kinds of CCRM.

2.4.1.1　CCRM Based on $\underline{Appr}$

The spectrum of chance-constrained model based on $\underline{Appr}$ is as follows:

$$\begin{cases} \max[\bar{f}_1, \bar{f}_2, \cdots, \bar{f}_m] \\ \text{s.t.} \begin{cases} \underline{Appr}\{f_i(\mathbf{x}, \boldsymbol{\xi}) \geq \bar{f}_i\} \geq \beta_i, i = 1, 2, \cdots, m \\ \underline{Appr}\{g_r(\mathbf{x}, \boldsymbol{\xi}) \leq 0\} \geq \alpha_r, r = 1, 2, \cdots, p \end{cases} \end{cases} \quad (2.46)$$

where α_i, β_r are predetermined confidence levels, $i = 1, 2, \cdots, m; r = 1, 2, \cdots, p$.

DEFINITION 2.19 $\mathbf{x}^*$ *is said to be a feasible solution at α_r-Appr levels to problem (2.46) if and only if it satisfies $\underline{Appr}\{g_r(\mathbf{x}^*, \boldsymbol{\xi}) \leq 0\} \geq \alpha_r, r = 1, 2, \cdots, p$.*

DEFINITION 2.20 *A feasible solution at α_r-Appr levels $\mathbf{x}^*$ is said to be a β_i-efficient solution for problem (2.46) if and only if there exists no other feasible solution at α_r-Appr levels $\mathbf{x}$ such that $\underline{Appr}\{f_i(\mathbf{x}, \boldsymbol{\xi}) \geq \bar{f}_i\} \geq \beta_i$ with $\bar{f}_i \geq \bar{f}_i^*$ for all i, and $f_{i_0} > f_{i_0}^*$ for at least one $i_0 \in \{1, 2, \cdots, m\}$.*

The model (2.46) is called maxmax model since it is equivalent to

$$\begin{cases} \max_{\mathbf{x}} \max_{\bar{f}_i}[\bar{f}_1, \bar{f}_2, \cdots, \bar{f}_m] \\ \text{s.t.} \begin{cases} \underline{Appr}\{f_i(\mathbf{x}, \boldsymbol{\xi}) \geq \bar{f}_i\} \geq \beta_i, i = 1, 2, \cdots, m \\ \underline{Appr}\{g_r(\mathbf{x}, \boldsymbol{\xi}) \leq 0\} \geq \alpha_r, r = 1, 2, \cdots, p \end{cases} \end{cases} \quad (2.47)$$

If we minimize objectives, the model may formulated as

$$\begin{cases} \min[\bar{f}_1, \bar{f}_2, \cdots, \bar{f}_m] \\ \text{s.t.} \begin{cases} \underline{Appr}\{f_i(\mathbf{x}, \boldsymbol{\xi}) \leq \bar{f}_i\} \geq \beta_i, i = 1, 2, \cdots, m \\ \underline{Appr}\{g_r(\mathbf{x}, \boldsymbol{\xi}) \leq 0\} \geq \alpha_r, r = 1, 2, \cdots, p \end{cases} \end{cases} \quad (2.48)$$

In addition, we have the maxmin model and minmax model, presented by (2.49) and (2.50), respectively:

$$\begin{cases} \max_{\mathbf{x}} \min_{\bar{f}_i}[\bar{f}_1, \bar{f}_2, \cdots, \bar{f}_m] \\ \text{s.t.} \begin{cases} \underline{Appr}\{f_i(\mathbf{x}, \boldsymbol{\xi}) \leq \bar{f}_i\} \geq \beta_i, i = 1, 2, \cdots, m \\ \underline{Appr}\{g_r(\mathbf{x}, \boldsymbol{\xi}) \leq 0\} \geq \alpha_r, r = 1, 2, \cdots, p \end{cases} \end{cases} \quad (2.49)$$

and

$$\begin{cases} \min_{\mathbf{x}} \max_{\bar{f}_i}[\bar{f}_1, \bar{f}_2, \cdots, \bar{f}_m] \\ \text{s.t.} \begin{cases} \underline{Appr}\{f_i(\mathbf{x}, \boldsymbol{\xi}) \leq \bar{f}_i\} \geq \beta_i, i = 1, 2, \cdots, m \\ \underline{Appr}\{g_r(\mathbf{x}, \boldsymbol{\xi}) \leq 0\} \geq \alpha_r, r = 1, 2, \cdots, p \end{cases} \end{cases} \quad (2.50)$$

2.4.1.2 CCRM Based on $\overline{Appr}$

The spectrum of the chance-constrained model based on $\overline{Appr}$ is as follows:

$$\begin{cases} \max[\bar{f}_1, \bar{f}_2, \cdots, \bar{f}_m] \\ \text{s.t.} \begin{cases} \overline{Appr}\{f_i(\mathbf{x}, \boldsymbol{\xi}) \geq \bar{f}_i\} \geq \beta_i, i = 1, 2, \cdots, m \\ \overline{Appr}\{g_r(\mathbf{x}, \boldsymbol{\xi}) \leq 0\} \geq \alpha_r, r = 1, 2, \cdots, p \end{cases} \end{cases} \quad (2.51)$$

where α_i, β_r are predetermined confidence levels, $i = 1, 2, \cdots, m; r = 1, 2, \cdots, p$.

DEFINITION 2.21 x^* *is said to be a feasible solution at α_r-$\overline{Appr}$ levels to problem (2.51) if and only if it satisfies $\overline{Appr}\{g_r(x^*, \xi) \leq 0\} \geq \alpha_r, r = 1, 2, \cdots, p$.*

DEFINITION 2.22 *A feasible solution at α_r-$\overline{Appr}$ levels x^* is said to be a β_i-efficient solution for problem (2.51) if and only if there exists no other feasible solution at α_r-$\overline{Appr}$ levels x such that $\overline{Appr}\{f_i(x, \xi) \geq \bar{f}_i\} \geq \beta_i$ with $\bar{f}_i \geq \bar{f}_i^*$ for all i, and $f_{i_0} > f_{i_0}^*$ for at least one $i_0 \in \{1, 2, \cdots, m\}$.*

2.4.1.3 CCRM Based on *Appr*

The spectrum of the chance-constrained model based on *Appr* is as follows:

$$\begin{cases} \max[\bar{f}_1, \bar{f}_2, \cdots, \bar{f}_m] \\ \text{s.t} \begin{cases} Appr\{f_i(x, \xi) \geq \bar{f}_i\} \geq \beta_i, i = 1, 2, \cdots, m \\ Appr\{g_r(x, \xi) \leq 0\} \geq \alpha_r, r = 1, 2, \cdots, p \end{cases} \end{cases} \tag{2.52}$$

where α_i, β_r are predetermined confidence levels, $i = 1, 2, \cdots, m; r = 1, 2, \cdots, p$.

DEFINITION 2.23 x^* *is said to be a feasible solution at α_r-Appr levels of problem (2.52) if and only if it satisfies $Appr\{g_r(x^*, \xi) \leq 0\} \geq \alpha_r, r = 1, 2, \cdots, p$.*

DEFINITION 2.24 *A feasible solution at α_r-Appr levels x^* is said to be a β_i-efficient solution to problem (2.52) if and only if there exists no other feasible solution at α_r-Appr levels x such that $Appr\{f_i(x, \xi) \geq \bar{f}_i\} \geq \beta_i$ with $\bar{f}_i \geq \bar{f}_i^*$ for all i, and $f_{i_0} > f_{i_0}^*$ for at least one $i_0 \in \{1, 2, \cdots, m\}$.*

REMARK 2.12 If $\eta = 1$, $Appr = \underline{Appr}$, then model (2.52) degenerates into model (2.46). If $\eta = 0$, $Appr = \overline{Appr}$, then model (2.52) degenerates into model (2.51).

2.4.2 L-CCRM and Ideal Point Method

Let us consider a class of multiple objective linear models as

$$\begin{cases} \max [\tilde{c}_1^T x, \tilde{c}_2^T x, \cdots, \tilde{c}_m^T x] \\ \text{s.t.} \begin{cases} \tilde{a}_r^T x \leq \tilde{b}_r, r = 1, 2, \cdots, p \\ x \geq 0 \end{cases} \end{cases} \tag{2.53}$$

where $\tilde{c}_i = (\tilde{c}_{i1}, \tilde{c}_{i2}, \cdots, \tilde{c}_{in})^T$, $\tilde{a}_r = (\tilde{a}_{r1}, \tilde{a}_{r2}, \cdots, \tilde{a}_{rn})^T$ are rough interval vectors, and $\tilde{b}_r$ are rough intervals, $i = 1, 2, \cdots, m; r = 1, 2, \cdots, p$.

The CCRM of (2.53) based on $\underline{Appr}, \overline{Appr}$, or $Appr$ can be formulated as the following three models, respectively:

$$\begin{cases} \max[\bar{f}_1, \bar{f}_2, \cdots, \bar{f}_m] \\ \text{s.t.} \begin{cases} \underline{Appr\{\tilde{\mathbf{c}}_i^T \mathbf{x}\} \geq \beta_i, i = 1, 2, \cdots, m} \\ \underline{Appr\{\tilde{\mathbf{a}}_r^T \mathbf{x} \leq \tilde{b}_r\} \geq \alpha_r, r = 1, 2, \cdots, p} \end{cases} \end{cases} \tag{2.54}$$

where max $\bar{f}_i$ is the β_i-return defined as

$$\max\{\bar{f}_i | \underline{Appr\{\tilde{\mathbf{c}}_i^T \mathbf{x} \geq \bar{f}_i\}} \geq \beta_i\}$$

and

$$\begin{cases} \max[\bar{f}_1, \bar{f}_2, \cdots, \bar{f}_m] \\ \text{s.t.} \begin{cases} \overline{Appr}\{\tilde{\mathbf{c}}_i^T \mathbf{x} \geq \bar{f}_i\} \geq \beta_i, i = 1, 2, \cdots, m \\ \overline{Appr}\{\tilde{\mathbf{a}}_r^T \mathbf{x} \leq \tilde{b}_r\} \geq \alpha_r, r = 1, 2, \cdots, p \end{cases} \end{cases} \tag{2.55}$$

where max $\bar{f}_i$ is the β_i-return defined as

$$\max\{\bar{f}_i | \overline{Appr}\{\tilde{c}_i^T x \geq \bar{f}_i\} \geq \beta_i\}$$

and

$$\begin{cases} \max \ [\bar{f}_1, \bar{f}_2, \cdots, \bar{f}_m] \\ \text{s.t.} \begin{cases} Appr\{\tilde{\mathbf{c}}_i^T \mathbf{x} \geq \bar{f}_i\} \geq \beta_i, i = 1, 2, \cdots, m \\ Appr\{\tilde{\mathbf{a}}_r^T \mathbf{x} \leq \tilde{b}_r\} \geq \alpha_r, r = 1, 2, \cdots, p \end{cases} \end{cases} \tag{2.56}$$

where max $\bar{f}_i$ is the β_i-return defined as

$$\max\{\bar{f}_i | Appr\{\tilde{c}_i^T x \geq \bar{f}_i\} \geq \beta_i\}.$$

All of them are called linear chance-constrained rough model (L-CCRM).

REMARK 2.13 If $\eta = 1$, $Appr = \underline{Appr}$, then model (2.56) degenerates into model (2.54). If $\eta = 0$, $Appr = \overline{Appr}$, then model (2.56) degenerates into model (2.55).

2.4.2.1 Crisp Equivalent Model

Traditional mathematical solution methods require conversion of the chance constraints to their respective deterministic equivalents. However, this process is usually hard to perform and only successful for some special cases. First we present some useful results.

THEOREM 2.8
Let $\tilde{c}_{ij} = ([c_{ij1}, c_{ij2}], [c_{ij3}, c_{ij4}])(c_{ij3} \leq c_{ij1} \leq c_{ij2} \leq c_{ij4})$ and $\tilde{c}_i^T \mathbf{x} \ ([c_{i1}, c_{i2}], [c_{i3}, c_{i4}])$. Then
1. $\underline{Appr}\{\tilde{c}_i^T \mathbf{x} \geq \bar{f}_i\} \geq \beta_i$ holds if and only if

$$\begin{cases} \Phi, & \text{if } c_{i2} \leq \bar{f}_i \\ \bar{f}_i \leq \beta_i c_{i1} + (1 - \beta_i)c_{i2}, & \text{if } c_{i1} \leq \bar{f}_i \leq c_{i2} \\ X, & \text{if } \bar{f}_i \leq c_{i1} \end{cases} \tag{2.57}$$

where Φ means that $\underline{Appr}\{\tilde{\mathbf{c}}_i^T \mathbf{x} \geq \bar{f}_i\} \geq \beta_i$ never holds for any $\mathbf{x}$. Meanwhile, X means that $\underline{Appr}\{\tilde{\mathbf{c}}_i^T \mathbf{x} \geq \bar{f}_i\} \geq \beta_i$ holds for every x as long as x satisfies other constraints.

2. $\overline{Appr}\{\tilde{\mathbf{c}}_i^T \mathbf{x} \geq \bar{f}_i\} \geq \beta_i$ holds if and only if

$$\begin{cases} \Phi, & \text{if } c_{i4} \leq \bar{f}_i \\ \bar{f}_i \leq \beta_i c_{i1} + (1 - \beta_i)c_{i4}, & \text{if } c_{i3} \leq \bar{f}_i \leq c_{i4} \\ X, & \text{if } \bar{f}_i \leq c_i \end{cases} \tag{2.58}$$

where Φ means that $\overline{Appr}\{\tilde{\mathbf{c}}_i^T \mathbf{x} \geq \bar{f}_i\} \geq \beta_i$ never holds for any $\mathbf{x}$. Meanwhile, X means that $\overline{Appr}\{\tilde{\mathbf{c}}_i^T \mathbf{xx} \geq \bar{f}_i\} \geq \beta_i$ holds for every $\mathbf{x}$ as long as $\mathbf{x}$ satisfies other constraints.

3. $Appr\{\tilde{\mathbf{c}}_i^T \mathbf{x} \geq \bar{f}_i\} \geq \beta_i$ holds if and only if

$$\begin{cases} \Phi, & \text{if } c_{i4} \leq \bar{f}_i \\ \bar{f}_i \leq c_{i4} - \frac{(c_{i4} - c_{i3})\beta_i}{1 - \eta}, & \text{if } c_{i2} \leq \bar{f}_i \leq c_{i4} \\ \bar{f}_i \leq \frac{\eta(c_{i4} - c_{i3})c_{i2} + (1 - \eta)(c_{i2} - c_{i1})c_{i4} - \beta_i(c_{i2} - c_{i1})(c_{i4} - c_{i3})}{\eta(c_{i4} - c_{i3}) + (1 - \eta)(c_{i1} - c_{i2})}, & \text{if } c_{i1} \leq \bar{f}_i \leq c_{i2} \\ \bar{f}_i \leq \beta_i c_{i1} + (1 - \beta_i)c_{i4}, & \text{if } c_{i3} \leq \bar{f}_i \leq c_{i4} \\ X, & \text{if } \bar{f}_i \leq c_i \end{cases} \tag{2.59}$$

where Φ means that $Appr\{\tilde{\mathbf{c}}_i^T \mathbf{x} \geq \bar{f}_i\} \geq \beta_i$ never holds for any $\mathbf{x}$. Meanwhile, X means that $Appr\{\tilde{\mathbf{c}}_i^T \mathbf{x} \geq \bar{f}_i\} \geq \beta_i$ holds for every x as long as x satisfies other constraints.

PROOF   Let us prove them in turn.

1. It follows from the definition of $\underline{Appr}$ that

$$\underline{Appr}\{\tilde{\mathbf{c}}_i^T \mathbf{x} \geq \bar{f}_i\} = \begin{cases} 0, & \text{if } c_{i2} \leq \bar{f}_i \\ \frac{c_{i2} - \bar{f}_i}{c_{i2} - c_{i1}}, & \text{if } c_{i1} \leq \bar{f}_i \leq c_{i2} \\ 1, & \text{if } \bar{f}_i \leq c_{i1} \end{cases}$$

Then $\underline{Appr}\{\tilde{\mathbf{c}}_i^T \mathbf{x} \geq \bar{f}_i\} \geq \beta_i$ holds if and only if

$$\begin{cases} \Phi, & \text{if } c_{i2} \leq \bar{f}_i \\ \bar{f}_i \leq \beta_i c_{i1} + (1 - \beta_i)c_{i2}, & \text{if } c_{i1} \leq \bar{f}_i \leq c_{i2} \\ X, & \text{if } \bar{f}_i \leq c_{i1} \end{cases}$$

2. It follows from the definition of $\overline{Appr}$ that

$$\overline{Appr}\{\tilde{\mathbf{c}}_i^T \mathbf{x} \geq \bar{f}_i\} = \begin{cases} 0, & \text{if } c_{i4} \leq \bar{f}_i \\ \frac{c_{i4} - \bar{f}_i}{c_{i4} - c_{i3}}, & \text{if } c_{i3} \leq \bar{f}_i \leq c_{i4} \\ 1, & \text{if } \bar{f}_i \leq c_{i3} \end{cases}$$

Then $\overline{Appr}\{\tilde{\mathbf{c}}_i^T \mathbf{x} \geq \bar{f}_i\} \geq \beta_i$ holds if and only if

$$
\begin{cases}
\Phi, & \text{if } c_{i4} \leq \bar{f}_i \\
\bar{f}_i \leq \beta_i c_{i1} + (1 - \beta_i) c_{i4}, & \text{if } c_{i3} \leq \bar{f}_i \leq c_{i4} \\
X, & \text{if } \bar{f}_i \leq c_i
\end{cases}
$$

3. It follows from the definition of *Appr* that

$$
Appr\{\tilde{\mathbf{c}}_i^T \mathbf{x} \geq \bar{f}_i\} =
\begin{cases}
0, & \text{if } c_{i4} \leq \bar{f}_i \\
(1 - \eta)\frac{c_{i4} - \bar{f}_i}{(c_{i4} - c_{i3})}, & \text{if } c_{i2} \leq \bar{f}_i \leq c_{i4} \\
\eta\frac{c_{i4} - \bar{f}_i}{c_{i4} - c_{i3}} + (1 - \eta)\frac{c_{i2} - \bar{f}_i}{c_{i2} - c_{i1}}), & \text{if } c_{i1} \leq \bar{f}_i \leq c_{i2} \\
(1 - \eta)\frac{c_{i4} - \bar{f}_i}{c_{i4} - c_{i3}} + \eta, & \text{if } c_{i3} \leq \bar{f}_i \leq c_{i1} \\
1, & \text{if } \bar{f}_i \leq c_{i3}
\end{cases}
$$

Then $Appr\{\tilde{\mathbf{c}}_i^T \mathbf{x} \geq \bar{f}_i\} \geq \beta_i$ holds if and only if

$$
\begin{cases}
\Phi, & \text{if } c_{i4} \leq \bar{f}_i \\
\bar{f}_i \leq c_{i4} - \frac{(c_{i4} - c_{i3})\beta_i}{1 - \eta}, & \text{if } c_{i2} \leq \bar{f}_i \leq c_{i4} \\
\bar{f}_i \leq \frac{\eta(c_{i4} - c_{i3})c_{i2} + (1-\eta)(c_{i2} - c_{i1})c_{i4} - \beta_i(c_{i2} - c_{i1})(c_{i4} - c_{i3})}{\eta(c_{i4} - c_{i3}) + (1-\eta)(c_{i1} - c_{i2})}, & \text{if } c_{i1} \leq \bar{f}_i \leq c_{i2} \\
\bar{f}_i \leq \beta_i c_{i1} + (1 - \beta_i)c_{i4}, & \text{if } c_{i3} \leq \bar{f}_i \leq c_{i4} \\
X, & \text{if } \bar{f}_i \leq c_i
\end{cases}
$$

Thus the theorem holds.

COROLLARY 2.1

If $\eta = 1/2$, then $Appr\{\tilde{\mathbf{c}}_i^T \mathbf{x} \geq \bar{f}_i\} \geq \beta_i$ holds if and only if

$$
\begin{cases}
0, & \text{if } c_{i4} \leq \bar{f}_i \\
\frac{c_{i4} - \bar{f}_i}{2(c_{i4} - c_{i3})}, & \text{if } c_{i2} \leq \bar{f}_i \leq c_{i4} \\
\frac{1}{2}(\frac{c_{i4} - \bar{f}_i}{c_{i4} - c_{i3}} + \frac{c_{i2} - \bar{f}_i}{c_{i2} - c_{i1}}), & \text{if } c_{i1} \leq \bar{f}_i \leq c_{i2} \\
\frac{1}{2}(\frac{c_{i4} - \bar{f}_i}{c_{i4} - c_{i3}} + 1), & \text{if } c_{i3} \leq \bar{f}_i \leq c_{i1} \\
1, & \text{if } \bar{f}_i \leq c_{i3}
\end{cases}
$$

THEOREM 2.9

Assume that $\tilde{\mathbf{a}}_r^T \mathbf{x} - \tilde{b}_r = ([a,b],[c,d]), (c \leq a \leq b \leq d)$. Then

1. $\underline{Appr}\{\tilde{\mathbf{a}}_r^T \mathbf{x} \leq \tilde{b}_r\} \geq \alpha_r$ holds if and only if

$$
\begin{cases}
\Phi, & \text{if } 0 \leq a \\
(1 - \alpha_r)b + \alpha_r a \geq 0, & \text{if } a \leq 0 \leq b \\
X, & \text{if } b \leq 0
\end{cases}
\tag{2.60}
$$

where Φ means that $\underline{Appr}\{\tilde{a}_r^T x \leq \tilde{b}_r\} \geq \alpha_r$ never holds for any *x*. Meanwhile, *X* means that $\underline{Appr}\{\tilde{\mathbf{a}}_r^T \mathbf{x} \geq \tilde{b}_r\} \leq \alpha_r$ holds for every $\mathbf{x}$ as long as $\mathbf{x}$ satisfies other constraints.

2. $\overline{Appr}\{\tilde{\mathbf{a}}_r^T \mathbf{x} \leq \tilde{b}_r\} \geq \alpha_r$ *holds if and only if*

$$\begin{cases} \Phi, & \text{if } 0 \leq c \\ (1 - \alpha_r)d + \alpha_r c \geq 0, & \text{if } c \leq 0 \leq d \\ X, & \text{if } d \leq 0 \end{cases} \tag{2.61}$$

where Φ *means that* $\overline{Appr}\{\tilde{\mathbf{b}}_r^T \mathbf{x} \leq \tilde{b}_r\} \geq \alpha_r$ *never holds for any* $\mathbf{x}$. *Meanwhile,* X *means that* $\overline{Appr}\{\tilde{\mathbf{b}}_r^T \mathbf{x} \geq \tilde{b}_r\} \leq \alpha_r$ *holds for every* $\mathbf{x}$ *as long as* $\mathbf{x}$ *satisfies other constraints.*

3. $Appr\{\tilde{\mathbf{a}}_r^T \mathbf{x} \leq \tilde{b}_r\} \geq \alpha_r$ *holds if and only if*

$$\begin{cases} \Phi, & \text{if } 0 \leq c \\ \alpha_r(d-c) + (1-\eta)c \leq 0, & \text{if } c \leq 0 \leq a \\ \alpha_r(b-a)(d-c) + \eta a(d-c) + (1-\eta)c(b-a) \leq 0, & \text{if } a \leq 0 \leq b \\ \alpha_r(d-c) - (1-\eta)d \leq 0, & \text{if } b \leq 0 \leq d \\ X, & \text{if } d \leq 0 \end{cases} \tag{2.62}$$

where Φ *means that* $Appr\{\tilde{\mathbf{a}}_r^T \mathbf{x} \leq \tilde{b}_r\} \geq \alpha_r$ *never holds for any* $\mathbf{x}$. *Meanwhile,* X *means that* $Appr\{\tilde{\mathbf{a}}_r^T \mathbf{x} \leq \tilde{b}_r\} \leq \alpha_r$ *holds for every* $\mathbf{x}$ *as long as* $\mathbf{x}$ *satisfies other constraints.*

PROOF Let us prove them one by one.

1. It follows from the definition of *Appr* that

$$\underline{Appr}\{\tilde{\mathbf{a}}_r^T \mathbf{x} \leq \tilde{b}_r\} = \begin{cases} 0, & \text{if } 0 \leq a \\ -\frac{a}{b-a}, & \text{if } a \leq 0 \leq b \\ 1, & \text{if } b \leq 0 \end{cases}$$

Then $\underline{Appr}\{\tilde{\mathbf{a}}_r^T \mathbf{x} \leq \tilde{b}_r\} \geq \alpha_r$ holds if and only if

$$\begin{cases} \Phi, & \text{if } 0 \leq a \\ (1 - \alpha_r)b + \alpha_r a \geq 0, & \text{if } a \leq 0 \leq b \\ X, & \text{if } b \leq 0 \end{cases}$$

2. It follows from the definition of $\overline{Appr}$ that

$$\overline{Appr}\{\tilde{\mathbf{a}}_r^T \mathbf{x} \leq \tilde{b}_r\} = \begin{cases} 0, & \text{if } 0 \leq c \\ -\frac{c}{d-c}, & \text{if } c \leq 0 \leq d \\ 1, & \text{if if } d \leq 0 \end{cases}$$

Then $\overline{Appr}\{\tilde{\mathbf{a}}_r^T \mathbf{x} \leq \tilde{b}_r\} \geq \alpha_r$ holds if and only if

$$\begin{cases} \Phi, & \text{if } 0 \leq c \\ (1 - \alpha_r)d + \alpha_r c \geq 0, & \text{if } c \leq 0 \leq d \\ X, & \text{if } d \leq 0 \end{cases}$$

3. It follows from the definition of *Appr* that

$$
Appr\{\tilde{\mathbf{a}}_r^T \mathbf{x} \leq \tilde{b}_r\} = \begin{cases} 0, & \text{if } 0 \leq c \\ \frac{-c(1-\eta)}{d-c}, & \text{if } c \leq 0 \leq a, \\ \frac{-a\eta}{b-a} + \frac{-c(1-\eta)}{d-c}, & \text{if } a \leq 0 \leq b \\ \eta + \frac{-c(1-\eta)}{d-c}, & \text{if } b \leq 0 \leq d, \\ 1, & \text{if } d \leq 0 \end{cases}
$$

Then $\overline{Appr}\{\tilde{\mathbf{a}}_r^T \mathbf{x} \leq \tilde{b}_r\} \geq \alpha_r$ holds if and only if

$$
\begin{cases} \Phi, & \text{if } 0 \leq c \\ \alpha_r(d-c) + (1-\eta)c \leq 0, & \text{if } c \leq 0 \leq a \\ \alpha_r(b-a)(d-c) + \eta a(d-c) + (1-\eta)c(b-a) \leq 0, & \text{if } a \leq 0 \leq b \\ \alpha_r(d-c) - (1-\eta)d \leq 0, & \text{if } b \leq 0 \leq d \\ X, & \text{if } d \leq 0 \end{cases}
$$

Thus the theorem holds.

2.4.2.2 Ideal Point Method

In this section, we make use of the ideal point method proposed in [283] to resolve the multiple objective problem with crisp parameters.

$$
\max_{\mathbf{x} \in X} [f_1(\mathbf{x}), f_2(\mathbf{x}), \cdots, f_m(\mathbf{x})] \tag{2.63}
$$

If the decision maker can first propose an estimated value $\bar{F}_i$ for each objective function $\Psi_i^c \mathbf{x}$ such that

$$
\bar{F}_i \geq \max_{\mathbf{x} \in X'} \Psi_1^c \mathbf{x} i = 1, 2, \cdots, m \tag{2.64}
$$

where $X' = \{\mathbf{x} \in X | \Psi_r^e \mathbf{x} \leq \Psi_r^b, r = 1, 2, \cdots, p, \mathbf{x} \geq 0\}$, then $\bar{F}_i = (\bar{F}_1, \bar{F}_2, \cdots, \bar{F}_m)^T$ is called the ideal point. Specially, if $\bar{F}_i \geq \max_{\mathbf{x} \in X'} \Psi_1^c \mathbf{x}$ for all i, we call $\bar{F}$ the most ideal point.

The basic theory of the ideal point method is to take an especial norm in the objective space $\mathbf{R}^m$ and obtain the feasible solution $\mathbf{x}$ that the objective value approaches the ideal point $\bar{F} = (\bar{F}_1, \bar{F}_2, \cdots, \bar{F}_m)^T$ under the norm distance, that is, to seek the feasible solution $\mathbf{x}$ satisfying

$$
\min_{\mathbf{x} \in X'} u(\Psi^c(\mathbf{x})) = \min_{\mathbf{x} \in X'} ||\Psi^c(\mathbf{x}) - \bar{F}||.
$$

Usually, the following norm functions are used to describe the distance:
1. *p*-mode function

$$
d_p(\Psi^c(\mathbf{x}), \bar{F}; \omega) = \left[\sum_{i=1}^m \omega_i |\Psi_i^c \mathbf{x} - \bar{F}_i|^p \right]^{\frac{1}{p}}, \quad 1 \leq p < +\infty \tag{2.65}
$$

2. The maximal deviation function

$$d_{+\infty}(\Psi^c(\mathbf{x}), \bar{F}; \omega) = \max_{1 \le i \le m} \omega_i |\Psi_i^c \mathbf{x} - \bar{F}_i| \tag{2.66}$$

3. Geometric mean function

$$d(\Psi^c(\mathbf{x}), \bar{F}) = \left[\prod_{i=1}^{m} |\Psi_i^c \mathbf{x} - \bar{F}_i|^p \right]^{\frac{1}{m}} \tag{2.67}$$

The weight parameter vector $\omega = (\omega_1, \omega_2, \cdots, \omega_m)^T > 0$ needs to be predetermined.

THEOREM 2.10
(F. Szidarovszky et al. [283]) Assume that $\bar{F}_i > \max_{\mathbf{x} \in X'} \Psi_1^c \mathbf{x} (i = 1, 2, \cdots, m)$. If $\mathbf{x}^*$ is the optimal solution to the following problem:

$$\min_{\mathbf{x} \in X'} d_p(\Psi^c(\mathbf{x}), \bar{F}; \omega) = \left[\sum_{i=1}^{m} \omega_i |\Psi_i^c \mathbf{x} - \bar{F}_i|^p \right]^{\frac{1}{p}} \tag{2.68}$$

then $\mathbf{x}^*$ is an efficient solution to problem (2.4.2.2). On the contrary, if $\mathbf{x}^*$ is an efficient solution of problem (2.4.2.2), then there exists a weight vector ω such that $\mathbf{x}^*$ is the optimal solution to problem (2.68).

Next, we take the p-mode function to describe the procedure for solving the problem (2.68).

Step 1. Find the ideal point. If the decision maker can give the ideal objective value satisfying the condition (2.64), the value will be considered as the ideal point. However, decision makers themselves do not know how to give the objective value. Then we can get the ideal point by solving the following problem:

$$\begin{cases} \max \Psi_i^c \mathbf{x} \\ \text{s.t.} \begin{cases} \Psi_r^e \mathbf{x} \le \Psi_r^b, r = 1, 2, \cdots, p \\ \mathbf{x} \in X \end{cases} \end{cases} \tag{2.69}$$

Then the ideal point $\bar{\mathbf{F}} = (\bar{F}_1, \bar{F}_2, \cdots, \bar{F}_m)^T$ can be fixed by $\bar{F}_i = \Psi_i^c \mathbf{x}^*$, where $\mathbf{x}^*$ is the optimal solution to problem (2.69).

Step 2. Fix the weight. The method of selecting the weight can be referred to in the literature, and interested readers can consult them. We usually use the following function to fix the weight:

$$\omega_i = \frac{\bar{F}_i}{\sum_{i=1}^{m} \bar{F}_i}.$$

Step 3. Construct the minimal distance problem. Solve the following single objective programming problem to obtain the efficient solution to problem :

$$\begin{cases} \min \left[\sum_{i=1}^{m} \omega_i |\Psi_i^c \mathbf{x} - \bar{F}_i|^t \right]^{\frac{1}{t}} \\ \text{s.t.} \begin{cases} \Psi_r^e \mathbf{x} \leq \Psi_r^b, r = 1, 2, \cdots, p \\ \mathbf{x} \in X \end{cases} \end{cases} \tag{2.70}$$

Usually we take $t = 2$ to compute it.

2.4.2.3 Numerical Example

Example 2.11

Consider the problem

$$\begin{cases} \max[\bar{f}_1, \bar{f}_2] \\ \text{s.t.} \begin{cases} Appr\{\xi_1 x_1 + \xi_2 x_2 + \xi_3 x_3 + \xi_4 x_4 + \xi_5 x_5 + \xi_6 x_6 \geq \bar{f}_1\} \geq 0.95 \\ Appr\{\xi_7 x_1 + \xi_8 x_2 + \xi_9 x_3 + \xi_{10} x_4 + \xi_{11} x_5 + \xi_{12} x_6 \geq \bar{f}_1\} \geq 0.90 \\ 2x_1 + 4x_2 + 6x_3 + 8x_4 + 10x_5 + 12x_6 \geq 300 \\ 3x_1 + 5x_2 + 7x_3 + 9x_4 + 11x_5 + 13x_6 \leq 600 \\ x_j \geq 0, j = 1, 2, \cdots, 12 \end{cases} \end{cases} \tag{2.71}$$

where $\xi_j (j = 1, 2, \cdots, 6)$ are rough intervals characterized as

$$\xi_1 = ([2,5],[1,6]), \quad \xi_2 = ([3,6],[2,7]), \quad \xi_3 = ([4,7],[3,8]),$$
$$\xi_4 = ([5,8],[4,9]), \quad \xi_5 = ([6,9],[5,10]), \quad \xi_6 = ([7,10],[6,11]),$$
$$\xi_7 = ([8,11],[7,12]), \quad \xi_8 = ([9,12],[8,13]), \quad \xi_9 = ([10,13],[9,14])$$
$$\xi_{10} = ([11,14],[10,15]), \quad \xi_{12} = ([12,15],[11,16]), \quad \xi_{12} = ([13,16],[12,17]).$$

From rough interval arithmetic, we have

$$\sum_{j=1}^{6} \xi_j x_j = ([\sum_{j=1}^{6} (j+1)x_j, \sum_{i=1}^{6} (j+4)x_j], [\sum_{j=1}^{6} jx_j, \sum_{j=1}^{6} (j+5)x_j])$$

$$\sum_{j=1}^{6} \xi_{j+7} x_j = ([\sum_{j=1}^{6} (j+7)x_j, \sum_{j=1}^{6} (j+10)x_j], [\sum_{j=1}^{6} (j+6)x_j, \sum_{j=1}^{6} (j+11)x_j]).$$

Let $\eta = 1/2$. It follows from Theorems 2.8 and 2.9 that problem (2.71) is equivalent to

$$\begin{cases} \max F_1(x) = -3.5x_1 - 2.5x_2 - 1.5x_3 - 0.5x_4 + 0.5x_5 + 1.5x_6 \\ \max F_2(x) = 3x_1 + 4x_2 + 5x_3 + 6x_4 + 7x_5 + 8x_6 \\ \text{s.t.} \begin{cases} 2x_1 + 4x_2 + 6x_3 + 8x_4 + 10x_5 + 12x_6 \geq 300 \\ 3x_1 + 5x_2 + 7x_3 + 9x_4 + 11x_5 + 13x_6 \leq 600 \\ x_j \geq 0, j = 1, 2, \cdots, 6 \end{cases} \end{cases} \tag{2.72}$$

By solving

$$\begin{cases} \max F_1(x) = -3.5x_1 - 2.5x_2 - 1.5x_3 - 0.5x_4 + 0.5x_5 + 1.5x_6 \\ \text{s.t.} \begin{cases} 2x_1 + 4x_2 + 6x_3 + 8x_4 + 10x_5 + 12x_6 \geq 300 \\ 3x_1 + 5x_2 + 7x_3 + 9x_4 + 11x_5 + 13x_6 \leq 600 \\ x_j \geq 0, j = 1, 2, \cdots, 6 \end{cases} \end{cases} \quad (2.73)$$

and

$$\begin{cases} \max F_2(x) = 3x_1 + 4x_2 + 5x_3 + 6x_4 + 7x_5 + 8x_6 \\ \text{s.t.} \begin{cases} 2x_1 + 4x_2 + 6x_3 + 8x_4 + 10x_5 + 12x_6 \geq 300 \\ 3x_1 + 5x_2 + 7x_3 + 9x_4 + 11x_5 + 13x_6 \leq 600 \\ x_j \geq 0, j = 1, 2, \cdots, 6 \end{cases} \end{cases} \quad (2.74)$$

the ideal point is calculated as $(69.23, 200)$. Solve the following problem:

$$\begin{cases} \max \sqrt{(69.23 - F_1(x)) + (200 - F_2(x))} \\ \text{s.t.} \begin{cases} 2x_1 + 4x_2 + 6x_3 + 8x_4 + 10x_5 + 12x_6 \geq 300 \\ 3x_1 + 5x_2 + 7x_3 + 9x_4 + 11x_5 + 13x_6 \leq 600 \\ x_j \geq 0, j = 1, 2, \cdots, 6 \end{cases} \end{cases} \quad (2.75)$$

We can obtain the efficient solution as follows: $(x_1, x_2, x_3, x_4, x_5, x_6) = (0, 0, 0, 0, 0, 46.39)$.

2.4.3 NL-CCRM and Rough Simulation-Based AWGA

Let us return to the nonlinear chance-constrained rough model (NL-CCRM).

$$\begin{cases} \max[\bar{f}_1, \bar{f}_2, \cdots, \bar{f}_m] \\ \text{s.t.} \begin{cases} Ch\{f_i(\mathbf{x}, \boldsymbol{\xi}) \geq \bar{f}_i\} \geq \beta_i, \ i = 1, 2, \cdots, m \\ Ch\{g_r(\mathbf{x}, \boldsymbol{\xi}) \leq b_r\} \geq \alpha_i, \ r = 1, 2, \cdots, p \\ \mathbf{x} \geq \mathbf{0}, 0 \leq \alpha_r, \beta_i \leq 1 \end{cases} \end{cases} \quad (2.76)$$

where $f_i(\mathbf{x}, \boldsymbol{\xi})$ or $g_r(\mathbf{x}, \boldsymbol{\xi})$ or both of them are nonlinear functions with respect to $\mathbf{x}$, and $\boldsymbol{\xi}$ is a rough intervals vector, $i = 1, 2, \cdots, m$; $r = 1, 2, \cdots, p$. Ch represents $Appr, \overline{Appr}, Appr$. For the class of problems that cannot be converted into crisp ones, the rough simulation is a useful tool to compute the chance measure of rough events.

2.4.3.1 Rough Simulation 2 for Critical Value

Let x be a fixed decision vector. Denote $(\underline{X}, \overline{X}) = f_i(\mathbf{x}, \boldsymbol{\xi})$. Here we employ rough simulation to estimate the maximal value $\bar{f}_i$ such that

$$Appr\{f_i(\mathbf{x}, \boldsymbol{\xi}) \geq \bar{f}_i\} \geq \beta_i \quad (2.77)$$

where β_i is a predetermined confidence level with $0 < \beta_i \leq 1$. We sample $\underline{\lambda}_1, \underline{\lambda}_2, \cdots, \underline{\lambda}_N$ from $\underline{X}$ and $\overline{\lambda}_1, \overline{\lambda}_2, \cdots, \overline{\lambda}_N$ from $\underline{X}$ randomly. For any number v, let $\underline{N}(v)$ denote the number of $\underline{\lambda}_n$ satisfying $f_i(\underline{\lambda}_n)x \geq v$ for $n = 1,2,\cdots,N$, and $\overline{N}(v)$ denote the number of $\overline{\lambda}_n$ satisfying $f_i(\overline{\lambda}_n)x \geq v$ for $n = 1,2,\cdots,N$. Then we may find the maximal value v such that

$$\frac{\eta \underline{N} + (1-\eta)\overline{N}}{N} \geq \beta_i \tag{2.78}$$

This value is an estimation of $\bar{f}_i$.

The process can be summarized as follows:

Step 1. Generate $\underline{\lambda}_1, \underline{\lambda}_2, \cdots \underline{\lambda}_N$ from $\underline{X}$ randomly.
Step 2. Generate $\overline{\lambda}_1, \overline{\lambda}_2, \cdots \overline{\lambda}_N$ from $\overline{X}$ randomly.
Step 3. Find the maximal value v such that (2.78) holds.
Step 4. Return v.

REMARK 2.14 If $\eta = 1$, the simulation process above is rough simulation of

$$\underline{Appr}\{f_i(\mathbf{x},\boldsymbol{\xi}) \geq \bar{f}_i\} \geq \beta_i \tag{2.79}$$

If $\eta = 0$, the simulation process above is rough simulation of

$$\overline{Appr}\{f_i(\mathbf{x},\boldsymbol{\xi}) \geq \bar{f}_i\} \geq \beta_i \tag{2.80}$$

Example 2.12
Assume that the rough intervals $\xi_1 = ([0,1],[-1,3]), \xi_2 = ([1,2],[0,3])$, and $\xi_3 = ([2,3],[1,5])$. Now we compute the maximal value $\bar{f}$ such that $Appr\{\xi_1 + \xi_2^2 + \xi_3^3 \geq \bar{f}\} \geq 0.8$. A run of the rough simulation with 2000 cycles shows that $\bar{f} = 12.7$.

2.4.3.2 AWGA

Next, we will introduce the adaptive weight GA (AWGA), which used in this section. This section attempts to apply rough simulation 1 to compute the expected value, convert the uncertain multiple objective problem into a deterministic one and make use of the adaptive weight genetic algorithm to solve this multiple objective problem.

R. Cheng and M. Gen [60] proposed an adaptive weight approach that utilizes some useful information from the current population to readjust weights to obtain a search pressure toward a positive ideal point. Without loss of generality, consider the maximization problem with q objectives. For the solutions examined in each generation, we define two extreme points: the maximum extreme point z^+ and the minimum point z^- in criteria space as follows: $z^+ = \{z_1^{\max}, z_2^{\max}, \cdots, z_q^{\max}\}$, and $z^- = \{z_1^{\min}, z_2^{\min}, \cdots, z_q^{\min}\}$, respectively, $k = 1,2,\cdots,q$, where $z_k^{\min}$ and $z_k^{\max}$ are the maximal value and minimal value, respectively of objective k in the current population. Let P denote the set of the current population. For a given individual x, the maximal value and minimal value for each objective are defined as $z_k^{\max} = \max\{f_k(x)|x \in P\}$ and $z_k^{\min} = \min\{f_k(x)|x \in P\}$, respectively, $k = 1,2,\cdots,q$.

The hyperparallelogram defined by the two extreme points is a minimal hyperparallelogram containing all current solutions. The two extreme points are renewed at each generation. The maximum extreme point will gradually approximate the positive ideal point. The adaptive weight for objective k is calculated by the following equation:

$$w_k = \frac{1}{z_k^{\max} - z_k^{\min}}, k = 1, 2, \cdots, q.$$

For a given individual x, the weighted-sum objective function is given by the following equation:

$$\begin{aligned}
z(x) &= \sum_{k=1}^{q} w_k (z_k - z_k^{\min}) \\
&= \sum_{k=1}^{q} \frac{z_k - z_k^{\min}}{z_k^{\max} - z_k^{\min}} \\
&= \sum_{k=1}^{q} \frac{f_k(x) - z_k^{\min}}{z_k^{\max} - z_k^{\min}}
\end{aligned} \tag{2.81}$$

As the extreme points are renewed at each generation, the weights are renewed accordingly. Equation (2.81) is hyperplane defined by the following extreme point in current solutions:

$$\begin{aligned}
&(z_1^{\max}, z_2^{\min}, \cdots, z_k^{\min}, \cdots, z_q^{\min}), \\
&(z_1^{\min}, z_2^{\max}, \cdots, z_k^{\min}, \cdots, z_q^{\min}), \\
&\quad\quad\quad\quad \cdots \\
&(z_1^{\min}, z_2^{\min}, \cdots, z_k^{\max}, \cdots, z_q^{\min}), \\
&\quad\quad\quad\quad \cdots \\
&(z_1^{\min}, z_2^{\min}, \cdots, z_k^{\min}, \cdots, z_q^{\max}).
\end{aligned}$$

The hyperplane divides the criteria space Z into two half-spaces: One half-space contains the positive ideal point, denoted as Z^+, and the other half-space contains the negative ideal point, denoted as Z^-. All Pareto solutions examined lie in the space Z^+, and all points lying in Z^+ have larger fitness values than those in the points in the space Z^-. As the maximum extreme point approximates the positive ideal point along with the evolutionary progress, the hyperplane will gradually approach the positive ideal point. Therefore, the adaptive weight method can readjust its weights according to the current population to obtain a search pressure toward the positive ideal point.

For the minimization case, we just need to transform the original problem into its equivalent maximization problem and then apply (2.81). For a maximization problem, (2.81) can be simplified as follows:

$$\begin{aligned}
z(x) &= \sum_{k=1}^{q} w_k z_k \\
&= \sum_{k=1}^{q} \frac{z_k}{z_k^{\max} - z_k^{\min}} \\
&= \sum_{k=1}^{q} \frac{f_k(x)}{z_k^{\max} - z_k^{\min}}.
\end{aligned}$$

Let us look at an example of a bicriteria maximization problem:

$$\begin{cases} \max\{z_1 = f_1(\mathbf{x}), z_2 = f_2(\mathbf{x})\} \\ \text{s.t. } g_i(\mathbf{x}) \leq 0, i = 1, 2, \cdots, m \end{cases}$$

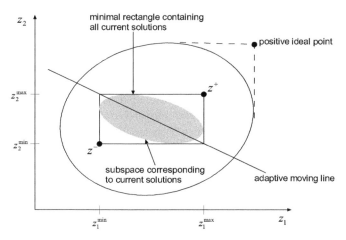

FIGURE 2.17 Adaptive weights and adaptive hyperplane.

For a given generation, two extreme points are identified as

$$\begin{cases} z_1^{\max} = \max\{z_1(\mathbf{x}^j), j = 1, 2, \cdots, pop - size\} \\ z_2^{\max} = \max\{z_2(\mathbf{x}^j), j = 1, 2, \cdots, pop - size\} \\ z_1^{\min} = \min\{z_1(\mathbf{x}^j), j = 1, 2, \cdots, pop - size\} \\ z_2^{\min} = \min\{z_2(\mathbf{x}^j), j = 1, 2, \cdots, pop - size\} \end{cases}$$

and the adaptive weights are calculated as

$$w_1 = \frac{1}{z_1^{\max} - z_1^{\min}}, \ w_2 = \frac{1}{z_2^{\max} - z_2^{\min}}.$$

The weighted-sum objective function is then given by

$$z(\mathbf{x}) = w_1 z_1 + w_2 z_2 = w_1 f_1(\mathbf{x}) + w_2 f_2(\mathbf{x}).$$

It is an adaptive moving line defined by the extreme points $(z_1^{\max}, z_2^{\min})$ and $(z_1^{\min}, z_2^{\max})$, as shown in Figure 2.17. The rectangle defined by the extreme point $(z_1^{\max}, z_2^{\max})$ and $(z_1^{\min}, z_2^{\min})$ is the minimal rectangle containing all current solutions.

One of important issues in genetic multiple objective optimization is how to handle constraints because genetic operators used to manipulate chromosomes often yield infeasible offspring. R. Cheng and M. Gen [60] suggested an adaptive penalty method to deal with infeasible individuals. Given an individual x in the current population $P(t)$, the adaptive penalty function is constructed as

$$p(x) = 1 - \frac{1}{m} \sum_{i=1}^{m} \left(\frac{\Delta b_i(\mathbf{x})}{\Delta b_i^{\max}} \right)^a,$$

where $\Delta b_i(\mathbf{x}) = \max\{0, g_i(x) - b_i\}$, $\Delta b_i^{\max} = \max\{\varepsilon, \Delta b_i(\mathbf{x}) | \mathbf{x} \in P(t)\}$.

Where $\Delta b_i(\mathbf{x})$ is the value of violation for constraint i for the ith chromosome, $\Delta b_i^{\max}$ the maximum of violation for constraint i among the current population, and ε a small positive number used to have penalty avoid from zero division. For highly constrained optimization problems, the infeasible solutions take a relatively big portion among population at each generation. The penalty approach adjusts the ratio of penalties adaptively at each generation to make a balance between the preservation of information and the pressure for infeasibility and to avoid overpenalty. With the penalty function, the fitness function then takes the following form:

$$eval(\mathbf{x}) = z(\mathbf{x})p(\mathbf{x}).$$

The overall procedure of adaptive weight approach is summarized as follows:

Step 1. Initialization. Generate the initial population randomly.

Step 2. Evaluation. Calculate the objective value, penalty value, and fitness value for each individual.

Step 3. Pareto set. Update the set of Pareto solutions.

Step 4. Selection. Select the next generation using the roulette wheel method.

Step 5. Production. Produce offspring with crossover and mutation.

Step 6. Termination. If the maximal generation is reached, stop; otherwise, go to Step 2.

2.4.3.3 Numerical Example

Example 2.13
Consider the problem

$$\begin{cases} \max F_1(x,\xi) = 2\xi_1^2 + 5\xi_1 x_2^2 - \xi_3 x_3 + \sqrt{\xi_4^2 + (10-x_5)^2} \\ \max F_2(x,\xi) = 3x_1 + 3x_2 - x_5 \\ \text{s.t.} \begin{cases} x_1 + x_2 + x_3 + x_4 + x_5 \le 70 \\ 5x_1 - 3x_2^2 - \sqrt{x_3} + x_4 + x_5^3 \le 160 \\ \xi_1^2 x_1 + 2\xi_2^2 x_2 - \xi_3^2 - 5\xi_4^2 x_4 \le 0 \\ x_i \ge 0, i = 1,2,3,4,5 \end{cases} \end{cases} \quad (2.82)$$

where $\xi_i = 1,2,3,4$ are rough intervals characterized as

$$\xi_1 = ([2,5],[1,6]), \xi_2 = ([3,6],[2,7]), \xi_3 = ([4,7],[3,8]), \xi_4 = ([5,8],[4,9]).$$

From the mathematical view, the problem (2.82) is not well defined because of the uncertain parameters. Then we apply the chance operator to deal with this uncertain programming as

$$
\begin{cases}
\max[\bar{f}_1, \bar{f}_2] \\
\text{s.t.}
\begin{cases}
Appr\{2\xi_1^2 + 5\xi_1 x_2^2 - \xi_3 x_3 + \sqrt{\xi_4^2 + (10 - x_5)^2} \geq \bar{f}_1\} \geq 0.8 \\
3x_1 + 3x_2 - x_5 \geq \bar{f}_2 \\
x_1 + x_2 + x_3 + x_4 + x_5 \leq 70 \\
Appr\{5x_1 - 3x_2^2 - \sqrt{x_3} + x_4 + x_5^3 \leq 160\} \geq 0.7 \\
Appr\{\xi_1^2 x_1 + 2\xi_2^2 x_2 - \xi_3^2 - 5\xi_4^2 x_4 \leq 0\} \geq 0.8 \\
x_i \geq 0, i = 1, 2, 3, 4, 5
\end{cases}
\end{cases}
\tag{2.83}
$$

Due to the existence of nonlinear objective function and constraint, we cannot transform it into its crisp equivalent model. In order to solve it, we use the rough simulation-based adaptive weight GA to deal with it. After running the algorithm, we get a solution as follows:

$$
(x_1^*, x_2^*, x_3^*, x_4^*, x_5^*) = (11.32, 0, 3.67, 22.77, 30.18), (\bar{f}_1^*, \bar{f}_2^*) = (120.58, 32.44).
$$

The Matlab® file is presented in A.2.

2.5 DCRM

Chance-constrained programming, pioneered by A. Charnes and W. Cooper [52], provides a means of handling uncertainty by specifying a confidence level at which it is desired that the stochastic constraint holds. Chance-constrained programming models can be converted into deterministic equivalents only for some special cases, and then solved by using solution methods of deterministic mathematical programming. In order to overcome this difficulty, B. Liu and K. Iwamura [195] provided a new stochastic programming framework called dependent-chance programming, in which a complex stochastic decision system undertakes multiple tasks called events, and the decision maker wishes to maximize the chance functions that are defined as the probabilities of satisfying these events.

Roughly speaking, dependent-chance programming is aimed at maximizing some chance functions of events in an uncertain environment. In deterministic mathematical programming, the feasible set is essentially assumed to be deterministic, and the optimal solution can always be implemented. However, when uncertainty is taken into account, the given solution may be infeasible if the realization of uncertain parameters is unfavorable. In other words, the feasible set of dependent-chance programming is described by a so-called uncertain environment. Although a deterministic solution is given by the dependent-chance programming model, this solution needs to be as flexible as possible with respect to the uncertain environment. This special feature of dependent-chance programming is very different from other existing stochastic programming frameworks. However, such problems do exist in the real world. Some applications of dependent chance programming have been presented L. Yang and L. Liu [324], and L. Liu et al. [197].

In this section we introduce the concept of chance function and discuss the DCM models. We also present some crisp equivalent models. Finally, some numerical examples are exhibited.

2.5.1   General Model for DCRM

We propose the general model of the DCRM as

$$
\begin{cases}
\max\left[Ch\{f_1(\mathbf{x},\boldsymbol{\xi})\leq 0\},Ch\{f_2(\mathbf{x},\boldsymbol{\xi})\leq 0\},\cdots,Ch\{f_m(\mathbf{x},\boldsymbol{\xi})\leq 0\}\right]\\
\text{s.t. } g_r(\mathbf{x},\boldsymbol{\xi})\leq 0, r=1,2,\cdots,p
\end{cases}
\tag{2.84}
$$

where $f_i(\mathbf{x},\boldsymbol{\xi})\leq 0, k=1,2,\cdots,m$ represent rough events, $\boldsymbol{\xi}=(\xi_1,\xi_2,\cdots,\xi_m)$. Ch are the chance of the rough event, and we could use $\underline{Appr},\overline{Appr}$, and $Appr$ measures according to different decision environments.

2.5.1.1   DCRM Based on *Appr* Measure

If we take Ch as $Appr$, the (2.84) can be rewritten as

$$
\begin{cases}
\max\left[Appr\{f_1(\mathbf{x},\boldsymbol{\xi})\leq 0\},Appr\{f_2(\mathbf{x},\boldsymbol{\xi})\leq 0\},\cdots,Appr\{f_m(\mathbf{x},\boldsymbol{\xi})\leq 0\}\right]\\
\text{s.t. } g_r(\mathbf{x},\boldsymbol{\xi})\leq 0, r=1,2,\cdots,p
\end{cases}
$$

$$\tag{2.85}$$

Thus, for a each given decision vector, we cannot judge whether or not a decision vector $\mathbf{x}$ is feasible before the realization of the rough intervals vector $\boldsymbol{\xi}$. Hence, problem (2.85) is not well defined mathematically. Motivated by the ideas of EVM or CCM, we may consider the expected value of $g_r(x,\xi)$ or require the possibility measure of $g_r(\mathbf{x},\boldsymbol{\xi})$ to be not less than the predetermined confidence levels, that is, we have the following models:

$$
\begin{cases}
\max[\underline{Appr}\{f_1(\mathbf{x},\boldsymbol{\xi})\leq 0\},\underline{Appr}\{f_2(\mathbf{x},\boldsymbol{\xi})\leq 0\},\cdots,\underline{Appr}\{f_m(\mathbf{x},\boldsymbol{\xi})\leq 0\}]\\
\text{s.t. } E[g_r(\mathbf{x},\boldsymbol{\xi})]\leq 0, r=1,2,\cdots,p
\end{cases}
\tag{2.86}
$$

or

$$
\begin{cases}
\max[\underline{Appr}\{f_1(\mathbf{x},\boldsymbol{\xi})\leq 0\},\underline{Appr}\{f_2(\mathbf{x},\boldsymbol{\xi})\leq 0\},\cdots,\underline{Appr}\{f_m(\mathbf{x},\boldsymbol{\xi})\leq 0\}]\\
\text{s.t. } \underline{Appr}\{g_r(\mathbf{x},\boldsymbol{\xi})\leq 0\}\geq \alpha_r, r=1,2,\cdots,p
\end{cases}
\tag{2.87}
$$

where α_r are the predetermined confidence levels, $r=1,2,\cdots,p$ level value.

Obviously, problem (2.86) and problem (2.87) are mathematically meaningful decision problems. In other words, we can judge whether or not a decision vector is feasible. Thus, they can be formulated as

$$
\max_{\mathbf{x}\in X}[\underline{Appr}\{f_1(\mathbf{x},\boldsymbol{\xi})\leq 0\},\underline{Appr}\{f_2(\mathbf{x},\boldsymbol{\xi})\leq 0\},\cdots,\underline{Appr}\{f_m(\mathbf{x},\boldsymbol{\xi})\leq 0\}]
\tag{2.88}
$$

where X is a fixed feasible set.

DEFINITION 2.25 $\mathbf{x}^* \in X$ *is said to be an Appr-efficient solution of problem (2.88) if and only if there exists no other rough feasible solution* $\mathbf{x}$ *such that* $\underline{Appr}\{f_i(\mathbf{x}, \boldsymbol{\xi}) \le 0\} \ge \underline{Appr}\{f_i(\mathbf{x}^*, \boldsymbol{\xi}) \le 0\}$ *for all i, and a strict inequality holds for at least one* $i_0 \in \{1, 2, \cdots, m\}$.

2.5.1.2 DCRM Based on $\overline{Appr}$ Measure

If we take Ch as $\overline{Appr}$, then problem (2.84) can be rewritten as

$$\begin{cases} \max \left[\overline{Appr}\{f_1(\mathbf{x}, \boldsymbol{\xi}) \le 0\}, \overline{Appr}\{f_2(\mathbf{x}, \boldsymbol{\xi}) \le 0\}, \cdots, \overline{Appr}\{f_m(\mathbf{x}, \boldsymbol{\xi}) \le 0\} \right] \\ \text{s.t. } g_r(\mathbf{x}, \boldsymbol{\xi}) \le 0, r = 1, 2, \cdots, p \end{cases}$$

(2.89)

Thus, for each given decision vector, we cannot judge whether or not a decision vector x is feasible before the realization of the rough intervals vector $\boldsymbol{\xi}$. Hence, problem (2.89) is not well defined mathematically. Motivated by the ideas of EVM or CCM, we may consider the expected value of $g_r(\mathbf{x}, \boldsymbol{\xi})$ or require the possibility measure of $g_r(\mathbf{x}, \boldsymbol{\xi})$ to be not less than the predetermined confidence levels, that is, we have the following models:

$$\begin{cases} \max \left[\overline{Appr}\{f_1(\mathbf{x}, \boldsymbol{\xi}) \le 0\}, \overline{Appr}\{f_2(\mathbf{x}, \boldsymbol{\xi}) \le 0\}, \cdots, \overline{Appr}\{f_m(\mathbf{x}, \boldsymbol{\xi}) \le 0\} \right] \\ \text{s.t. } E[g_r(\mathbf{x}, \boldsymbol{\xi})] \le 0, r = 1, 2, \cdots, p \end{cases}$$

(2.90)

or

$$\begin{cases} \max \left[\overline{Appr}\{f_1(\mathbf{x}, \boldsymbol{\xi}) \le 0\}, \overline{Appr}\{f_2(\mathbf{x}, \boldsymbol{\xi}) \le 0\}, \cdots, \overline{Appr}\{f_m(\mathbf{x}, \boldsymbol{\xi}) \le 0\} \right] \\ \text{s.t. } \overline{Appr}\{g_r(\mathbf{x}, \boldsymbol{\xi}) \le 0\} \ge \alpha_r, r = 1, 2, \cdots, p \end{cases}$$

(2.91)

where $\alpha_r, r = 1, 2, \cdots, p$ are the predetermined confidence level value.

Obviously, problem (2.90) and problem (2.91) are mathematically meaningful decision problems. In other words, we can judge whether or not a decision vector is feasible. Thus, they can be formulated as

$$\max_{\mathbf{x} \in X} [\overline{Appr}\{f_1(\mathbf{x}, \boldsymbol{\xi}) \le 0\}, \overline{Appr}\{f_2(\mathbf{x}, \boldsymbol{\xi}) \le 0\}, \cdots, \overline{Appr}\{f_m(\mathbf{x}, \boldsymbol{\xi}) \le 0\}] \quad (2.92)$$

where X is a fixed feasible set.

DEFINITION 2.26 $\mathbf{x}^* \in X$ *is said to be an Appr-efficient solution of problem (2.92) if and only if there exists no other rough feasible solution* x *such that* $\overline{Appr}\{f_i(\mathbf{x}, \boldsymbol{\xi}) \le 0\} \ge \overline{Appr}\{f_i(\mathbf{x}^*, \boldsymbol{\xi}) \le 0\}$ *for all i, and a strict inequality holds for at least one* $i_0 \in \{1, 2, \cdots, m\}$.

2.5.1.3 DCRM Based on $Appr$ Measure

If we take Ch as $Appr$, problem (2.84) can be rewritten as

$$\begin{cases} \max \left[Appr\{f_1(\mathbf{x}, \boldsymbol{\xi}) \le 0\}, Appr\{f_2(\mathbf{x}, \boldsymbol{\xi}) \le 0\}, \cdots, Appr\{f_m(\mathbf{x}, \boldsymbol{\xi}) \le 0\} \right] \\ \text{s.t. } g_r(\mathbf{x}, \boldsymbol{\xi}) \le 0, r = 1, 2, \cdots, p \end{cases}$$

(2.93)

Thus, for each given decision vector, we cannot judge whether or not a decision vector $\mathbf{x}$ is feasible before the realization of the rough interval vector $\boldsymbol{\xi}$. Hence, problem (2.93) is not well defined mathematically. Motivated by the ideas of EVM or CCM, we may consider the expected value of $g_r(\mathbf{x}, \boldsymbol{\xi})$ or require the possibility measure of $g_r(\mathbf{x}, \boldsymbol{\xi})$ to be not less than a predetermined confidence levels, that is, we have the following models:

$$\begin{cases} \max \left[Appr\{f_1(\mathbf{x}, \boldsymbol{\xi}) \leq 0\}, Appr\{f_2(\mathbf{x}, \boldsymbol{\xi}) \leq 0\}, \cdots, Appr\{f_m(\mathbf{x}, \boldsymbol{\xi}) \leq 0\} \right] \\ \text{s.t. } E[g_r(\mathbf{x}, \boldsymbol{\xi})] \leq 0, r = 1, 2, \cdots, p \end{cases}$$

(2.94)

or

$$\begin{cases} \max \left[Appr\{f_1(\mathbf{x}, \boldsymbol{\xi}) \leq 0\}, Appr\{f_2(\mathbf{x}, \boldsymbol{\xi}) \leq 0\}, \cdots, Appr\{f_m(\mathbf{x}, \boldsymbol{\xi}) \leq 0\} \right] \\ \text{s.t. } Appr\{g_r(\mathbf{x}, \boldsymbol{\xi}) \leq 0\} \geq \alpha_r, r = 1, 2, \cdots, p \end{cases}$$

(2.95)

where $\alpha_r, r = 1, 2, \cdots, p$ are the predetermined confidence level value.

Obviously, problem (2.94) and problem (2.95) are mathematically meaningful decision problems. In other words, we can judge whether or not a decision vector is feasible. Thus, they can be formulated as

$$\max_{\mathbf{x} \in X} [Appr\{f_1(\mathbf{x}, \boldsymbol{\xi}) \leq 0\}, Appr\{f_2(\mathbf{x}, \boldsymbol{\xi}) \leq 0\}, \cdots, Appr\{f_m(\mathbf{x}, \boldsymbol{\xi}) \leq 0\}] \quad (2.96)$$

where X is a fixed feasible set.

DEFINITION 2.27 $\mathbf{x}^* \in X$ *is said to be an Appr-efficient solution of problem (2.96) if and only if there exists no other rough feasible solution $\mathbf{x}$ such that $Appr\{f_i(\mathbf{x}, \boldsymbol{\xi}) \leq 0\} \geq Appr\{f_i(\mathbf{x}^*, \boldsymbol{\xi}) \leq 0\}$ for all i, and a strict inequality holds for at least one $i_0 \in \{1, 2, \cdots, m\}$.*

2.5.2 L-DCRM and Goal Programming Method

If the objective functions and constraints are linear and with rough interval coefficients, we propose the following linear dependent chance constraints rough models (L-DCRM) (2.97) and (2.98):

$$\begin{cases} \max \left[Ch\{\tilde{\mathbf{c}}_1^T \mathbf{x} \geq \bar{f}_1\}, Ch\{\tilde{\mathbf{c}}_2^T \mathbf{x} \geq \bar{f}_2\}, \cdots, Ch\{\tilde{\mathbf{c}}_m^T \mathbf{x} \geq \bar{f}_m\} \right] \\ \text{s.t. } \begin{cases} E[\tilde{\mathbf{a}}_r^T \mathbf{x}] \leq E[\tilde{b}_r], r = 1, 2, \cdots, p \\ \mathbf{x} \geq \mathbf{0} \end{cases} \end{cases}$$

(2.97)

and

$$\begin{cases} \max \left[Ch\{\tilde{\mathbf{c}}_1^T \mathbf{x} \geq \bar{f}_1\}, Ch\{\tilde{\mathbf{c}}_2^T \mathbf{x} \geq \bar{f}_2\}, \cdots, Ch\{\tilde{\mathbf{c}}_m^T \mathbf{x} \geq \bar{f}_m\} \right] \\ \text{s.t. } \begin{cases} Ch\{\tilde{\mathbf{a}}_r^T \mathbf{x} \leq \tilde{b}_r\} \geq \alpha_r, r = 1, 2, \cdots, p \\ \mathbf{x} \geq \mathbf{0} \end{cases} \end{cases}$$

(2.98)

where $\tilde{\mathbf{c}}_i = (\tilde{c}_{i1}, \tilde{c}_{i2}, \cdots, \tilde{c}_{in})^T$, $\tilde{\mathbf{a}}_r = (\tilde{a}_{r1}, \tilde{a}_{r2}, \cdots, \tilde{a}_{rn})^T$ are rough interval vectors and $\tilde{b}_r$ are rough intervals, $\bar{f}_i$ are the predetermined ideal objective value for each objective function. $i = 1, 2, \cdots, m, r = 1, 2, \cdots, p$, and Ch represent $\underline{Appr}, \overline{Appr}$, or $Appr$.

By introducing variables β_i, model (2.97) and (2.98) can be rewritten as (2.99) and (2.100), respectively,

$$
\begin{cases}
\max[\beta_1,\beta_2,\cdots,\beta_m] \\
\text{s.t.}
\begin{cases}
Ch\{\tilde{\mathbf{c}}_i^T\mathbf{x} \geq \bar{f}_1\} \geq \beta_i, i = 1,2,\cdots,p \\
E[\tilde{\mathbf{c}}_r^T\mathbf{x}] \leq E[\tilde{b}_r], r = 1,2,\cdots,p \\
\mathbf{x} \geq \mathbf{0}
\end{cases}
\end{cases}
\tag{2.99}
$$

and

$$
\begin{cases}
\max[\beta_1,\beta_2,\cdots,\beta_m] \\
\text{s.t.}
\begin{cases}
Ch\{\tilde{\mathbf{c}}_i^T\mathbf{x} \geq \bar{f}_1\} \geq \beta_i, i = 1,2,\cdots,p \\
Ch\{\tilde{\mathbf{a}}_r^T\mathbf{x} \leq \tilde{b}_r\} \geq \alpha_r, r = 1,2,\cdots,p \\
\mathbf{x} \geq \mathbf{0}
\end{cases}
\end{cases}
\tag{2.100}
$$

2.5.2.1 Crisp Equivalent Model

One way of solving the dependent-chance multiple objective programming model is to convert the objectives and constraints of problems (2.97) and (2.98) into their respective crisp equivalents. As we know, this process is usually very hard and only successful in some special cases. Next we will consider a special case and present the result in this section.

THEOREM 2.11
Let $\tilde{c}_{ij} = ([c_{ij1},c_{ij2}],[c_{ij3},c_{ij4}])(c_{ij3} \leq c_{ij1} \leq c_{ij2} \leq c_{ij4})$, $\tilde{\mathbf{c}}_i^T\mathbf{x} = ([c_{i1},c_{i2}],[c_{i3},c_{i4}])$ and $\tilde{\mathbf{a}}_r^T\mathbf{x} - \tilde{b}_r = ([a,b],[c,d])(c \leq a \leq b \leq d)$. If $Ch = Appr$, then
1. Model (2.99) is equivalent to the following multiple objective decision-making problems:

$$
\begin{cases}
\max\left[(1-\eta)\frac{c_{14}-\bar{f}_1}{(c_{14}-c_{13})},(1-\eta)\frac{c_{24}-\bar{f}_2}{(c_{24}-c_{23})},\cdots,(1-\eta)\frac{c_{m4}-\bar{f}_m}{(c_{m4}-c_{m3})}\right] \\
\text{s.t.}
\begin{cases}
\eta/2(a+b)+(1-\eta)/2(c+d) \leq 0 \\
\mathbf{x} \geq \mathbf{0}
\end{cases}
\end{cases}
\tag{2.101}
$$

$$
\begin{cases}
\max\left[\eta\frac{c_{14}-\bar{f}_1}{c_{14}-c_{13}}+(1-\eta)\frac{c_{12}-\bar{f}_1}{c_{12}-c_{11}},\eta\frac{c_{24}-\bar{f}_2}{c_{24}-c_{23}}+(1-\eta)\frac{c_{22}-\bar{f}_2}{c_{22}-c_{21}},\cdots,\eta\frac{c_{m4}-\bar{f}_m}{c_{m4}-c_{m3}}+\right. \\
\left.(1-\eta)\frac{c_{m2}-\bar{f}_m}{c_{m2}-c_{m1}}\right] \\
\text{s.t.}
\begin{cases}
\eta/2(a+b)+(1-\eta)/2(c+d) \leq 0 \\
\mathbf{x} \geq \mathbf{0}
\end{cases}
\end{cases}
\tag{2.102}
$$

and

$$
\begin{cases}
\max\left[(1-\eta)\frac{c_{14}-\bar{f}_1}{c_{14}-c_{13}}+\eta\frac{c_{12}-\bar{f}_1}{c_{12}-c_{11}},(1-\eta)\frac{c_{24}-\bar{f}_2}{c_{24}-c_{23}}+\eta\frac{c_{22}-\bar{f}_2}{c_{22}-c_{21}},\cdots,(1-\eta)\frac{c_{m4}-\bar{f}_m}{c_{m4}-c_{m3}}+\right. \\
\left.\eta\frac{c_{m2}-\bar{f}_m}{c_{m2}-c_{m1}}\right] \\
\text{s.t.}
\begin{cases}
\eta/2(a+b)+(1-\eta)/2(c+d) \leq 0 \\
\mathbf{x} \geq \mathbf{0}
\end{cases}
\end{cases}
\tag{2.103}
$$

2. *Model (2.100) is equivalent to the following multiple objective decision-making problems:*

$$
\begin{cases}
\max \left[(1-\eta)\frac{c_{14}-\bar{f}_1}{(c_{14}-c_{13})}, (1-\eta)\frac{c_{24}-\bar{f}_2}{(c_{24}-c_{23})}, \cdots, (1-\eta)\frac{c_{m4}-\bar{f}_m}{(c_{m4}-c_{m3})} \right] \\
\text{s.t.} \begin{cases}
\alpha_r(d-c)+(1-\eta)c \leq 0, & \text{if } c \leq 0 \leq a \\
\alpha_r(b-a)(d-c)+\eta a(d-c)+(1-\eta)c(b-a) \leq 0, & \text{if } a \leq 0 \leq b \\
\alpha_r(d-c)-(1-\eta)d \leq 0, & \text{if } b \leq 0 \leq d \\
\mathbf{x} \geq \mathbf{0}
\end{cases}
\end{cases}
\tag{2.104}
$$

$$
\begin{cases}
\max \left[\eta\frac{c_{14}-\bar{f}_1}{c_{14}-c_{13}}+(1-\eta)\frac{c_{12}-\bar{f}_1}{c_{12}-c_{11}}, \eta\frac{c_{24}-\bar{f}_2}{c_{24}-c_{23}}+(1-\eta)\frac{c_{22}-\bar{f}_2}{c_{22}-c_{21}}, \cdots, \eta\frac{c_{m4}-\bar{f}_m}{c_{m4}-c_{m3}}+ \right. \\
\left. (1-\eta)\frac{c_{m2}-\bar{f}_m}{c_{m2}-c_{m1}} \right] \\
\text{s.t.} \begin{cases}
\alpha_r(d-c)+(1-\eta)c \leq 0, & \text{if } c \leq 0 \leq a \\
\alpha_r(b-a)(d-c)+\eta a(d-c)+(1-\eta)c(b-a) \leq 0, & \text{if } a \leq 0 \leq b \\
\alpha_r(d-c)-(1-\eta)d \leq 0, & \text{if } b \leq 0 \leq d \\
\mathbf{x} \geq \mathbf{0}
\end{cases}
\end{cases}
\tag{2.105}
$$

and

$$
\begin{cases}
\max \left[(1-\eta)\frac{c_{14}-\bar{f}_1}{c_{14}-c_{13}}+\eta\frac{c_{12}-\bar{f}_1}{c_{12}-c_{11}}, (1-\eta)\frac{c_{24}-\bar{f}_2}{c_{24}-c_{23}}+\eta\frac{c_{22}-\bar{f}_2}{c_{22}-c_{21}}, \cdots, (1-\eta)\frac{c_{m4}-\bar{f}_m}{c_{m4}-c_{m3}}+ \right. \\
\left. \eta\frac{c_{m2}-\bar{f}_m}{c_{m2}-c_{m1}} \right] \\
\text{s.t.} \begin{cases}
\alpha_r(d-c)+(1-\eta)c \leq 0, & \text{if } c \leq 0 \leq a \\
\alpha_r(b-a)(d-c)+\eta a(d-c)+(1-\eta)c(b-a) \leq 0, & \text{if } a \leq 0 \leq b \\
\alpha_r(d-c)-(1-\eta)d \leq 0, & \text{if } b \leq 0 \leq d \\
\mathbf{x} \geq \mathbf{0}
\end{cases}
\end{cases}
\tag{2.106}
$$

PROOF   The proof is direct from Theorem 2.8 and Theorem 2.9.

2.5.2.2   Goal Programming Method

The goal programming method was initialized by A. Charnes and W. Cooper [51] in 1961. After that, Y. Ijiri [142], S. Lee [176], K. Kendall and S. Lee [161], J. Ignizio [141], R. Narasimhan [223], J. Dyer [95], J. Lee and S. Kim [175], N. Freed and F. Glover [109], E. Karsak et al. [158], M. Schniederjans [266], F. Buffa and W. Jackson [41], V. Barichard [20], H. Calvete et al. [43], S. Chu [62] deeply researched and widely developed it. When dealing with many multiple objective decision-making problems, the goal programming method is widely applied since it could provide with a technique that is accepted by many decision makers, that is, it could point out the preference information and harmoniously inosculate it into the model.

The basic idea of the goal programming method is that, for the objective function $\mathbf{f}(\mathbf{x})=(f_1(\mathbf{x}),f_2(\mathbf{x}),\cdots,f_m(\mathbf{x}))^T$, decision makers give a goal value $\mathbf{f}^o=(f_1^o,f_2^o,\cdots,f_m^o)^T$ such that every objective function $f_i(\mathbf{x})$ approximates the goal value f_i^o as closely as possible. Let $d_p(\mathbf{f}(\mathbf{x}),\mathbf{f}^o) \in \mathbf{R}^m$ be the deviation between $\mathbf{f}(\mathbf{x})$ and $\mathbf{f}^o$, and then consider the following problem:

$$\min_{\mathbf{x} \in X} d_p(\mathbf{f}(\mathbf{x}), \mathbf{f}^o) \qquad (2.107)$$

where the goal value $\mathbf{f}^o$ and the weight vector $\mathbf{w}$ are predetermined by the decision maker. The weight w_i expresses the important factor that the objective function $f_i(\mathbf{x})$ $(i = 1, 2, \cdots, m)$ approximates the goal value f_i^o, $1 \leq p \leq \infty$.

When $p = 1$, it is recalled the simple goal programming method, which is most widely used. Then we have

$$d_p(\mathbf{f}(\mathbf{x}), \mathbf{f}^o) = \sum_{i=1}^{m} w_i |\mathbf{f}(\mathbf{x}), \mathbf{f}^o)|.$$

Since there is the notation $|\cdot|$ in $d_p(\mathbf{f}(\mathbf{x}), \mathbf{f}^o)$, it is not a differentiable function anymore. Therefore, denote that $d_i^+ = \frac{1}{2}(|f_i(\mathbf{x}) - f_i^o| + (f_i(\mathbf{x}) - f_i^o))$ and $d_i^- = \frac{1}{2}(|f_i(\mathbf{x}) - f_i^o| - (f_i(\mathbf{x}) - f_i^o))$, where d_i^+ expresses the quantity that $f_i(\mathbf{x})$ exceeds f_i^o, and d_i^- expresses the quantity that $f_i(\mathbf{x})$ is less than f_i^o. It is easy to prove that

$$d_i^+ + d_i^- = |f_i(\mathbf{x}) - f_i^o|, d_i^+ - d_i^- = f_i(\mathbf{x}) - f_i^o, d_i^+ d_i^- = 0, \ d_i^+, d_i^- \geq 0 \qquad (2.108)$$

When $p = 1$, problem (2.107), can be rewritten as

$$\begin{cases} \min \sum_{i=1}^{m} w_i(d_i^+ + d_i^-) \\ \text{s.t.} \begin{cases} f_i(\mathbf{x}) + d_i^+ - d_i^- = f_i^o \\ d_i^+ d_i^- = 0 \\ d_i^+, d_i^- \geq 0 \\ i = 1, 2, \cdots, m \\ \mathbf{x} \in X \end{cases} \end{cases} \qquad (2.109)$$

In order to easily solve problem (2.109), abandon the constraint $d_i^+ d_i^- = 0$ $(i = 1, 2, \cdots, m)$, and we have

$$\begin{cases} \min \sum_{i=1}^{m} w_i(d_i^+ + d_i^-) \\ \text{s.t.} \begin{cases} f_i(\mathbf{x}) + d_i^+ - d_i^- = f_i^o \\ d_i^+, d_i^- \geq 0 \\ i = 1, 2, \cdots, m \\ \mathbf{x} \in X \end{cases} \end{cases} \qquad (2.110)$$

THEOREM 2.12

If $(\mathbf{x}, \bar{\mathbf{d}}^+, \bar{\mathbf{d}}^-)$ is the optimal solution to problem (2.110), then $\bar{\mathbf{x}}$ is doubtlessly the optimal solution to problem (2.107), where $\bar{\mathbf{d}}^+ = (\bar{d}_1^+, \bar{d}_2^+, \cdots, \bar{d}_m^+)$ and $\bar{\mathbf{d}}^- = (\bar{d}_1^-, \bar{d}_2^-, \cdots, \bar{d}_m^-)$.

PROOF Since $(\mathbf{x}, \bar{\mathbf{d}}^+, \bar{\mathbf{d}}^-)$ is the optimal solution to problem (2.110), we have $\mathbf{x} \in X$, $\bar{\mathbf{d}}^+ \geq 0$, $\bar{\mathbf{d}}^- \geq 0$ and

$$f_i(\mathbf{x}) + \bar{d}_i^+ - \bar{d}_i^- = f_i^o, \ i = 1, 2, \cdots, m \qquad (2.111)$$

1. If $\bar{d}_i^+ = \bar{d}_i^- = 0$, we have $f_i(\mathbf{x}) = f_i^o$, which means $\mathbf{x}$ is the optimal solution problem (2.107).

2. If there exists $i_0 \in \{1, 2, \cdots, m\}$ such that $f_i(\mathbf{x}) \neq f_i^o$, $\bar{d}_i^+ \bar{d}_i^- = 0$ doubtlessly holds. If not, we have $\bar{d}_i^+ > 0$ and $\bar{d}_i^- > 0$. We discuss them respectively, as follows:

(1) If $\bar{d}_i^+ - \bar{d}_i^- > 0$, for $i \in \{1, 2, \cdots, m\}$, let

$$\tilde{d}_i^+ = \begin{cases} \bar{d}_i^+ - \bar{d}_i^-, & \text{if } i = i_0 \\ \bar{d}_i^+, & \text{if } i \neq i_0 \end{cases} \quad \text{and} \quad \tilde{d}_i^- = \begin{cases} 0, & \text{if } i = i_0 \\ \bar{d}_i^-, & \text{if } i \neq i_0 \end{cases} \tag{2.112}$$

Thus, $\tilde{d}_{i_0}^+ < \bar{d}_{i_0}^+$ and $\tilde{d}_{i_0}^- < \bar{d}_{i_0}^-$ both hold. It follows from equations. (2.111) and (2.112) that

$$f_i(\mathbf{x}) + \tilde{d}_i^+ - \tilde{d}_i^- = \begin{cases} f_i(\mathbf{x}) + 0 - (\bar{d}_i^+ - \bar{d}_i^-) = f_i^o, & i = i_0 \\ f_i(\mathbf{x}) + \bar{d}_i^+ - \bar{d}_i^- = f_i^o, & i \neq i_0 \end{cases}$$

We also know that $\mathbf{x} \in X$, $\tilde{d}_i^+ \geq 0$ and $\tilde{d}_i^- \geq 0$. Denote $\tilde{\mathbf{d}}^+ = (\tilde{d}_1^+, \tilde{d}_2^+, \cdots, \tilde{d}_m^+)$ and $\tilde{\mathbf{d}}^- = (\tilde{d}_1^-, \tilde{d}_2^-, \cdots, \tilde{d}_m^-)$, and then we have $(\mathbf{x}, \tilde{\mathbf{d}}^+, \tilde{\mathbf{d}}^-)$ as a feasible solution to problem (2.110). It follows from $\tilde{d}_{i_0}^+ < \bar{d}_{i_0}^+$ and $\tilde{d}_{i_0}^- < \bar{d}_{i_0}^-$ that

$$\sum_{i=1}^{m} (\tilde{d}_{i_0}^+ + \tilde{d}_{i_0}^-) < \sum_{i=1}^{m} (\bar{d}_{i_0}^+ + \bar{d}_{i_0}^-) \tag{2.113}$$

This conflicts with the assumption that $(\mathbf{x}, \bar{\mathbf{d}}^+, \bar{\mathbf{d}}^-)$ is the optimal solution to problem (2.110).

(2) If $\bar{d}_i^+ - \bar{d}_i^- < 0$, for $i \in \{1, 2, \cdots, m\}$, let

$$\tilde{d}_i^+ = \begin{cases} 0, & \text{if } i = i_0 \\ \bar{d}_i^+, & \text{if } i \neq i_0 \end{cases} \quad \text{and} \quad \tilde{d}_i^- = \begin{cases} -(\bar{d}_i^+ - \bar{d}_i^-), & \text{if } i = i_0 \\ \bar{d}_i^-, & \text{if } i \neq i_0 \end{cases} \tag{2.114}$$

We can similarly prove that this conflicts with the assumption that $(\mathbf{x}, \bar{\mathbf{d}}^+, \bar{\mathbf{d}}^-)$ is the optimal solution to problem (2.110).

So far we have proved that $(\mathbf{x}, \bar{\mathbf{d}}^+, \bar{\mathbf{d}}^-)$ is the optimal solution to problem (2.109). Since the feasible region of problem (2.109) is included in problem (2.110), $(\mathbf{x}, \bar{\mathbf{d}}^+, \bar{\mathbf{d}}^-)$ is the optimal solution to problem (2.110). Next, we will prove that $(\mathbf{x}, \bar{\mathbf{d}}^+, \bar{\mathbf{d}}^-)$ is the optimal solution to problem (2.107). For any feasible solution $(\mathbf{x}, \bar{\mathbf{d}}^+, \bar{\mathbf{d}}^-)$, it follows from (2.108) that

$$|f_i(\mathbf{x}) - f_i^o| = d_i^+ + d_i^-, \quad |f_i(\bar{\mathbf{x}}) - f_i^o| = \bar{d}_i^+ + \bar{d}_i^-, \quad i = 1, 2, \cdots, m.$$

For any $\mathbf{x} \in X$, since

$$\sum_{i=1}^{m} |f_i(\bar{\mathbf{x}}) - f_i^o| = \sum_{i=1}^{m} (\bar{d}_i^+ + \bar{d}_i^-) \leq \sum_{i=1}^{m} (d_i^+ + d_i^-) = \sum_{i=1}^{m} |f_i(\mathbf{x}) - f_i^o|,$$

it means that $\bar{\mathbf{x}}$ is the optimal solution to problem (2.107).

2.5.2.3 Numerical Example

Example 2.14
Consider the problem

$$\begin{cases} \max f_1(\mathbf{x}, \boldsymbol{\xi}) = \xi_1 x_1 + \xi_2 x_2 + \xi_3 x_3 + \xi_4 x_4 + \xi_5 x_5 + \xi_6 x_6 \\ \max f_2(\mathbf{x}, \boldsymbol{\xi}) = \xi_7 x_1 + \xi_8 x_2 + \xi_9 x_3 + \xi_{10} x_4 + \xi_{11} x_5 + \xi_{12} x_6 \\ \text{s.t.} \begin{cases} 2x_1 + 4x_2 + 6x_3 + 8x_4 + 10x_5 + 12x_6 \geq 300 \\ 3x_1 + 5x_2 + 7x_3 + 9x_4 + 11x_5 + 13x_6 \leq 600 \\ x_j \geq 0, j = 1, 2, \cdots, 6 \end{cases} \end{cases} \quad (2.115)$$

where $\xi_j (j = 1, 2, \cdots, 12)$ are rough intervals characterized as

$$\begin{aligned} &\xi_1 = ([2,5],[1,6]), \quad &&\xi_2 = ([3,6],[2,7]), \quad &&\xi_3 = ([4,7],[3,8]), \\ &\xi_4 = ([5,8],[4,9]), \quad &&\xi_5 = ([6,9],[5,10]), \quad &&\xi_6 = ([7,10],[6,11]), \\ &\xi_7 = ([8,11],[7,12]), \quad &&\xi_8 = ([9,12],[8,13]), \quad &&\xi_9 = ([10,13],[9,14]) \\ &\xi_{10} = ([11,14],[10,15]), &&\xi_{12} = ([12,15],[11,16]), &&\xi_{12} = ([13,16],[12,17]). \end{aligned}$$

We use the *Appr* operator to deal with the objective functions of the model, and then we can obtain the following DCRM:

$$\begin{cases} \max Appr\{\xi_1 x_1 + \xi_2 x_2 + \xi_3 x_3 + \xi_4 x_4 + \xi_5 x_5 + \xi_6 x_6 \geq \bar{f}_1\} \\ \max Appr\{\xi_7 x_1 + \xi_8 x_2 + \xi_9 x_3 + \xi_{10} x_4 + \xi_{11} x_5 + \xi_{12} x_6 \geq \bar{f}_2\} \\ \text{s.t.} \begin{cases} 2x_1 + 4x_2 + 6x_3 + 8x_4 + 10x_5 + 12x_6 \geq 300 \\ 3x_1 + 5x_2 + 7x_3 + 9x_4 + 11x_5 + 13x_6 \leq 600 \\ x_j \geq 0, j = 1, 2, \cdots, 6 \end{cases} \end{cases} \quad (2.116)$$

or

$$\begin{cases} \max [\beta_1, \beta_2] \\ \text{s.t.} \begin{cases} Appr\{\xi_1 x_1 + \xi_2 x_2 + \xi_3 x_3 + \xi_4 x_4 + \xi_5 x_5 + \xi_6 x_6 \geq \bar{f}_1\} \geq \beta_1 \\ Appr\{\xi_7 x_1 + \xi_8 x_2 + \xi_9 x_3 + \xi_{10} x_4 + \xi_{11} x_5 + \xi_{12} x_6 \geq \bar{f}_2\} \geq \beta_2 \\ 2x_1 + 4x_2 + 6x_3 + 8x_4 + 10x_5 + 12x_6 \geq 300 \\ 3x_1 + 5x_2 + 7x_3 + 9x_4 + 11x_5 + 13x_6 \leq 600 \\ r_j \geq 0, j - 1, 2, \cdots, 6 \end{cases} \end{cases} \quad (2.117)$$

By Theorem 2.11, if the decision maker gives aspiration levels $\bar{f}_1, \bar{f}_2$ for each objective function and sets $\eta = 0.5$, we have the following equivalent model:

$$\begin{cases} \max \frac{1}{2} \left(\frac{5x_1+6x_2+7x_3+8x_4+9x_5+10x_6-\bar{f}_1}{3(x_1+x_2+x_3+x_4+x_5+x_6)} + \frac{6x_1+7x_2+8x_3+9x_4+10x_5+11x_6-\bar{f}_1}{5(x_1+x_2+x_3+x_4+x_5+x_6)} \right) \\ \max \frac{1}{2} \left(\frac{11x_1+12x_2+13x_3+14x_4+15x_5+16x_6-\bar{f}_2}{3(x_1+x_2+x_3+x_4+x_5+x_6)} + \frac{12x_1+13x_2+14x_3+15x_4+16x_5+17x_6-\bar{f}_2}{5(x_1+x_2+x_3+x_4+x_5+x_6)} \right) \\ \text{s.t.} \begin{cases} 2x_1 + 4x_2 + 6x_3 + 8x_4 + 10x_5 + 12x_6 \geq 300 \\ 3x_1 + 5x_2 + 7x_3 + 9x_4 + 11x_5 + 13x_6 \leq 600 \\ x_j \geq 0, j = 1, 2, \cdots, 6 \end{cases} \end{cases}$$

$$(2.118)$$

Suppose the risk tolerance given by the decision maker is $\beta_1^* = 0.8, \beta_1^* = 0.6$, and $\bar{f}_1 = 120, \bar{f}_2 = 480$, we use the goal programming method to handle model (2.118), and we can get the following goal programming model:

$$
\begin{cases}
\min \ \{w_1(d_1^- + d_1^+) + w_2(d_2^- + d_2^+)\} \\
\text{s.t.} \begin{cases}
\frac{1}{2}\left(\frac{5x_1+6x_2+7x_3+8x_4+9x_5+10x_6-120}{3(x_1+x_2+x_3+x_4+x_5+x_6)} + \frac{6x_1+7x_2+8x_3+9x_4+10x_5+11x_6-120}{5(x_1+x_2+x_3+x_4+x_5+x_6)}\right) \\
\quad +d_1^- - d_1^+ = 0.8 \\
\frac{1}{2}\left(\frac{11x_1+12x_2+13x_3+14x_4+15x_5+16x_6-480}{3(x_1+x_2+x_3+x_4+x_5+x_6)} + \frac{12x_1+13x_2+14x_3+15x_4+16x_5+17x_6-480}{5(x_1+x_2+x_3+x_4+x_5+x_6)}\right) \\
\quad +d_2^- - d_2^+ = 0.6 \\
2x_1 + 4x_2 + 6x_3 + 8x_4 + 10x_5 + 12x_6 \geq 300 \\
3x_1 + 5x_2 + 7x_3 + 9x_4 + 11x_5 + 13x_6 \leq 600 \\
x_j \geq 0, j = 1, 2, \cdots, 6
\end{cases}
\end{cases}
$$

$$(2.119)$$

We take $w_1 = w_2 = 0.5$, and by solving the above model, we can obtain the efficient solution as follows:

$$(x_1^*, x_2^*, x_3^*, x_4^*, x_5^*, x_6^*) = (0, 2.1, 12, 5, 40.1, 20.7, 29.4).$$

2.5.3 NL-DCRM and Rough Simulation-Based STGA

Similar to EVRM and CCRM, for problems that are easily converted into crisp ones, we deal with them by the chance measure, and stochastic simulation is used to deal with those that cannot be converted into crisp ones. Next, let us introduce the process of the stochastic simulation dealing with dependent-chance models. Consider the nonlinear dependent-chance rough model (NL-DCRM) as follows:

$$
\begin{cases}
\max \ \left[Ch\{f_1(\mathbf{x}, \boldsymbol{\xi}) \geq \bar{f}_1\}, Ch\{f_2(\mathbf{x}, \boldsymbol{\xi}) \geq \bar{f}_2\}, \cdots, f_m(\mathbf{x}, \boldsymbol{\xi}) \geq \bar{f}_m\}\right] \\
\text{s.t.} \begin{cases} Ch\{g_r(\mathbf{x}, \boldsymbol{\xi}) \leq 0\} \geq \alpha_r, r = 1, 2, \cdots, p \\ \mathbf{x} \geq \mathbf{0} \end{cases}
\end{cases}
\qquad (2.120)
$$

where $f_i(\mathbf{x}, \boldsymbol{\xi})$ and $g_r(\mathbf{x}, \boldsymbol{\xi})$ are nonlinear functions with respect to $\mathbf{x}$, and $\boldsymbol{\xi}$ is a rough interval vector.

2.5.3.1 Rough simulation 3 for *Appr*

In order to verify the feasibility of a fixed decision vector in rough multiple objective decision-making problem, it is crucial to calculate the trust of rough event $\{g_r(\mathbf{x}, \boldsymbol{\xi}) \leq b_r\}$. Let x be fixed decision vector. Denote $(\underline{X}, \overline{X}) = g_r(\mathbf{x}, \boldsymbol{\xi})$. The process can be summarized as follows:

Step 1. Set $\underline{N} = 0$ and $\overline{N} = 0$.
Step 2. Generate $\underline{\lambda}$ and $\overline{\lambda}$ from $\underline{X}$ and $\overline{X}$ randomly.
Step 3. If $g_r(x, \underline{\lambda}) \leq \overline{b}_r(\underline{\lambda})$ for $i = 1, 2, \cdots, m$, then $\underline{N}++$.

Step 4. If $g_r(x,\overline{\lambda}) \le \overline{b}_r(\overline{\lambda})$ for $i = 1, 2, \cdots, m$, then $\overline{N}++$.

Step 5. Repeat the second to fourth steps N times.

Step 6. $L = (\eta \overline{N} + (1-\eta)\underline{N})/N$.

Example 2.15

Assume that the rough interval $\xi = ([1,2],[0,5])$ and $\xi_2 = ([2,3],[1,4])$. Let $\eta = 1/2$. In order to calculate $L = Appr\{\xi_1^2 + \xi_2^2 \le 18\}$, we perform the rough simulation with 2000 cycles and obtain $L = 0.87$.

2.5.3.2 STGA

The spanning tree-based genetic algorithms (STGA) plays an important role within many fields. It generally arises in one of two ways, directly or indirectly. In some direct applications, we wish to connect a set of points using the least cost or least length collection of arcs. Frequently, the points represent physical entities such as components of a computer chip or users of a system who need to be connected to each other or to a central service such as a central processor in a computer system [4]. In indirect applications, we either (1) wish to connect some set of points using a measure of performance that, on the surface, bears little resemblance to the minimum spanning tree objective (sum of arc costs), or (2) the problem itself bears little resemblance to an "optimal tree" problem. In these instances, we often need to be creative in modeling the problem so that it becomes a minimum spanning tree problem. Next, we introduce the steps of spanning tree-based genetic algorithm in detail, and interested readers can refer to the relate literatures [115, 163, 282].

Representation and initialization. Genetic representation is a kind of data structure that represents the candidate solutions to the problem in coding space. Usually, different problems have different data structures or genetic representations. Here, we employ the sub-tree I-J and the sub-tree J-k to represent the transport pattern from plants to distribution centers (DCs), and from DCs to customers, respectively. Each chromosome in this problem consists of three parts. The first part is J binary digits to represent the opened/closed DCs. The last two parts are two Prüfer numbers representing the distribution pattern from plants to DCs, and from DCs to customers, respectively.

In 1889, Cayley proved that there are p^{p-2} distinct labeled trees for a complete graph with p nodes. Prüfer presented the simplest proof of Cayley's formula by establishing a one-to-one correspondence between the set of spanning tree and a set of $p-2$ digit with an integer between 1 and p inclusive [114]. For the sub-tree I-J, denote the plants $1, 2, \cdots, I$ as the component of set $I = \{1, 2, \cdots, I\}$ and define DCs $1, 2, \cdots, J$ as the component of the set $D = \{I+1, I+2, \cdots, I+J\}$. Obviously, this distribution graph has $I+J$ nodes, which means that we need $I+J-2$ digit Prüfer numbers in the range $[1, I+J]$ to uniquely represent the subtree I-J. By using the similar ways, we produce another $J+K-2$ digit Prüfer numbers representing subtrees J-K. An illustration of a feasible chromosome representation is given in Figure2.18. The first substring is 4 binary digits, representing opened/closed DCs.

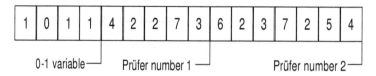

FIGURE 2.18   An illustration of chromosome.

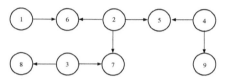

FIGURE 2.19   Spanning tree.

The last two sub-strings are Prüfer numbers consisting of 5 and 7 digits to represent distribution patterns.

Then we take the third substring for example to explain the encoding, feasibility check, and decoding algorithm.

Encoding. The transport tree illustrated in Figure 2.19, converted into the corresponding Prüfer number, is shown in Figure 2.20.

The encoding procedure is as follows:

Step 1. Let j be the lowest-numbered leaf node in the tree. Let k be the node that is the incident to node j. Then k becomes the leftmost digit of the number $P(T)$. $P(T)$ is built up by appending digits to the right; thus $P(T)$ is built and read from left to right.

Step 2. Remove j and edge (j, k) from further consideration. Thus, j is no longer considered at all, and if j is the only successor to k, then j becomes a leaf node.

Step 3. If only two nodes remain to be considered, $P(T)$ has been formed with $J + K - 2$ digits between 1 and $J + K$ inclusive, so stop; otherwise, return to Step 1.

Feasibility check for Prüfer number. The process of initializing a chromosome(a Prüfer number) is by choosing $J + K - 2$ digits from the range $[1, J + K]$ at random. Thus, it is possible to generate some infeasible chromosomes that cannot be adapted into the transport network graph. Due to this reason, feasibility should be checked before decoding the Prüfer number into the spanning tree. As we know, Prüfer number encoding explicitly contains information of a node degree such that any node with degree d will appear exactly $d - 1$ times in the encoding. Thus, when a node appears d times in the Prüfer number, the node has exactly $d + 1$ connections with other nodes.

Then we create the handling for the feasibility of the chromosome with the following criterion: Denote that L_J^j and L_K^k are the number of appearance of nodes j and k which are included in the set J and K, respectively, from $P(T)$. Also we denote that $\overline{L_J^j}$ and $\overline{L_K^k}$ are the number of appearances of nodes j and k in $\overline{P}(T)$ that are included

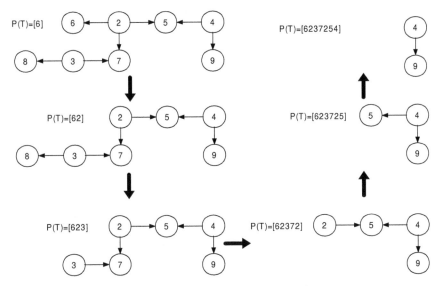

FIGURE 2.20 Encoding procedure.

in sets J and K, respectively. If $\sum_{j\in J}(L_J^j+1)+\sum_{j\in J}\overline{L_J^j}=\sum_{k\in K}(L_K^k+1)+\sum_{k\in K}\overline{L_K^k}$, then $P(T)$ is feasibile; otherwise, $P(T)$ is infeasible. Here, we design the feasibility check and repairing procedure for the Prüfer number to be decoded into spanning tree as follows:

Produce: feasibility check by using the rough simulation and repairing procedure for the Prüfer number. Repeat the following steps until $\sum_{j\in J}(L_J^j+1)+\sum_{j\in J}\overline{L_J^j}=\sum_{k\in K}(L_K^k+1)+\sum_{k\in K}\overline{L_K^k}$.

Step 1. Determine L_J^j and L_K^k from $P(T)$, and $\overline{L_J^j}$ and $\overline{L_K^k}$ from $\overline{P}(T)$.

Step 2. If $\sum_{j\in J}(L_J^j+1)+\sum_{j\in J}\overline{L_J^j}>\sum_{k\in K}(L_K^k+1)+\sum_{k\in K}\overline{L_K^k}$, then select one digit in $P(T)$ that contains node $j(j\in J)$ and replace it with the number $k(k\in K)$. Otherwise, select one digit in $P(T)$ that contains node $k(k\in K)$ and replace it with the number $j(j\in J)$.

Decoding. After checking the feasibility of the chromosome, the chromosome of this problem can be decoded into spanning trees in order to determine the transport pattern. Considering that the total capacity of DCs that will be opened has to satisfy the total demanded by customers, the chromosome is decoded in the backward direction. First, the transport tree between opened DCs and customers is obtained by changing the capacity of the closed DCs to be zero and decoding of the last segment of the chromosome. After that, the total amount required for a product on each DC is determined. Last, the transport tree between suppliers and opened DCs is obtained by decoding the second segment of the chromosome.

The decoding procedure of the second Prüfer number shown in Figure 2.18 and its trace table are given in Figure 2.21. We also give the step-by-step procedure for

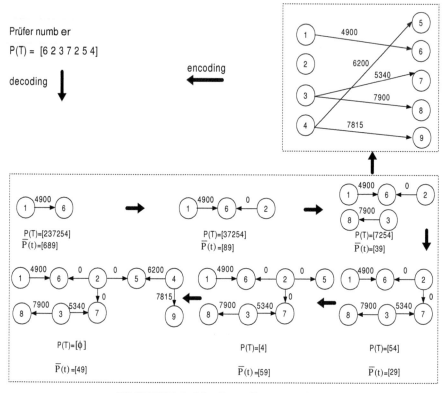

FIGURE 2.21   Decoding procedure.

decoding the second Prüfer number as follow: First, Let $P(T) = [\,6\,2\,3\,7\,2\,5\,4\,]$ be the original Prüfer number, and we have $\overline{P}(T) = [\,1\,8\,9\,]$ as being the set of all nodes that are not part of $P(T)$ and are designed as eligible for consideration. Node 1 is the lowest-numbered eligible node in $\overline{P}(T)$, and node 6 is the leftmost digit of $P(T)$. However, since these two nodes are not in the same set, we add an edge $(1, 6)$ to the tree, remove node 1 from $\overline{P}(T)$ and node 6 from $P(T)$, leaving $P(T) = [\,2\,3\,7\,2\,5\,4\,]$, and since node 6 no longer appears in the remaining part of $P(T)$, add it to $\overline{P}(T)$, so $\overline{P}(T) = [\,6\,8\,9\,]$. Assign the available amount of units to $x_{16} = \min\{m_1, b_6\} = 4900$, which satisfies the defined constraints. Update availability $m_1 = m_1 - x_{16} = 3851$ and $b_6 = b_6 - x_{16} = 0$. Second, node 6, lowest-numbered eligible node in $\overline{P}(T)$, and node 2 is the leftmost digit of $P(T)$, and these two nodes are not in the same set, so we add $(2, 6)$ to the tree, remove node 6 from $\overline{P}(T)$ and node 2 from $P(T)$, leaving $P(T) = [\,3\,7\,2\,5\,4\,]$ and $\overline{P}(T) = [\,8\,9\,]$. Assign $x_{26} = \min\{m_2, b_6\} = 0$. Update $m_2 = m_2 - x_{26} = 0$ and $b_6 = b_6 - x_{26} = 0$. Repeat this process until, finally, $P(T)$ is empty and there are left with only node 4 and 9 in $\overline{P}(T)$. Since there is still an available source in node 4 and demand in node 9, we add the edge $(4, 9)$ to the tree. The decoding procedures for the first Prüfer number is similar.

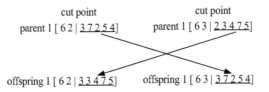

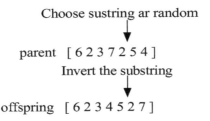

FIGURE 2.22 One-point crossover process.

Choose sustring ar random

$\downarrow$

parent [6 2 3 7 2 5 4]

Invert the substring

$\downarrow$

offspring [6 2 3 4 5 2 7]

FIGURE 2.23 Inversion mutation process.

Genetic operators. Genetic operators mimic the process hereditary to genes to create new offspring in each generation. The operators are used to alter the genetic composition of individuals during representation. There are two common genetic operators: crossover and mutation.

Crossover is the main genetic operator that is employed to explore a new solution space. It operates on two chromosomes at a time and generates offspring by combining both chromosomes' features. As a simple way to achieve crossover, a one-cut point crossover operation is used to choose a random cut-point and to generate the offspring by combining the segment of one parent to the left of the cut-point with the segment of the other parent to the right of the cut-point as shown by Figure 2.22.

Mutation is a background operator that produces spontaneous random changes in various chromosomes to explore a new solution space. A simple way to achieve mutation would be to alter one or more genes. In this book, we adopt inversion mutation by selecting two positions within a chromosome at random and then inverting the substring between these two positions as illustrated in Figure 2.23. The Prüfer numbers resulting by this mutation operation are always feasible in the sense that they can be decoded into a corresponding transport tree due to the feasibility criteria $L_s + \overline{L_s} = L_D + \overline{L_D}$ being unchanged after these operations.

Fitness evaluation and selection. Fitness evaluation is to check the solution value of the objective function subjected to the problem constraints. The single objective problem can be easily manipulated by calculating the fitness value of each chromosome according to the objective function. However, with the multiple objective problem, we can only calculate each objective value, and we cannot simply evaluate its fitness value when in practice the objective functions conflict with each other. In other words, we cannot obtain the absolute optimal solutions, but can only get the Pareto optimal solution.

As to the fitness function for evaluating chromosomes, we employ the weighted-sums method to contract the fitness function. Then we use the following evaluation to combine the multiple objective functions into one overall fitness function and evaluate each chromosome.

Step 1. Calculate the objective values $(F_i, i = 1, 2)$.

Step 2. Chose the solution points that contain the maximum F_1 (or the minimum F_2) corresponding to each objective function value, and then compare them with the stored solution points in the previous generation and select the best points to save again.

$$
\begin{cases}
F_i^{max(t)} = \max_k \left\{ F_i^{max(t-1)}, F_i^{(t)}(X_k) | k = 1, 2, \cdots, i_{size} \right\} \\
F_i^{min(t)} = \min_k \left\{ F_i^{min(t-1)}, F_i^{(t)}(X_k) | k = 1, 2, \cdots, i_{size} \right\}
\end{cases}
\tag{2.121}
$$

where $F_i^{max(t)}$, $F_i^{min(t)}$ are the maximum and minimum values of the ith objective function at generation t, respectively; $F_i^{(t)}(X_k))$ is the ith objective function value of the kth chromosome at generation t; and i_{size} is equal to the pop_{size} plus the offsprings generated after genetic operations.

Step 3. Solve the following equation to get weight for the fitness function:

$$
\varepsilon_i = F_i^{max(t)} - F_i^{min(t)}, \omega_i = \varepsilon_i / \sum_{i=1}^{2} \varepsilon_i, i = 1, 2
\tag{2.122}
$$

Step 4. Fitness function is obtained by combining the objective functions as follows:

$$
eval(X_k) = \sum_{i=1}^{2} \omega_i F_i(X_k), k = 1, 2, \cdots, i_{size}
\tag{2.123}
$$

Selection provides the driving force in a GA. It directs the genetic search toward promising regions in the search space. During the past two decades, many selection methods have been proposed, examined, and compared. Roulette wheel selection, proposed by Holland, is the best-known selection type. The basic idea is to determine selection probability or survival probability for each chromosome proportional to the fitness value. Then a model roulette wheel can be made displaying these probabilities. The selection process is based on spinning the wheel the number of times equal to population size, each selecting a single chromosome for the new procedure.

The overall procedure can be summarized as follows:

Step 1. Calculate a cumulative probability q_k for each chromosome $X_k, k = 1, 2, \cdots, i_{size})$.

Step 2. Generate a random real number $r \in [0, 1]$.

Step 3. If $r \leq q_1$, then select the first chromosome X_1; otherwise select the kth chromosome $X_k(2 \leq k \leq i_{size})$ such that $q_{k-1} < r \leq q_k$.

Step 4. Repeat steps 2 and 3 i_{size} times and obtain i_{size} copies of chromosomes.

Step 5. If the best chromosome is not selected in the next generation, replace one from the new population randomly by the best one.

Using this selection process, we can keep the best chromosome from the current generation for the next generation.

2.5.3.3 Numerical Example

Example 2.16

Consider the problem

$$
\begin{cases}
\max F_1(x,\xi) = 2\xi_1^2 + 5\xi_1 x_2^2 - \xi_3 x_3 + \sqrt{\xi_4^2 + (10 - x_5)^2} \\
\max F_2(x,\xi) = 3x_1 + 3x_2 - x_5 + \xi_2^2 \\
\text{s.t.}
\begin{cases}
x_1 + x_2 + x_3 + x_4 + x_5 \leq 70 \\
5x_1 - 3x_2^2 - \sqrt{x_3} + x_4 + x_5^3 \leq 160 \\
\xi_1^2 x_1 + 2\xi_2^2 x_2 - \xi_3^2 - 5\xi_4^2 x_4 \leq 0 \\
x_i \geq 0, i = 1,2,3,4,5
\end{cases}
\end{cases}
\tag{2.124}
$$

where $\xi_i, = 1,2,3,4$ are rough intervals as follows:

$$\xi_1 = ([2,5],[1,6]), \xi_2 = ([3,6],[2,7]), \xi_3 = ([4,7],[3,8]), \xi_4 = ([5,8],[4,9]).$$

From the mathematical point of view, problem (2.82) is not well defined because of uncertain parameters. Then we apply the *Appr* operator to deal with this uncertain programming.

$$
\begin{cases}
\max \ [\beta_1, \beta_2] \\
\text{s.t.}
\begin{cases}
Appr\{2\xi_1^2 + 5\xi_1 x_2^2 - \xi_3 x_3 + \sqrt{\xi_4^2 + (10 - x_5)^2} \geq \bar{f}_1\} \geq \beta_1 \\
Appr\{3x_1 + 3x_2 - x_5 + \xi_2^2 \geq \bar{f}_2\} \geq \beta_2 \\
x_1 + x_2 + x_3 + x_4 + x_5 \leq 70 \\
Appr\{5x_1 - 3x_2^2 - \sqrt{x_3} + x_4 + x_5^3 \leq 160\} \geq 0.7 \\
Appr\{\xi_1^2 x_1 \mid 2\xi_2^2 x_2 \quad \xi_3^2 - 5\xi_4^2 x_4 \leq 0\} \geq 0.8 \\
x_i \geq 0, i = 1,2,3,4,5
\end{cases}
\end{cases}
\tag{2.125}
$$

Due to the existence of nonlinear objective function and constraint, we cannot transform it into its crisp equivalent model. In order to solve it, we use the rough simulation-based GA to deal with it. After running the algorithm, we get a solution as follows:

$$(x_1^*, x_2^*, x_3^*, x_4^*, x_5^*) = (10.32, 3.25, 0, 0, 10.18), (\beta_1^*, \beta_2^*) = (0.73, 0.59).$$

2.6   The Longtan Construction Site Layout Planning

Next we will develop a class of multiple objective decision-making models with rough interval parameters for the construction site layout planning problem. In addition, the construction site layout planning at the Longtan large-scale water conservancy and hydropower construction project is illustrated to show the application of the proposed models and algorithms.

2.6.1   Background Statement

Construction site layout deals with the assignment of appropriate site locations for temporary facilities such as warehouses, site offices, workshops, and batch plants. The site layout problem involves multiple sources of expertise with often conflicting goals, requiring the scheduling of activities with multiple resources and space considerations. Project managers, superintendents, and subcontractors may jointly agree upon the location of major material pieces and equipment, by using their past experience, trial and error, insight, preference, common sense, and intuition [173]. Construction site layout can be delimited as a design problem of arranging a set of predetermined facilities on the site while satisfying a set of constraints and optimizing layout objectives. Effective placement of facilities within the site is significantly influenced by the movement of resources or the interaction among the facilities [227].

Four quantitative factors, namely, material flows (MF), information flows (IF), personnel flows (PF), and equipment flows (EF) and two qualitative factors, namely, safety/environment concerns (SE) and users' preference (UP) are usually considered in the construction site layout planning problem. According to [94, 157, 227], these six factors are defined as

1. Material flows (MF): the flow of parts, raw materials, works-in- process and finished products between departments. MF can be measured by unit per time unit.

2. Information flows (IF): the communication (oral or reports) between facilities. IF can be measured via a survey of involved personnel, and it can be expressed by the number of communications per time unit.

3. Personnel flows (PF): the number of employees from one or both facilities that perform tasks across facilities.

4. Equipment flows (EF): EF is defined by the number of material-handling equipment (trucks, mixers, etc.) used to transfer materials between facilities.

5. Safety/environment concerns (SE): SE represents the level of safety and environment hazards, measured by safety concerns, which may arise when the two facilities are close to each other and may affect site workers by increasing the likelihood of accidents, noise, uncomfortable temperatures, and pollution.

6. Users' preference (UP) represents the project manager's desire to have the facilities close to or apart from each other.

Good layout demands fulfills a number of competing and yet often conflicting design objectives. Some objectives are to maximize operational efficiency by promoting worker productivity, shortening project time, reducing costs, and maintaining good employee morale by providing for the employees' safety and job satisfaction; to minimize travel distance and time for movement of resources and to decrease material handling time. Besides this, legal obligations may impose safety and permit constraints, where technical and physical limitations cannot be neglected.

In dynamic construction site layout planning problems, facilities serviced in the different construction phases in accordance with the requirements of the construction work during the whole progress of a construction project are different. For example, in Figure 2.1, the construction project lasts for n years, and according to the required facilities, the whole project can be divided into 3 phases. In phase 1, six facilities are set-up, located in six of the eight locations; in phase 2, facilities 1, 2, 6 are still open, facility 4 is closed, while facility 8 changes its location, and so on. In the last phase, there are only four facilities.

Next we will consider the factors, MF, IF, PF, EF, and SE in the mathematical model while considering the UP, namely, users' preferences, in the algorithm design.

In practice, we often face dual-layer information in construction projects, especially large-scale construction projects. The imprecision and complexity in large-scale construction projects cannot be dealt with by simple fuzzy variables or random variables. Compared with fuzzyness and randomness, rough intervals describe dual-layer information over some ranges of imprecision. For instance, in one phase, the operation costs of the Labor Residence is "between 170 thousand China Yuan (CNY) and 180 thousand CNY" in normal cases, and "between 160 thousand CNY and 190 thousand CNY" in special cases. For this type of information, the existing methods can neither reflect its dual-layer features nor entirely pass it to the resulting decisions. However, we can use rough interval $([170, 180], [160, 190])$ to deal with these kind of uncertain parameters. The situation is similar with the interactive cost of different facilities.

2.6.2 Modeling and Analysis

This problem can be modeled as a Quadratic Assignment Problem (QAP). This formulation requires an equal number of facilities and locations. If the number m of facilities is less than the number n of locations, $n - m$ dummy facilities can be created and zero setup cost and transportation cost assigned to them. However, if there are fewer locations than facilities, the problem is not feasible. The goal in solving the dynamic construction site layout problem is to minimize the cost of each single facility, and the interactive cost between different facilities, and to maximize the distance between the facilities that may cause safety and environmental accidents. The facilities and locations are interrelated by two kinds of constraints. The first is the area constraint. The location area has to meet the facility requirement. The second is the logical constraints, which force a feasible solution. For example, different facilities must not appear in one location at the same time, and one facility can be located in only one location at one period.

2.6.2.1   Assumptions

To model the dynamic construction site layout planning problem of large-scale construction projects with a mixed uncertain environment, in this book, the flowing assumptions are used:

1. All the possible locations for the different facilities can be identified [209].

2. F facilities are to be positioned on a site, L locations are available for each facility to position, $L \geq F$.

3. The interactive cost of different facilities and the operational cost of facilities are regarded as fuzzy random variables.

4. All demands for resources at each facility during every period can be satisfied.

5. For each assignment of a facility to a candidate location, there are different set-up and removal costs.

6. The possibility of safety or environment accidents and the possible loss are proportional to the reciprocal of the distance between "high-risk" facilities and "high-protection" facilities.

2.6.2.2   Notations

In order to formulate the model, indices, variables, certain parameters, rough interval parameters, and decision variables are introduced as follows:

Indices

x, y different types of site-level facilities index, $x, y \in \{1, 2, \cdots, F\}$. Among those, "high-risk" facilities that may cause safety or environmental accidents, are marked as x^k

k "high-risk" facilities index, $k \in \{1, 2, \cdots, K\}$

r "high-protection" facilities index, $r \in \{1, 2, \cdots, R\}$

i, j locations index, $i, j \in \{1, 2, \cdots, L\}$

t periods of the problem index, $t \in \{1, 2, \cdots, T\}$

Variables

η_{xit} denotes the opening of facility x at location i during period t

ω_{xit} denotes the closure of facility x at location i during period t

Certain parameters

α_t an appropriate discount rate

A_{ij} the distance from location i to location j

C^s_{xit} the startup cost of facility x at location i during time period t

C^c_{xit} the closure cost of facility x at location i during time period t

S_{xt} the area required for facilities x during time period t

D_i the area of the location i

w_{kr} the risk weight of "high-risk" facility x_k near "high-protection" facility y_r

Uncertain variables

$\tilde{C}_{xyt}$ the interactive cost per unit distance of facility x and facility y for unit-distance during time period t

$\tilde{C}_{xit}^z$ the cost of operation at location i during time period t for facility type x

$\tilde{B}_{xyt}$ the penalty cost of safety accidents or environmental accidents

Decision variables

δ_{xit} denotes the existence of facility x at location i during period t, with the initial condition of $\delta_{xi0} = 0$, $\delta_{xj0} = 0$

δ_{yit} denotes the existence of facility y at location i during period t, with the initial condition of $\delta_{yi0} = 0$, $\delta_{yj0} = 0$

Based on the assumptions and notations foregoing, we propose a multiple objective decision-making model with rough interval parameters for the dynamic construction site layout planning problem of large-scale construction projects.

2.6.2.3 Objective Functions

Various layout plans should be evaluated to determine the optimum one according to the six factors in construction site layout planning. There are three objectives, which are explained as follows: (1) The total cost of site layout planning The cost of each single facility is composed of three parts: the setup cost, the closure cost, and the operating cost. Since η_{xit} is the opening of facility x at location i during phase t, C_{xit}^s is the startup cost. $\eta_{xit} = 1$ denotes that the facility x is setup at location i during phase t, and C_{xit}^s is a certain variable that can be forecast by construction site layout managers, so that $C_{xit}^s \eta_{xit}$ denotes the opening cost of all the facilities. Similarly, $C_{xit}^c \omega_{xit}$ denotes the closure cost, and $\tilde{C}_{xit}^z \delta_{xit}$ denotes the operating cost of facilities. In addition, in large-scale construction projects, there could be a very long time duration. Some construction projects last for several years or even more than 10 years. It is necessary, therefore, to consider the time value of the fund. a_t denotes an appropriate discount rate to determine the net present value. Thereupon, $a_t(C_{xit}^s \eta_{xit} + C_{xit}^c \omega_{xit} + \tilde{C}_{xit}^z \delta_{xit})$ is the cost of facility x located in location i.

It is not enough to just minimize the cost of each single facility. There are imperative material flows, information flows, personnel flows, and equipment flows between different facilities. An optimum dynamic site layout plan has to minimize the interaction cost between these facilities. In phase t, if facility x is housed in location i and facility y is housed in location j, there are activities between x and y. Considering the complicated flows of materials, information, personnel, and equipment between these facilities, and the future condition being hard to predict given the above assumptions, the interactive cost of x and y for unit-distance is a rough interval, marked as $\tilde{C}_{xyt}$. Since A_{ij} is the distance from i to j, the interactive cost between x and y during phase t is $\delta_{xit} \delta_{yjt} A_{ij} \tilde{C}_{xyt}$. So, considering the time value of the fund, the total cost of site layout C is

$$\sum_{t=1}^{T} \sum_{x,y=1}^{F} \sum_{i,j=1}^{L} \alpha_t \left(C_{xit}^s \eta_{xit} + C_{xit}^c \omega_{xit} + \tilde{C}_{xit}^z \delta_{xit} + \delta_{xit} \delta_{yjt} A_{ij} \tilde{C}_{xyt} \right) \tag{2.126}$$

In the construction industry, the risk of a fatality is five times more likely than in a manufacturing, based industry, while the risk of a major injury is two and a half times higher [262]. Safety and environmental issues are an inescapable aspect when designing construction site layouts. A well-planned and well-run project will be both safe and efficient. It will save lives and money, and present injury and ill-health.

The "high-risk" facilities that may cause safety and environmental accidents are marked as x^k, for example, oil depots, explosive storage, dangerous chemical storage, etc. The "high-protection" ones are vulnerable facilities and may suffer great loss once safety and environmental accidents happen. It is known that the nearer these two kinds of facilities, the greater chance of an accident. Therefore, the farther these "high-risk" facilities are from "high-protection" facilities, the better. $\delta_{x^k i t}$ denotes facility x^k located in location i, and $\delta_{y^r j t}$ denotes facility y^r housed in location j. Thereupon, the distance between "high-risk" and other facilities D is

$$D = \sum_{t=1}^{T} \sum_{i,j=1}^{L} \sum_{k=1}^{K} \sum_{r=1}^{R} (w_{kr} \delta_{x^k i t} \delta_{y^r jt} A_{ij})^{-1} \qquad (2.127)$$

2.6.2.4 Constraints

The location area has to meet the requirements of the facility. S_{xt} is the area required for facilities x during time period t, while the location area i is D_i. So, we have

$$\delta_{xit} S_{xt} < D_i, \forall x \in \{1,2,\cdots,F\}, t \in \{1,2,\cdots,T\}, i \in \{1,2,\cdots,L\} \qquad (2.128)$$

The use ratio of the area cannot reach 100%, so $S_{xt} < D_i$ instead of $S_{xt} \leq D_i$.

In order to get feasible solution, there are some logical constraints. At most only one facility can be housed in one location, namely, $\delta_{xit} + \delta_{yit} \leq 1, \delta_{xjt} + \delta_{yjt} \leq 1$, while one type of facility can be housed in more than one locations, namely $\delta_{xit} + \delta_{xjt} \geq 0, \delta_{yit} + \delta_{yjt} \geq 0$. S_{xrt}, D_{yrt} and A_{ij} are nonnegative parameters in the model for practicality. $\delta_{xit}, \delta_{xjt}, \delta_{yit}, \delta_{yjt}$ denote the existence of the facility; 1 denotes, in this phase, that this facility exists, while 0 denotes that it does not exist.

η_{xit} is the opening of facility x at location i in phase t, and ω_{xit} is the closure of facility x at location i or j during phase t. Practically, if $\delta_{xi,t-1} = 0$, while $\delta_{xit} = 1$, namely, $\delta_{xit} - \delta_{xi,t-1} = 1$, then facility x is opened in location i in phase t, so that $\eta_{xit} = 1$. $\delta_{xit} - \delta_{xi,t-1} = 0$ denotes that facility x is neither opened nor closed in phase t, so that $\eta_{xit} = 0, \omega_{xit} = 0$. $\delta_{xit} - \delta_{xi,t-1} = -1$ denotes facility x is closed at location i during period t, so that $\omega_{xit} = 1$, namely,

$$\begin{cases} \eta_{xit} = \begin{cases} 1 \text{ if } \delta_{xit} - \delta_{xi,t-1} = 1 \\ 0 \text{ otherwise} \end{cases} \\ \omega_{xit} = \begin{cases} 1 \text{ if } \delta_{xit} - \delta_{xi,t-1} = -1 \\ 0 \text{ otherwise} \end{cases} \\ x,y \in \{1,2,\cdots,F\}, i,j \in \{1,2,\cdots,L\}, t \in \{1,2,\cdots,T\} \end{cases} \qquad (2.129)$$

2.6.2.5 Formulation

From the above discussions, by the integration of equations (2.126) $\sim$ (2.129), the mathematical model of the dynamic construction site layout planning problem for large-scale construction projects could be stated as follows:

$$
\begin{cases}
\min C = \sum\limits_{t=1}^{T} \sum\limits_{x,y=1}^{F} \sum\limits_{i,j=1}^{L} \alpha_t \left(C_{xit}^s \eta_{xit} + C_{xit}^c \omega_{xit} + \tilde{C}_{xit}^z \delta_{xit} + \delta_{xit} \delta_{yjt} A_{ij} \tilde{C}_{xyt} \right) \\
\max D = \sum\limits_{t=1}^{T} \sum\limits_{i,j=1}^{L} \sum\limits_{k=1}^{K} \sum\limits_{r=1}^{R} \left(w_{kr} \delta_{x^k it} \delta_{y^r jt} A_{ij} \right)^{-1} \\
\text{s.t.} \begin{cases}
\delta_{xit} S_{xt} < D_i \\
\delta_{xit} + \delta_{yit} \leq 1 \\
\delta_{xjt} + \delta_{yjt} \leq 1 \\
\delta_{xit} + \delta_{xjt} \geq 0 \\
\delta_{yit} + \delta_{yjt} \geq 0 \\
S_{xt} \geq 0 \\
D_i > 0 \\
\delta_{xit}, \delta_{xjt}, \delta_{yit}, \delta_{yjt} = 0 \text{ or } 1 \\
\eta_{xit} = \begin{cases} 1 \text{ if } \delta_{xit} - \delta_{xi,t-1} = 1 \\ 0 \text{ otherwise} \end{cases} \\
\omega_{xit} = \begin{cases} 1 \text{ if } \delta_{xit} - \delta_{xi,t-1} = -1 \\ 0 \text{ otherwise} \end{cases} \\
\alpha_t = \frac{1}{(1+\alpha)^{(t-1)}} \\
A_{ij} \geq 0 \\
x \neq y, i \neq j
\end{cases}
\end{cases} \tag{2.130}
$$

Since model (2.130) includes uncertain parameters, for each given decision, it is meaningless to maximize the objective functions before we know the exact value of the relative parameters, just like we cannot maximize a fuzzy variable in possibility programming. Also, we cannot judge whether or not a decision is feasible before we know the exact value of the relative parameters. Hence, both the objectives and the constraints in model (2.130) are not well defined mathematically. It is a challenge for a decision maker to select the "optimal decision" with multiple objectives. Thus, rough interval goal programming is proposed in the following text to overcome these problems.

Goal programming proposed by A. Charnes et al. [54] and then detailed by A. Charnes and W. Cooper [51] is an effective tool to deal with multiple objective programming problems. Goal programming can be regarded as a compromise method for multiple objective programming problems and has been applied to a wide range of practical problems.

For the general multiple objective programming problem

$$
\max_{x \in X} [f_1, f_2, \cdots, f_m] \tag{2.131}
$$

the goal programming model can be employed as

$$
\begin{cases}
\min \sum\limits_{i=1}^{m} (w_i^+ d_i^+ + w_i^- d_i^-) \\
\text{s.t.} \begin{cases} f_i(x) + d_i^- - d_i^+ = b_i, i = 1, 2, \cdots, m \\ x \in X \\ d_i^-, d_i^+ \geq 0, i = 1, 2, \cdots, m \end{cases}
\end{cases} \tag{2.132}
$$

where b_i is the goal value for the ith objective. d_i^+ and d_i^- positive are defined as $d_i^+ = [f_i(x) - b_i] \vee 0, d_i^- = [b_i - f_i(x)] \vee 0$, w_i^+ and w_i^- are weighting factors corresponding to the positive deviation and negative deviation for the ith goal, respectively.

In addition, the goal programming method has been applied to solve the multiobjective interval programming problem proposed by H. Ishihuchi and H. Tanaka [149]. Taguchi et al. have presented a typical formulation of lexicographic interval integer goal programming.

If the decision maker can give the targets for the objectives and preference structure, the rough interval goal programming model for problem (2.130) is analogous to the crisp goal programming model (2.132) as follows:

$$
\begin{cases}
\min \ (w_1^+ d_1^+ + w_1^- d_1^-) + (w_2^+ d_2^+ + w_2^- d_2^-)) \\
\text{s.t.} \begin{cases}
\sum\limits_{t=1}^{T} \sum\limits_{x,y=1}^{F} \sum\limits_{i,j=1}^{L} \alpha_t \left(C_{xit}^s \eta_{xit} + C_{xit}^c \omega_{xit} + \tilde{C}_{xit}^z \delta_{xit} + \delta_{xit} \delta_{yjt} A_{ij} \tilde{C}_{xyt} \right) + d_1 - d_1^+ \\
\quad = \tilde{G}_1 \\
\sum\limits_{t=1}^{T} \sum\limits_{i,j=1}^{L} \sum\limits_{k=1}^{K} \sum\limits_{r=1}^{R} (w_{kr} \delta_{x^k it} \delta_{y^r jt} A_{ij})^{-1} + d_2 - d_2^+ = G_2 \\
\delta_{xit} S_{xt} < D_i \\
\delta_{xit} + \delta_{yit} \leq 1 \\
\delta_{xjt} + \delta_{yjt} \leq 1 \\
\delta_{xit} + \delta_{xjt} \geq 0 \\
\delta_{yit} + \delta_{yjt} \geq 0 \\
S_{xt} \geq 0 \\
D_i > 0 \\
\delta_{xit}, \delta_{xjt}, \delta_{yit}, \delta_{yjt} = 0 \text{ or } 1 \\
\eta_{xit} = \begin{cases} 1 \text{ if } \delta_{xit} - \delta_{xi,t-1} = 1 \\ 0 \text{ otherwise} \end{cases} \\
\omega_{xit} = \begin{cases} 1 \text{ if } \delta_{xit} - \delta_{xi,t-1} = -1 \\ 0 \text{ otherwise} \end{cases} \\
\alpha_t = \frac{1}{(1+\alpha)^{(t-1)}} \\
A_{ij} \geq 0 \\
x \neq y, i \neq j
\end{cases}
\end{cases}
$$

$$(2.133)$$

where $\tilde{G}_1 = ([\underline{G}_{1,1}, \overline{G}_{1,1}], [\underline{G}_{1,2}, \overline{G}_{1,2}])$ is the rough interval goal for the first objective, and G_2 is the goal for the second objective. It is noted that model (2.133) is still meaningless due to uncertain parameters. However, it may be converted into its crisp equivalent formulation in some sense. The result is shown by Theorem 2.13.

THEOREM 2.13

If the decision maker adopts the order relation "$\preceq$" (Defintion 1.5) for rough intervals as a preference structure, then model (2.133) is equivalent to the *following crisp multiple objective decision-making model:*

$$
\begin{cases}
\min \ (w_1^+ d_1^+ + w_1^- d_1^-) + (w_2^+ d_2^+ + w_2^- d_2^-)) \\
\text{s.t.} \begin{cases}
\displaystyle\sum_{t=1}^{T}\sum_{x,y=1}^{F}\sum_{i,j=1}^{L} \alpha_t \left(C_{xit}^s \eta_{xit} + C_{xit}^c \omega_{xit} + \underline{C}_{xit,k}^z \delta_{xit} + \delta_{xit}\delta_{yjt}A_{ij}\underline{C}_{xyt,k} \right) + d_1 - d_1^+ \\
= \underline{G}_{1,k}, k=1,2 \\
\displaystyle\sum_{t=1}^{T}\sum_{x,y=1}^{F}\sum_{i,j=1}^{L} \alpha_t \left(C_{xit}^s \eta_{xit} + C_{xit}^c \omega_{xit} + \bar{C}_{xit,k}^z \delta_{xit} + \delta_{xit}\delta_{yjt}A_{ij}\bar{C}_{xyt,k} \right) + d_1 - d_1^+ \\
= \bar{G}_{1,k}, k=1,2 \\
\displaystyle\sum_{t=1}^{T}\sum_{i,j=1}^{L}\sum_{k=1}^{K}\sum_{r=1}^{R} (w_{kr}\delta_{x^k it}\delta_{y^r jt}A_{ij})^{-1} + d_2 - d_2^+ = G_2 \\
\delta_{xit}S_{xt} < D_i \\
\delta_{xit} + \delta_{yit} \le 1 \\
\delta_{xjt} + \delta_{yjt} \le 1 \\
\delta_{xit} + \delta_{xjt} \ge 0 \\
\delta_{yit} + \delta_{yjt} \ge 0 \\
S_{xt} \ge 0 \\
D_i > 0 \\
\delta_{xit}, \delta_{xjt}, \delta_{yit}, \delta_{yjt} = 0 \text{ or } 1 \\
\eta_{xit} = \begin{cases} 1 \text{ if } \delta_{xit} - \delta_{xi,t-1} = 1 \\ 0 \text{ otherwise} \end{cases} \\
\omega_{xit} = \begin{cases} 1 \text{ if } \delta_{xit} - \delta_{xi,t-1} = -1 \\ 0 \text{ otherwise} \end{cases} \\
\alpha_t = \frac{1}{(1+\alpha)^{(t-1)}} \\
A_{ij} \ge 0 \\
x \ne y, i \ne j
\end{cases}
\end{cases}
$$
(2.134)

where $\tilde{C}_{xyt} = ([\underline{C}_{xyt,1}, \bar{C}_{xyt,1}], [\underline{C}_{xyt,2}, \bar{C}_{xyt,2}]), \tilde{C}_{xit}^z = ([\underline{C}_{xit,1}^z, \bar{C}_{xit,1}^z], [\underline{C}_{xit,2}^z, \bar{C}_{xit,2}^z]), \tilde{G}_1 = ([\underline{G}_{1,1}, \bar{G}_{1,1}], [\underline{G}_{1,2}, \bar{G}_{1,2}]).$

PROOF It follows from rough interval arithmetic that

$$
\sum_{t=1}^{T}\sum_{x,y=1}^{F}\sum_{i,j=1}^{L} \alpha_t \left(C_{xit}^s \eta_{xit} + C_{xit}^c \omega_{xit} + \tilde{C}_{xit}^z \delta_{xit} + \delta_{xit}\delta_{yjt}A_{ij}\tilde{C}_{xyt} \right) + d_1 - d_1^+
$$

$$
= ([\sum_{t=1}^{T}\sum_{x,y=1}^{F}\sum_{i,j=1}^{L} \alpha_t \left(C_{xit}^s \eta_{xit} + C_{xit}^c \omega_{xit} + \underline{C}_{xit,1}^z \delta_{xit} + \delta_{xit}\delta_{yjt}A_{ij}\underline{C}_{xyt,1} \right) + d_1 - d_1^+,
$$

$$
\sum_{t=1}^{T}\sum_{x,y=1}^{F}\sum_{i,j=1}^{L} \alpha_t \left(C_{xit}^s \eta_{xit} + C_{xit}^c \omega_{xit} + \bar{C}_{xit,1}^z \delta_{xit} + \delta_{xit}\delta_{yjt}A_{ij}\bar{C}_{xyt,1} \right) + d_1 - d_1^+]
$$

$$
, [\sum_{t=1}^{T}\sum_{x,y=1}^{F}\sum_{i,j=1}^{L} \alpha_t \left(C_{xit}^s \eta_{xit} + C_{xit}^c \omega_{xit} + \underline{C}_{xit,1}^z \delta_{xit} + \delta_{xit}\delta_{yjt}A_{ij}\underline{C}_{xyt,1} \right) + d_1 - d_1^+
$$

$$
, \sum_{t=1}^{T}\sum_{x,y=1}^{F}\sum_{i,j=1}^{L} \alpha_t \left(C_{xit}^s \eta_{xit} + C_{xit}^c \omega_{xit} + \bar{C}_{xit,1}^z \delta_{xit} + \delta_{xit}\delta_{yjt}A_{ij}\bar{C}_{xyt,1} \right) + d_1 - d_1^+]).
$$

Thus

$$\sum_{t=1}^{T}\sum_{x,y=1}^{F}\sum_{i,j=1}^{L} \alpha_t \left(C_{xit}^s \eta_{xit} + C_{xit}^c \omega_{xit} + \tilde{C}_{xit}^z \delta_{xit} + \delta_{xit}\delta_{yjt}A_{ij}\tilde{C}_{xyt} \right) + d_1 - d_1^+ = \tilde{G}_1$$

is equivalent to

$$\sum_{t=1}^{T}\sum_{x,y=1}^{F}\sum_{i,j=1}^{L} \alpha_t \left(C_{xit}^s \eta_{xit} + C_{xit}^c \omega_{xit} + \underline{C}_{xit,1}^z \delta_{xit} + \delta_{xit}\delta_{yjt}A_{ij}\underline{C}_{xyt,1} \right) + d_1 - d_1^+ = \underline{G}_{1,1}$$

$$\sum_{t=1}^{T}\sum_{x,y=1}^{F}\sum_{i,j=1}^{L} \alpha_t \left(C_{xit}^s \eta_{xit} + C_{xit}^c \omega_{xit} + \bar{C}_{xit,1}^z \delta_{xit} + \delta_{xit}\delta_{yjt}A_{ij}\bar{C}_{xyt,1} \right) + d_1 - d_1^+ = \bar{G}_{1,1}$$

$$\sum_{t=1}^{T}\sum_{x,y=1}^{F}\sum_{i,j=1}^{L} \alpha_t \left(C_{xit}^s \eta_{xit} + C_{xit}^c \omega_{xit} + \underline{C}_{xit,2}^z \delta_{xit} + \delta_{xit}\delta_{yjt}A_{ij}\underline{C}_{xyt,2} \right) + d_1 - d_1^+ = \underline{G}_{1,2}$$

$$\sum_{t=1}^{T}\sum_{x,y=1}^{F}\sum_{i,j=1}^{L} \alpha_t \left(C_{xit}^s \eta_{xit} + C_{xit}^c \omega_{xit} + \bar{C}_{xit,2}^z \delta_{xit} + \delta_{xit}\delta_{yjt}A_{ij}\bar{C}_{xyt,2} \right) + d_1 - d_1^+ = \bar{G}_{1,2}.$$

The theorem is proved.

2.6.2.6 Solution Method

In order to solve problem (2.134), we used st-GA. Proposed as a help to escape from repair mechanisms in the search process of GA by M. Gen and R. Cheng [113], it has had some successful applications [282, 309].

The overall procedure of GA is stated as follows:

Step 1. Set the initial values and the parameters of the genetic algorithms: population size *popsize*, crossover rate p_c, and maximum generation, *max_gen*.

Step 2. Generate the initial population.

Step 3. Generic operation: crossover and mutation.

Step 4. Check the feasibility of the offspring and repair the infeasible offspring.

Step 5. Evaluate and select the chromosomes.

Step 6. Repeat the second to fifth steps for the given number of max gen.

Step 7. Report the best chromosome as the optimal solution.

Now let us state the main aspects of st-GA for our problem step by step in more detail.

Representation. For the problems in this book, we use three subtrees, *I-J*, *J-K* and *K-L* to represent the transportation strategy for customer-collection center stage, collection center-disassembly center stage, and disassembly center-manufactory zone stage respectively. Each chromosome in this problem consists of five parts . The first part are *J* binary digits to represent the opened/closed collection centers. The second part are *K* binary digits to represent opened/closed disassembly centers. The last three parts are three Prüfer numbers representing the transportation pattern of each stage, respectively. It is well known that there exists a one-to-one correspondence between the spanning tree set and a set of $p-2$ digit with an integer between 1 and p inclusive, where p is the number of nodes. Then the chromosome of our problem has the form as shown by Figure 2.24. Next, we will take the first stage as an example to illustrate the conversion process of the transportation tree and the Prüfer number.

Procedure: convert a tree to a Prüfer number for $I - J$.

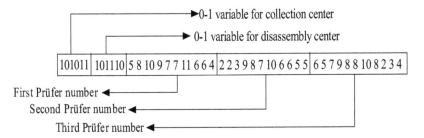

FIGURE 2.24 Illustration of the chromosome.

Step 1. Let i be the lowest-numbered leaf node in tree T. Let j be the node that is the predecessor of i. Then j becomes the rightmost digit of the Prüfer number $P(T)$. $P(T)$ is built up by appending digits to the right; thus $P(T)$ is built and read from left to right.

Step 2. Remove i and edge (i, j) from further consideration. Thus, i is no longer considered at all, and if i is the only successor of j, then j becomes a leaf node.

Step 3. If only two nodes remain to be considered, $P(T)$ has been formed with $I+J-2$ digits between 1 and $I-J$ inclusive, so stop; otherwise, return to Step 1.

Procedure: convert a Prüfer number to a transportation tree for I-J.

Step 1. Let $P(T)$ be the original Prüfer number, and let $\bar{P}(T)$ be the set of all nodes that are not part of $\bar{P}(T)$ and are designed as eligible for consideration.

Step 2. Repeat steps (1)-(5) until no digits are left in $P(T)$.

1. Let i be the lowest-numbered eligible nodes in $\bar{P}(T)$ and j be the leftmost digit of $P(T)$.

2. If i and j are not in the same set I or J, add edge (i, j) to tree T. Otherwise, select the next digit k from $P(T)$ that is not included in the same set with i, exchange j with k, and add edge (i, k) to the tree T.

3. Remove j (or k) from $P(T)$ and i form $\bar{P}(T)$. If j (or k) does not occur anywhere in the remaining part of $P(T)$, put it into $\bar{P}(T)$. Designate i as no longer eligible.

4. Assign the available amount of units to $\gamma_{i,j} = \min\{a_i, b_k\}$(or $\gamma_{i,k} = \min\{a_i, b_k\}$) to edge (i, j) (or (j, k)), where $i \in I$ and $j, k \in J$.

5. Update availability $a_i = a_i - \gamma_{i,j}$ and $b_j = b_j - \gamma_{i,j}$ (or $b_k = b_k - \gamma_{i,k}$).

Step 3. If no digits remain in $P(T)$, there are exactly two nodes, i and j, still eligible in $\bar{P}(T)$ for consideration. Add edge $(i, j)$to tree T and form a tree with $I+J-1$ edges.

Step 4. If there no available units to assign, stop; otherwise, supply r and demand s remain. Add edge (r, s) to the tree and assign the available amount $\gamma_{r,s} = a_r = b_s$ to the edge. If there exists a cycle, remove the edge that is assigned zero flow. A new spanning tree is formed with $I+J-1$ edges.

Similarly, we can obtain the other two Prüfer numbers.

Feasibility check for Prüfer number. When generating the Prüfer number, it is also possible for us to generate an infeasible Prüfer number that cannot be adapted to generate the transportation subtree. For this reason, we need to check its feasibility by using the following feasibility condition.

Repeat the following steps, until $\sum_{i=1}^{I} L_i = \sum_{i=1}^{I} L_i$ (feasibility condition) [114]:

Step 1. Determine R_i for $i \in I \cup J$ from $P(T)$.

Step 2. $R_i = L_i + 1$.

Step 3. If $\sum_{i=1}^{I} L_i > \sum_{i=1}^{I} L_i$, then select one digit in $P(T)$ that contains node $i \in I$ and replace it with the number $j \in D$ generated randomly. Otherwise, select one digit in $P(T)$ that contains node $\in I$ and replace it with the number $j \in J$ generated randomly.

Initialization. Generate chromosome X randomly and check its feasibility. If X is feasible, we take it as an initial chromosome. Otherwise, we repeat this step until the chromosome x is checked to be a feasible chromosome. Repeat the above process *popsize* times, then we have *popsize* initial feasible chromosomes $X_1, X_2, \cdots, X_{popsize}$.

Selection. Determining the fitness value of the chromosome for survival is an important issue in optimization problems. Each chromosome is converted into spanning trees and the objective function are calculated. Then we give an order relation among chromosomes $X_1, X_2, \cdots, X_{popsize}$. Based on the values of their objective functions, the *popsize* chromosomes can be rearranged from good to bad. For the sake of convenience, the rearranged chromosomes are still denoted by $X_1, X_2, \cdots, X_{popsize}$. Finally, we define the rank-based evaluation function as follows:

$$eval(X_i) = a(1-a)^{i-1}, i = 1, 2, \cdots, popsize,$$

where $a \in (0, 1)$ is a parameter in the genetic system.

Crossover. Crossover is the main genetic operator. It operates on two parents (chromosomes) at a time and generates offspring by combining both chromosomes' features. The crossover is done to explore new solution space, and the crossover operator corresponds to the exchanging of parts of strings between selected parents. In this book we employ a one-cut-point crossover operation, which randomly selects a one-cut-point and exchanges the right parts of two parents to generate offspring (see Figure 2.25). To avoid unnecessary decoding from which an infeasible chromosome (a Prüfer number) may be generated after using the crossover operator, the feasibility criteria is embedded within the crossover operation to guarantee the production of feasible offspring, that is, by this checking procedure, offspring always correspond to a transportation tree.

Mutation. Similar to crossover, mutation is used to prevent premature convergence and explore new solution space. However, unlike crossover, mutation is usually done by modifying genes within a chromosome. We use the inversion mutation operation illustrated in Figure 2.26, which randomly selects two positions within a chromosome and then inverts the substring between these two positions. Prüfer numbers resulting from this mutation operation are always feasible in the sense that they can be decoded into a corresponding transportation tree as the feasibility criteria is unchanged after these operations.

The following is an illustrative example of construction site layout planning at Longtan large-scale water conservancy and hydropower construction project to show the application of the proposed models and algorithms.

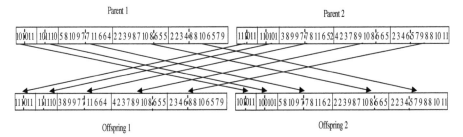

FIGURE 2.25 Illustration of the one-point crossover process.

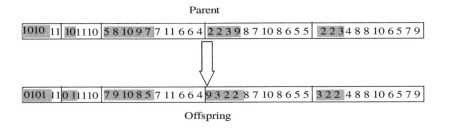

FIGURE 2.26 The illustration of the inversion mutation process.

The problem considered in this book is the Longtan hydropower station project, which is a large-scale water conservancy and hydropower construction project in the southwest region of China. The Longtan hydropower station, located in Tiane county of Guangxi Zhuang Autonomous Region, is one of ten major indicative projects of the national strategies of the Great Western Development in China and the Power Transmission from West to East, and it is a control project in the cascade development on the Hong Shui River. The primary mission of the project with integrative efficiency is flood control and navigation. This hydropower project is of type I large-scale. Layout is as follows: roller compacted concrete gravity dam, flood building with seven outlets and two bottom outlets are arranged in the river bed, and the power capacity of the stream systems with nine installations arranged on the left bank with navigation structures arranged on right bank, and a 2-stage vertical ship lift is used for navigation. The project is designed 400 m in accordance with the normal pool level, and installed plant capacity is 5400 MW. Tunnel diversion is applied in river diversion during construction, and two diversion openings are arranged on the left and right bank. The standard of diversion is a 10-year flood, and the corresponding flow is 14,700 m^3/s.

There are 10 bid sections in the principal part of the Longtan large-scale water conservancy and hydropower construction project. I bid is the excavation treatment

TABLE 2.7   Facilities to be distributed and their required area during each phase

Index	Facility	Phase 1	Phase 2	Phase 3	Phase 4	Phase 5	Phase 6	Phase 7	Phase 8
F_1	Reinforcing Steel Shop	1800	1800	1800	2000	2000	1800	1000	0
F_2	Carpentry shop	0	2500	2500	3000	3000	2500	2000	2000
F_3	Concrete precast shop	0	0	0	1360	1360	0	0	0
F_4	Drill tools repair shop	800	800	800	800	800	0	0	0
F_5	Equipment repairing workshop	3500	3500	3500	3500	3500	3000	3000	3000
F_6	Truck maintenance shop	4000	4000	4000	4000	4000	3500	3000	3000
F_7	Metal and electrical installing workshop	0	0	0	0	0	3000	3300	3300
F_8	Oil depot	1000	1000	1000	1000	1000	1000	1000	1000
F_9	Explosive storage	750	750	750	0	750	750	0	0
F_{10}	Rebar storage	2000	2000	2500	3510	3510	3000	2500	0
F_{11}	Steel storage	2000	2000	2000	2500	2500	2000	2000	2000
F_{12}	Integrated warehouse	800	1000	1000	1000	1000	800	1000	1000
F_{13}	Office	4500	4500	4500	4500	4500	4500	4500	4500
F_{14}	Labor residence	8000	8000	8000	8000	10000	8000	8000	8000

of the left bank slope, and the diversion tunnel of the left bank. II bid is the excavation treatment of the right bank slope, a diversion tunnel, and a navigation structure. I bid and II bid can be done simultaneously. III1 bid is the river closure, cofferdam, and excavation treatment of the riverbed. III2 bid is concrete pouring of the diversion, and the retaining dam on the left bank. IV bid is construction of plant, water diversion and tailrace system in underground powerhouse. V bid is the second excavation treatment of navigation structure and concrete pouring. VI bid is steel tube processing. VII, VIII, and IX are to install metal structures for the generating system, the flood system, and the ship lift system respectively. X bid is the electrical installation bid. Among those, I bid is followed by III2 bid.; II bid is followed by III1 bid and V bid.

There are 14 facilities involved in the principal part of the Longtan hydropower construction project. These facilities as well as their required area during each phase are listed in Table 2.7.

Based on the plan of each bid construction section, the principal part of the Longtan hydropower construction project can be divided into eight phases. The facilities required in each phase are shown in Figure 2.27.

fourteen candidate locations were identified after a comprehensive consideration of the carrying capacity, slope stability, as well as every kinds of topographic, geological, and traffic conditions. The area of these locations are denoted as $D_i, i = \{1, 2, \cdots, 14\}$, $D_i = (6500, 6000, 5750, 7800, 7200, 25000, 26000, 6600, 7200, 4600, 7200, 7200, 7400, 4800)$. The distances between these 17 locations are indicated in Table 2.8.

The closure costs of facilities include dismantling costs, transport costs, material loss costs, and function loss costs, and is in proportion to the distance between facilities that have series flows of materials and equipment. In this book, the closure costs of facilities at different locations is counted as 10% startup costs (see Table 2.9).

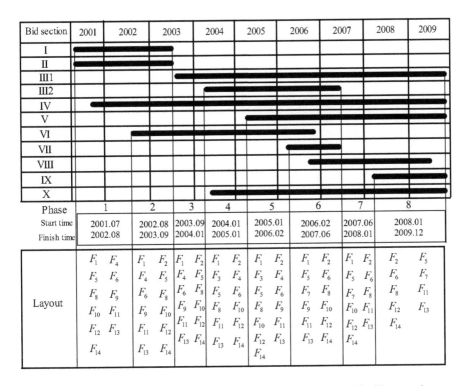

FIGURE 2.27 Construction schedule as well as phase and facility requirement.

TABLE 2.8 The distances between locations (m)

	L_1	L_2	L_3	L_4	L_5	L_6	L_7	L_8	L_9	L_{10}	L_{11}	L_{12}	L_{13}	L_{14}
								A_{ij}						
L_1	0	1440	1584	1740	1908	2148	2256	1956	1800	2172	1680	1620	1476	1452
L_2	1440	0	144	300	468	708	816	636	516	732	348	264	72	1140
L_3	1584	144	0	156	324	564	672	492	372	1092	204	120	180	1284
L_4	1740	300	156	0	168	408	480	336	216	1248	120	144	228	1440
L_5	1908	468	324	168	0	240	348	108	132	1416	288	264	348	1608
L_6	2148	708	564	408	240	0	108	240	312	1656	420	504	588	1848
L_7	2256	816	672	480	348	108	0	264	360	1680	600	708	828	2028
L_8	1956	636	492	336	108	240	264	0	120	1500	336	444	564	1656
L_9	1800	516	372	216	132	312	360	120	0	1380	168	324	444	1776
L_{10}	2172	732	1092	1248	1416	1656	1680	1500	1380	0	1212	1128	888	1872
L_{11}	1680	348	204	120	288	420	600	336	168	1212	0	84	276	1488
L_{12}	1620	264	120	144	264	504	708	444	324	1128	84	0	192	1404
L_{13}	1476	72	180	228	348	588	828	564	444	888	276	192	0	1212
L_{14}	1452	1140	1284	1440	1608	1848	2028	1656	1776	1872	1488	1404	1212	0

TABLE 2.9   The startup costs of facilities during the first period
(1000 CNY)

x	C^s_{xit}													
	i													
	L_1	L_2	L_3	L_4	L_5	L_6	L_7	L_8	L_9	L_{10}	L_{11}	L_{12}	L_{13}	L_{14}
F_1	367.1	264.5	250.8	264.1	267.5	214.4	223.5	255.5	170.2	159.6	131.0	105.3	174.0	360.9
F_2	157.3	126.4	130.4	84.7	85.5	150.0	138.6	127.7	106.3	180.1	97.3	117.4	98.8	174.9
F_3	126.1	196.4	192.6	267.8	270.6	235.4	223.5	255.5	287.8	160.1	130.4	108.9	218.3	229.6
F_4	105.9	116.1	120.0	127.6	128.3	182.2	95.2	159.6	148.8	138.1	131.4	179.1	163.7	131.2
F_5	126.1	105.2	170.7	95.9	96.5	139.6	127.3	180.9	191.3	159.6	130.0	95.2	109.1	98.4
F_6	126.4	100.6	80.2	74.1	74.6	85.7	63.1	95.8	127.5	148.4	140.7	95.2	131.8	109.3
F_7	84.7	106.0	180.6	169.4	171.1	160.0	85.0	95.8	106.3	116.7	195.6	137.8	142.1	87.4
F_8	96.0	107.6	100.3	105.8	106.4	192.6	191.4	202.8	148.8	138.1	151.1	190.9	87.4	98.4
F_9	105.9	97.0	80.2	95.2	96.5	150.8	212.1	170.4	180.7	169.9	98.3	95.5	109.1	98.3
F_{10}	84.7	95.0	90.3	105.8	106.4	128.6	116.9	106.6	85.0	95.2	98.0	63.0	81.9	109.7
F_{11}	63.0	73.9	103.8	95.9	96.5	117.9	127.3	95.8	97.2	84.9	98.3	63.5	76.2	90.1
F_{12}	63.5	74.9	90.0	63.5	64.6	96.4	106.6	95.8	95.6	74.7	86.9	74.0	98.8	89.1
F_{13}	115.0	126.4	140.7	148.3	149.2	117.9	106.7	95.2	96.8	106.4	97.3	76.7	131.8	164.5
F_{14}	167.3	73.3	60.2	74.0	74.6	107.2	116.9	117.1	85.0	95.6	97.3	63.0	76.5	89.6

Empirically, F_8 (Oil Depot) and F_9 (Explosive Storage) are most likely to cause safety or environmental accidents, and they are identified as "high-risk" facilities, namely $\{x^1, x^2\} = \{F_8, F_9\}$. F_2 (Carpentry Shop), F_{13} (Office), and F_{14} (Labor Residence) are identified as "high-protection" facilities, namely, $\{y^1, y^2, y^3\} = \{F_{12}, F_{13}, F_{14}\}$. Decision makers can set the weight by the possible severity of the safety or environmental accidents that may happen, or by preference. In the Longtan case, we set $\{w_{11}, w_{12}, w_{13}\} = \{w_{21}, w_{22}, w_{23}\} = \{0.2, 0.3, 0.5\}$.

The rough intervals of the penalty cost of safety or environmental accidents are given in Table 2.10.

As for the operational cost, because of scanty and imprecise data, the value of $\tilde{C}^z_{xit}$ is difficult to confirm, especially, when facility x has not been located in location i, and there are complicated material flows, information flows, personnel flows, and equipment flows between facilities and working area in the construction site during the process of operating facilities. However, based on years of statistical data and statistical analysis, $\tilde{C}^z_{xit}$ can be identified as rough intervals that can also be estimated based on the specific situation of this project, including scale, scheme, schedule, etc., as well as the experience of decision maker. The operational costs of each facility located in L_2, L_3, L_4, L_9, L_{11}, L_{12}, and L_{13} are listed in Table 2.11. The operational costs of facilities located in L_1 and L_{14} will be 5.8% more because of far away from the workshop. Similarly, the operational costs of facilities located in L_5, L_6, L_7, L_8, and L_{10} will be 7.5% more.

As the decision maker is required to give his goals, he set two goals as follows:

1. Goal 1: The total cost objective should not exceed ($[5.6 \times 10^6, 6.6 \times 10^6], [6.2 \times 10^6, 7.2 \times 10^6]$) as far as possible.

2. Goal 2: The safety and environmental objective should exceed 1×10^{-3} as far as possible.

Then $d_1^- = d_2^+ = 0$. In addition, the weight are assigned as $w_1^- = 0.6, w_3^- = w_3^- = 0.2$. The algorithm parameters for the case problem were set as follows: *popsize* =

TABLE 2.10 The rough intervals of the interactive costs (1000 CNY/km)

	$\tilde{C}_{1,3,t}$	$\tilde{C}_{1,10,t}$	$\tilde{C}_{2,3,t}$	$\tilde{C}_{5,8,t}$	$\tilde{C}_{6,8,t}$	$\tilde{C}_{7,11,t}$
Phase 1	0	([5,6],[4,7])	0	([10,12],[8,16])	([12,16],[9,18])	0
Phase 2	0	([4,6],[3,8])	0	([10,13],[8,15])	([11,15],[8,18])	0
Phase 3	0	([4,5],[4,8])	0	([9,14],[8,15])	([10,14],[8,18])	0
Phase 4	([4,5],[3,7])	([6,10],[5,12])	([8,10],[5,13])	([5,7],[4,8])	([10,13],[8,19])	0
Phase 5	([4,5],[4,8])	([6,10],[5,13])	([8,10],[5,14])	([5,7],[4,10])	([10,12],[8,17])	0
Phase 6	0	([5,8],[4,9])	0	([10,13],[8,15])	([12,18],[8,22])	([5,6],[4,7])
Phase 7	0	([5,7],[4,8])	0	([10,13],[9,16])	([12,20],[9,24])	([5,7],[4,8])
Phase 8	0	0	0	([10,13],[9,16])	([14,18],[9,26]	([5,6],[4,8])

TABLE 2.11 The operational costs of facilities at different locations (1000 CNY)

	Phase 1	Phase 2	Phase 3	Phase 4
F_1	([72,78],[80,90])	([78,84],[76,86])	([100,106],[97,110])	([106,118],[103,120])
F_2		([60,63],[57,65])	([41,44],[40,45])	([70,75],[68,77])
F_3				([140,150],[134,155])
F_4	([41,43],[40,44])	([42,43],[40,44])	([41,42],[40,44])	([42,43],[40,44])
F_5	([55,56],[53,57])	([55,57],[54,58])	([55,56],[54,58])	([60,62],[58,63])
F_6	([43,44],[42,45]	([43,44],[42,45]	([43,44],[42,45]	([45,46],[44,47]])
F_7				
F_8	([36,37],[35,38])	([36,37],[35,38])	([37,39],[36,40])	([38,39],[36,40])
F_9	([41,42],[40,46])	([41,43],[40,46])	([42,45],[40,46])	
F_{10}	([30,32],[29,33])	[31,32],[29,33])	[30,32],[29,33])	([33,35],[32,36])
F_{11}	([30,32],[29,33])	([30,31],[29,33])	([30,33],[29,34])	([34,36],[35,38])
F_{12}	([34,35],[33,36])	([34,35],[33,36])	([34,35],[33,36])	([34,35],[33,36])
F_{13}	([42,43],[41,44])	([42,43],[41,44])	([42,43],[41,44])	([42,43],[41,44])
F_{14}	([54,55],[52,57])	([54,56],[52,58])	([60,64],[58,65])	([61,66],[59,67])
	Phase 5	Phase 6	Phase 7	Phase 8
F_1	([95,96],[93,97])	([87,88],[86,89])	([73,76],[72,78])	
F_2	(78,79[],[76,80])	([78,79],[77,80])	([69,72],[67,73])	([60,62],[58,63])
F_3	([140,143],[135,146])			
F_4	([40,41],[39,42])			
F_5	([60,62],[58,63])	([55,56],[54,57])	([53,54],[52,55])	([51,52],[50,54])
F_6	([45,46],[44,47])	([43,44],[42,45])	([43,44],[42,45])	([40,42],[38,43])
F_7		([66,67],[65,68])	([66,67],[65,68])	([66,67],[65,68])
F_8	([37,39],[36,40])	([37,38],[36,39])	([34,35],[33,36])	([34,35],[33,36])
F_9	([42,44],[41,46])	([42,45],[41,46])		
F_{10}	([33,35],[32,36])	([33,34],[32,36])	([29,30],[28,32])	
F_{11}	([36,37],[35,38])	([35,36],[34,37])	([32,34],[31,35])	([30,32],[29,33])
F_{12}	([34,35],[33,36])	([34,35],[33,36])	([34,35],[33,36])	([34,35],[33,36])
F_{13}	([42,43],[41,44])	([42,43],[41,44])	([42,43],[41,44])	([42,43],[41,44])
F_{14}	([60,62],[58,65])	([58,60],[56,63])	([54,55][53,56])	([52,54],[51,55])

TABLE 2.12   The safest dynamic site layout plan in the Longtan case

Phase	F_1	F_2	F_3	F_4	F_5	F_6	F_7	F_8	F_9	F_{10}	F_{11}	F_{12}	F_{13}	F_{14}
1	3	-	-	1	4	5	-	10	14	9	2	12	6	7
2	3	13	-	1	4	5	-	10	14	9	2	12	6	7
3	3	13	-	1	4	9	-	10	14	11	2	8	6	7
4	12	3	1	13	4	9	-	10	-	11	2	8	6	7
5	12	3	1	13	4	9	-	10	14	11	2	8	6	7
6	12	13	-	-	4	9	5	10	14	11	2	8	6	7
7	12	13	-	-	4	9	5	14	-	11	2	8	6	7
8	-	13	-	-	4	9	5	14	-	-	11	8	6	7

TABLE 2.13   The minimum cost of dynamic site layout plan in the Longtan case

Phase	F_1	F_2	F_3	F_4	F_5	F_6	F_7	F_8	F_9	F_{10}	F_{11}	F_{12}	F_{13}	F_{14}
1	8	-	-	1	4	5	-	3	10	11	2	12	6	7
2	8	13	-	1	4	5	-	3	10	11	2	12	6	7
3	12	13	-	14	4	9	-	3	10	11	2	8	6	7
4	12	13	1	14	4	9	-	3	-	11	2	8	6	7
5	12	13	1	14	4	9	-	3	10	11	2	8	6	7
6	12	13	-	-	4	9	5	3	10	11	2	8	6	7
7	12	13	-	-	4	9	5	10	-	11	2	8	6	7
8	-	13	-	-	4	9	5	10	-	-	11	8	6	7

$40, P_c = 0.4, P_m = 0.6, a = 0.05$. The computer running environment is an inter core 2 Duo 2.00 GHz clock pulse with 2048 MB memory. The case problem is solved by the proposed algorithm with satisfactory solutions within 27 minutes on average, which time is acceptable.

Decision maker can choose the plan from these Pareto optimal solutions by his or her preference. For example, the decision maker thinks that the safety and environmental objective is the most important objective, he may sacrifice more cost for the safest site layout plan, and he or she will choose the absolute right of the Pareto optimal solution, which means the site layout plan shown in Table 2.12. On the contrary, if the decision maker thinks the cost objective is the most important, he or she may choose the minimum total cost of the dynamic site layout plan and sacrifice the safety and environmental objective. The minimum total cost plan is shown in Table 2.13.

By the goal programming technique and arithmetic and order of rough intervals, the multiple objective mixed-integer nonlinear decision-making model of the Longtan large-scale water conservancy and hydropower construction project problem has been degenerated into deterministic a single objective model, and a genetic algorithm is proposed to solve the model. Till now, none has formulated a layout planning problem in the above manner. Here, the techniques illustrated in this book can easily be applied to other layout planning problems. Therefore, these techniques are the appropriate tools to tackle other layout planning problems in realistic environments.

Chapter 3

Bilevel Multiple Objective Rough Decision Making

Organizational decision making often involves two levels of decision makers, uncertain information, and multiple conflicting objectives. With the complex decision-making environment, knowledge-based intelligent systems, including fuzzy sets and fuzzy logic, neural networks, optimization algorithms, etc., provide effective assistance for decision problem recognition, modeling, and solving. In this chapter we introduce the bilevel multiple objective decision making model with rough interval parameters, which is also called bilevel multiple objective rough decision-making (BL-MORDM). The first section is about the crisp bilevel multiple objective decision-making model. After that, we introduce the BL-MORDM problem and the group of models, that is, the expected value model, the chance-constrained model, and the dependent-chance model to deal with it. For each kind of model, we give the equivalent model based on some special cases with good properties. However, for general models, it is usually difficult to give the equivalent form, so we present the rough simulations of these three kinds of equivalent models in another section. In order to solve these models, we adopt the rough simulation-based tabu search algorithms. In the end, an application to an optimization problem is presented as an illustration.

3.1 Bilevel Allocation Problem

Water scarcity is a worldwide occurrence. With the development of society and economy, there is a growing demand for fresh water. However, the limited availability of water does not meet the increasing water demand. The contradiction between the supply and the demand of water resources has become a major constraining factor in economic development in many areas. What is more, water shortage becomes more severe due to the geographically and temporally uneven distribution of regional water resources, (see [5]). Conflicts often arise when different water users (including the environment) compete for a limited water supply for their own benefit. In order to achieve sustainable development and a secure society, institutions and methodologies for water allocation should be reformed, especially for regions having water resource shortages. Water allocation among the regions should consider three key principles: equity, efficiency, and sustainability (see [104]). However, it is not easy to fulfill all three of the principles or goals for a water allocation problem at the basin scale. Effective management and rational allocation is to be a key way to settle water resource shortage. So, higher requirements on the optimal allocation of regional water resources should be put forward.

Research on the optimization of water resources began in the mid-1950s and developed rapidly in the 1960s. Z. Shangguan et al. [270] studied a model for the regional allocation of irrigation water resources under deficit irrigation. B. Abolpour et al. [3] developed the Adaptive Neural Fuzzy Reinforcement Learning approach for water allocation improvement in river basins. G. Wang et al. [293] proposed an improved multiobjective optimization model that considers ecoenvironmental water demand (EWD) for allocating water resources in a river basin over the long term. The above studies are focused on providing an optimal water distribution program for water managers, and some considered regional sustainable development. However, they lack effective incentive mechanisms for water resources allocation, and there is no motivation to save water in each subarea; the phenomenon of wasting water resources is very serious, which aggravates the contradiction between the supply of and demand for water resources.

Economic measures such as water rights and the water market have become the primary elements to optimize the allocation of water resources. Water rights are the rights of using water resources for purposes such as navigation, water withdrawal, and even the right to trade water for personal and financial gain. Water users' original Water rights are assigned by the water resources administrator. After users get their water rights, such rights become their personal property, which they can use or trade in a water market to gain financial profit. Using market means through the water rights trading mechanisms has become an important economic tool to achieve the optimization of regional water resources allocation as it can improve water use efficiency and allocation efficiency (K. Easter and R. Hearne [97]). H. Bjornlund and J. Mckey [33] separately studied the concentration mechanism and market mechanism of water resources allocation based on water rights. After the 2000 Ministerial Meeting of the World Water Forum held in the Hague, water rights have been actively explored in areas such as theoretical frameworks, policies and regulations, and reform and practice (see L. Wang et al. [294, 295] and A. Garrido [110]). Water rights management is a new resources management system and is a management mechanism combining government macro-control with market regulation. As water is different from general commodities due to its unique nature, the water market is not a complete market but a quasi-market. It cannot be separated from macro-control based on basin unified management, while the government's macro-control is aimed at the optimal allocation of water resources based on ensuring social equity.

Generally, models are increasingly being relied upon to inform and support natural resources management; it is particularly important that uncertainty is considered in developing any model (L. Jakeman et al. [153]). In practice, there are uncertain factors in regional water allocation problems; a large number of inexact optimization methods were developed for dealing with uncertainties in water resources management (D. Morgan et al. [218], C. Revelle [253], M. Schluter and N. Rüger [265], N. Isendahl et al. [147]). For example, N. Edirisinghe et al. [98] studied a mathemati-

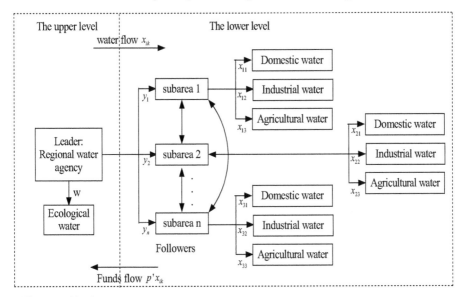

The upper objective :
maximize social total benefit

The lower objective: maximize economic benefit in each subarea

FIGURE 3.1   Model structure for allocation in regional water planning.

cal programming model for the planning of reservoir capacity under random stream flows, which was based on a chance-constrained programming method with a special target-priority policy being considered according to given system reliabilities. S. Chanas and D. Kuchta [49] presented fuzzy-stochastic linear programming in water resources engineering. I. Maqsood et al. [208] developed an interval-parameter fuzzy two-stage stochastic programming (IFTSP) method for the planning of water resources management systems under uncertainty. J. Xu and Y. Tu [310] considered a regional water resource allocation problem whose structure can be expressed by Figure 3.1. Y. Li et al. [190] proposed a multistage fuzzy-stochastic programming model for supporting sustainable water resources allocation and management. Y. Li et al. [191] developed and applied a robust interval-based minimax-regret analysis (RIMA) method to the identification of optimal water resources allocation strategies under uncertainty. The developed RIMA approach can address uncertainties with multiple presentations. Moreover, it can be used for analyzing all possible scenarios associated with different system costs/benefits and risk levels without making assumptions on probabilistic distributions for random variables.

Now let us consider the uncertainty from another viewpoint. If the decision maker classifies the decision environment into two categories, the normal case and special case. The decision parameters are assumed as intervals in each case. For this type of information, the existing methods can neither reflect its dual-layer features nor entirely pass it to the resulting decisions. Two challenges thus emerge: one is to find an effective expression that could reflect dual-layer information (i.e., not only the parameters' most possible value but also its most reliable value), and the other is to use an appropriate method to generate decisions with dual-layer information directly corresponding to the most possible and reliable conditions of the system. Rough intervals can be a suitable concept to express such information. Obviously, the essence of regional water resources allocation based on water rights under a market mechanisms is a two-story structure, game decision-making problem. We will make some contribution to the discussion about regional water allocation by constructing a bilevel decision-making model with rough interval parameters under a market mechanism and to obtaining solutions that can optimize the total benefit to the society and the economic benefit of each subarea, that is, to satisfy the upper and lower objectives of their possibilities in practice. Once these actual representations are available, the real-life regional water allocation problems can be solved realistically.

There exist many methods for solving multilevel programming problems (B. Colson et al. [68]), which can be classified roughly into four categories: the vertex enumeration (J. Bard and J. Moore [19]); the Kuhn–Tucker Methods (W. Bialns and M. Karwan [29]); the penalty function approach (D. White and G. Anandalingam [301], Y. Ishizuka and E. Aiyoshi [150]); methods based on meta heuristics (S. Hejazi et al. [130], H. Calvete et al. [42]) However, these methods must be designed according to different problems, and it is very difficult to find a usual or normal pattern. G. Zhang et al. [341] discussed a decentralized multiobjective bilevel decision making with fuzzy demands that can be modeled by

$$
\begin{cases}
\max_{\mathbf{x} \in X} F(\mathbf{x}, \mathbf{y}) = [\tilde{\mathbf{c}}_{11}^T \mathbf{x} + \tilde{\mathbf{d}}_{11}^T \mathbf{y}, \tilde{\mathbf{c}}_{21}^T \mathbf{x} + \tilde{\mathbf{d}}_{21}^T \mathbf{y}, \cdots, \tilde{\mathbf{c}}_{s1}^T \mathbf{x} + \tilde{\mathbf{d}}_{s1}^T \mathbf{y}] \\
\text{s.t. } \tilde{\mathbf{A}}_1 \mathbf{x} + \tilde{\mathbf{B}}_1 \mathbf{y} \lesssim \tilde{\mathbf{b}}_1 \\
\text{where } y \text{ solves} \\
\quad \begin{cases}
\max_{\mathbf{y} \in Y} f(\mathbf{x}, \mathbf{y}) = [\tilde{\mathbf{c}}_{12}^T \mathbf{x} + \tilde{\mathbf{d}}_{12}^T \mathbf{y}, \tilde{\mathbf{c}}_{22}^T \mathbf{x} + \tilde{\mathbf{d}}_{22}^T \mathbf{y}, \cdots, \tilde{\mathbf{c}}_{t2}^T \mathbf{x} + \tilde{\mathbf{d}}_{t2}^T \mathbf{y}] \\
\text{s.t. } \tilde{\mathbf{A}}_2 \mathbf{x} + \tilde{\mathbf{B}}_2 \mathbf{y} \lesssim \tilde{\mathbf{b}}_2
\end{cases}
\end{cases}
\tag{3.1}
$$

where $\tilde{\mathbf{c}}_{i1}, \tilde{\mathbf{c}}_{i2} \in \mathscr{F}^*(\mathfrak{R}^n), \tilde{\mathbf{d}}_{i1} \in \mathscr{F}^*(\mathfrak{R}^n), \tilde{\mathbf{d}}_{i2} \in \mathscr{F}^*(\mathfrak{R}^n), \tilde{\mathbf{b}}_1 \in \mathscr{F}^*(\mathfrak{R}^p), \tilde{\mathbf{b}}_2 \in \mathscr{F}^*(\mathfrak{R}^q),$ $\tilde{\mathbf{A}}_1 = (\tilde{a}_{ij})_{pn}, \tilde{a}_{ij} \in \mathscr{F}^*(\mathfrak{R}), \tilde{\mathbf{B}}_1 = (\tilde{t}_{ij})_{pm}, \tilde{p}_{ij} \in \mathscr{F}^*(\mathfrak{R}), \tilde{\mathbf{A}}_2 = (\tilde{e}_{ij})_{qn}, \tilde{e}_{ij} \in \mathscr{F}^*(\mathfrak{R}),$ and $\tilde{\mathbf{B}}_2 = (\tilde{s}_{ij})_{qm}, \tilde{s}_{ij} \in \mathscr{F}^*(\mathfrak{R}), i = 1, 2, \cdots, n; j = 1, 2, \cdots, t.$ More details about this model can be found in [341].

Interactive programming techniques are applied in this book to convert the bilevel model into the single-level one, and optimal alternative solutions are obtained by a tabu search that is used to solve the complex nonlinear decision-making problem generated by the decision-making technique.

If DMs considered multiple objectives, and the parameters are assumed as rough intervals, then we can propose BL-MORDM. By using the expected value operator and the chance operator, we present three treatment method classes. A real application is illustrated in the last section of this chapter.

3.2 Bilevel Decision Making Model

In this section we give a brief introduction to (classical) bilevel optimization. In doing so, we follow basically the definitions of [75]. Comprehensive overviews on bilevel optimization can be found in [17, 67, 75, 76, 292]. Bilevel optimization (or bi-level programming) problems constitute a particular kind of hierarchical optimization problem where a part of the constraints for the upper-(or higher-) level problem is defined by another parametric optimization problem, that is, the lower-level problem.

The problems we want to consider have the following common characteristics: (1) The system has interacting decision-making units within a hierarchical structure. (2) Each subordinate level executes its policies after, and in view of, the decisions of superior levels. (3) Each unit maximizes net benefits independently of other units, but may be affected by the actions and reactions of those units. (4) The external effect on a decision-maker's problem can be reflected in both his or her objective function and set of feasible decisions.

3.2.1 Bilevel Single Objective Decision Making Model

A general formulation of bilevel decision-making (BLDM) problems is shown below:

$$
\begin{cases}
\max\limits_{\mathbf{x}} F(\mathbf{x},\mathbf{y}) \\
\text{s.t. } G(\mathbf{x},\mathbf{y}) \leq 0 \\
\text{where } y \text{ solves} \\
\quad \begin{cases}
\max\limits_{\mathbf{y}} f(\mathbf{x},\mathbf{y}) \\
\text{s.t. } \begin{cases} g(\mathbf{x},\mathbf{y}) \leq 0 \\ \mathbf{x} \in \Re^{n_1} \\ \mathbf{y} \in \Re^{n_2} \end{cases}
\end{cases}
\end{cases}
\tag{3.2}
$$

where $\mathbf{x}$ and $\mathbf{y}$ are n_1-dimensional and n_2-dimensional variable column vectors and $\mathbf{x}$ and $\mathbf{y}$ are the decision variables of the leader and the follower. The outer objective function, $F(\mathbf{x},\mathbf{y})$, is constrained by inequality constraints $G(\mathbf{x},\mathbf{y})$. In addition, the outer problem is constrained by an inner problem that minimizes $f(\mathbf{x},\mathbf{y})$ with respect to $\mathbf{y}$, inequality constraints $g(\mathbf{x},\mathbf{y})$, and equality. This inner problem is always active. Each decision maker completely knows the objective functions and constraints of self and the opponent; first, the leader makes a decision, then the follower makes a decision to minimize the objective function with full knowledge of the decision of the leader. That is, let $\hat{\mathbf{x}}$ be the decision specified by the leader; then, the follower solves a linear decision-making problem

$$
\begin{cases}
\max\limits_{\mathbf{y}} f(\hat{\mathbf{x}},\mathbf{y}) \\
\text{s.t. } g(\hat{\mathbf{x}},\mathbf{y}) \leq 0
\end{cases}
\tag{3.3}
$$

and chooses a rational response as an optimal solution $y(x)$ to problem (3.3). Assuming that the follower chooses the rational response, the leader also makes the decision to maximize the objective function $F(\hat{x}, y(\hat{x}))$. Then, a solution defined by the above- mentioned procedure is called a Stackelberg solution.

When the Stackelberg solution is employed, it is assumed that there is no communication between the two decision makers, or that they do not make any binding agreement even if there such communication exists. In other words, there is no cooperative relationship between the two decision makers.

We define the following concepts related to the Stackelberg solution to problem (3.3):

1. Constraint region S, $\{(\mathbf{x}, \mathbf{y}) | G(\mathbf{x}, \mathbf{y}) \leq 0, g(\mathbf{x}, \mathbf{y}) \leq 0\}$.
2. Feasible region of the follower, $S(\mathbf{x}), S(\mathbf{x}) = \{\mathbf{y} | G(\mathbf{x}, \mathbf{y}) \leq 0, g(\mathbf{x}, \mathbf{y}) \leq 0\}$.
3. Set of the follower rational responses $R(\mathbf{x}), R(\mathbf{x}) = \{\mathbf{y} | \mathbf{y} \in arg \max\limits_{\mathbf{y} \in S(\mathbf{x})} f(\mathbf{x}, \mathbf{y})\}$.

The set $R(\mathbf{x})$ is often assumed to be a singleton.

4. Inducible region $IR, IR = \{(\mathbf{x}, \mathbf{y}) | (\mathbf{x}, \mathbf{y}) \in S, \mathbf{y} \in R(\mathbf{x})\}$.
5. Stackelberg solutions. $\{(\mathbf{x}, \mathbf{y}) | (\mathbf{x}, \mathbf{y}) \in arg \max\limits_{(\mathbf{x}, \mathbf{y}) \in IR} F(\mathbf{x}, \mathbf{y})\}$.

The majority of research on BLDM has centered on the linear version of the problem. For $\mathbf{x} \in X \subseteq \Re^n, \mathbf{y} \in Y \subseteq \Re^m, F : X \times Y \to \Re$, and $f : X \times Y \to \Re$, a linear BLDM problem is given by J. Bard [17].

$$
\begin{cases}
\max\limits_{\mathbf{x}} F(\mathbf{x}, \mathbf{y}) = \mathbf{c}_1 \mathbf{x} + \mathbf{d}_1 \mathbf{y} \\
\text{s.t. } \mathbf{A}_1 \mathbf{x} + \mathbf{B}_1 \mathbf{y} \leq \mathbf{b}_1 \\
\text{where } \mathbf{y} \text{ solves} \\
\quad \begin{cases}
\max\limits_{y} f(\mathbf{x}, \mathbf{y}) = \mathbf{c}_2 \mathbf{x} + \mathbf{d}_2 \mathbf{y} \\
\text{s.t. } \begin{cases} \mathbf{A}_2 \mathbf{x} + \mathbf{B}_2 \mathbf{y} \leq \mathbf{b}_2 \\ \mathbf{x} \in \Re^{n_1}, \mathbf{y} \in \Re^{n_2} \end{cases}
\end{cases}
\end{cases} \tag{3.4}
$$

where $\mathbf{x}$ and $\mathbf{y}$ are n_1-dimensional and n_2-dimensional variable column vectors, $\mathbf{c}_i$ and $\mathbf{d}_i, i = 1, 2$ are n_1-dimensional and n_2-dimensional constant row vectors, and $\mathbf{b}_i, i = 1, 2$ is an m-dimensional constant column vector. There have been nearly two dozen algorithms [18, 29, 45, 127, 301] proposed for solving linear BLDM problems since the field caught the attention of researchers in the mid-1970s.

For linear BLDM (3.4), we have the following concepts:

1. Constraint region $S = \{(\mathbf{x}, \mathbf{y}) | \mathbf{x} \in X, \mathbf{y} \in Y, \mathbf{A}_1 \mathbf{x} + \mathbf{B}_1 \mathbf{y} \leq \mathbf{b}_1, \mathbf{A}_2 \mathbf{x} + \mathbf{B}_2 \mathbf{y} \leq \mathbf{b}_2\}$.
2. Feasible region of the follower $S(\mathbf{x})$ for each fixed $\mathbf{x} \in X$, $S(\mathbf{x}) = \{\mathbf{y} \in Y | \mathbf{B}_2 \mathbf{y} \leq \mathbf{b}_2 - \mathbf{A}_2 \mathbf{x}\}$.
3. Projection of S onto the leader's decision space, $S(X)\{\mathbf{x} \in X | \exists \mathbf{y} \in Y, \mathbf{A}_1 \mathbf{x} + \mathbf{B}_1 \mathbf{y} \leq \mathbf{b}_1, \mathbf{A}_2 \mathbf{x} + \mathbf{B}_2 \mathbf{y} \leq \mathbf{b}_2\}$.
4. The follower's rational reaction set for $\mathbf{x} \in S(X)$, $P(\mathbf{x}) = \{\mathbf{y} \in Y | \mathbf{y} \in arg \max\{f(\mathbf{x}, \hat{\mathbf{y}}), \hat{\mathbf{y}} \in S(\mathbf{x})\}\}$ where $arg \max\{f(\mathbf{x}, \hat{\mathbf{y}}), \hat{\mathbf{y}} \in S(\mathbf{x})\} = \{\mathbf{y} \in S(\mathbf{x}) | f(\mathbf{x}, \mathbf{y}) \leq f(\mathbf{x}, \hat{\mathbf{y}}), \hat{\mathbf{y}} \in S(\mathbf{x})\}$.
5. Inducible region $IR = \{(\mathbf{x}, \mathbf{y}) | (\mathbf{x}, \mathbf{y}) \in S, \mathbf{y} \in P(\mathbf{x})\}$.

To ensure that (3.4) has an optimal solution, J. Bard gave the following assumption:

1. S is nonempty and compact. For decisions taken by the leader, the follower has some room to respond, that is $P(x) \neq \emptyset$.

2. $P(\mathbf{x})$ is a point-to-point map.

3.2.2 Bilevel Multiple Objective Decision Making Model

In bilevel optimization, also called two-level optimization, problems are considered where the set of feasible points of the so-called upper-level problem is given by the solution set of a so-called lower-level parametric optimization problem. The variables of the upper-level are the parameters of the lower-level problem and again the solutions of the optimization problem on the lower-level influence the upper level objective function value. With a vector-valued objective on one or both of the levels we speak of a multiobjective bilevel optimization problem.

Despite multiobjective bilevel optimization problems having not yet received a broad attention in the literature, they are very interesting in their view of possible applications (see, e.g., [289]).We illustrate this with an example. Let us consider a city bus transportation system financed by the public authorities. They have as target the reduction of the money losses in this nonprofitable business. As a second target they want to bring as many people as possible to use the buses instead of their own cars, as it is a public mission to reduce the overall traffic. The public authorities can decide about the bus ticket price, but this will influence the customers in their use of the buses. The public has maybe several competing objectives, too, such as to minimize their transportation time and costs. Hence, the use of the public transportation system can be modeled on the lower level with the bus ticket price as parameter and with the solutions influencing the objective values of the public authorities on the upper level again, and thus such a problem can be modeled by bilevel multiobjective optimization.

For solving such a problem, we show that the set of feasible points of the upper-level problem can be expressed completely as the solution set of a multiobjective optimization problem. Then we can determine an approximation of this solution set based on a secularization approach by Pascoletti and Serafini using results from multiobjective optimization.

Here our aim is not a good approximation of the image set of the minimal solutions, the so called efficient set, as it is the aim generally in multiobjective optimization (see, e.g., [72, 108, 264, 280]), but of the solution set itself. A good approximation of this minimal solution set and thus of the set of feasible points of the upper-level problem is meant in the sense of an approximation with almost equidistant points. This is important for getting a representative approximation of the set of feasible points with a given accuracy and for avoiding neglect of larger parts.

For generating such an approximation of the set of feasible points, we use sensitivity results for controlling the parameters of the secularization problem adaptively. We will use these sensitivity results again for solving the upper-level problem in an iterative process: the approximation of the set of feasible points is refined adaptively around the approximated minimal solutions of the upper level. Thereby, not only one minimal solution but an approximation of the whole efficient set of the multiobjective bilevel optimization problem is determined.

Based on the work of J. Bard [15, 16] and T. Shi and S. Xia ([271], we develop a mathematical model of bilevel multiple objective decision-making problem with multiple interconnected lower-level decision makers. The model can be stated as follows:

$$
\begin{cases}
\max_{\mathbf{x}}[F_1(\mathbf{x},\mathbf{y}),F_2(\mathbf{x},\mathbf{y}),\cdots,F_M(\mathbf{x},\mathbf{y})] \\
\text{s.t. } G_r(\mathbf{x},\mathbf{y})) \le 0, r=1,2,\cdots,P \\
\text{where } \mathbf{y} \text{ solves} \\
\quad \begin{cases}
\max_{\mathbf{y}}[f_1(\mathbf{x},\mathbf{y}),f_2(\mathbf{x},\mathbf{y}),\cdots,f_m(\mathbf{x},\mathbf{y})] \\
\text{s.t. } \begin{cases} g_r(\mathbf{x},\mathbf{y}) \le 0, r=1,2,\cdots,p \\ \mathbf{x} \in \Re^{n_1}, \mathbf{y} \in \Re^{n_2} \end{cases}
\end{cases}
\end{cases} \tag{3.5}
$$

DEFINITION 3.1 *Let $\mathbf{x}$ be given by the upper-level decision maker (ULDM) satisfying $G_r(\mathbf{x},\mathbf{y}) \le 0, r=1,2,\cdots,P$. Then $(\mathbf{x},\mathbf{y})$ is said to be a feasible solution to the bilevel multiple objective decision-making problem (3.5) if and only if $\mathbf{y}$ at the lower level is the efficient solution to the lower-level decision maker (LLDM).*

DEFINITION 3.2 *Let $(\mathbf{x}^*,\mathbf{y}^*)$ be a feasible solution to the problem (3.5). Then $(\mathbf{x}^*,\mathbf{y}^*)$ is the efficient solution to problem (3.5) if and only if there is no other feasible solution $(\mathbf{x},\mathbf{y})$ exists, such that $F_i(\mathbf{x},\mathbf{y}) \ge F_i(\mathbf{x}^*,\mathbf{y}^*)$, and at least one $i=1,2,\cdots,M$ is strict inequality.*

The multiobjective two-level linear programming problem is proposed to take into account the diversity of evaluation of outcomes yielded by the decisions of the leader and the follower. It is represented by

$$
\begin{cases}
\max_{\mathbf{x}}[\mathbf{c}_{11}\mathbf{x}+\mathbf{d}_{11}\mathbf{y},\mathbf{c}_{12}\mathbf{x}+\mathbf{d}_{12}\mathbf{y},\cdots,\mathbf{c}_{1M}\mathbf{x}+\mathbf{d}_{1M}\mathbf{y}] \\
\text{s.t. } \mathbf{A}_1\mathbf{x}+\mathbf{B}_1\mathbf{y} \le \mathbf{b}_1 \\
\text{where } \mathbf{y} \text{ solves} \\
\quad \begin{cases}
\max_{\mathbf{y}}[\mathbf{c}_{21}\mathbf{x}+\mathbf{d}_{21}\mathbf{y},\mathbf{c}_{22}\mathbf{x}+\mathbf{d}_{22}\mathbf{y},\cdots,\mathbf{c}_{2m}\mathbf{x}+\mathbf{d}_{2m}\mathbf{y}] \\
\text{s.t. } \begin{cases} \mathbf{A}_2\mathbf{x}+\mathbf{B}_2\mathbf{y} \le \mathbf{b}_1 \\ \mathbf{x} \in \Re^{n_1}, \mathbf{y} \in \Re^{n_2} \end{cases}
\end{cases}
\end{cases} \tag{3.6}
$$

where $\mathbf{x}$ and $\mathbf{y}$ are n_1-dimensional and n_2-dimensional variable column vectors; $\mathbf{c}_i$ and $\mathbf{d}_i, i=1,2$ are n_1-dimensional and n_2-dimensional constant row vectors; and $b_i, i=1,2$ is an m-dimensional constant column vector. If there are more than one followers in the bilevel decision-making problem, we have the model as follows:

$$
\begin{cases}
\max_{\mathbf{x}} F(\mathbf{x}, \mathbf{y}_1, \mathbf{y}_2, \cdots, \mathbf{y}_m) \\
\text{s.t. } G(\mathbf{x}, \mathbf{y}_1, \mathbf{y}_2, \cdots, \mathbf{y}_m) \le 0, \\
\text{where } \mathbf{y}_1, \mathbf{y}_2, \cdots, \mathbf{y}_m \text{ solves} \\
\quad \begin{cases}
\max_{\mathbf{y}_i} f_i(\mathbf{x}, \mathbf{y}_1, \mathbf{y}_2, \cdots, \mathbf{y}_m) \\
\text{s.t. } \begin{cases} g_r(\mathbf{x}, \mathbf{y}_1, \mathbf{y}_2, \cdots, \mathbf{y}_m) \le 0, r = 1, 2, \cdots, p \\ \mathbf{x} \in \Re^{n_1}, y \in \Re^{n_2} \end{cases}
\end{cases}
\end{cases}
\tag{3.7}
$$

where $\mathbf{x}_i$ and $\mathbf{y}_i$ be the control vector of the leader and the ith followers, $i = 1, 2, \cdots, p$, respectively, $F(\mathbf{x}, \mathbf{y}_1, \mathbf{y}_2, \cdots, \mathbf{y}_m)$, and $f_i(\mathbf{x}, \mathbf{y}_1, \mathbf{y}_2, \cdots, \mathbf{y}_m)$ are the objective functions (without loss of generality, all are to be maximized) of the leader and ith followers. For each decision x chosen by the leader, the feasible set Y of control array $(\mathbf{y}_1, \mathbf{y}_2, \cdots, \mathbf{y}_m)$ of the followers should be dependent on $\mathbf{x}$, and is generally represented by

$$
Y(\mathbf{x}) = \{(\mathbf{y}_1, \mathbf{y}_2, \cdots, \mathbf{y}_m) | g(\mathbf{x}, \mathbf{y}_1, \mathbf{y}_2, \cdots, \mathbf{y}_m)\}.
$$

This bilevel programming is a multiperson noncooperative game with leader–follower strategy. When $m = 1$ and $f_l(\mathbf{x}, \mathbf{y}_l) = -F(\mathbf{x}, \mathbf{y}_l)$, that is, the objective function of the follower is in direct opposition to the objective function of the leader, the bilevel programming is a classical minimax problem. For each follower $i, (i = 1, 2, \cdots, m)$, if $\mathbf{x}, \mathbf{y}_1, \cdots, \mathbf{y}_{i-1}, \mathbf{y}_{i+1}, \cdots, \mathbf{y}_m$ are revealed by the leader and other followers, then the reaction $\mathbf{y}_i^*$ of the ith follower must be the optimal solution to the optimization problem

$$
\begin{cases}
\max_{\mathbf{y}_i} f_i(\mathbf{x}, \mathbf{y}_1, \mathbf{y}_2, \cdots, \mathbf{y}_m) \\
\text{s.t. } \begin{cases} g(\mathbf{x}, \mathbf{y}_1, \mathbf{y}_2, \cdots, \mathbf{y}_m) \le 0 \\ \mathbf{x} \in \Re^{n_1} \\ \mathbf{y}_k \in \Re^{n_k}, k = 1, 2, \cdots, m \end{cases}
\end{cases}
\tag{3.8}
$$

whose set of optimal solutions may consist of multiple points, and such a set is also called the rational reaction set of the ith follower.

However, all the followers are of equal status, and they must reveal their strategies simultaneously. So, for all followers, a popular solution concept is the so-called Nash equilibrium defined as the array $(\mathbf{y}_1^*, \mathbf{y}_2^*, \cdots, \mathbf{y}_m^*) \in Y(\mathbf{x})$ with respect to $\mathbf{x}$, that is, $f_i(\mathbf{x}, \mathbf{y}_1^*, \cdots, \mathbf{y}_{i-1}^*, \mathbf{y}_i, \mathbf{y}_{i+1}^*, \cdots, \mathbf{y}_m^*) \le f_i(\mathbf{x}, \mathbf{y}_1^*, \cdots, \mathbf{y}_{i-1}^*, \mathbf{y}_i^*, \mathbf{y}_{i+1}^*, \cdots, \mathbf{y}_m^*)$ for any $(\mathbf{x}, \mathbf{y}_1^*, \cdots, \mathbf{y}_{i-1}^*, \mathbf{y}_i, \mathbf{y}_{i+1}^*, \cdots, \mathbf{y}_m^*) \in Y(\mathbf{x})$ and $i = 1, 2, \cdots, m$. If there is a unique Nash equilibrium, perhaps all the followers might make such an equilibrium because any follower cannot improve his or her own objective by altering his or her strategy unilaterally. It must be also noted that, in general, the Nash equilibrium is neither unique nor Pareto optimal, and any follower can deviate from the Nash equilibrium and move to a better solution.

In the following text, we will discuss a class of bilevel multiple objective rough decision-making problem, in which the parameters are rough intervals. In order to handle this class of problems, we present the bilevel expected value rough model

(BL-EVRM), bilevel chance-constrained rough model (BL-CCRM) and bilevel dependent-chance rough model (BL-DCRM). Traditional solution methods and rough simulation-based intelligent algorithms are proposed to solve the corresponding models. These contents are described in detail in the following three sections.

3.3 BL-EVRM

We give the general form of the bilevel multiple objective rough decision-making model with rough interval parameters as follows, and the discussion in this section is based on this model:

$$
\begin{cases}
\max_{\mathbf{x}}[F_1(\mathbf{x},\mathbf{y},\boldsymbol{\xi}),F_2(\mathbf{x},\mathbf{y},\boldsymbol{\xi}),\cdots,F_M(\mathbf{x},\mathbf{y},\boldsymbol{\xi})] \\
\text{s.t. } G_r(\mathbf{x},\mathbf{y},\boldsymbol{\xi}) \le 0, r=1,2,\cdots,P \\
\text{where } \mathbf{y} \text{ solves} \\
\quad \begin{cases}
\max_{y}[f_1(\mathbf{x},\mathbf{y},\boldsymbol{\xi}),f_2(\mathbf{x},\mathbf{y},\boldsymbol{\xi}),\cdots,f_m(\mathbf{x},\mathbf{y},\boldsymbol{\xi})] \\
\text{s.t. } \begin{cases} g_r(\mathbf{x},\mathbf{y},\boldsymbol{\xi}) \le 0, r=1,2,\cdots,p \\ \mathbf{x}\in\Re^{n_1}, y\in\Re^{n_2} \end{cases}
\end{cases}
\end{cases}
\tag{3.9}
$$

where x, F_i, and G_r are the upper-level decision variable, objective functions, and constraint functions respectively, and $\mathbf{y}, f_i$, and g_r are the lower-level decision variable, objective function and constraint functions, respectively, and $\boldsymbol{\xi}$ is a rough interval vectors.

3.3.1 General Model for BL-EVRM

We know that the expected value operator is a tool that can transform rough model into crisp one. The spectrum of the rough expected value model is as follows:

$$
\begin{cases}
\max_{\mathbf{x}}[E[F_1(\mathbf{x},\mathbf{y},\boldsymbol{\xi})],E[F_2(\mathbf{x},\mathbf{y},\boldsymbol{\xi})],\cdots,F_M[(\mathbf{x},\mathbf{y},\boldsymbol{\xi})]] \\
\text{s.t. } E[G_r(\mathbf{x},\mathbf{y},\boldsymbol{\xi})] \le 0, r=1,2,\cdots,P \\
\text{where } \mathbf{y} \text{ solves} \\
\quad \begin{cases}
\max_{y}[E[f_1(\mathbf{x},\mathbf{y},\boldsymbol{\xi})],E[f_2(\mathbf{x},\mathbf{y},\boldsymbol{\xi})],\cdots,E[f_m(\mathbf{x},\mathbf{y},\boldsymbol{\xi})]] \\
\text{s.t. } \begin{cases} E[g_r(\mathbf{x},\mathbf{y},\boldsymbol{\xi})] \le 0, r=1,2,\cdots,p \\ \mathbf{x}\in\Re^{n_1}, y\in\Re^{n_2}. \end{cases}
\end{cases}
\end{cases}
\tag{3.10}
$$

Model (3.10) is called BL-EVRM.

DEFINITION 3.3 *Let $\mathbf{x}$ be given by the upper-level decision maker. Then $(\mathbf{x},\mathbf{y})$ is said to be a feasible solution to problem (3.10) if and only if the decision making variable $y(E[g_r(\mathbf{x},\mathbf{y},\boldsymbol{\xi})] \le 0, r=1,2,\cdots,p)$ at the lower level is the efficient solution to the upper-level decision maker.*

DEFINITION 3.4 *Let* $(\mathbf{x}^*, \mathbf{y}^*)$ *be a feasible solution to problem (3.10). Then* $(\mathbf{x}^*, \mathbf{y}^*)$ *is the efficient solution to problem (3.10) if and only if there is no other feasible solution* $(\mathbf{x}, \mathbf{y})$ *exists, such that* $E[F_i(\mathbf{x}, \mathbf{y})] \geq E[F_i(\mathbf{x}^*, \mathbf{y}^*)]$ *and at least one* $i = 1, 2, \cdots, M$ *is strict inequality.*

3.3.2 BL-LEVRM and KKT Transformation Approach

The general bilevel linear expected value rough model (BL-LEVRM) can be formulated as

$$
\begin{cases}
\max \left[E[\tilde{\mathbf{C}}_1^T \mathbf{x} + \tilde{\mathbf{D}}_1^T \mathbf{y}], E[\tilde{\mathbf{C}}_2^T \mathbf{x} + \tilde{\mathbf{D}}_2^T \mathbf{y}], \cdots, E[\tilde{\mathbf{C}}_{m_1}^T \mathbf{x} + \tilde{\mathbf{D}}_{m_1}^T \mathbf{y}] \right] \\
\text{s.t. } \tilde{\mathbf{A}}_r^T \mathbf{x} + \tilde{\mathbf{B}}_r^T \mathbf{y} \leq \tilde{E}_r, r = 1, 2, \cdots, p_1 \\
\text{where } \mathbf{y} \text{ solves} \\
\quad \begin{cases}
\max \left[E[\tilde{\mathbf{c}}_1^T \mathbf{x} + \tilde{\mathbf{d}}_1^T \mathbf{y}], E[\tilde{\mathbf{c}}_2^T \mathbf{x} + \tilde{\mathbf{d}}_2^T \mathbf{y}], \cdots, E[\tilde{\mathbf{c}}_{m_2}^T \mathbf{x} + \tilde{\mathbf{d}}_{m_2}^T \mathbf{y}] \right] \\
\text{s.t. } E[\tilde{\mathbf{a}}_r^T \mathbf{x} + \tilde{\mathbf{b}}_r^T \mathbf{y}] \leq E[\tilde{e}_r], r = 1, 2, \cdots, p_2 \\
\mathbf{x} \in \mathfrak{R}^{n_1}, \mathbf{y} \in \mathfrak{R}^{n_2}
\end{cases}
\end{cases} \tag{3.11}
$$

where $\tilde{\mathbf{C}}_i = (\tilde{C}_{i1}, \tilde{C}_{i2}, \cdots, \tilde{C}_{in_1})^T, \tilde{\mathbf{D}}_i = (\tilde{D}_{i1}, \tilde{D}_{i2}, \cdots, \tilde{D}_{in_1})^T, \tilde{\mathbf{A}}_r = (\tilde{A}_{r1}, \tilde{A}_{r2}, \cdots, \tilde{A}_{rn_1})^T,$
$\tilde{\mathbf{B}}_r = (\tilde{B}_{r1}, \tilde{B}_{r2}, \cdots, \tilde{B}_{rn_1})^T,$ are rough interval vectors and $\tilde{E}_r$ are rough intervals, $i = 1, 2, \cdots, m_1, r = 1, 2, \cdots, p_1.$ At the same time, $\tilde{\mathbf{c}}_i = (\tilde{c}_{i1}, \tilde{c}_{i2}, \cdots, \tilde{c}_{in_2})^T, \tilde{\mathbf{d}}_i = (\tilde{d}_{i1}, \tilde{d}_{i2}, \cdots, \tilde{d}_{in_2})^T, \tilde{\mathbf{a}}_r = (\tilde{a}_{r1}, \tilde{a}_{r2}, \cdots, \tilde{a}_{rn_1})^T, \tilde{\mathbf{b}}_r = (\tilde{b}_{r1}, \tilde{b}_{r2}, \cdots, \tilde{b}_{rn_1})^T$ are rough interval vectors and $\tilde{e}_r$ are rough intervals, $i = 1, 2, \cdots, m_2, r = 1, 2, \cdots, p_2.$

3.3.2.1 Crisp Equivalent Model

One way of solving problem (3.11) is to convert the objectives and constraints into their respective crisp equivalents and then solve them with traditional bilevel multiple objective decision-making methods.

THEOREM 3.1
Let $\tilde{C}_{ij} = ([C_{ij1}, C_{ij2}], [C_{ij3}, C_{ij4}])(C_{ij3} \leq C_{ij1} \leq C_{ij2} \leq C_{ij4}), \tilde{D}_{ij} = ([D_{ij1}, D_{ij2}], [D_{ij3}, D_{ij4}])(D_{ij3} \leq D_{ij1} \leq D_{ij2} \leq D_{ij4}), \tilde{A}_{rj} = ([A_{rj1}, A_{rj2}], [A_{rj3}, A_{rj4}])(A_{rj3} \leq A_{rj1} \leq A_{rj2} \leq A_{rj4}), \tilde{A}_{rj} = ([A_{rj1}, A_{rj2}], [B_{rj3}, B_{rj4}])(B_{rj3} \leq B_{rj1} \leq B_{rj2} \leq B_{rj4}), \tilde{E}_r = ([E_{r1}, E_{r2}], [E_{r3}, E_{r4}])(E_{r3} \leq E_{r1} \leq E_{r2} \leq E_{r4})$ for $i = 1, 2, \cdots, m_1, r = 1, 2, \cdots, p_1, j = 1, 2, \cdots, n_1.$ And $\tilde{c}_{ij} = ([c_{ij1}, c_{ij2}], [c_{ij3}, c_{ij4}])(c_{ij3} \leq c_{ij1} \leq c_{ij2} \leq c_{ij4}), \tilde{d}_{ij} = ([d_{ij1}, d_{ij2}], [d_{ij3}, d_{ij4}])(d_{ij3} \leq d_{ij1} \leq d_{ij2} \leq d_{ij4}), \tilde{a}_{rj} = ([a_{rj1}, a_{rj2}], [a_{rj3}, a_{rj4}])(a_{rj3} \leq a_{rj1} \leq a_{rj2} \leq a_{rj4}), \tilde{b}_{rj} = ([b_{rj1}, b_{rj2}], [b_{rj3}, b_{rj4}])(b_{rj3} \leq b_{rj1} \leq b_{rj2} \leq b_{rj4}), \tilde{e}_r = ([e_{r1}, e_{r2}], [e_{r3}, e_{r4}])(e_{r3} \leq e_{r1} \leq e_{r2} \leq e_{r4})$ for $i = 1, 2, \cdots, m_2, r = 1, 2, \cdots, p_2, j = 1, 2, \cdots, n_2.$ Then problem (3.11) is equivalent to the conventional multiple objective*

decision-making problem:

$$
\begin{cases}
\max[\frac{1}{2}\sum\limits_{j=1}^{n_1}(\eta(C_{1j1}+C_{1j2})+(1-\eta)(C_{1j3}+C_{1j4}))x_j+\sum\limits_{j=1}^{n_2}(\eta(D_{1j1}+D_{1j2})\\
+(1-\eta)(D_{1j3}+D_{1j4}))y_j,\frac{1}{2}\sum\limits_{j=1}^{n_1}(\eta(C_{2j1}+C_{2j2})+(1-\eta)(C_{2j3}+C_{2j4}))x_j+\\
\sum\limits_{j=1}^{n_2}(\eta(D_{2j1}+D_{2j2})+(1-\eta)(D_{2j3}+D_{2j4}))y_j,\cdots,\frac{1}{2}\sum\limits_{j=1}^{n_1}(\eta(C_{m_1j1}+C_{m_1j2})+\\
(1-\eta)(C_{m_1j3}+C_{m_1j4}))x_j+\sum\limits_{j=1}^{n_2}((\eta(D_{m_1j1}+D_{m_1j2})+(1-\eta)(D_{m_1j3}+D_{m_1j4})\\
))y_j]\\
\text{s.t. }\sum\limits_{j=1}^{n_1}\eta(A_{rj1}+A_{rj2})+(1-\eta)(A_{rj3}+A_{rj4})+\sum\limits_{j=1}^{n_2}\eta(B_{rj1}+B_{rj2})+(1-\eta)\\
(B_{rj3}+B_{rj4})\le\eta(E_{r1}+E_{r2})+(1-\eta)(E_{r3}+E_{r4}),r=1,2,\cdots,p_1\\
\text{where }\mathbf{y}\text{ solves}\\
\quad\begin{cases}
\max[\frac{1}{2}\sum\limits_{j=1}^{n_2}(\eta(c_{1j1}+c_{1j2})+(1-\eta)(c_{1j3}+c_{1j4}))x_j+\sum\limits_{j=1}^{n_2}(\eta(d_{1j1}+d_{1j2})\\
+(1-\eta)(d_{1j3}+d_{1j4}))y_j,\frac{1}{2}\sum\limits_{j=1}^{n_1}(\eta(c_{2j1}+c_{2j2})+(1-\eta)(c_{2j3}+c_{2j4}))x_j+\\
\sum\limits_{j=1}^{n_2}(\eta(d_{2j1}+d_{2j2})+(1-\eta)(d_{2j3}+d_{2j4}))y_j,\cdots,\frac{1}{2}\sum\limits_{j=1}^{n_2}(\eta(c_{m_2j1}+c_{m_2j2})+\\
(1-\eta)(c_{m_2j3}+c_{m_2j4}))x_j+\sum\limits_{j=1}^{n_2}(\eta(d_{m_1j1}+d_{m_1j2})+(1-\eta)(d_{m_2j3}+d_{m_2j4}\\
))]\\
\text{s.t. }\begin{cases}
\sum\limits_{j=1}^{n_2}\eta(a_{rj1}+a_{rj2})+(1-\eta)(a_{rj3}+a_{rj4})+\sum\limits_{j=1}^{n_2}\eta(b_{rj1}+b_{rj2})\\
+(1-\eta)(b_{rj3}+b_{rj4})\le\eta(e_{r1}+e_{r2})+(1-\eta)(e_{r3}+e_{r4})\\
r=1,2,\cdots,p_2\\
\mathbf{x}\in\Re^{n_1},\mathbf{y}\in\Re^{n_2}
\end{cases}
\end{cases}
\end{cases}
$$

$$(3.12)$$

where $0\le\eta\le1$ is predetermined by the decision maker.

PROOF It follows from the linearity of the rough expected value operator that

$$
\begin{aligned}
&E[\tilde{\mathbf{C}}_i^T\mathbf{x}+\tilde{\mathbf{D}}_i^T\mathbf{y}]\\
&=E[\tilde{\mathbf{C}}_i^T\mathbf{x}]+E[\tilde{\mathbf{D}}_i^T\mathbf{y}]\\
&=E[\sum_{j=1}^{n_1}\tilde{C}_{ij}x_j]+E[\sum_{j=1}^{n_2}\tilde{D}_{ij}y_j]\\
&=\sum_{j=1}^{n_1}E[\tilde{C}_{ij}x_j]+\sum_{j=1}^{n_2}E[\tilde{D}_{ij}y_j]\\
&=\frac{1}{2}\sum_{j=1}^{n_1}(\eta(C_{1j1}+C_{1j2})+(1-\eta)(C_{1j3}+C_{1j4}))x_j\\
&\quad+\sum_{j=1}^{n_2}(\eta(D_{1j1}+D_{1j2})+(1-\eta)(D_{1j3}+D_{1j4}))y_j
\end{aligned}
$$

for $i = 1, 2, \cdots, m_1$.

$$
\begin{aligned}
E[\tilde{\mathbf{A}}_r^T \mathbf{x} + \tilde{\mathbf{B}}_r^T \mathbf{x}] \\
= E[\tilde{\mathbf{A}}_r^T \mathbf{x}] + E[\tilde{\mathbf{B}}_r^T \mathbf{x}] \\
= E[\sum_{j=1}^{n_1} \tilde{A}_{rj} x_j] + E[\sum_{j=1}^{n_2} \tilde{B}_{rj} y_j] \\
= \sum_{j=1}^{n_1} E[\tilde{A}_{rj} x_j] + \sum_{j=1}^{n_2} E[\tilde{B}_{rj} y_j] \\
= \sum_{j=1}^{n_1} \eta(A_{rj1} + A_{rj2}) + (1 - \eta)(A_{rj3} + A_{rj4}) \\
+ \sum_{j=1}^{n_2} \eta(B_{rj1} + B_{rj2}) + (1 - \eta)(B_{rj3} + B_{rj4})
\end{aligned}
$$

and

$$
E[\tilde{E}_r] = \eta(E_{r1} + E_{r2}) + (1 - \eta)(E_{r3} + E_{r4})
$$

for $r = 1, 2, \cdots, p_1$.

Similarly, for the lower-level decision making, we have

$$
\begin{aligned}
E[\tilde{\mathbf{c}}_i^T \mathbf{x} + \tilde{\mathbf{d}}_i^T \mathbf{y}] = E[\tilde{\mathbf{c}}_i^T \mathbf{y}] + E[\tilde{\mathbf{d}}_i^T \mathbf{x}] \\
= E[\sum_{j=1}^{n_1} \tilde{c}_{ij} x_j] + E[\sum_{j=1}^{n_2} \tilde{d}_{ij} y_j] \\
= \sum_{j=1}^{n_1} E[\tilde{c}_{ij} x_j] + \sum_{j=1}^{n_2} E[\tilde{d}_{ij} y_j] \\
= \tfrac{1}{2} \sum_{j=1}^{n_1} (\eta(c_{1j1} + c_{1j2}) + (1 - \eta)(c_{1j3} + c_{1j4})) x_j \\
+ \sum_{j=1}^{n_2} (\eta(d_{1j1} + d_{1j2}) + (1 - \eta)(d_{1j3} + d_{1j4})) y_j
\end{aligned}
$$

for $i = 1, 2, \cdots, m_1$

$$
\begin{aligned}
E[\tilde{\mathbf{a}}_r^T \mathbf{x} + \tilde{\mathbf{b}}_i^T \mathbf{y}] = E[\tilde{a}_i^T \mathbf{r}] + E[\tilde{\mathbf{b}}_r^T \mathbf{y}] - E[\sum_{j=1}^{n_1} \tilde{a}_{rj} x_j] + E[\sum_{j=1}^{n_2} \tilde{b}_{rj} y_j] \\
= \sum_{j=1}^{n_1} E[\tilde{a}_{rj} x_j] + \sum_{j=1}^{n_2} E[\tilde{b}_{rj} y_j] = \sum_{j=1}^{n_1} \eta(a_{rj1} + a_{rj2}) + (1 - \eta)(a_{rj3} + a_{rj4}) \\
+ \sum_{j=1}^{n_2} \eta(b_{rj1} + b_{rj2}) + (1 - \eta)(b_{rj3} + b_{rj4})
\end{aligned}
$$

and

$$
E[\tilde{e}_r] = \eta(e_{r1} + e_{r2}) + (1 - \eta)(e_{r3} + e_{r4})
$$

for $r = 1, 2, \cdots, p_2$. Thus the theorem is proved.

REMARK 3.1   If $\eta = 1/2$, the equivalent model can be rewritten as

$$
\left\{
\begin{aligned}
& \max\left[\tfrac{1}{4} \sum_{j=1}^{n_1} (C_{1j1}+C_{1j2}+C_{1j3}+C_{1j4})x_j + \sum_{j=1}^{n_2} D_{1j1}+D_{1j2}+D_{1j3}+D_{1j4})y_j, \right. \\
& \tfrac{1}{4} \sum_{j=1}^{n_1} (C_{2j1}+C_{2j2}+C_{2j3}+C_{2j4})x_j + \sum_{j=1}^{n_2} (D_{2j1}+D_{2j2}+D_{2j3}+D_{2j4})y_j, \cdots, \tfrac{1}{4} \\
& \sum_{j=1}^{n_1} (C_{m_1 j1}+C_{m_1 j2}+C_{m_1 j3}+C_{m_1 j4})x_j + \sum_{j=1}^{n_2} (D_{m_1 j1}+D_{m_1 j2}+D_{m_1 j3}+D_{m_1 j4})y_j] \\
& \text{s.t. } \sum_{j=1}^{n_1} (A_{rj1}+A_{rj2}+A_{rj3}+A_{rj4})x_j + \sum_{j=1}^{n_2} (B_{rj1}+B_{rj2}+B_{rj3}+B_{rj4})y_j \leq E_{r1}+ \\
& E_{r2}+E_{r3}+E_{r4}, r=1,2,\cdots,p_1 \\[4pt]
& \text{where } y \text{ solves} \\[4pt]
& \left\{
\begin{aligned}
& \max\left[\tfrac{1}{4} \sum_{j=1}^{n_1} (c_{1j1}+c_{1j2}+c_{1j3}+c_{1j4})x_j + \sum_{j=1}^{n_2} d_{1j1}+d_{1j2}+d_{1j3}+d_{1j4})y_j, \right. \\
& \tfrac{1}{4} \sum_{j=1}^{n_1} (c_{2j1}+c_{2j2}+c_{2j3}+c_{2j4})x_j + \sum_{j=1}^{n_2} (d_{2j1}+d_{2j2}+d_{2j3}+d_{2j4})y_j, \cdots, \tfrac{1}{4} \\
& \sum_{j=1}^{n_1} (c_{m_2 j1}+c_{m_2 j2}+c_{m_2 j3}+c_{m_2 j4})x_j + \sum_{j=1}^{n_2} (d_{m_2 j1}+d_{m_2 j2}+d_{m_2 j3}+d_{m_2 j4})y_j] \\
& \text{s.t. } \left\{
\begin{aligned}
& \sum_{j=1}^{n_1} (a_{rj1}+a_{rj2}+a_{rj3}+a_{rj4})x_j + \sum_{j=1}^{n_2} (b_{rj1}+b_{rj2}+b_{rj3}+b_{rj4})y_j \leq \\
& e_{r1}+e_{r2}+e_{r3}+e_{r4}, r=1,2,\cdots,p_2 x \in Re^{n_1}, y \in Re^{n_2}.
\end{aligned} \right.
\end{aligned} \right.
\end{aligned}
\right.
$$

$$\tag{3.13}$$

3.3.2.2   KKT Transformation Approach

Decision making with p linear objectives and m linear constraints defined by n decision variables is a common problem encountered in operation reasearch. This constrained optimization problem is transformed to an unconstrained one by employing the Lagrange multipliers of the constraints. Assuming the functions (objectives and constraints) are differentiable, the Karush–Kuhn–Tucker (KKT) optimality conditions are obtained as the first-order derivative of the Lagrangian. These are the necessary conditions that give a local optimal solution.

    Linear MODM was addressed as far back as the early 1950s by Kuhn and Tucker via their formulation of the so-called vector maximum model. The same problem was addressed by Karush independently. The problem was to recognize the optimal solution of the nonlinear programming problem (with differentiable functions).

    A multilevel programming problem is very similar to the multiobjective decision-making problem since conflicting objectives are present in both cases but furthermore, in an multilevel programming problem the decision makers are placed in hierarchical order and each decision maker controls only a subset of the decision variables. Therefore, some of the multiobjective decision-making methods are applicable to solvemultilevel programming problems. A multilevel programming problem consists of two or more hierarchical levels in descending order of importance while in a multiobjective decision-making problem all objectives may or may not be equally important, but are not placed in any level. multilevel programming problem are exemplified by typical pyramidal structure where more than one division at any level forms a decentralized system.

J. Bard [18] has extended the open-loop Stackelberg game to players by multilevel programming problem. and thus used it as a model for a variety of hierarchical systems in which sequential planning is the norm. G. Anandalingam [8] has considered hierarchical organizations, identifying the manufacturing enterprise with three-level programming problem and used the KKT transformation approach. M. Rijckaert and E. Walraven [255] present a method for estimating Lagrange multipliers for generalized geometric programming. O. Mangasarian [207] has shown that the satisfaction of a standard constraint qualification of mathematical programming at a stationary point of a nonconvex differentiable nonlinear program provides explicit numerical bounds for the set of all Lagrange multipliers associated with the stationary point. Sufficient Fritz John optimality criteria were derived by T. Weir and B. Mond [297] for nonlinear programs in which the objective is pseudo-convex, and the constraints are quasi-convex or semi-strictly pseudo-convex. A. Ruszczynski [257] proposed a new method for large LPPs of angular structure. It combines the ideas of augmented Lagrangians, simplical decomposition, column generation, and dual ascent. The paper by P. Fetterolf and G. Anandalingam [107] addresses the problem of interconnecting a group of LANs with bridges. First-order approximations as abstract generalized derivatives are studied by S. Komlosi [168]. In the case of quasi-convex first-order approximations, the quasi-sub differential is introduced, and Kuhn–Tucker type necessary optimality conditions are formulated by its help. Several interactive procedures for solving multiple criteria NLPPs have been developed by A. Sadagopan [258]. The paper by Prasad [246]shows that it is enough to consider the extreme points for minimizing over a bounded polyhedron, a function that is bounded and quasi-concave on the polyhedron. A. Msilti and P. Tolla [220] have given a method for an interactive multiobjective nonlinear programming procedure. P. Toint [287] presents two new trust-region methods for solving nonlinear optimization problems over convex feasible domains. These methods are distinguished by the fact that they do not enforce strict monotonicity of the objective function values at successive iterates.

Let f and $g_i, i = 1, 2, \cdots, n$, be real-valued functions defined on $\Re^n$. We consider the following optimization problem:

$$(P) \quad \begin{cases} \min f(\mathbf{x}) = f(x_1, x_2, \cdots, x_n) \\ \text{s.t. } g_i(\mathbf{x}) \leq 0, i = 1, 2, \cdots, m \end{cases}$$

Suppose that the constraint functions g_i are convex on $\Re^n$ for each $i = 1, 2, \cdots, m$. Then the feasible set $X = \{x \in \Re^n | g_i(\mathbf{x}) \leq 0, i = 1, 2, \cdots, m\}$ is a convex subset of $\Re^n$. The well-known KKT condition for problem (P) (e.g., see [136]) is stated as follows.

THEOREM 3.2

(H. Wu [304]) Assume that the constraint functions $g_i : \Re^n \rightarrow \Re$ are convex on $\Re^n$ for $i = 1, 2, \cdots, m$. Let $X = \{\mathbf{x} \in \Re^n | g_i(\mathbf{x}) \leq 0, i = 1, 2, \cdots, m\}$ be a feasible set and a point $\mathbf{x}^* \in X$. Suppose that the objective function $f : \Re^n \rightarrow \Re$ is convex at $\mathbf{x}^*$, and $f, g_i, i = 1, 2, \cdots, m$, are continuously differentiable at $\mathbf{x}^*$. If there exist (Lagrange) multipliers $0 \leq \mu_i \in \Re^n$, such that

1. $\nabla f(\mathbf{x}^*) + \sum_{i=1}^m \mu_i \nabla g_i(\mathbf{x}^*) = \mathbf{0}$;

2. $\mu_i g_i(\mathbf{x}^*) = 0$ *for all $i = 1, 2, \cdots, m$, then $\mathbf{x}^*$ is an optimal solution to problem (P).*

3.3.2.3 Numerical Example

In this section, a numerical example is given to illustrate the effectiveness of the application of the proposed models and algorithms above.

Example 3.1

Consider the problem

$$
\begin{cases}
\max\limits_{x_1, x_2, x_3} F_1 = 2\xi_1 x_1 + \xi_2 x_2 + \xi_3 x_3 + 2\xi_4 y_1 + \xi_5 y_2 + \xi_6 y_3 \\
\max\limits_{x_1, x_2, x_3} F_2 = x_1 + x_2 + x_3 + y_1 + y_2 + y_3 \\
\text{where } (y_1, y_2, y_3) \text{ solves} \\
\quad \begin{cases}
f_1 = \max\limits_{y_1, y_2, y_3} 7\xi_1 x_1 + 3\xi_2 x_2 + 8\xi_3 x_3 + 12\xi_4 y_1 + 11\xi_5 y_2 + 5\xi_6 y_3 \\
f_2 = \max\limits_{y_1, y_2, y_3} x_1 + 5x_2 + 7x_3 + y_1 + 11y_2 - 8y_3 \\
\quad \begin{cases}
9x_1 \xi_1 + 3x_2 \xi_2 + 5x_3 \xi_3 + 4y_4 \xi_4 + 17y_3 \xi_6 \leq 1500 \\
3x_1 + 6x_+ 12x_9 + 14x_{10} + 19x_{11} \leq 2500 \\
\text{s.t.} \quad x_1 + x_2 + x_3 + y_1 + y_2 + y_3 \geq 120 \\
x_1 + x_2 + x_3 + y_1 + y_2 + y_3 \leq 200 \\
x_1, x_2, x_3, y_1, y_2, y_3 \geq 0
\end{cases}
\end{cases}
\end{cases}
\tag{3.14}
$$

where $\xi_j, j = 1, 2, \cdots, 6$ are rough intervals characterized as

$$
\xi_1 = ([8, 10], [7, 11]), \quad \xi_2 = ([9, 11], [8, 12]), \quad \xi_3 = ([10, 12], [9, 13]),
$$
$$
\xi_4 = ([11, 13], [10, 14]), \quad \xi_5 = ([12, 14], [11, 15]), \quad \xi_6 = ([13, 15], [12, 16]).
$$

In order to solve it, we use the expected operator to deal with rough objectives and rough constraints, and then we can obtain the model

$$
\begin{cases}
\max\limits_{x_1,x_2,x_3} E[2\xi_1 x_1 + \xi_2 x_2 + \xi_3 x_3 + 2\xi_4 y_1 + \xi_5 y_2 + \xi_6 y_3] \\
\max\limits_{x_1,x_2,x_3} \{x_1 + x_2 + x_3 + y_1 + y_2 + y_3\} \\
\text{where } (y_1,y_2,y_3) \text{ solves} \\
\quad
\begin{cases}
\max\limits_{y_1,y_2,y_3} E[7\xi_1 x_1 + 3\xi_2 x_2 + 8\xi_3 x_3 + 12\xi_4 y_1 + 11\xi_5 y_2 + 5\xi_6 y_3] \\
\max\limits_{y_1,y_2,y_3} \{x_1 + 5x_2 + 7x_3 + y_1 + 11y_2 - 8y_3\} \\
\quad \text{s.t.}
\begin{cases}
E[9x_1\xi_1 + 3x_2\xi_2 + 5x_3\xi_3 + 4y_4\xi_4 + 17y_3\xi_6] \le 1540 \\
3x_1 + 6x_+ 12x_2 + 14x_3 + 9y_1 + 20y_1 + 15y_2 + 10y_3 \le 2400 \\
x_1 + x_2 + x_3 + y_1 + y_2 - y_3 \ge 150 \\
x_1 + x_2 + x_3 + y_1 + y_2 + y_3 \le 200 \\
x_1,x_2,x_3,y_1,y_2,y_3 \ge 0
\end{cases}
\end{cases}
\end{cases}
\tag{3.15}
$$

Here we suppose that $\eta = 1/2$, and by Theorem 3.1, we know that the problem (3.15) is equivalent to model

$$
\begin{cases}
\max\limits_{x_1,x_2,x_3} \{18x_1 + 10x_2 + 11x_3 + 24y_1 + 13y_2 + 14y_3\} \\
\max\limits_{x_1,x_2,x_3} \{x_1 + x_2 + x_3 + y_1 + y_2 + y_3\} \\
\text{where } (y_1,y_2,y_3) \text{ solves} \\
\quad
\begin{cases}
\max\limits_{y_1,y_2,y_3} \{63x_1 + 30x_2 + 88x_3 + 144y_1 + 143y_2 + 70y_3\} \\
\max\limits_{y_1,y_2,y_3} \{x_1 + 5x_2 + 7x_3 + y_1 + 11y_2 - 8y_3\} \\
\quad \text{s.t.}
\begin{cases}
9x_1 + 3x_2 + 5x_3 + 4y_4 + 17y_3 \le 1500 \\
3x_1 + 6x_+ 12x_9 + 14x_{10} + 19x_{11} \le 2500 \\
x_1 + x_2 + x_3 + y_1 + y_2 - y_3 \ge 120 \\
x_1 + x_2 + x_3 + y_1 + y_2 + y_3 \le 200 \\
x_1,x_2,x_3,y_1,y_2,y_3 \ge 0
\end{cases}
\end{cases}
\end{cases}
\tag{3.16}
$$

By using KKT method, we obtain $(x_1^*,x_2^*,x_3^*,y_1^*,y_2^*,y_3^*) = (12.3.16.7.12, 4, 20.5, 20.7, 15, 3)$ and $(F_1^*,F_2^*) = (126.7, 138.5), (f_1^*,f_2^*) = (116.9, 128.9)$.

3.3.3 BL-NLEVRM and Rough Simulation-Based ECTS

It is difficult to transform the bilevel nonlinear expected value rough model (BL-NLEVRM) into its equivalent form, so we introduce the rough simulation-based enhanced continuous tabu search (ECTS) algorithm to solve it.

3.3.3.1 ECTS

Local search employs the idea that a given solution x may be improved by making small changes. Those solutions obtained by modifying solution x are called neighbors of x. The local search algorithm starts with some initial solution and moves from neighbor to neighbor as long as possible while decreasing the objective function value. The main problem with this strategy is to escape from local minima where the search cannot find any further neighborhood solution that decreases the objective function value. Different strategies have been proposed to solve this problem. One of the most efficient strategies is tabu search. Tabu search allows the search to explore solutions that do not decrease the objective function value only in those cases where these solutions are not forbidden. This is usually obtained by keeping track of the last solutions in terms of the action used to transform one solution to the next. When an action is performed, it is considered tabu for the next T iterations, where T is the tabu status length. A solution is forbidden if it is obtained by applying a tabu action to the current solution. The Tabu Search metaheuristic has been defined by F. Glover [118]. The basic ideas of tabu search (TS) have also been sketched by P. Hansen [126]. After that, TS has achieved widespread success in solving practical optimization problems in different domains (such as resource management, process design, logistic, and telecommunications).

A tabu list is a set of solutions determined by historical information from the last t iterations of the algorithm, where t is fixed or is a variable that depends on the state of the search, or a particular problem. At each iteration, given the current solution x and its corresponding neighborhood $N(x)$, the procedure moves to the solution in the neighborhood $N(x)$ that most improves the objective function. However, moves that lead to solutions on the tabu list are forbidden, or are tabu. If there are no improving moves, TS chooses the move that least changes the objective function value. The tabu list avoids returning to the local optimum from which the procedure has recently escaped. A basic element of tabu search is the aspiration criterion, which determines when a move is admissible despite being on the tabu list. One termination criterion for the tabu procedure is a limit in the number of consecutive moves for which no improvement occurs. Given an objective function $f(\mathbf{x})$ over a feasible domain D, a generic tabu search for finding an approximation of the global minimum of $f(\mathbf{x})$ is given as follows:

Step 1. Initialization: (a) Generate an initial solution $\mathbf{x}$ and set $\mathbf{x}^* = \mathbf{x}$. (b) Initialize the tabu list $T = \Phi$. (c) Set iteration counters $k = 0$ and $l = 0$.

Step 2. While $(N(\mathbf{x})\ T \neq \Phi)$, do (1) $k = k + 1$, $l = l + 1$. (2) Select $\mathbf{x}$ as the best solution from the set $N(\mathbf{x})\ T$. (3) If $f(\mathbf{x}) < f(\mathbf{x}^*)$, then update $\mathbf{x}^* = \mathbf{x}$ and set $l = 0$. (4) If $k = \bar{k}$ or if $l = \bar{l}$, go to Step 3.

Step 3. Output the best found solution $\mathbf{x}^*$.

In the following part, we will introduce detail steps on how to apply the special TS algorithm–ECTS proposed by R. Chelouah and P. Siarry [56] based on the rough simulation to solve a bilevel multiple objective expected value model with rough interval parameters.

Setting of parameters. Two parameters must be set before any execution of ECTS, that is, initialization and control parameters.

For each of these categories, some parameter values must be chosen by the user and some parameter values must be calculated. These four subsets of parameters are listed in Table 3.1.

Initialization. In this stage, we will list the representation of the solution. We have resumed and adapted the method described in detail in [274]. Randomly generate a solution $\mathbf{x}$ and check its feasibility by the random rough simulation such that

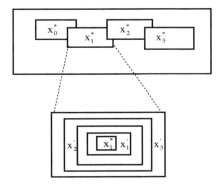

FIGURE 3.2 Partition of current solution neighborhood.

$E[g_r(\mathbf{x},\mathbf{x})] \le 0 (r = 1,2,\cdots,p)$. Then generate its neighborhood by the concept of "ball" defined in [274]. A ball $B(\mathbf{x},r)$ is centered on $\mathbf{x}$ with radius r, which contains all points $\mathbf{x}'$ such that $||\mathbf{x}' - \mathbf{x}|| \le 4$ (the symbol $||\cdot||$ denotes the Euclidean norm). To obtain a homogeneous exploration of the space, we consider a set of balls centered on the current solution $\mathbf{x}$, with $h_0, h_1, \cdots, h_\eta$. Hence the space is partitioned into concentric "crowns" $C_i(\mathbf{x}, h_{i-1}, h_i)$, such that

$$C_i(\mathbf{x}, h_{i-1}, h_i) = \{\mathbf{x}' | h_{i-1} \le ||\mathbf{x}' - \mathbf{x}|| \le h_i\}.$$

The η neighbors of s are obtained by random selection of one point inside each crown C_i, for i varying from 1 to η. Finally, we select the best neighbor of $\mathbf{x}$ among these η neighbors, even if it is worse than $\mathbf{x}$. In ECTS, we replace the balls by hyperrectangles for the partition of the current solution neighborhood (see Figure 3.2), and we generate neighbors in the same way. The reason for using a hyperrectangular neighborhood instead of crown "balls" is the following: it is mathematically much easier to select a point inside a specified hyperrectangular zone than to select a point inside a specified crown ball. Therefore, in the first case, we only have to compare the coordinates of the randomly selected point with the bounds that define the hyperrectangular zone at hand. Next, we will describe the initialization of some parameters and the tuning of the control parameters. In other words, we give the "definition" of all the parameters of ECTS. The parameters in part A of Table 3.1 are automatically built by using the parameters fixed at the beginning. The parameters in part B of Table 3.1 are valued in the following way:

1. The search domain of analytical test functions is set as prescribed in the literature, the initial solution $\mathbf{x}^*$ is randomly chosen and checked if it is feasible by the rough simulation.

2. The tabu list is initially empty.

3. To complete the promising list, the algorithm randomly draws a point. This point is accepted as the center of an initial promising ball if it does not belong to an already generated ball. In this way, the algorithm generates N_p sample points, which are uniformly dispersed in the whole space solution S.

TABLE 3.1   Listing of the ECTS parameters

A. *Initialization parameters chosen by the user*
Search domain of each function variable
Starting point
Content of the tabu list
Content of the promising list

B. *Initialization parameters calculated*
Length δ of the smallest edge of the initial hyperrectangular search domain
Initial threshold for the acceptance of a promising area
Initial best point
Number η of neighbors of the current solution investigated at each iteration
Maximum number of successive iterations without any detection of a promising
   area
Maximum number of successive iterations without any improvement of the
   objective function value
Maximum number of successive reductions of the hyperrectangular neighborhood
   and of the radius of tabu balls without any improvement
Maximum number of iterations

C. *Control parameters chosen by the user*
Length N_t of the tabu list
Length N_p of the promising list
Parameter ρ_t allowing calculation of the initial radius of tabu balls
Parameter ρ_{neigh} allowing to calculate the initial size of the
hyperrectangular neighborhood

D. *Control parameters calculated*
Initial radius ε_t of tabu balls
Initial radius ε_p of promising balls
Initial size of the hyperrectangular neighborhood

---

4. The initial threshold for the acceptance of a promising area is taken equal to the average of the objective function values over the previous N_p sample points.

5. The best point found is taken equal to the best point among the previous N_p.

6. The number η of neighbors of the current solution investigated at each iteration is set to twice the number of variables, if this number is equal or smaller than five, otherwise η is set to 10.

7. The maximum number of successive iterations without any detection of a new promising area is equal to twice the number of variables.

8. The maximum number of successive iterations without any improvement of the objective function value is equal to five times the number of variables.

9. The maximum number of successive reductions of the hyperrectangular neighborhood and of the radius of tabu balls without any improvement of the objective function value is set to twice the number of variables.

10. The maximum number of iterations is equal to 50 times the number of variables.

Two types of control parameters exist. Some parameters are chosen by the user; others are deduced from the chosen parameters. The fixed parameters are the length of the tabu list (set to 7, which is the usual tuning advocated by Glover), the length of the promising list (set to 10, as in [69]), and the parameters ρ_t, ρ_p and ρ_{neigh} (set to 100, 50, and 5, respectively). The expressions of ε_t and ε_p are δ/ρ_t and δ/ρ_p, respectively, and the initial size of the hyperrectangular neighborhood of the current solution (the more external hyperrectangle) is obtained by dividing δ by the factor ρ_{neigh}.

Diversification. At this stage, the process starts with the initial solution that used as the current one. ECTS generates a specified number of neighbors: one point is selected inside each hyperrectangular zone around the current solution. Each neighbor is accepted only if it does not belong to the tabu list. The best of these neighbors becomes the new current solution even if it is worse than the previous one. A new promising solution is detected and generated according to the procedure described above. This promising solution defines a new promising area if it does not already belong to a promising ball. If a new promising area is accepted, the worst area of the promising list is replaced by the newly accepted promising area. The use of the promising and tabu lists stimulates the search for solutions far from the starting one and the identified promising areas. The diversification process stops after a given number of successive iterations without any detection of a new promising area. Then the algorithm determines the most promising area among those present in the promising list.

Search for the most promising area. In order to determine the most promising area, we proceed in three steps. First, we calculate the average value of the objective function over all the solutions present in the promising list. Second, we eliminate all the solutions for which the function value is higher than this average value. Third, we deal with the thus-reduced list in the following way. We halve the radius of the tabu balls and the size of the hyperrectangular neighborhood. For each remaining promising solution, we perform the generation of the neighbors and selection of the best. We replace the promising solution by the best neighbor located, yet only if this neighbor is better than that solution. After having scanned the whole promising list, the algorithm removes the least promising solution. This process is reiterated after halving again the above two parameters. It stops when just one promising area remains.

Intensification. The first step of the intensification stage is the resetting of the tabu list. The remaining promising area allows the definition of a new search domain. The center of this area is taken as the current point, and the tabu search starts again: generation of neighbors not belonging to the tabu list, selection of the best, and insertion of the best solution into the tabu list. This selected neighbor becomes the new current solution even if it is worse than the previous one. After a predetermined number of successive iterations without any improvement of the objective function value (e.g., quadratic error between two successive solutions less than 10^{-3}), the size of the hyperrectangular neighborhood and the radius of the tabu balls are halved, the tabu list is reset, and we restart the procedure from the best point found until now. To stop the algorithm, we use two criteria: a specified number of successive reductions of the two parameters mention earlier without any significant improvement of the objective function value and a specified maximum number of iterations.

3.3.3.2 Numerical Example

In this section, a numerical example is given to illustrate the effectiveness of the application of the proposed models and algorithms.

Example 3.2
Consider the problem

$$
\begin{cases}
\max\limits_{x_1,x_2,x_3} F_1 = 2\xi_1^2 x_1 + \sqrt{\xi_2 x_2 + \xi_3 x_3} + \xi_4 y_1 + \sqrt{\xi_5} y_2 + \sqrt{\xi_6} y_3 \\
\max\limits_{x_1,x_2,x_3} F_2 = x_1 + x_2 + x_3 + y_1 + y_2 + y_3 \\
\text{where } (y_1, y_2, y_3) \text{ solves} \\
\quad \begin{cases}
\max\limits_{y_1,y_2,y_3} f_1 = \sqrt{7\xi_1 x_1 + 3\xi_2 x_2 + 8\xi_3 x_3} + (12\xi_4 y_1 + 11\xi_5 y_2 + 5\xi_6 y_3)^2 \\
\max\limits_{y_1,y_2,y_3} f_2 = x_1 + 5x_2 + 7x_3 + y_1 + 11y_2 - 8y_3 \\
\quad \text{s.t.} \begin{cases}
\sqrt{x_1 \xi_1 + x_2 \xi_2} + x_3^2 \xi_3 + \xi_4 y_4 + 17 y_3 \xi_6 \le 1540 \\
3x_1 + 6x_+ 12x_9 + 14x_{10} + 19x_{11} \le 2400 \\
x_1 + x_2 + x_3 + y_1 + y_2 - y_3 \ge 150 \\
x_1 + x_2 + x_3 + y_1 + y_2 + y_3 \le 200 \\
x_1, x_2, x_3, y_1, y_2, y_3 \ge 0
\end{cases}
\end{cases}
\end{cases}
$$

(3.17)

where $\xi_j (j = 1, 2, \cdots, 6)$ are rough intervals characterized as

$$\xi_1 = ([8,10],[7,11]), \quad \xi_2 = ([9,11],[8,12]), \quad \xi_3 = ([10,12],[9,13]),$$
$$\xi_4 = ([11,13],[10,14]), \quad \xi_5 = ([12,14],[11,15]), \quad \xi_6 = ([13,15],[12,16]).$$

In order to solve it, we use the expected operator to deal with rough objectives and rough constraints, then we can obtain the model

$$
\begin{cases}
\max\limits_{x_1,x_2,x_3} E[2\xi_1^2 x_1 + \sqrt{\xi_2 x_2 + \xi_3 x_3} + \xi_4 y_1 + \sqrt{\xi_5} y_2 + \sqrt{\xi_6} y_3] \\
\max\limits_{x_1,x_2,x_3} \{x_1 + x_2 + x_3 + y_1 + y_2 + y_3\} \\
\text{where } (y_1,y_2,y_3) \text{ solves} \\
\quad \begin{cases}
\max\limits_{y_1,y_2,y_3} E[\sqrt{7\xi_1 x_1 + 3\xi_2 x_2 + 8\xi_3 x_3} + (12\xi_4 y_1 + 11\xi_5 y_2 + 5\xi_6 y_3)^2] \\
\max\limits_{y_1,y_2,y_3} \{x_1 + 5x_2 + 7x_3 + y_1 + 11y_2 - 8y_3\} \\
\text{s.t.} \begin{cases}
E[\sqrt{x_1\xi_1 + x_2\xi_2} + x_3^2\xi_3 + \xi_4 y_4 + 17y_3\xi_6] \le 1540 \\
3x_1 + 6x_+ 12x_2 + 14x_3 + 9y_1 + 20y_1 + 15y_2 + 10y_3 \le 2400 \\
x_1 + x_2 + x_3 + y_1 + y_2 - y_3 \ge 150 \\
x_1 + x_2 + x_3 + y_1 + y_2 + y_3 \le 200 \\
x_1,x_2,x_3,y_1,y_2,y_3 \ge 0
\end{cases}
\end{cases}
\end{cases}
\tag{3.18}
$$

Here we suppose that $\eta = 1/2$. After running the rough simulate-based TS, we obtain $(x_1^*,x_2^*,x_3^*,y_1^*,y_2^*,y_3^*) = (12.3, 16.7, 12.4, 20.5, 20.7, 15, 3)$ and $(F_1^*,F_2^*) = (106.9, 118.9)$, $(f_1^*,f_2^*) = (86.5, 78.3)$.

3.4   BL-CCRM

In order to maximize the optimistic return with a given confidence level subject to some chance constraint, we may employ bilevel chance constrained rough model (BL-CCRM).

3.4.1   General model for BL-CCRM

For the following bilevel multiple objective rough model

$$
\begin{cases}
\max\limits_{\mathbf{x}} [F_1(\mathbf{x},\mathbf{y},\xi), F_2(\mathbf{x},\mathbf{y},\xi), \cdots, F_{m_1}(\mathbf{x},\mathbf{y},\xi)] \\
\text{s.t. } G_r(\mathbf{x},\mathbf{y},\xi) \le 0, r = 1,2,\cdots,p_1 \\
\text{where } \mathbf{y} \text{ solves} \\
\quad \begin{cases}
\max\limits_{\mathbf{y}} [f_1(\mathbf{x},\mathbf{y},\xi), f_2(\mathbf{x},\mathbf{y},\xi), \cdots, f_{m_2}(\mathbf{x},\mathbf{y},\xi)] \\
\text{s.t.} \begin{cases}
g_r(\mathbf{x},\mathbf{y},\xi) \le 0, r = 1,2,\cdots,p_2 \\
\mathbf{x} \in \mathfrak{R}^{n_1}, \mathbf{y} \in \mathfrak{R}^{n_2}
\end{cases}
\end{cases}
\end{cases}
\tag{3.19}
$$

where ξ is a rough interval vector.

The general BL-CCRM is formulated as

$$
\begin{cases}
\max_{\mathbf{x}}[\bar{F}_1,\bar{F}_2,\cdots,\bar{F}_{m_1}] \\
\text{s.t.} \begin{cases} Ch\{F_i(\mathbf{x},\mathbf{y},\boldsymbol{\xi}) \geq \bar{F}_i\} \geq \alpha_i, i = 1,2,\cdots,m_1 \\ Ch\{G_r(\mathbf{x},\mathbf{y},\boldsymbol{\xi}) \leq 0\} \geq \beta_r, r = 1,2,\cdots,p_1 \end{cases} \\
\text{where } \mathbf{y} \text{ solves} \\
\quad \begin{cases} \max_{\mathbf{y}}[\bar{f}_1,\bar{f}_2,\cdots,\bar{f}_{m_2}] \\ \text{s.t.} \begin{cases} Ch\{f_i(\mathbf{x},\mathbf{y},\boldsymbol{\xi}) \geq \bar{f}_i\} \geq \gamma_i, i = 1,2,\cdots,m_2 \\ Ch\{g_r(\mathbf{x},\mathbf{y},\boldsymbol{\xi}) \leq 0\} \geq \delta_r, r = 1,2,\cdots,p_2 \\ \mathbf{x} \in \Re^{n_1}, \mathbf{y} \in \Re^{n_2} \end{cases} \end{cases}
\end{cases} \tag{3.20}
$$

where $\alpha_i, \beta_r, \gamma_i, \delta_r$ are predetermined confidence levels. We adopt $\underline{Appr}$, $\overline{Appr}$, or $Appr$ measures, respectively, to measure the rough event, and then we have three kinds of BL-CCRM.

3.4.1.1   BL-CCRM Based on $\underline{Appr}$

The spectrum of the chance-constrained model based on $\underline{Appr}$ is formulated as

$$
\begin{cases}
\max_{\mathbf{x}}[\bar{F}_1,\bar{F}_2,\cdots,\bar{F}_{m_1}] \\
\text{s.t.} \begin{cases} \underline{Appr}\{F_i(\mathbf{x},\mathbf{y},\boldsymbol{\xi}) \geq \bar{F}_i\} \geq \alpha_i, i = 1,2,\cdots,m_1, \\ \underline{Appr}\{G_r(\mathbf{x},\mathbf{y},\boldsymbol{\xi}) \leq 0\} \geq \beta_r, r = 1,2,\cdots,p_1 \end{cases} \\
\text{where } y \text{ solves} \\
\quad \begin{cases} \max_{\mathbf{y}}\{\bar{f}_1,\bar{f}_2,\cdots,\bar{f}_{m_2}\} \\ \text{s.t.} \begin{cases} \underline{Appr}\{f_i(\mathbf{x},\mathbf{y},\boldsymbol{\xi}) \geq \bar{f}_i\} \geq \gamma_i, i = 1,2,\cdots,m_1 \\ \underline{Appr}\{g_r(\mathbf{x},\mathbf{y},\boldsymbol{\xi}) \leq 0\} \geq \delta_r, r = 1,2,\cdots,p_2 \\ \mathbf{x} \in \Re^{n_1}, \mathbf{y} \in \Re^{n_2} \end{cases} \end{cases}
\end{cases} \tag{3.21}
$$

DEFINITION 3.5   *Let* $\mathbf{x}$ *be given by the upper-level decision maker (ULDM) satisfying* $\underline{Appr}\{G_r(\mathbf{x},\mathbf{y},\boldsymbol{\xi}) \leq 0\} \geq \beta_r, r = 1,2,\cdots,p_1$. *Then* $(\mathbf{x},\mathbf{y})$ *is said to be a feasible solution to model (3.21) at* (β_r,δ_r)-$\underline{Appr}$ *levels if and only if* $\mathbf{y}(\underline{Appr}\{g_r(\mathbf{x},\mathbf{y},\boldsymbol{\xi}) \leq 0\} \geq \delta_r, r = 1,2,\cdots,p_2$ *at the lower level is the efficient solution to the lower-level decision maker (LLDM).*

DEFINITION 3.6   *Let* $(\mathbf{x}^*,\mathbf{y}^*)$ *be a feasible solution at* (β_r,δ_r)-$\underline{Appr}$ *levels. Then* $(\mathbf{x}^*,\mathbf{y}^*)$ *is said to be an* (α_i,γ_i) *efficient solution to problem (3.21) if and only if there exists no other feasible at* (β_r,δ_r)-$\underline{Appr}$ *levels,* $(\mathbf{x},\mathbf{y})$ *such that* $\underline{Appr}\{F_i(\mathbf{x},\mathbf{y},\boldsymbol{\xi}) \geq \bar{F}_i\} \geq \alpha_i$ *with* $\bar{F}_i \geq \bar{F}_i$ *for all* i *and* $F_{i_0} > \bar{F}_{i_0}$ *for at least one* $i_0 \in \{1,2,\cdots,m_1\}$.

3.4.1.2 BL-CCRM Based on $\overline{Appr}$

The spectrum of the chance-constrained model based on $\overline{Appr}$ is formulated as

$$
\begin{cases}
\max\limits_{\mathbf{x}}[\bar{F}_1,\bar{F}_2,\cdots,\bar{F}_{m_1}] \\
\text{s.t.}\begin{cases} \overline{Appr}\{F_i(\mathbf{x},\mathbf{y},\boldsymbol{\xi}) \geq \bar{F}_i\} \geq \alpha_i, i = 1,2,\cdots,m_1 \\ \overline{Appr}\{G_r(\mathbf{x},\mathbf{y},\boldsymbol{\xi}) \leq 0\} \geq \beta_r, r = 1,2,\cdots,p_1 \end{cases} \\
\text{where } \mathbf{y} \text{ solves} \\
\quad \begin{cases} \max\limits_{\mathbf{y}}\{\bar{f}_1,\bar{f}_2,\cdots,\bar{f}_{m_2}\} \\ \text{s.t.}\begin{cases} \overline{Appr}\{f_i(\mathbf{x},\mathbf{y},\boldsymbol{\xi}) \geq \bar{f}_i\} \geq \gamma_i, i = 1,2,\cdots,m_2 \\ \overline{Appr}\{g_r(\mathbf{x},\mathbf{y},\boldsymbol{\xi}) \leq 0\} \geq \delta_r, r = 1,2,\cdots,p_2 \\ \mathbf{x} \in \Re^{n_1}, \mathbf{y} \in \Re^{n_2} \end{cases} \end{cases}
\end{cases}
\tag{3.22}
$$

DEFINITION 3.7 *Let* $\mathbf{x}$ *be given by the ULDM satisfying* $\overline{Appr}\{G_r(\mathbf{x},\mathbf{y},\boldsymbol{\xi}) \leq 0\} \geq \beta_r, r = 1,2,\cdots,p_1$. *Then* $(\mathbf{x},\mathbf{y})$ *is said to be a feasible solution to model (3.22) at* (β_r,δ_r)-$\overline{Appr}$ *levels if and only if* $\mathbf{y}(\overline{Appr}\{g_r(\mathbf{x},\mathbf{y},\boldsymbol{\xi}) \leq 0\} \geq \delta_r, r = 1,2,\cdots,p_2$ *at the upper level is the efficient solution to the LLDM.*

DEFINITION 3.8 *Let* $(\mathbf{x}^*,\mathbf{y}^*)$ *be a feasible solution at* (β_r,δ_r)-$\overline{Appr}$ *levels. Then* $(\mathbf{x}^*,\mathbf{y}^*)$ *is said to be an* (α_i,γ_i) *efficient solution to problem (3.22) if and only if there exists no other feasible at* (β_r,δ_r)-$\overline{Appr}$ *levels,* $(\mathbf{x},\mathbf{y})$ *such that* $\overline{Appr}\{F_i(\mathbf{x},\mathbf{y},\boldsymbol{\xi}) \geq \bar{F}_i\} \geq \alpha_i$ *with* $\bar{F}_i \geq \bar{F}_i$ *for all* i *and* $F_{i_0} > \bar{F}_{i_0}$ *for at least one* $i_0 \in \{1,2,\cdots,m_1\}$.

3.4.1.3 BL-CCRM Based on $Appr$

The spectrum of the chance-constrained model based on $Appr$ is formulated as

$$
\begin{cases}
\max\limits_{\mathbf{x}}[\bar{F}_1,\bar{F}_2,\cdots,\bar{F}_{m_1}] \\
\text{s.t.}\begin{cases} Appr\{F_i(\mathbf{x},\mathbf{y},\boldsymbol{\xi}) \geq \bar{F}_i\} \geq \alpha_i, i = 1,2,\cdots,m_1 \\ Appr\{G_r(\mathbf{x},\mathbf{y},\boldsymbol{\xi}) \leq 0\} \geq \beta_r, r = 1,2,\cdots,p_1 \end{cases} \\
\text{where } y \text{ solves} \\
\quad \begin{cases} \max\limits_{\mathbf{y}}[\bar{f}_1,\bar{f}_2,\cdots,\bar{f}_{m_2}] \\ \text{s.t.}\begin{cases} Appr\{f_i(\mathbf{x},\mathbf{y},\boldsymbol{\xi}) \geq \bar{f}_i\} \geq \gamma_i, i = 1,2,\cdots,m_2 \\ Appr\{g_r(\mathbf{x},\mathbf{y},\boldsymbol{\xi}) \leq 0\} \geq \delta_r, r = 1,2,\cdots,p_2 \\ \mathbf{x} \in \Re^{n_1}, \mathbf{y} \in \Re^{n_2} \end{cases} \end{cases}
\end{cases}
\tag{3.23}
$$

DEFINITION 3.9 *Let* $\mathbf{x}$ *be given by the ULDM satisfying* $Appr\{G_r(\mathbf{x},\mathbf{y},\boldsymbol{\xi}) \leq 0\} \geq \beta_r, r = 1,2,\cdots,p_1$. *Then* $(\mathbf{x},\mathbf{y})$ *is said to be a feasible solution to model (3.23) at* (β_r,δ_r)-$Appr$ *levels if and only if* $\mathbf{y}(Appr\{g_r(\mathbf{x},\mathbf{y},\boldsymbol{\xi}) \leq 0\} \geq \delta_r, r = 1,2,\cdots,p_2$ *at the lower level is the efficient solution to the LLDM.*

DEFINITION 3.10 *Let* $(\mathbf{x}^*,\mathbf{y}^*)$ *be a feasible solution at* (β_r,δ_r)-$Appr$

levels. Then $(\mathbf{x}^*, \mathbf{y}^*)$ *is said to be an* (α_i, γ_i) *efficient solution to problem (3.23) if and only if there exists no other feasible at* (β_r, δ_r)-*Appr levels,* $(\mathbf{x}, \mathbf{y})$ *such that* $Appr\{F_i(\mathbf{x}, \mathbf{y}, \boldsymbol{\xi}) \geq \bar{F}_i\} \geq \alpha_i$ *with* $\bar{F}_i \geq \bar{F}_i$ *for all i and* $F_{i_0} > \bar{F}_{i_0}$ *for at least one* $i_0 \in \{1, 2, \cdots, m_1\}$.

3.4.2 BL-LCCRM and Fuzzy Decision Approach

Let us still consider a class of linear models as follows:

$$
\begin{cases}
\max \left[\tilde{\mathbf{C}}_1^T \mathbf{x} + \tilde{\mathbf{D}}_1^T \mathbf{y}, \tilde{\mathbf{C}}_2^T \mathbf{x} + \tilde{\mathbf{D}}_2^T \mathbf{y}, \cdots, \tilde{\mathbf{C}}_{m_1}^T \mathbf{x} + \tilde{\mathbf{D}}_{m_1}^T \mathbf{y} \right] \\
\text{s.t. } \tilde{\mathbf{A}}_r^T \mathbf{x} + \tilde{\mathbf{B}}_r^T \mathbf{y} \leq \tilde{E}_r, r = 1, 2, \cdots, p_1 \\
\text{where } \mathbf{y} \text{ solves} \\
\quad \begin{cases}
\max \left[\tilde{\mathbf{c}}_1^T \mathbf{x} + \tilde{\mathbf{d}}_1^T \mathbf{y}, \tilde{\mathbf{c}}_2^T \mathbf{x} + \tilde{\mathbf{d}}_2^T \mathbf{y}, \cdots, \tilde{\mathbf{c}}_{m_2}^T \mathbf{x} + \tilde{\mathbf{d}}_{m_2}^T \mathbf{y} \right] \\
\text{s.t. } \tilde{\mathbf{a}}_r^T \mathbf{x} + \tilde{\mathbf{b}}_r^T \mathbf{y} \leq \tilde{e}_r, r = 1, 2, \cdots, p_2 \\
\mathbf{x} \in \Re^{n_1}, \mathbf{y} \in \Re^{n_2}
\end{cases}
\end{cases}
\tag{3.24}
$$

where $\tilde{\mathbf{C}}_i = (\tilde{C}_{i1}, \tilde{C}_{i2}, \cdots, \tilde{C}_{in_1})^T$, $\tilde{\mathbf{D}}_i = (\tilde{D}_{i1}, \tilde{D}_{i2}, \cdots, \tilde{D}_{in_1})^T$, $\tilde{\mathbf{A}}_r = (\tilde{A}_{r1}, \tilde{A}_{r2}, \cdots, \tilde{A}_{rn_1})^T$, $\tilde{\mathbf{B}}_r = (\tilde{B}_{r1}, \tilde{B}_{r2}, \cdots, \tilde{B}_{rn_1})^T$, are rough interval vectors and $\tilde{E}_r$ are rough intervals, $i = 1, 2, \cdots, m_1; r = 1, 2, \cdots, p_1$. At the same time, $\tilde{\mathbf{c}}_i = (\tilde{c}_{i1}, \tilde{c}_{i2}, \cdots, \tilde{c}_{in_2})^T$, $\tilde{\mathbf{d}}_i = (\tilde{d}_{i1}, \tilde{d}_{i2}, \cdots, \tilde{d}_{in_2})^T$, $\tilde{\mathbf{a}}_r = (\tilde{a}_{r1}, \tilde{a}_{r2}, \cdots, \tilde{a}_{rn_2})^T$, $\tilde{\mathbf{b}}_r = (\tilde{b}_{r2}, \tilde{b}_{r2}, \cdots, \tilde{b}_{rn_2})^T$ are rough interval vectors and $\tilde{e}_r$ are rough intervals, $i = 1, 2, \cdots, m_2; r = 1, 2, \cdots, p_2$.

Considering model (3.24), constraints can no longer give a crisp feasible set due to the existence of rough interval. Naturally, one requires constraints helds at a predetermined confidence level. The chance constraints with fuzzy parameters are expressed as

$$
\underline{Appr}\{\tilde{\mathbf{a}}_r^T \mathbf{x} \leq \bar{b}_r\} \geq \alpha_r,
$$

where $\underline{Appr}$ denotes the lower approximation of the event $\{\cdot\}$. If the decision maker hopes that the lower approximation of the objectives are not less than β_i, and that the lower approximation of the constraints being no greater than $\bar{b}_r$ is not less than β_r, then the chance-constrained linear programming model with rough interval parameters is formulated as

$$
\begin{cases}
\max \left[\bar{F}_1, \bar{F}_2, \cdots, \bar{F}_{m_1} \right] \\
\text{s.t. } \begin{cases}
\underline{Appr}\{\tilde{\mathbf{C}}_i^T \mathbf{x} + \tilde{\mathbf{D}}_i^T \mathbf{y} \geq \bar{F}_i\} \geq \alpha_i, i = 1, 2, \cdots, m_1 \\
\underline{Appr}\{\tilde{\mathbf{A}}_r^T \mathbf{x} + \tilde{\mathbf{B}}_r^T \mathbf{y} \leq \tilde{E}_r\} \geq \beta_r, r = 1, 2, \cdots, p_1
\end{cases} \\
\text{where } \mathbf{y} \text{ solves} \\
\quad \begin{cases}
\max \left[\bar{f}_1, \bar{f}_2, \cdots, \bar{f}_{m_2} \right] \\
\text{s.t. } \begin{cases}
\underline{Appr}\{\tilde{\mathbf{c}}_i^T \mathbf{x} + \tilde{\mathbf{d}}_i^T \mathbf{y} \geq \bar{f}_i\} \geq \gamma_i, i = 1, 2, \cdots, m_2 \\
\underline{Appr}\{\tilde{\mathbf{a}}_r^T \mathbf{x} + \tilde{\mathbf{b}}_r^T \mathbf{y} \leq \tilde{e}_r\} \geq \delta_r, r = 1, 2, \cdots, p_2 \\
\mathbf{x} \in \Re^{n_1}, \mathbf{y} \in \Re^{n_2}
\end{cases}
\end{cases}
\end{cases}
\tag{3.25}
$$

Similarly, we have

$$
\begin{cases}
\max [\bar{F}_1, \bar{F}_2, \cdots, \bar{F}_{m_1}] \\
\text{s.t.} \begin{cases} \overline{Appr}\{\tilde{\mathbf{C}}_i^T \mathbf{x} + \tilde{\mathbf{D}}_i^T \mathbf{x} \geq \bar{F}_i\} \geq \alpha_i, i = 1, 2, \cdots, m_1 \\ \overline{Appr}\{\tilde{\mathbf{A}}_r^T \mathbf{x} + \tilde{\mathbf{B}}_r^T \mathbf{y} \leq \tilde{E}_r\} \geq \beta_r, r = 1, 2, \cdots, p_1 \end{cases} \\
\text{where } \mathbf{y} \text{ solves} \\
\quad \begin{cases} \max [\bar{f}_1, \bar{f}_2, \cdots, \bar{f}_{m_2}] \\ \text{s.t.} \begin{cases} \overline{Appr}\{\tilde{\mathbf{c}}_i^T \mathbf{x} + \tilde{\mathbf{d}}_i^T \mathbf{y} \geq \bar{f}_i\} \geq \gamma_i, i = 1, 2, \cdots, m_1 \\ \overline{Appr}\{\tilde{\mathbf{a}}_r^T \mathbf{x} + \tilde{\mathbf{b}}_r^T \mathbf{x} \leq \tilde{e}_r\} \geq \delta_r, r = 1, 2, \cdots, p_2 \\ \mathbf{x} \in \Re^{n_1}, \mathbf{y} \in \Re^{n_2} \end{cases} \end{cases}
\end{cases}
\tag{3.26}
$$

and

$$
\begin{cases}
\max [\bar{F}_1, \bar{F}_2, \cdots, \bar{F}_{m_1}] \\
\text{s.t.} \begin{cases} Appr\{\tilde{\mathbf{C}}_i^T \mathbf{x} + \tilde{\mathbf{D}}_i^T \mathbf{y} \geq \bar{F}_i\} \geq \alpha_i, i = 1, 2, \cdots, m_1 \\ Appr\{\tilde{\mathbf{A}}_r^T \mathbf{x} + \tilde{\mathbf{B}}_r^T \mathbf{x} \leq \tilde{E}_r\} \geq \beta_r, r = 1, 2, \cdots, p_1 \end{cases} \\
\text{where } \mathbf{y} \text{ solves} \\
\quad \begin{cases} \max [\bar{f}_1, \bar{f}_2, \cdots, \bar{f}_{m_2}] \\ \text{s.t.} \begin{cases} Appr\{\tilde{\mathbf{c}}_i^T \mathbf{x} + \tilde{\mathbf{d}}_i^T \mathbf{y} \geq \bar{f}_i\} \geq \gamma_i, i = 1, 2, \cdots, m_1 \\ Appr\{\tilde{\mathbf{a}}_r^T \mathbf{x} + \tilde{\mathbf{b}}_r^T \mathbf{y} \leq \tilde{e}_r\} \geq \delta_r, r = 1, 2, \cdots, p_2 \\ \mathbf{x} \in \Re^{n_1}, \mathbf{y} \in \Re^{n_2} \end{cases} \end{cases}
\end{cases}
\tag{3.27}
$$

All of (3.25), (3.26) and (3.27) are called bilevel linear chance-constrained rough model (BL-LCCRM).

REMARK 3.2 If $\eta = 1$, $Appr = \underline{Appr}$, then model (3.27) degenerates into model (3.25). If $\eta = 0$, $Appr = \overline{Appr}$, then model (3.27) degenerates into model (3.26).

3.4.2.1 Crisp Equivalent Model

The mathematical traditional solution methods require conversion of the chance constraints to their respective deterministic equivalents. However, this process is usually hard to perform and only successful in some special cases. First, we present some useful results

THEOREM 3.3
Let $\tilde{\mathbf{C}}_i^T \mathbf{x} + \tilde{\mathbf{D}}_i^T \mathbf{y} = ([c_{i1}, c_{i2}], [c_{i3}, c_{i4}])$. Then $Appr\{\tilde{\mathbf{C}}_i^T \mathbf{x} + \tilde{\mathbf{D}}_i^T \mathbf{y} \geq \bar{F}_i\} \geq \alpha_i$ holds if and only if

$$
\begin{cases}
\Phi, & \text{if } c_{i4} \leq \bar{F}_i \\
\bar{F}_i \leq c_{i4} - \frac{(c_{i4} - c_{i3})\alpha_i}{1 - \eta}, & \text{if } c_{i2} \leq \bar{F}_i \leq c_{i4} \\
\bar{F}_i \leq \frac{\eta(c_{i4} - c_{i3})c_{i2} + (1 - \eta)(c_{i2} - c_{i1})c_{i4} - \alpha_i(c_{i2} - c_{i1})(c_{i4} - c_{i3})}{\eta(c_{i4} - c_{i3}) + (1 - \eta)(c_{i1} - c_{i2})}, & \text{if } c_{i1} \leq \bar{F}_i \leq c_{i2} \\
\bar{F}_i \leq \alpha_i c_{i1} + (1 - \alpha_i)c_{i4}, & \text{if } c_{i3} \leq \bar{F}_i \leq c_{i4} \\
X, & \text{if } \bar{F}_i \leq c_i
\end{cases}
\tag{3.28}
$$

where Φ means that $Appr\{\tilde{\mathbf{C}}_i^T \mathbf{x} + \tilde{\mathbf{D}}_i^T \mathbf{y} \geq \bar{F}_i\} \geq \alpha_i$ never holds for any (x,y). Meanwhile, X means that $Appr\{\tilde{\mathbf{C}}_i^T \mathbf{x} + \tilde{\mathbf{D}}_i^T \mathbf{y} \geq \bar{F}_i\} \geq \alpha_i$ holds for every $\mathbf{x}, \mathbf{y}$ as long as $\mathbf{x}, \mathbf{y}$ satisfies other constraints.

PROOF   It follow from the definition of *Appr* that

$$Appr\{\tilde{\mathbf{C}}_i^T \mathbf{x} + \tilde{\mathbf{D}}_i^T \mathbf{y} \geq \bar{F}_i\} = \begin{cases} 0, & \text{if } c_{i4} \leq \bar{F}_i \\ (1-\eta)\frac{c_{i4}-\bar{F}_i}{(c_{i4}-c_{i3})}, & \text{if } c_{i2} \leq \bar{F}_i \leq c_{i4} \\ \eta\frac{c_{i4}-\bar{F}_i}{c_{i4}-c_{i3}} + (1-\eta)\frac{c_{i2}-\bar{F}_i}{c_{i2}-c_{i1}}), & \text{if } c_{i1} \leq \bar{F}_i \leq c_{i2} \\ (1-\eta)\frac{c_{i4}-\bar{F}_i}{c_{i4}-c_{i3}} + \eta, & \text{if } c_{i3} \leq \bar{F}_i \leq c_{i1} \\ 1, & \text{if } \bar{F}_i \leq c_{i3} \end{cases}$$

Then $Appr\{\tilde{\mathbf{C}}_i^T \mathbf{x} + \tilde{\mathbf{D}}_i^T \mathbf{y} \geq \bar{F}_i\} \geq \alpha_i$ holds if and only if

$$\begin{cases} \Phi, & \text{if } c_{i4} \leq \bar{F}_i \\ \bar{F}_i \leq c_{i4} - \frac{(c_{i4}-c_{i3})\alpha_i}{1-\eta}, & \text{if } c_{i2} \leq \bar{F}_i \leq c_{i4} \\ \bar{F}_i \leq \frac{\eta(c_{i4}-c_{i3})c_{i2}+(1-\eta)(c_{i2}-c_{i1})c_{i4}-\alpha_i(c_{i2}-c_{i1})(c_{i4}-c_{i3})}{\eta(c_{i4}-c_{i3})+(1-\eta)(c_{i1}-c_{i2})}, & \text{if } c_{i1} \leq \bar{F}_i \leq c_{i2} \\ \bar{F}_i \leq \alpha_i c_{i1} + (1-\alpha_i)c_{i4}, & \text{if } c_{i3} \leq \bar{F}_i \leq c_{i4} \\ X, & \text{if } \bar{F}_i \leq c_i \end{cases}$$

Thus this theorem holds.

THEOREM 3.4
Let $\tilde{\mathbf{A}}_r^T \mathbf{x} + \tilde{\mathbf{B}}_r^T \mathbf{y} \leq \tilde{E}_r = ([a,b],[c,d]), (c \leq a \leq b \leq d)$. Then $Appr\{\tilde{\mathbf{A}}_r^T \mathbf{x} + \tilde{\mathbf{B}}_r^T \mathbf{y} \leq \tilde{E}_r\} \geq \beta_r$ holds if and only if

$$\begin{cases} \Phi, & \text{if } 0 \leq c \\ \beta_r(d-c)+(1-\eta)c \leq 0, & \text{if } c \leq 0 \leq a \\ \beta_r(b-a)(d-c)+\eta a(d-c)+(1-\eta)c(b-a) \leq 0, & \text{if } a \leq 0 \leq b \\ \beta_r(d-c)-(1-\eta)d \leq 0, & \text{if } b \leq 0 \leq d \\ X, & \text{if } d \leq 0 \end{cases} \qquad (3.29)$$

where Φ means that $Appr\{\tilde{\mathbf{A}}_r^T \mathbf{x} + \tilde{\mathbf{B}}_r^T \mathbf{y} \leq \tilde{E}_r\} \geq \beta_r$ never holds for any $\mathbf{x}$. Meanwhile, X means that $Appr\{\tilde{\mathbf{A}}_r^T \mathbf{x} + \tilde{\mathbf{B}}_r^T \mathbf{y} \leq \tilde{E}_r\} \leq \beta_r$ holds for every $\mathbf{x}$ as long as x satisfies other constraints.

PROOF   It follow from the definition of *Appr* that

$$Appr\{\tilde{\mathbf{A}}_r^T \mathbf{x} + \tilde{\mathbf{B}}_r^T \mathbf{y} \leq \tilde{E}_r\} = \begin{cases} 0, & \text{if } 0 \leq c \\ \frac{-c(1-\eta)}{d-c}, & \text{if } c \leq 0 \leq a \\ \frac{-a\eta}{b-a} + \frac{-c(1-\eta)}{d-c}, & \text{if } a \leq 0 \leq b \\ \eta + \frac{-c(1-\eta)}{d-c}, & \text{if } b \leq 0 \leq d \\ 1, & \text{if } d \leq 0 \end{cases}$$

Then $Appr\{\tilde{\mathbf{A}}_r^T\mathbf{x} + \tilde{\mathbf{B}}_r^T\mathbf{y} \leq \tilde{E}_r\} \geq \beta_r$ holds if and only if

$$\begin{cases} \Phi, & \text{if } 0 \leq c \\ \beta_r(d-c) + (1-\eta)c \leq 0, & \text{if } c \leq 0 \leq a \\ \beta_r(b-a)(d-c) + \eta a(d-c) + (1-\eta)c(b-a) \leq 0, & \text{if } a \leq 0 \leq b \\ \beta_r(d-c) - (1-\eta)d \leq 0, & \text{if } b \leq 0 \leq d \\ X, & \text{if } d \leq 0 \end{cases}$$

Thus the proof is complete.

In a similar way, the lower-level programming problem of (3.27) can be transformed into crisp equivalent forms.

3.4.2.2 Fuzzy Decision Approach

To solve the BLMOP (3.5) by adopting, the two-planner Stakelberg (see [174, 273]), and the well-known fuzzy decision model of M. Sakawa [260], one first gets the satisfactory solution that is acceptable to ULDM, and then give the ULDM decision variables and goals with some leeway to the lower level decision make LLDM for him or her to seek the satisfactory solution, and to arrive at the solution that is closest to the satisfactory solution of the ULDM. This due to the LLDM, who should not only optimize his or her objective functions but also try to satisfy the ULDM goals and preferences as much as possible.

In this way, the solution method simplifies a BLDM by transforming it into separate MODM problems at two levels, and by that means the difficulty associated with nonconvexity to arrive at an optimal solution is avoided.

First, the ULDM solves the following MODM problem:

$$\max_{(\mathbf{x},\mathbf{y})\in D} [F_1(\mathbf{x},\mathbf{y}), F_2(\mathbf{x},\mathbf{y}), \cdots, F_{m_1}(\mathbf{x},\mathbf{y})] \qquad (3.30)$$

where $D := \{(\mathbf{x},\mathbf{y})|G_r(\mathbf{x},\mathbf{y}) \leq 0, r = 1, 2, \cdots, p_1, g_r(x,y) \leq 0, r = 1, 2, \cdots, p_2\}$. To build membership functions, goals and tolerances should be determined first. However, they can hardly be determined without meaningful supporting data.

We should first find the individual best solutions F_i^+ and individual worst solutions F_i^- for each objective of (3.30), where

$$F_i^+ = \max_{x \in D} F_i(\mathbf{x},\mathbf{y}), F_i^- = \min_{x \in D} F_i(\mathbf{x},\mathbf{y}), i = 1, 2, \cdots, P \qquad (3.31)$$

Goals and tolerances can then be reasonably set for individual solutions and the differences of the best and worst solutions, respectively. This data can then be formulated as the following membership functions of fuzzy set theory:

$$\mu_{F_i}(F_i(\mathbf{x},\mathbf{y})) = \begin{cases} 1, & \text{if } F_i(\mathbf{x},\mathbf{y}) > F_i^+ \\ \frac{F_i(\mathbf{x},\mathbf{y})-F_i^-}{F_i^+-F_i^-}, & \text{if } F_i^- \leq F_i(\mathbf{x},\mathbf{y}) \leq F_i^+ \\ 0, & \text{if } F_i(x,y) < F_i^- \end{cases} \qquad (3.32)$$

Now we can get the solution to the ULDM problem by solving the following Tchebycheff problem (see [2, 271]):

$$
\begin{cases}
\max \lambda \\
\text{s.t.} \begin{cases}
(\mathbf{x}, \mathbf{y}) \in D \\
\mu_{F_i}(F_i(\mathbf{x}, \mathbf{y})) \geq \lambda, i = 1, 2, \cdots, m_1 \\
\lambda \in [0, 1]
\end{cases}
\end{cases}
\tag{3.33}
$$

whose solution is assumed to be $(\mathbf{x}^U, \mathbf{y}^U, \lambda^U, F_i^U, i = 1, 2, \cdots, m_1)$.

Second, in the same way, the LLDM independently solves

$$
\max_{(\mathbf{x}, \mathbf{y}) \in D} [f_1(\mathbf{x}, \mathbf{y}), f_2(\mathbf{x}, \mathbf{y}), \cdots, f_{m_2}(\mathbf{x}, \mathbf{y})]
\tag{3.34}
$$

where $D := \{(\mathbf{x}, \mathbf{y}) | G_r(\mathbf{x}, \mathbf{y}) \leq 0, r = 1, 2, \cdots, P, g_r(\mathbf{x}, \mathbf{y}) \leq 0, r = 1, 2, \cdots, p\}$. To build membership functions, goals and tolerances should be determined first. However, they could hardly be determined without meaningful supporting data.

We should first find the individual best solutions f_i^+ and individual worst solutions f_i^- for each objective of (3.34), where

$$
f_i^+ = \max_{(\mathbf{x}, \mathbf{y}) \in D} f_i(\mathbf{x}, \mathbf{y}) \text{ and } f_i^- = \min_{(\mathbf{x}, \mathbf{y}) \in D} f_i(\mathbf{x}, \mathbf{y}), i = 1, 2, \cdots, p
\tag{3.35}
$$

Goals and tolerances can then be reasonably set for individual solutions and the differences of the best and worst solutions, respectively. This data can then be formulated as the following membership functions:

$$
\mu_{f_i}(f_i(\mathbf{x}, \mathbf{y})) = \begin{cases}
1, & \text{if } f_i(\mathbf{x}, \mathbf{y}) > f_i^+ \\
\frac{f_i(\mathbf{x}, \mathbf{y}) - f_i^-}{f_i^+ - f_i^-}, & \text{if } f_i^- \leq f_i(\mathbf{x}, \mathbf{y}) \leq f_i^+ \\
0, & \text{if } f_i(\mathbf{x}, \mathbf{y}) < f_i^-
\end{cases}
\tag{3.36}
$$

Now, we can get the solution of the LLDM problem by solving the following Tchebycheff problem:

$$
\begin{cases}
\max \delta \\
\text{s.t.} \begin{cases}
(\mathbf{x}, \mathbf{y}) \in D \\
\mu_{f_i}(f_i(\mathbf{x}, \mathbf{y})) \geq \lambda, i = 1, 2, \cdots, m \\
\lambda \in [0, 1]
\end{cases}
\end{cases}
\tag{3.37}
$$

whose solutions is assumed to be $(\mathbf{x}^L, \mathbf{y}^L, f_i^L, \lambda^L, i = 1, 2, \cdots, m)$.

Now the solution to the ULDM and LLDM are disclosed. However, the two solutions are usually different because of the difference in nature between two level objective functions. The ULDM knows that using the optimal decisions $\mathbf{x}^U$ as a control factors for the LLDM are not practical. It is more reasonable to have some tolerance that gives the LLDM an extent of feasible region to search for his or her optimal solution, and also to reduce searching time or interactions.

In this way, the range of decision variables $\mathbf{x}$ and $\mathbf{y}$ should be around $\mathbf{x}^U$ with maximum tolerance t and the following membership function specifies x_1^U as

$$\mu_{\mathbf{x}^U} = \begin{cases} \frac{\mathbf{x}-(\mathbf{x}^U-t)}{t} & \mathbf{x}^U - t \leq \mathbf{x} \leq \mathbf{x}^U \\ \frac{(\mathbf{x}^U+t)-\mathbf{x}}{t} & \mathbf{x}^U \leq \mathbf{x} \leq \mathbf{x}^U + t \end{cases} \tag{3.38}$$

where $\mathbf{x}^U$ is the most preferred solution; $(\mathbf{x}^U - t)$ and $(\mathbf{x}^U + t)$ are the worst acceptable decisions; and that satisfaction is linearly increasing with the interval of $[\mathbf{x}^U - t, x]$ and linearly decreasing with $[\mathbf{x}_1, \mathbf{x}^U + t]$, and other decision are not acceptable. In order to supervise the LLDM to search for solutions in the right direction, we will take the following steps.

First, the ULDM goals may reasonably consider that all $F_i \geq F_i^U, k = 1, 2, \cdots, N_1$ are absolutely acceptable and all $F_i \geq F_i' = F_i(\mathbf{x}, \mathbf{y}), k = 1, 2, \cdots, N_1$ are absolutely unacceptable, and that the preference with $[F_i', F_i^U], i = 1, 2, \cdots, M$ is linearly increasing. This is due to the fact that the LLDM obtained the optimum at $(\mathbf{x}^L, \mathbf{y}^L)$, which in turn provides the ULDM the objective function values F_i', which makes any $F_i < F_i', i = 1, 2, \cdots, M$ unattractive in practice.

The following membership functions of the ULDM can be stated as

$$\mu_{\mathbf{x}^U}' = \begin{cases} 1, & \text{if } F_i(\mathbf{x}, \mathbf{y}) > F_i^U \\ \frac{F_i(\mathbf{x},\mathbf{y})-F_i'}{F_i^U - F_i'}, & \text{if } F_i' \leq F_i(\mathbf{x}, \mathbf{y}) \leq F_i^U \\ 0, & \text{if } F_i(\mathbf{x}, \mathbf{y}) \leq F_i' \end{cases} \tag{3.39}$$

Second, the LLDM may be willing to build a membership function for his or her objective functions, so that he or she can rate the satisfaction of each potential solution. In this way, the LLDM has the following membership functions for his/her goals:

$$\mu_{\mathbf{x}^U}' = \begin{cases} 1, & \text{if } f_i(x,y) > f_i^L \\ \frac{f_i(\mathbf{x},\mathbf{y})-f_i'}{f_i^L - f_i'}, & \text{if } f_i' \leq f_i(\mathbf{x}, \mathbf{y}) \leq f_i^L \\ 0, & \text{if } f_i(\mathbf{x}, \mathbf{y}) \leq f_i' \end{cases} \tag{3.40}$$

where $f_i' = f_i(x^U, y^U)$.

Finally, in order to generate the satisfactory solution, which is also a Pareto-optimal solution with overall satisfaction for all DMs, we can solve the following Tchebycheff problem:

$$\begin{cases} \max \delta \\ \text{s.t.} \begin{cases} \frac{[(\mathbf{x}^U+t)-\mathbf{x}]}{t} \geq \delta I \\ \frac{[\mathbf{x}-(\mathbf{x}^U-t)]}{t} \geq \delta I \\ \mu_{F_i}'[F_i(\mathbf{x}, \mathbf{y})] \geq \delta, i = 1, 2, \cdots, M \\ \mu_{f_i}'[f_i(\mathbf{x}, \mathbf{y})] \geq \delta, i = 1, 2, \cdots, m \\ (\mathbf{x}, \mathbf{y}) \in D \\ t \geq 0 \\ \delta \in [0, 1] \end{cases} \end{cases} \tag{3.41}$$

where δ is the overall satisfaction, and I is the column vector with all elements equal to 1s. Equation (3.41) is actually a fuzzy problem by M. Sakawa [260].

By solving problem (3.41), if the ULDM is satisfied with solution, then a satisfactory solution is reached. Otherwise, he or she should provide a new membership function for the control variables and objectives to the LLDM until a satisfactory solution is reached.

3.4.2.3 Numerical Example

In this section, a numerical example is given to illustrate the effectiveness of the application of the models and algorithms proposed above.

Example 3.3
Consider the problem

$$\begin{cases} \max_{x_1,x_2} F_1 = \xi_1 x_1 + 2\xi_2 x_2 + 3\xi_3 y_1 + 4\xi_4 y_2 \\ \max_{x_1,x_2} F_2 = 5\xi_1 x_1 + 3\xi_2 x_2 + 2\xi_3 y_1 + \xi_4 y_2 \\ \text{where } (y_1,y_2,y_3) \text{ solves} \\ \quad \begin{cases} \max_{y_1,y_2} f_1 = 7\xi_1 x_1 + 5\xi_2 x_2 + 4\xi_3 y_1 + \xi_4 y_2 \\ \max_{y_1,y_2} f_2 = 3\xi_1 x_1 + 4\xi_2 x_2 + 5\xi_3 y_1 + 6\xi_4 y_2 \\ \quad \text{s.t.} \begin{cases} x_1 + x_2 + y_1 + y_2 \geq 120 \\ x_1 + x_2 + y_1 + y_2 \leq 200 \\ x_1,x_2,x_3,y_1,y_2,y_3 \geq 10 \end{cases} \end{cases} \end{cases} \quad (3.42)$$

where $\xi_j, j = 1,2,3,4$ are rough intervals characterized as

$$\xi_1 = ([4,6],[2,8]), \xi_2 = ([9,11],[8,12]), \xi_3 = ([7,9],[6,10]), \xi_4 = ([3,9],[2,10])).$$

In order to solve it, we use the chance operator to deal with rough objectives and rough constraints, and then we can obtain the BL-CCRM as

$$\begin{cases} \max_{x_1,x_2} [\bar{F}_1, \bar{F}_2] \\ \text{s.t.} \begin{cases} Appr\{\xi_1 x_1 + 2\xi_2 x_2 + 3\xi_3 y_1 + 4\xi_4 y_2 \geq \bar{F}_1\} \geq 0.6 \\ Appr\{5\xi_1 x_1 + 3\xi_2 x_2 + 2\xi_3 y_1 + \xi_4 y_2 \geq \bar{F}_2\} \geq 0.8 \end{cases} \\ \text{where } (y_1,y_2) \text{ solves} \\ \quad \begin{cases} \max_{y_1,y_2} [\bar{f}_1, \bar{f}_2] \\ \quad \text{s.t.} \begin{cases} Appr\{7\xi_1 x_1 + 5\xi_2 x_2 + 4\xi_3 y_1 + \xi_4 y_2 \geq \bar{f}_1\} \geq 0.8 \\ Appr\{3\xi_1 x_1 + 4\xi_2 x_2 + 5\xi_3 y_1 + 6\xi_4 y_2 \geq \bar{f}_2\} \geq 0.6 \\ x_1 + x_2 + y_1 + y_2 \geq 120 \\ x_1 + x_2 + y_1 + y_2 \leq 200 \\ x_1,x_2,x_3,y_1,y_2,y_3 \geq 10 \end{cases} \end{cases} \end{cases} \quad (3.43)$$

Here we suppose that $\eta = 1/2$, and, by Theorem 3.3 and Theorem 3.4, we know that problem (3.43) is equivalent to the model

$$
\begin{cases}
\max[0.8x_1 + 16.8x_2 + 19.6y_1 + 7.6y_2, -8x_1 + 16.8x_2 + 7.2y_1 - 2.8y_2] \\
\text{where } (y_1, y_2) \text{ solves} \\
\max[-14.2x_1 + 28x_2 + 14.4y_1 - 2.8y_2, 2.4x_1 + 28.8x_2 + 26y_1 + 2.4y_2] \\
\text{s.t.} \begin{cases}
x_1 + x_2 + y_1 + y_2 \geq 120 \\
x_1 + x_2 + y_1 + y_2 \leq 200 \\
x_1, x_2, y_1, y_2 \geq 10
\end{cases}
\end{cases} \tag{3.44}
$$

Without causing ambiguity, we still use F_1, F_2, f_1, f_2 to represent $0.8x_1 + 16.8x_2 + 19.6y_1 + 7.6y_2, -8x_1 + 16.8x_2 + 7.2y_1 - 2.8y_2, -14.2x_1 + 28x_2 + 14.4y_1 - 2.8y_2$, and $2.4x_1 + 28.8x_2 + 26y_1 + 2.4y_2$, respectively.

First, the ULDM solves his or her problem as follows:

1. Finding individual optimal solutions, we get $F_1^+ = 3584, F_1^- = 512, F_2^+ = 2820, F_2^- = -1148$.

2. By using (3.34), build the membership functions μ_{F_1}, μ_{F_2}, hen solve (3.35) as follows:

$$
\begin{cases}
\max \lambda \\
\text{s.t.} \begin{cases}
0.8x_1 + 16.8x_2 + 19.6y_1 + 7.6y_2 \geq 3072\lambda + 512 \\
-8x_1 + 16.8x_2 + 7.2y_1 - 2.8y_2 \geq 3968\lambda - 1148 \\
x_1 + x_2 + y_1 + y_2 \geq 120 \\
x_1 + x_2 + y_1 + y_2 \leq 200 \\
x_1, x_2, y_1, y_2 \geq 10 \\
\lambda \in [0, 1]
\end{cases}
\end{cases}
$$

whose solution is $(x_1^U, x_2^U, y_1^U, y_2^U) = (10, 126.2, 53.8, 10), (F_1^U, F_2^U) = (3258.6, 2399.7), \lambda^U = 0.894$.

Then, in the same way, the LLDM solves his/her problem as follows:

1. Finding individual optimal solutions, we get $f_1^+ = 4734, f_1^- = 2018, f_2^+ = 5204, f_2^- = 788$.

By solving

$$
\begin{cases}
\max \lambda \\
\text{s.t.} \begin{cases}
-14.2x_1 + 28x_2 + 14.4y_1 - 2.8y_2 \geq 6752\lambda - 2018 \\
2.4x_1 + 28.8x_2 + 26y_1 + 2.4y_2 \geq 4416\lambda + 788 x_1 + x_2 + y_1 + y_2 \geq 120 \\
x_1 + x_2 + y_1 + y_2 \leq 200 \\
x_1, x_2, y_1, y_2 \geq 10 \lambda \in [0, 1]
\end{cases}
\end{cases}
$$

whose solution is $(x_1^L, x_2^L, y_1^L, y_2^L) = (10, 170, 10, 10), (f_1^L, f_2^L) = (4734, 5204), \lambda^L = 1$.

Finally, we assume that the ULDMs control decision (x_1^U, x_2^U) is around 0 with tolerance 1.

By using following problem

$$
\begin{cases}
\max \delta \\
\text{s.t.} \begin{cases}
x_1 - 9 \geq \delta \\
11 - x_1 \geq \delta \\
x_2 - 125.2 \geq \delta \\
127.2 - x_2 \geq \delta \\
0.8x_1 + 16.8x_2 + 19.6y_1 + 7.6y_2 \geq 122.6\delta + 3136 \\
-8x_1 + 16.8x_2 + 7.2y_1 - 2.8y_2 \leq -420.38\delta + 2820 \\
-14.2x_1 + 28x_2 + 14.4y_1 - 2.8y_2 \geq 592.68\delta + 4138.32 \\
2.4x_1 + 28.8x_2 + 26y_1 + 2.4y_2 \geq 122.64\delta + 5081.36 \\
x_1 + x_2 + y_1 + y_2 \geq 120 \\
x_1 + x_2 + y_1 + y_2 \leq 200 \\
x_1, x_2, y_1, y_2 \geq 10 \\
\delta \in [0,1]
\end{cases}
\end{cases}
$$

whose compromise solution is $(x_1^*, x_2^*, y_1^*, y_2^*) = (10, 127.2, 52.8, 10)$, $(F_1^*, F_2^*) = (3255.9, 2408.9)$, $(f_1^*, f_2^*) = (4151.6, 5084, 1)$, $\delta^* = 0.962$.

If the ULDM is satisfied with the above solution, then a satisfactory solution is obtained. Otherwise, he or she should provide new membership functions for the control variable and objectives to the LLDM until a satisfactory solution is reached. It is easy to see that there is an inverse correlation between t and δ.

3.4.3 BL-NLCCRM and Rough Simulation-Based PTS

It is difficult to transform the bilevel nonlinear chance-constrained rough model (BL-NLCCRM) into its equivalent form, so we introduce the rough simulation-based parametric tabu search (PTS) algorithm to solve it.

3.4.3.1 PTS

Next let us recall the detail of the PTS introduced by F. Glover [120]. The solution approach consists of a parametric form of tabu search utilizing moves based on the approach of parametric branch and bound [119]. The tabu search framework amends parametric branch and bound by replacing its tree search memory structure with an adaptive memory structure that provides greater flexibility and facilitates the use of strategies outside the scope of tree search.

Let N^+ and N^- denote selected subsets of $N = \{1, 2, \cdots, n\}$, the index set for x. Also, let $N' = N^+ \cup N^-$, and let x' denote an associated trial vector satisfying $x' \in X$. For convenience of terminology, we understand the vector x' to be the partial vector consisting of the components x'_j for $j \in N'$, disregarding the components of $j \in N - N'$. Parametric branch and bound uses a variant of absolute value norm minimization to maintain linear programming (LP) feasibility while seeking to enforce the following conditions:

$$\begin{cases} \text{(UP)} & x_j \geq x_j' \text{ for } j \in N^+ \text{(provided } x_j' > 0) \\ \text{(DN)} & x_j \leq x_j' \text{ for } j \in N^- \text{(provided } x_j' < U_j) \end{cases}$$

Stipulating $x_j' > 0$ in (UP) and $x_j' < U_j$ in (DN) avoids consideration of redundant inequalities. The condition $x_j = x_j'$ is handled by allowing j to belong to both N^+ and N^-. As light variation on the foregoing representation introduces two different instances of x_j', one for N^+ and the other for N^-, thereby permitting (UP) and (DN) to bracket x_j between two different values. However, by our present formulation, j can be an element of both N^+ and N^- only in the case where the preceding inequalities constrain x_j to the single value x_j'. By convention, throughout this book, we may speak of a variable belonging to a set as shorthand for saying that its index belongs to the set (as in referring to x_j as an element of some specified subset of N').

We refer to (UP) and (DN) as goal conditions and call x_j' the goal value in these conditions, because we do not seek to enforce (UP) and (DN) directly by imposing them as constraints in the manner of customary branch-and-bound procedures but rather indirectly by incorporating them into the objective function of LP relaxation . This can be done by means of the following linear program, where M denotes a large positive number:

$$(LP') \begin{cases} \min u_0 = cx + dy + M(\sum_{j \in N^-} u_j + \sum_{j \in N^+} v_j) \\ \text{s.t.} \begin{cases} x_j = x_j' + u_j - v_j \\ u_j, v_j \geq 0, j \in N' \end{cases} \end{cases}$$

We say that (LP') targets the inequalities of (UP) and (DN). Evidently, the inequalities will be satisfied in an optimal solution to (LP') if and only if the problem (LP) has a feasible solution when expanded to include these inequalities.

The representation of (LP') simplifies in the special cases where $x_j' = U_j'$ for $j \in N^+$ and $x_j' = 0$ for $j \in N^-$ since we may then modify the objective by respectively replacing u_j by $x_j, j \in N^-$ and v_j by $U_j - x_j, j \in N^+$ and do not need to introduce the constraints associated with u_j and v_j. Thus, let N_U denote the subset of N^+ such that $x_j = U_j'$, and N_0 denote the subset of N^- such that $x_j = 0$, (LP') can be written more precisely in the form as

$$(LP') \begin{cases} \min u_0 = cx + dy + M(\sum_{j \in N_0} x_j + \sum_{j \in N_U} (U_j - x_j) + \sum_{j \in N^- - N_0} u_j + \sum_{j \in N^+ - N_U} v_j) \\ \text{s.t.} \begin{cases} x_j = x_j' + u_j - v_j \\ u_j, v_j \geq 0, j \in N' - N_0 - N_U \end{cases} \end{cases}$$

The terms U_j in the objective can be removed by introducing a single constant term $c_o = MU_j, j \in N_U$. Evidently, in the case of 0-1 multiobjective integer prgramming problems, no u_j or v_j variables are required.

From an implementation standpoint, a two-phase approach can be used for optimizing (LP') as an option to using the explicit form of the objective. That is, the objective of (LP') can be equivalently handled by first creating a primary objective that sets $\mathbf{M} = 1$ and disregards the $cx + dy$ component. Then, at optimality, all non-basic variables are held constant at their currently assigned (lower or upper) bounds, removing them from further consideration. The second phase is then implemented by optimizing $cx + dy$ over the residual system.

3.4.3.2   Numerical Example

In this section, a numerical example is given to illustrate the effectiveness of the application of the models and algorithms proposed earlier.

Example 3.4

Consider the problem

$$
\begin{cases}
\max\limits_{x_1,x_2,x_3} F_1 = 2\xi_1^2 x_1 + \sqrt{\xi_2} x_2 + \xi_3 x_3 + \xi_4 y_1 + \sqrt{\xi_5} y_2 + \sqrt{\xi_6} y_3 \\
\max\limits_{x_1,x_2,x_3} F_2 = x_1 + x_2 + x_3 + y_1 + y_2 + y_3 \\
\text{where } (y_1, y_2, y_3) \text{ solves} \\
\quad
\begin{cases}
\max\limits_{y_1,y_2,y_3} f_1 = \sqrt{7\xi_1 x_1 + 3\xi_2 x_2 + 8\xi_3 x_3} + (12\xi_4 y_1 + 11\xi_5 y_2 + 5\xi_6 y_3)^2 \\
\max\limits_{y_1,y_2,y_3} f_2 = x_1 + 5x_2 + 7x_3 + y_1 + 11y_2 - 8y_3 \\
\quad
\text{s.t.}
\begin{cases}
\sqrt{x_1 \xi_1 + x_2 \xi_2} + x_3^2 \xi_3 + \xi_4 y_4 + 17 y_3 \xi_6 \le 1540 \\
3x_1 + 6x_+ 12x_9 + 14x_{10} + 19x_{11} \le 2400 \\
x_1 + x_2 + x_3 + y_1 + y_2 - y_3 \ge 150 \\
x_1 + x_2 + x_3 + y_1 + y_2 + y_3 \le 200 \\
x_1, x_2, x_3, y_1, y_2, y_3 \ge 0
\end{cases}
\end{cases}
\end{cases}
$$

$$\tag{3.45}$$

where $\xi_j, j = 1, 2, \cdots, 6$ are rough intervals characterized as

$$
\xi_1 = ([8, 10], [7, 11]), \quad \xi_2 = ([9, 11], [8, 12]), \quad \xi_3 = ([10, 12], [9, 13]),
$$
$$
\xi_4 = ([11, 13], [10, 14]), \quad \xi_5 = ([12, 14], [11, 15]), \quad \xi_6 = ([13, 15], [12, 16]).
$$

In order to solve it, we use the *Appr* operator to deal with rough objectives and rough constraints, and then we can obtain the model

$$
\begin{cases}
\max\limits_{x_1,x_2,x_3} \{\bar{F}_1, x_1 + x_2 + x_3 + y_1 + y_2 + y_3\} \\
\text{s.t. } Appr\{\xi_1^2 x_1 + \sqrt{\xi_2} x_2 + \xi_3 x_3 + \xi_4 y_1 + \sqrt{\xi_5} y_2 + \sqrt{\xi_6} y_3 \ge \bar{F}_1\} \ge 0.8 \\
\text{where } (y_1, y_2, y_3) \text{ solves} \\
\quad
\begin{cases}
\max\limits_{y_1,y_2,y_3} [\bar{f}_1, x_1 + 5x_2 + 7x_3 + y_1 + 11y_2 - 8y_3] \\
\quad
\begin{cases}
Appr\{\sqrt{7\xi_1 x_1 + 3\xi_2 x_2 + 8\xi_3 x_3} + (12\xi_4 y_1 + 11\xi_5 y_2 + 5\xi_6 y_3)^2 \ge \bar{f}_1\} \\
\quad \ge 0.8 \\
Appr\{\sqrt{x_1 \xi_1 + x_2 \xi_2} + x_3^2 \xi_3 + \xi_4 y_4 + 17 y_3 \xi_6 \le 1400\} \ge 0.6 \\
\text{s.t.} \; 3x_1 + 6x_+ 12x_2 + 14x_3 + 9y_1 + 20y_1 + 15y_2 + 10y_3 \le 2400 \\
x_1 + x_2 + x_3 + y_1 + y_2 - y_3 \ge 150 \\
x_1 + x_2 + x_3 + y_1 + y_2 + y_3 \le 200 \\
x_1, x_2, x_3, y_1, y_2, y_3 \ge 0
\end{cases}
\end{cases}
\end{cases}
$$

$$\tag{3.46}$$

Here we suppose that $\eta = 1/2$. After running of the above algorithm, we obtain $(x_1^*, x_2^*, x_3^*, y_1^*, y_2^*, y_3^*) = (10.3.6.7.11.4, 20.3, 10.7, 25, 3)$ and $(F_1^*, F_2^*) = (136.9, 158.9), (f_1^*, f_2^*) = (106.9, 78.9)$. The Matlab® file is presented in A.3.

3.5 BL-DCRM

In this section, we introduce the concept of chance function and discuss the bilevel dependent-chance rough models (BL-DCRM). We also present some crisp equivalent models. Finally, a few numerical examples are exhibited.

3.5.1 General model of BL-DCRM

We propose the general model of BL-DCRM:

$$
\begin{cases}
\max \left[Ch\{F_1(\mathbf{x}, \mathbf{y}, \boldsymbol{\xi}) \le 0\}, Ch\{F_2(\mathbf{x}, \mathbf{y}, \boldsymbol{\xi}) \le 0\}, \cdots, Ch\{F_{m_1}(\mathbf{x}, \mathbf{y}, \boldsymbol{\xi}) \le 0\} \right] \\
\text{s.t. } Ch\{G_r(\mathbf{x}, \mathbf{y}, \boldsymbol{\xi}) \le 0\} \ge \alpha_r, r = 1, 2, \cdots, p_1 \\
\text{where } \mathbf{y} \text{ solves} \\
\quad \begin{cases}
\max \left[Ch\{f_1(\mathbf{x}, \mathbf{y}, \boldsymbol{\xi}) \le 0\}, Ch\{f_2(\mathbf{x}, \mathbf{y}, \boldsymbol{\xi}) \le 0\}, \cdots, Ch\{f_{m_2}(\mathbf{x}, \mathbf{y}, \boldsymbol{\xi}) \le 0\} \right] \\
\text{s.t. } \begin{cases} Ch\{g_r(\mathbf{x}, \mathbf{y}, \boldsymbol{\xi}) \le 0\} \ge \beta_r, r = 1, 2, \cdots, p_2 \\ \mathbf{x} \in \mathfrak{R}^{n_1}, \mathbf{y} \in \mathfrak{R}^{n_2} \end{cases}
\end{cases}
\end{cases}
\tag{3.47}
$$

where α_r, β are predetermined confidence levels, Ch are the chance of the rough event, and we could use $\underline{Appr}, \overline{Appr}$, and $Appr$ measures according different decision environments.

3.5.1.1 BL-DCRM Based on $\underline{Appr}$

If we take Ch as $\underline{Appr}$, then (3.47) can be rewritten as

$$
\begin{cases}
\max[\underline{Appr}\{F_1(\mathbf{x}, \mathbf{y}, \boldsymbol{\xi}) \le 0\}, \underline{Appr}\{F_2(\mathbf{x}, \mathbf{y}, \boldsymbol{\xi}) \le 0\}, \cdots, \underline{Appr}\{F_{m_1}(\mathbf{x}, \mathbf{y}, \boldsymbol{\xi}) \le 0\}] \\
\text{s.t. } \underline{Appr}\{G_r(\mathbf{x}, \mathbf{y}, \boldsymbol{\xi}) \le 0\} \ge \alpha_r, r = 1, 2, \cdots, p_1 \\
\text{where } \mathbf{y} \text{ solves} \\
\quad \begin{cases}
\max[\underline{Appr}\{f_1(\mathbf{x}, \mathbf{y}, \boldsymbol{\xi}) \le 0\}, \underline{Appr}\{f_2(\mathbf{x}, \mathbf{y}, \boldsymbol{\xi}) \le 0\}, \cdots, \underline{Appr}\{f_{m_2}(\mathbf{x}, \mathbf{y}, \boldsymbol{\xi}) \\
\le 0\}] \\
\text{s.t. } \begin{cases} \underline{Appr}\{g_r(\mathbf{x}, \mathbf{y}, \boldsymbol{\xi}) \le 0\} \ge \beta_r, r = 1, 2, \cdots, p_2 \\ \mathbf{x} \in \mathfrak{R}^{n_1}, \mathbf{y} \in \mathfrak{R}^{n_2} \end{cases}
\end{cases}
\end{cases}
\tag{3.48}
$$

DEFINITION 3.11 *Let $\mathbf{x}$ be given by the ULDM satisfying $\underline{Appr}\{G_r(\mathbf{x}, \mathbf{y}, \boldsymbol{\xi}) \le 0\} \ge \alpha_r, r = 1, 2, \cdots, p_1$. Then $(\mathbf{x}, \mathbf{y})$ is said to be an $\underline{Appr}$-feasible solution to problem (3.48) if and only $\mathbf{y}$ is an efficient solution of $\underline{LLDM}$.*

DEFINITION 3.12   *Let* $(\mathbf{x}^*, \mathbf{y}^*)$ *be a <u>Appr</u>-feasible solution of problem (3.48). Then* $(\mathbf{x}^*, \mathbf{y}^*)$ *is said to be an efficient solution of (3.48) if and only if no other feasible solution* $(\mathbf{x}, \mathbf{y})$ *exists, such that* $\underline{Appr}\{F_i(\mathbf{x}, \mathbf{y}) \geq 0\} \geq \underline{Appr}\{F_i(\mathbf{x}^*, \mathbf{y}^*)\}$, *and at least one* $i = 1, 2, \cdots, m_1$ *is strict inequality.*

3.5.1.2   BL-DCRM Based on $\overline{Appr}$

If we take *Ch* as $\overline{Appr}$, then (3.47) can be rewritten as

$$
\begin{cases}
\max[\overline{Appr}\{F_1(\mathbf{x}, \mathbf{y}, \boldsymbol{\xi}) \leq 0\}, \overline{Appr}\{F_2(\mathbf{x}, \mathbf{y}, \boldsymbol{\xi}) \leq 0\}, \cdots, \overline{Appr}\{F_{m_1}(\mathbf{x}, \mathbf{y}, \boldsymbol{\xi}) \leq 0\}] \\
\text{s.t. } \overline{Appr}\{G_r(\mathbf{x}, \mathbf{y}, \boldsymbol{\xi}) \leq 0\} \geq \alpha_r, r = 1, 2, \cdots, p_1 \\
\text{where } \mathbf{y} \text{ solves} \\
\quad \begin{cases}
\max[\overline{Appr}\{f_1(\mathbf{x}, \mathbf{y}, \boldsymbol{\xi}) \leq 0\}, \overline{Appr}\{f_2(\mathbf{x}, \mathbf{y}, \boldsymbol{\xi}) \leq 0\}, \cdots, \overline{Appr}\{f_{m_2}(\mathbf{x}, \mathbf{y}, \boldsymbol{\xi}) \\
\leq 0\}] \\
\text{s.t. } \begin{cases} \overline{Appr}\{g_r(\mathbf{x}, \mathbf{y}, \boldsymbol{\xi}) \leq 0\} \geq \beta_r, r = 1, 2, \cdots, p_2 \\ \mathbf{x} \in \mathfrak{R}^{n_1}, \mathbf{y} \in \mathfrak{R}^{n_2} \end{cases}
\end{cases}
\end{cases}
\tag{3.49}
$$

DEFINITION 3.13   *Let* $\mathbf{x}$ *be given by the ULDM satisfying* $\overline{Appr}\{G_r(\mathbf{x}, \mathbf{y}, \boldsymbol{\xi}) \leq 0\} \geq \alpha_r, r = 1, 2, \cdots, p_1$. *Then* $(\mathbf{x}, \mathbf{y})$ *is said to an* $\overline{Appr}$-*feasible solution to problem (3.49) if and only* $\mathbf{y}$ *is an efficient solution of LLDM.*

DEFINITION 3.14   *Let* $(\mathbf{x}^*, \mathbf{y}^*)$ *be a* $\overline{Appr}$-*feasible solution to problem (3.49). Then* $(\mathbf{x}^*, \mathbf{y}^*)$ *is said to be an efficient solution to (3.48) if and only if no other feasible solution* $(\mathbf{x}, \mathbf{y})$ *exists, such that* $\overline{Appr}\{F_i(\mathbf{x}, \mathbf{y}) \geq 0\} \geq \overline{Appr}\{F_i(\mathbf{x}^*, \mathbf{y}^*)\}$, *and at least one* $i = 1, 2, \cdots, m_1$ *is strict inequality.*

3.5.1.3   BL-DCRM Based on $Appr$

If we take *Ch* as $\overline{Appr}$, then (3.47) can be rewritten as

$$
\begin{cases}
\max[Appr\{F_1(\mathbf{x}, \mathbf{y}, \boldsymbol{\xi}) \leq 0\}, Appr\{F_2(\mathbf{x}, \mathbf{y}, \boldsymbol{\xi}) \leq 0\}, \cdots, Appr\{F_{m_1}(\mathbf{x}, \mathbf{y}, \boldsymbol{\xi}) \leq 0\}] \\
\text{s.t. } Appr\{G_r(\mathbf{x}, \mathbf{y}, \boldsymbol{\xi}) \leq 0\} \geq \alpha_r, r = 1, 2, \cdots, p_1 \\
\text{where } \mathbf{y} \text{ solves} \\
\quad \begin{cases}
\max[Appr\{f_1(\mathbf{x}, \mathbf{y}, \boldsymbol{\xi}) \leq 0\}, \overline{Appr}\{f_2(\mathbf{x}, \mathbf{y}, \boldsymbol{\xi}) \leq 0\}, \cdots, Appr\{f_{m_2}(\mathbf{x}, \mathbf{y}, \boldsymbol{\xi}) \\
\leq 0\}] \\
\text{s.t. } \begin{cases} Appr\{g_r(\mathbf{x}, \mathbf{y}, \boldsymbol{\xi}) \leq 0\} \geq \beta_r, r = 1, 2, \cdots, p_2 \\ \mathbf{x} \in \mathfrak{R}^{n_1}, \mathbf{y} \in \mathfrak{R}^{n_2} \end{cases}
\end{cases}
\end{cases}
\tag{3.50}
$$

DEFINITION 3.15   *Let* $\mathbf{x}$ *be given by the ULDM satisfying* $Appr\{G_r(\mathbf{x}, \mathbf{y}, \boldsymbol{\xi}) \leq 0\} \geq \alpha_r, r = 1, 2, \cdots, p_1$. *Then* $(\mathbf{x}, \mathbf{y})$ *is said to an* $\overline{Appr}$-*feasible solution to problem (3.50) if and only* $\mathbf{y}$ *is an efficient solution to LLDM.*

DEFINITION 3.16   *Let* $(\mathbf{x}^*, \mathbf{y}^*)$ *be an Appr-feasible solution to problem (3.50). Then* $(\mathbf{x}^*, \mathbf{y}^*)$ *is said to be efficient solution to (3.50) if and only if no other feasible solution* $(\mathbf{x}, \mathbf{y})$ *exists, such that* $Appr\{F_i(\mathbf{x}, \mathbf{y}) \geq 0\} \geq Appr\{F_i(\mathbf{x}^*, \mathbf{y}^*)\}$, *and at least one* $i = 1, 2, \cdots, m_1$ *is strict inequality.*

3.5.2 BL-LDCRM and Interactive Method

If the objective functions and the constraints are linear and with rough interval parameters, we propose the following bilevel linear dependent-chance rough model (BL-LDCRM):

$$
\begin{cases}
\max\,[\,Appr\{\widetilde{\mathbf{C}}_1^T\mathbf{x} + \widetilde{\mathbf{D}}_1^T\mathbf{y} \geq \overline{F}_1\}, Appr\{\widetilde{\mathbf{C}}_2^T\mathbf{x} + \widetilde{\mathbf{D}}_2^T\mathbf{y} \geq \overline{F}_2\}, \cdots, \\
Appr\{\widetilde{\mathbf{C}}_{m_1}^T\mathbf{x} + \widetilde{\mathbf{D}}_{m_1}^T\mathbf{y} \geq \overline{F}_{m_1}\}\,] \\
\text{s.t. } Appr\{\widetilde{\mathbf{A}}_r^T\mathbf{x} + \widetilde{\mathbf{B}}_r^T\mathbf{x} \leq \widetilde{E}_r\} \geq \alpha_r, r = 1, 2, \cdots, p_1 \\
\text{where } \mathbf{y} \text{ solves} \\
\quad
\begin{cases}
\max\,[\,Appr\{\tilde{\mathbf{c}}_1^T\mathbf{x} + \tilde{\mathbf{d}}_1^T\mathbf{y} \geq \bar{f}_1\}, Appr\{\tilde{\mathbf{c}}_2^T\mathbf{x} + \tilde{\mathbf{d}}_2^T\mathbf{y} \geq \bar{f}_2\}, \cdots, \\
Appr\{\tilde{\mathbf{c}}_{m_2}^T\mathbf{x} + \tilde{\mathbf{d}}_{m_2}^T\mathbf{y} \geq \bar{f}_{m_2}\}\,] \\
\text{s.t. }
\begin{cases}
Appr\{\tilde{\mathbf{a}}_r^T\mathbf{x} + \tilde{b}_r^T\mathbf{x} \leq \widetilde{E}_r\} \geq \beta_r, r = 1, 2, \cdots, p_2 \\
\mathbf{x} \in \Re^{n_1}, \mathbf{y} \in \Re^{n_2}
\end{cases}
\end{cases}
\end{cases}
\tag{3.51}
$$

where $\widetilde{\mathbf{C}}_i = (\widetilde{C}_{i1}, \widetilde{C}_{i2}, \cdots, \widetilde{C}_{in_1})^T, \widetilde{\mathbf{D}}_i = (\widetilde{D}_{i1}, \widetilde{D}_{i2}, \cdots, \widetilde{D}_{in_1})^T, \widetilde{\mathbf{A}}_r = (\widetilde{A}_{r1}, \widetilde{A}_{r2}, \cdots, \widetilde{A}_{rn_1})^T$, and $\widetilde{\mathbf{B}}_r = (\widetilde{B}_{r1}, \widetilde{B}_{r2}, \cdots, \widetilde{B}_{rn_1})^T$ are rough interval vectors, and $\widetilde{E}_r$ are rough intervals, $i = 1, 2, \cdots, m_1; r = 1, 2, \cdots, p_1$. At the same time, $\tilde{\mathbf{c}}_i = (\tilde{c}_{i1}, \tilde{c}_{i2}, \cdots, \tilde{c}_{in_2})^T, \tilde{\mathbf{d}}_i = (\tilde{d}_{i1}, \tilde{d}_{i2}, \cdots, \tilde{d}_{in_2})^T, \tilde{\mathbf{a}}_r = (\tilde{a}_{r1}, \tilde{a}_{r2}, \cdots, \tilde{a}_{rn_1})^T, \tilde{b}_r = (\tilde{b}_{r1}, \tilde{b}_{r2}, \cdots, \tilde{b}_{rn_1})^T$ are rough interval vectors, and $\tilde{e}_r$ are rough intervals, $i = 1, 2, \cdots, m; r = 1, 2, \cdots, p$. Ch represent $Appr, \overline{Appr}$, or $\underline{Appr}$.

By introducing variables $\gamma_i, i = 1, 2, \cdots, m_1; \delta_i = 1, 2, \cdots, m_2$, model (3.51) can be rewritten as

$$
\begin{cases}
\max\,[\gamma_1, \gamma_2, \cdots, \gamma_{m_1}] \\
\text{s.t. }
\begin{cases}
Appr\{\overline{\mathbf{C}}_i^T\mathbf{x} + \overline{\mathbf{D}}_i^T\mathbf{y} \geq \overline{F}_i\} \geq \gamma_i, i = 1, 2, \cdots, m_1 \\
Appr\{\overline{\mathbf{A}}\mathbf{x} + \overline{\mathbf{B}}_i\mathbf{y} \leq \overline{E}_r\} \geq \alpha_r, r = 1, 2, \quad , p_1
\end{cases} \\
\text{where } \mathbf{y} \text{ solves} \\
\quad
\begin{cases}
\max\,[\delta_1, \delta_2, \cdots, \delta_{m_2}] \\
\text{s.t. }
\begin{cases}
Appr\{\tilde{\mathbf{c}}_i^T\mathbf{x} + \tilde{\mathbf{d}}_i^T\mathbf{y} \geq \bar{f}_i\} \geq \delta_i, i = 1, 2, \cdots, m_2 \\
Appr\{\tilde{\mathbf{a}}_r^T\mathbf{x} + \tilde{\mathbf{b}}_r^T\mathbf{y} \leq \tilde{e}_r\} \geq \alpha_r, r = 1, 2, \cdots, p_2
\end{cases} \\
\mathbf{x} \in \Re^{n_1}, \mathbf{y} \in \Re^{n_2}
\end{cases}
\end{cases}
\tag{3.52}
$$

3.5.2.1 Crisp Equivalent Model

In this section, we consider a special case of the linear model and present the crisp equivalent model.

THEOREM 3.5

Let $\bar{C}_i^T x + \bar{D}_i^T y = ([c_{i1}, c_{i2}], [c_{i3}, c_{i4}]), \bar{A}_r^T x + \bar{B}_r^T y - \bar{E}_r = ([a,b],[c,d])$, and $Ch = Appr$. Then the upper decision making model of model (3.52) is equivalent to the following multiple objective decision-making problems:

$$
\begin{cases}
\max \left[(1-\eta)\frac{c_{14}-\bar{f}_1}{(c_{14}-c_{13})}, (1-\eta)\frac{c_{24}-\bar{f}_2}{(c_{24}-c_{23})}, \cdots, (1-\eta)\frac{c_{m4}-\bar{f}_{m_1}}{(c_{m_14}-c_{m_13})} \right] \\
\text{s.t.} \begin{cases}
\alpha_r(d-c) + (1-\eta)c \le 0, & \text{if } c \le 0 \le a \\
\alpha_r(b-a)(d-c) + \eta a(d-c) + (1-\eta)c(b-a) \le 0, & \text{if } a \le 0 \le b \\
\alpha_r(d-c) - (1-\eta)d \le 0, & \text{if } b \le 0 \le d \\
x \ge 0
\end{cases}
\end{cases}
$$

$$\tag{3.53}$$

$$
\begin{cases}
\max \left[\eta\frac{c_{14}-\bar{f}_1}{c_{14}-c_{13}} + (1-\eta)\frac{c_{12}-\bar{f}_1}{c_{12}-c_{11}}, \eta\frac{c_{24}-\bar{f}_2}{c_{24}-c_{23}} + (1-\eta)\frac{c_{22}-\bar{f}_2}{c_{22}-c_{21}}, \cdots, \eta\frac{c_{m_14}-\bar{f}_{m_1}}{c_{m_14}-c_{m_13}} + \right. \\
\left. (1-\eta)\frac{c_{m_12}-\bar{f}_{m_1}}{c_{m_12}-c_{m_11}} \right] \\
\text{s.t.} \begin{cases}
\alpha_r(d-c) + (1-\eta)c \le 0, & \text{if } c \le 0 \le a \\
\alpha_r(b-a)(d-c) + \eta a(d-c) + (1-\eta)c(b-a) \le 0, & \text{if } a \le 0 \le b \\
\alpha_r(d-c) - (1-\eta)d \le 0, & \text{if } b \le 0 \le d \\
x \ge 0
\end{cases}
\end{cases}
$$

$$\tag{3.54}$$

and

$$
\begin{cases}
\max \left[(1-\eta)\frac{c_{14}-\bar{f}_1}{c_{14}-c_{13}} + \eta\frac{c_{12}-\bar{f}_1}{c_{12}-c_{11}}, (1-\eta)\frac{c_{24}-\bar{f}_2}{c_{24}-c_{23}} + \eta\frac{c_{22}-\bar{f}_2}{c_{22}-c_{21}}, \cdots, (1-\eta)\frac{c_{m_14}-\bar{f}_{m_1}}{c_{m_14}-c_{m_13}} \right. \\
\left. +\eta\frac{c_{m_12}-\bar{f}_{m_1}}{c_{m_12}-c_{m_11}} \right] \\
\text{s.t.} \begin{cases}
\alpha_r(d-c) + (1-\eta)c \le 0, & \text{if } c \le 0 \le a \\
\alpha_r(b-a)(d-c) + \eta a(d-c) + (1-\eta)c(b-a) \le 0, & \text{if } a \le 0 \le b \\
\alpha_r(d-c) - (1-\eta)d \le 0, & \text{if } b \le 0 \le d \\
x \ge 0
\end{cases}
\end{cases}
$$

$$\tag{3.55}$$

PROOF   The proof is direct from Theorem 3.3 and the definition of *Appr*.

Similarly, we can obtain the crisp equivalents for the lower level model.

3.5.2.2   Interactive Method

In the section we introduce the interactive method for the crisp equivalent model. The process can be summarized by five steps as follows:

Step 1.   Set $k = 0$, solve the upper-level decision-making problem to obtain a set of preferred solutions that are acceptable to the ULDM; the ULDM then puts the solutions in order in the format as follows:

Preferred solution

$$(x^{(k)}, y^{(k)}), (x^{(k+1)}, y^{(k+1)}), \cdots, (x^{(k+p)}, y^{(k+p)}),$$

Preferred ranking

$$(x^{(k)}, y^{(k)}) \succ (x^{(k+1)}, y^{(k+1)}) \succ \cdots \succ (x^{(k+p)}, y^{(k+p)}).$$

Step 2. Given $x^{(k)}$ to the lower level, solve the lower-level decision-making problem and obtain $\bar{y}^{(k)}$.

Step 3. If $\frac{\|F_0(x^{(k)}, y^{(k)}) - F_0(x^{(k)}, \bar{y}^{(k)})\|}{\|F_0(x^{(k)}, y^{(k)})\|} \leq \sigma$, (where σ is a fairly small positive number) go to Step 4. Otherwise, go to Step 5.

Step 4. If The ULDM is satisfied with $(x^{(k)}, \bar{y}^{(k)})$ and $F_0(x^{(k)}, \bar{y}^{(k)})$, then $(x^{(k)}, \bar{y}^{(k)})$ is the preferred solution to the bilevel decision-making problem. Otherwise, go to Step 5.

Step 5. Let $k = k + 1$, and go to Step 2.

3.5.2.3 Numerical Example

In this section, a numerical example is given to illustrate the effectiveness of the application of the models and algorithms proposed earlier.

Example 3.5

Consider the problem

$$\begin{cases} \max_{x_1, x_2} F_1 = 2\xi_1 x_1 + \xi_2 x_2 + 2\xi_3 y_1 + \xi_4 y_2 \\ \max_{x_1, x_2} F_2 = \xi_1 x_1 + 2\xi_2 x_2 + 2\xi_3 y_1 + 4\xi_4 y_2 \\ \text{where } (y_1, y_2) \text{ solves} \\ \quad \begin{cases} \max_{y_1, y_2} f_1 = \xi_1 x_1 + 2\xi_2 x_2 + \xi_3 y_1 + 2\xi_4 y_2 \\ \max_{y_1, y_2} f_2 = 2\xi_1 x_1 + \xi_2 x_2 + 2\xi_3 y_1 + 2\xi_4 y_2 \\ \quad \text{s.t.} \begin{cases} x_1 + x_2 + y_1 + y_2 \geq 100 \\ x_1 + x_2 + y_1 + y_2 \leq 200 \\ x_1, x_2, y_1, y_2 \geq 10 \end{cases} \end{cases} \end{cases} \quad (3.56)$$

where $\xi_j, j = 1, 2, \cdots, 4$ are rough intervals characterized as

$$\xi_1 = ([2,3], [1,4]), \xi_2 = ([3,4], [2,5]), \xi_3 = ([4,5], [3,6]), \xi_4 = ([5,6], [4,7]).$$

In order to solve it, we use the chance operator to deal with rough objectives and rough constraints, and then we can obtain the BL-DCRM

$$
\begin{cases}
\max\limits_{x_1,x_2}[\alpha_1,\alpha_2] \\
\text{s.t.} \begin{cases} Appr\{2\xi_1 x_1 + \xi_2 x_2 + 2\xi_3 y_1 + \xi_4 y_2 \geq 1500\} \geq \alpha_1 \\ Appr\{\xi_1 x_1 + 2\xi_2 x_2 + 2\xi_3 y_1 + 4\xi_4 y_2 \geq 3000\} \geq \alpha_2 \end{cases} \\
\text{where } (y_1,y_2) \text{ solves} \\
\quad \begin{cases} \max\limits_{y_1,y_2}[\beta_1,\beta_2] \\ \text{s.t.} \begin{cases} Appr\{\xi_1 x_1 + 2\xi_2 x_2 + \xi_3 y_1 + 2\xi_4 y_2 \geq 1800\} \geq \beta_1 \\ Appr\{2\xi_1 x_1 + \xi_2 x_2 + 2\xi_3 y_1 + 2\xi_4 y_2 \geq 1800\} \geq \beta_2 \\ x_1 + x_2 + y_1 + y_2 \geq 100 \\ x_1 + x_2 + y_1 + y_2 \leq 200 \\ x_1, x_2, y_1, y_2 \geq 10 \end{cases} \end{cases}
\end{cases} \quad (3.57)
$$

Here we suppose that $\eta = 1/2$. and by Theorem 3.5 , we know that the problem (3.56) is equivalent to model

$$
\begin{cases}
\max\limits_{x_1,x_2}\left[\dfrac{8x_1+5x_2+12y_1+7y_2-1500}{6(2x_1+x_2+2y_1+y_2)}, \dfrac{2x_1+5x_2+6_1+14y_2-1500}{3(x_1+2x_2+2y_1+4y_2)}\right] \\
\text{where } (y_1,y_2,y_3) \text{ solves} \\
\quad \begin{cases} \max\limits_{y_1,y_2}\left[\dfrac{4x_1+5x_2+12y_1+14y_2-1800}{6(x_1+2x_2+3y_1+2y_2)}, \dfrac{8x_1+5x_2+12y_1+14y_2-1800}{6(2x_1+x_2+2y_1+2y_2)}\right] \\ \text{s.t.} \begin{cases} x_1 + x_2 + y_1 + y_2 \geq 100 \\ x_1 + x_2 + y_1 + y_2 \leq 200 \\ x_1, x_2, y_1, y_2 \geq 10 \end{cases} \end{cases}
\end{cases} \quad (3.58)
$$

After running of interactive method, we obtain $(x_1^*, x_2^*, y_1^*, y_2^*,) = (10.3, 6.7, 12, 4, 21.5, 23.8, 15, 3.8)$ and $(\alpha_1^*, \alpha_2^*) = (0.72, 0.93), (\beta_1^*, \beta_2^*) = (0.69, 0.89)$.

3.5.3   BL-NLDCRM and Rough Simulation-Based RTS

It is difficult to transform the bilevel nonlinear dependent-chance rough model (BL-NLDCRM) into its equivalent form, so we introduce the rough simulation-based reactive tabu search (RTS) algorithm to solve it.

The RTS is an improved version of TS. It was first proposed by R. Battiti and G. Tecchiolli [22]. The tabu list length of RTS can be self-adapted to balance intensification and diversification. Furthermore, an escape mechanism is introduced to avoid recycling. The RTS method has been successfully applied in many fields [21, 35, 46, 61, 222, 231].

The RTS algorithm goes further in the direction of robustness by proposing a simple mechanism for adapting the list size to the properties of the optimization problem. The configurations visited during the search and the corresponding iteration numbers are stored in memory so that, after the last movement is chosen, one can check for the repetition of configurations and calculate the interval between two visits. The basic fast reaction mechanism increases the list size when configurations are repeated. This is accompanied by a slower reduction mechanism so that the size is reduced in regions of the search space that do not need large sizes.

The RTS algorithm uses the reactive mechanism to adjust the length of the tabu list as well as balance the centralized strengthening search strategy and decentralized diversification search strategy. RTS involves increasing the adjustment coefficient ($N_{IN} > 1$) and decreasing the adjustment coefficient ($0 < N_{DE} < 1$). In the searching process, all the solutions visited are stored. As operating a step of move, the current solution is checked to see if it has been visited. If it has been visited, which shows that it is entering a cycle, then the length of tabu list become N_{IN} times of the original length. If there are no repeated solutions after several iterations, then the length of tabu list become N_{DE} times the original length.

In order to cycle, RTS presents escape mechanism. In the searching process, when repeated times of a large number of repeated solutions exceed the given times R_{EP}, the escape mechanism is activated. The escape strategy is based on the execution of a series of random exchanges. Their number is random. To avoid an immediate return to the old region of the search space, all random steps executed are made tabu.

The tabu search algorithm uses historical memory optimization, and search optimization with the tabu list, which combined with the level of desire of the system, is achieved through an intensive search and distributed search for the balance of diversification. The RTS uses active feedback strategies and uses the escape mechanism to strengthen the balance. Thus, in theory, it is superior than the general tabu search algorithm better, search for higher quality.

The key idea of RTS is feedback strategies and escape mechanism. In practical application, there are many ways to achieve these. For example, then length of tabu list become the original list *num − dec* times if there is no repeated solutions in the *num − dec* iteration. The escape mechanism is implemented if the times of repetition achieve *num − esc*.

The basic procedures of RTS are summarized as follows:

Step 1. Initialize two counters: *num − dec = num − esc = 0*.

Step 2. Initialize other parameters, and present the initial solution.

Step 3. Propose candidate solutions set according to the current solution.

Step 4. According to the tabu list situation and desired level, select a solution as the initial solution for next iteration update record list (including the tabu list and all the normal solutions visited).

Step 5. If the selected solution occurred earlier, then the length of the tabu list $t = tN_{IN}, n − esc = n − esc + 1, n − dec = 0$; otherwise $n − dec = n − dec + 1$.

Step 6. If $n − dec = num − dec$, then $t = tN_{DE}, n − dec = 0$.

Step 7. If $n − esc = num − esc$, then implement escape mechanism, $n − esc = 0, n − dec = 0$.

A flowchart is shown in Figure 3.3.

3.5.3.1 Numerical Example

In this section, a numerical example is given to illustrate the effectiveness of the application of the models and algorithms proposed earlier.

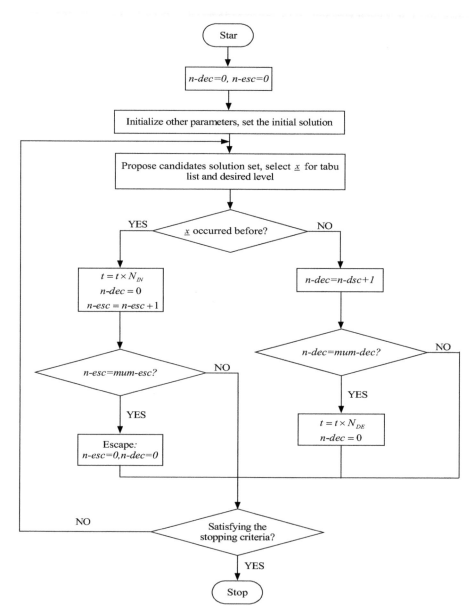

FIGURE 3.3 Flowchart of RTS algorithm.

Example 3.6

Consider the problem

$$
\begin{cases}
\max\limits_{x_1,x_2,x_3} F_1 = 2\xi_1^2 x_1 + \sqrt{\xi_2 x_2 + \xi_3 x_3} + \xi_4 y_1 + \sqrt{\xi_5} y_2 + \sqrt{\xi_6} y_3 \\
\max\limits_{x_1,x_2,x_3} F_2 = \xi_1^2 x_1 + \xi_2^2 x_2 + x_3 + y_1 + y_2 + y_3 \\
\text{where } (y_1, y_2, y_3) \text{ solves} \\
\quad \begin{cases}
\max\limits_{y_1,y_2,y_3} f_1 = \sqrt{7\xi_1 x_1 + 3\xi_2 x_2 + 8\xi_3 x_3} + (12\xi_4 y_1 + 11\xi_5 y_2 + 5\xi_6 y_3)^2 \\
\max\limits_{y_1,y_2,y_3} f_2 = x_1 + 5x_2 + 7x_3 + y_1 + 11y_2 - 8\xi_6^3 y_3 \\
\quad \text{s.t.} \begin{cases}
\sqrt{x_1 \xi_1 + x_2 \xi_2} + x_3^2 \xi_3 + \xi_4 y_4 + 17y_3 \xi_6 \leq 1540 \\
3x_1 + 6x_+ 12x_9 + 14x_{10} + 19x_{11} \leq 2400 \\
x_1 + x_2 + x_3 + y_1 + y_2 - y_3 \geq 150 \\
x_1 + x_2 + x_3 + y_1 + y_2 + y_3 \leq 200 \\
x_1, x_2, x_3, y_1, y_2, y_3 \geq 0
\end{cases}
\end{cases}
\end{cases} \tag{3.59}
$$

where $\xi_j, j = 1, 2, \cdots, 6$ are rough intervals characterized as

$$
\begin{aligned}
&\xi_1 = ([8, 10], [7, 11]), \quad \xi_2 = ([9, 11], [8, 12]), \quad \xi_3 = ([10, 12], [9, 13]), \\
&\xi_4 = ([11, 13], [10, 14]), \quad \xi_5 = ([12, 14], [11, 15]), \quad \xi_6 = ([13, 15], [12, 16]).
\end{aligned}
$$

In order to solve it, we use the *Appr* operator to deal with rough objectives and rough constraints, and then we can obtain the model

$$
\begin{cases}
\max\limits_{x_1,x_2,x_3} [\alpha_1, \alpha_2] \\
\text{s.t.} \begin{cases}
Appr\{\xi_1^2 x_1 + \sqrt{\xi_2 x_2 + \xi_3 x_3} + \xi_4 y_1 + \sqrt{\xi_5} y_2 + \sqrt{\xi_6} y_3 \geq \bar{F}_1\} \geq \alpha_1 \\
Appr\{x_1 + 5x_2 + 7x_3 + y_1 + 11y_2 - 8\xi_6^3 y_3 \geq 680\} \geq \alpha_2
\end{cases} \\
\text{where } (y_1, y_2, y_3) \text{ solves} \\
\quad \begin{cases}
\max\limits_{y_1,y_2,y_3} [\beta_1, \beta_2] \\
\quad \text{s.t.} \begin{cases}
Appr\{\sqrt{7\xi_1 x_1 + 3\xi_2 x_2 + 8\xi_3 x_3} + (12\xi_4 y_1 + 11\xi_5 y_2 + 5\xi_6 y_3)^2 \\
\geq 460\} \geq \beta_1 \\
Appr\{x_1 + 5x_2 + 7x_3 + y_1 + 11y_2 - 8\xi_6^3 y_3 \geq 390\} \geq \beta_2 \\
Appr\{\sqrt{x_1 \xi_1 + x_2 \xi_2} + x_3^2 \xi_3 + \xi_4 y_4 + 17y_3 \xi_6 \leq 1400\} \geq 0.6 \\
3x_1 + 6x_+ 12x_2 + 14x_3 + 9y_1 + 20y_1 + 15y_2 + 10y_3 \leq 2400 \\
x_1 + x_2 + x_3 + y_1 + y_2 - y_3 \geq 150 \\
x_1 + x_2 + x_3 + y_1 + y_2 + y_3 \leq 200 \\
x_1, x_2, x_3, y_1, y_2, y_3 \geq 0
\end{cases}
\end{cases}
\end{cases} \tag{3.60}
$$

Here we suppose that $\eta = 1/2$. After running the rough simulate-based TS, we obtain $(x_1^*, x_2^*, x_3^*, y_1^*, y_2^*, y_3^*) = (10.3, 26.7, 12.4, 20.5, 10.7, 0)$ and $(\alpha_1^*, \alpha_2^*) = (0.56, 0.70)$, $(\beta_1^*, \beta_2^*) = (0.67, 0.49)$.

3.6   Water Resource Management of Gan-Fu Plain

A practical application of regional water allocation is presented in this section. In this chapter, we take the irrigation district of Gan-Fu plain an example to explain the bilevel model with rough interval parameters and use the methods presented in the proceding sections to solve the proposed model.

3.6.1   Background Introduction

Efficiency in regional water planning is very important for national development. Attention should therefore be paid to effectiveness of planning in practice. In regional water planning, there is always more than one sub-area water managers, and they cannot compromise each other. At the same time, water managers must face an uncertain environment in the course of their work. All of these factors should be considered.

This section considers the design of regional water planning among subareas beside the same river basin. Therefore, regional water planning under market mechanisms, that is, the amount of water resources allocation among different subareas is no longer directly determined by the water resources scheduler of the river basin. Each subarea is an independent decision maker, and the regional water resources agency distributes the initial water rights to the various subareas. After obtaining the initial rights, each subarea water manager makes watertrading decisions based on water use and water allocation to promote equitable cooperation in a river basin and to achieve the efficient use of water. Also, water users pay for the allocated water. Therefore, the regional water resources agency influences the decision making of the water manager of each subarea through adjusting the initial water rights, while the water managers of the subareas intend to meet their individual economic benefit goals through making rational decisions about the water allocated to the water user on the basis of the decision making by the regional water resources agency. Basin water resources are mainly used as ecological water, agricultural water, industrial water, and domestic water, with ecological water taking priority. Thus, the regional water resources agency distributes certain public water rights as the ecological water to ensure the sustainable development of the river basin.

Therefore, the regional water resources optimization problems in this book can be abstracted as a bilevel programming problem, with the upper-level decision maker being the regional water resources agency who makes decision regarding the initial water rights to subareas to maximize the total social benefits and the lower-level decision makers are the subarea water managers, who decide the amount of allocated water to maximize their own economic benefits. A model structure of the regional water allocation problem can be seen in Figure 3.1.

3.6.2　Modeling and Analysis

The problem under consideration is how to make decisions about regional water resources allocation to maximize the total benefit to the society and the economic benefit to each subarea over the planning horizon. A bilevel decision-making model for regional water planning, considering roughness, will be constructed in this section, starting with the mathematical description of the problem.

Before constructing the model for regional water resources allocation, the following assumptions are taken account:

1. Available water for distribution is provided by a single river basin.

2. The total amount of available water for distribution in the basin, actual water demand of user in each subarea, and the proportion of wastewater water user in each subarea are known.

3. The decision maker at the upper level assumes that the rational responses of the decision makers at the lower level with respect to the decision he or she makes are a Nash Equilibrium Solution, and the decision maker at the upper level selects a decision that optimizes his or her objective function.

4. There is perfect exchange of information between all the participants such that the objectives and constraints are known.

5. Actual water demand of user is not less than the water allocated to user. This part of the water shortage can be dealt with saving water and by improving water efficiency. When the allocated water to the subarea is more than its initial water right, it needs to buy water from the water market. Conversely, it can sell excess water in the water market.

6. Water demand of user are considered as rough intervals.

The following notations will be used in the process of constructing the model.

Indices
i　subarea in the river basin, $i \in \Psi = \{1, 2, \cdots, n\}$
k　water user in the subarea, $k \in \Phi = \{1, 2, \cdots, m\}$

Functions
$h(w)$　ecological benefit function of the river basin
$p(r)$　water transaction price function, and r represents the total available water
　　　for trading in the water market

Certain parameters
Q　　　total amount of available water for distribution in basin ($10^6 \ m^3$)
w　　　public water right, which ensures ecological water ($10^6 \ m^3$)
η　　　minimum ecological water requirement in the river basin ($10^6 \ m^3$)
θ_i　　　minimum amount that should be allocated to subarea i ($10^6 \ m^3$)
$T_{ik\min}$　minimum water requirements of user k in subarea i($10^6 \ m^3$)
$T_{ik\max}$　maximum targets of taken water of user k in subarea i ($10^6 \ m^3$)
κ, δ　water transaction price function coefficients
λ_{ik}　　wastewater emission proportion of water user k in subarea i
q_i　　　maximum amount of wastewater in subarea i ($10^6 \ m^3$)

p_w cost parameter of treating waste water (CNY/m^3)

p'_k the charge of per unit of water resources (CNY/m^3)

b_{ik} benefit parameter for user k in subarea i per unit of water allocated (CNY/m^3)

c_{ik} the expenditure of saving per unit of water to user k in sub-area i (CNY/m^3)

e ecoefficiency parameter for the whole basin per unit of public water supply (CNY/m^3)

Rough interval parameters

$\tilde{d}_{ik}$ water demand of user k in subarea i (10^6 m^3), which presents in terms of roughness

Decision variables x_i initial water right of subarea i (10^6 m^3), which is one of the resource decision variables of the upper decision maker

y_{ik} the allocated water to user k in subarea i (10^6 m^3), which is the resource decision variable of the lower decision makers

3.6.2.1   Model Formulation

The lower-level model for water allocation problem in subareas. Lower levels can be discussed first. From the preceding description, it is known that the lower level model is to maximize the economic benefits of water allocation in each subarea. When the mathematical description of the lower level is founded, it is easy to give the mathematical description of the upper level.

For the lower level of the bilevel decision making model, first, the water manager of each subarea distributes water to the municipal water user, industrial water user, and agricultural water user, which can generate economic benefits: $\sum_{k=1}^{m} b_{ik} y_{ik}$ for $i = 1, 2, \cdots, n$.

However, the promised water cannot be delivered due to insufficient supply sometimes, so it leads to the acquisition of water from higher-priced alternatives and/or negative consequences generated from the curbing of regional development plans (D. Loucks et al. [199], Y. Li et al. [192]). For example, municipal residents may have to curtail the watering of lawns, industries may have to reduce production levels or increase water recycling rates, and farmers may not be able to conduct irrigation as planned. These actions will result in increased costs or decreased benefits for regional development, that is, $\sum_{k=i}^{m} c_{ik}(\tilde{d}_{ik} - y_{ik})$ for $i = 1, 2, \cdots, n$.

As the list of mentioned in assumptions, when the allocated water to the subarea is more than its initial water right, the shortfall needs to be bought from the water market. Conversely, when its initial water rights exceed the water allocated to the subarea, the water manager in each subarea can sell the excess water in the water market, which can generate the transaction price $(x_{ik} - \sum_{k=1}^{m} y_{ik})p(r)$, $\forall i \in \Psi$, where $\Psi = \{1, 2, \cdots, n\}$.

Water managers in each subarea need to pay water charges to get water from the regional water resources agency, and this part of the cost can be denoted as $p'_k \sum_{k=1}^{m} y_{ik}$

for $i = 1, 2, \cdots, n$.

What is more, the cost of treating wastewater is also considered in the objective function. Different water users in different subareas discharge different proportions of wastewater. so the cost can be given as

$$\sum_{k=1}^{m} p_w \lambda_{ik} y_{ik}$$

for $i = 1, 2, \cdots, n$.

The lower objective function is to measure and maximize the economic benefit of each subarea, Thus benefits should be maximized and costs can be minimized, and so the objective function can be described as follows:

$$
\begin{aligned}
V_i &= \sum_{k=1}^{m} b_{ik} y_{ik} - \sum_{k=1}^{m} c_{ik}(\tilde{d}_{ik} - y_{ik}) - p'_k \sum_{k=1}^{m} y_{ik} - \sum_{k=1}^{m} p_w \lambda_{ik} y_{ik} + (x_i - \sum_{k=1}^{m} y_{ik}) p(r) \\
&= \sum_{k=1}^{m} \left(b_{ik} y_{ik} - c_{ik}(\tilde{d}_{ik} - y_{ik}) - p'_k y_{ik} - p_w \lambda_{ik} y_{ik} \right) + (x_i - \sum_{k=1}^{m} y_{ik}) p(r)
\end{aligned}
\tag{3.61}
$$

for $i = 1, 2, \cdots, n$.

The constraints of the lower-level problem is comprised of four parts: constraints on water demand, constraints on water transportation, water transaction price function, and nonnegative constraints. The details of the four parts are explained as follows:

Constraints on water demand. The allocated water to each user must satisfy the users' minimum requirements, and not exceed the maximum allocation targets. So the constraints are given as (3.62), which ensures that none of the constraints are violated.

$$T_{ik\,\min} \le y_{ik} \le T_{ik\,\max} \tag{3.62}$$

for $i = 1, 2, \cdots, n$.

Constraints on water transportation. As is generally known, any production activities can produce different degrees of wastewater. Thus the wastewater emission proportion of water user k in subarea i is assumed to be λ_{ik}. From the perspective of environmental protection, the wastewater of each subarea cannot exceed the maximum amount. Thus the following constraints can be employed:

$$\sum_{k=1}^{m} \lambda_{ik} y_{ik} \le q_i \tag{3.63}$$

for $i = 1, 2, \cdots, n$.

Water transaction price function. In water allocation under the water market, water trade price is the basic term. Although the water market is not a completely competitive market, the market prices are still influenced by the relation of supply and demand, namely, the price rises when demand is more than the supply, and conversely, the price drops. According to the theoretical analysis of the oligarch competition model, the water rights transaction price function in the water market can be expressed as $p(r) = \kappa - \delta r (a > 0, b > 0)$, where r represents the total available water for trading in the water market, which is denoted by $\sum\limits_{i=1}^{n}(x_i - \sum\limits_{k=1}^{m} y_{ik})$. So the water transaction price function can be obtained as

$$p(r) = \kappa - \delta \left(\sum_{i=1}^{n} \left(x_i - \sum_{k=1}^{m} y_{ik} \right) \right), \quad \kappa > 0, \quad \delta > 0 \tag{3.64}$$

Non-negative constraints. Additionally, in order to describe some practical non-negative variables in the lower-level model, the logical constraints are presented as follows:

$$y_{ik} \geq 0 \tag{3.65}$$

for $i = 1, 2, \cdots, n$ and $k = 1, 2, \cdots, m$.

The upper-level model for the water allocation problem in the overall region. After the mathematical description of the lower level model is given, we can now find the objective function of the upper level.

The upper objective function defines the total benefits to society of water resources allocation. The aim of the regional water resources agency is to maximize the total benefits, which are comprised of three components: the ecological benefit, the charge of water taken from the subareas and the complete economic benefit of the water output. If f_0 is denoted as an ecological benefit, we can have $f_0 = h(w) + \sum\limits_{i=1}^{n} p'_k \sum\limits_{k=1}^{m} y_{ik}$, where $h(w) = ew$, which means the ecological benefit function of the river basin. So, the upper objective function can be represented by $V_0 = h(w) + \sum\limits_{i=1}^{n} \sum\limits_{k=1}^{m} p'_k y_{ik} + (\sum\limits_{i=1}^{n} V_i - \sum\limits_{i=1}^{n} \sum\limits_{k=1}^{m} p'_k y_{ik})$. Then the upper objective can be transformed into

$$V_0 = ew + \sum_{i=1}^{n} \left[\sum_{k=1}^{m} \left(b_{ik} y_{ik} - c_{ik}(\tilde{d}_{ik} - y_{ik}) + p'_k y_{ik} - p_w \lambda_{ik} y_{ik} \right) + \left(x_i - \sum_{k=1}^{m} y_{ik} \right) p(r) \right] \tag{3.66}$$

In regard to the constraints of the upper level, water supply should be limited. First the total amount of available water for distribution in the river basin has a capacity limit and the actual water allocated to the users must not exceed this capacity. Further, the initial water rights of water users, and public water rights should be equal to the total amount of available water for allocation. Thus we need a restriction on capacity as

$$\sum_{i=1}^{n} x_i + w = Q, \quad \sum_{i=1}^{n} \sum_{k=1}^{m} y_{ik} + w \leq Q \tag{3.67}$$

As mentioned above, if the ecological water requirements cannot be met, the ecological environment will be destroyed as this does not conform to the requirement of sustainable development. So public water rights w should not be lower than the minimum ecological water requirement α in the river basin, and therefore, we have the following inequalities:

$$w \geq \eta \ (\eta > 0) \tag{3.68}$$

The initial water rights x_i must ensure the lowest amount of water supply β_i in each subarea, and thus we have

$$x_i \geq \theta_i \ (\theta_i > 0) \tag{3.69}$$

for $i = 1, 2, \cdots, n$.

The regional water resources agency influences the decision variables y_{ik} of water managers in subareas through its control of the decision variables, namely, the initial water right x_i, which indicates that it is appropriate to use the bilevel programming model to describe the hierarchical relationships between the water resources agency and the water managers of subareas. Based on the description and notations above, by the integration of (3.61) $\sim$ (3.69), the following global model of bilevel programming can now be formulated to pose the regional water allocation problem:

$$
\begin{cases}
\max\limits_{x_i} V_0 = ew + \sum\limits_{i=1}^{n}\left[\sum\limits_{k=1}^{m}\left(b_{ik}y_{ik} - c_{ik}(\tilde{d}_{ik} - y_{ik}) + p_k' y_{ik} - p_w \lambda_{ik} y_{ik}\right) + \left(x_i - \sum\limits_{k=1}^{m} y_{ik}\right)p(r)\right] \\
\text{s.t.} \begin{cases}
\sum\limits_{i=1}^{n} x_i + w = Q \\
x_i \geq \theta_i \ (\theta_i > 0) \\
w \geq \eta \ (\eta > 0) \\
\max\limits_{y_{ik}} V_i = \sum\limits_{k=1}^{m}\left(b_{ik}y_{ik} - c_{ik}(\tilde{d}_{ik} - y_{ik}) - p_k' y_{ik} - p_w \lambda_{ik} y_{ik}\right) + \left(x_i - \sum\limits_{k=1}^{m} y_{ik}\right)p(r) \\
\text{s.t.} \begin{cases}
\sum\limits_{i=1}^{n}\sum\limits_{k=1}^{m} y_{ik} + w \leq Q \\
T_{ik\min} \leq y_{ik} \leq T_{ik\max} \\
\sum\limits_{k=1}^{m} \lambda_{ik} y_{ik} \leq q_i \\
p(r) = \kappa - \delta\left(\sum\limits_{i=1}^{n}\left(x_i - \sum\limits_{k=1}^{m} y_{ik}\right)\right) \ (\kappa > 0, \delta > 0) \\
y_{ik} \geq 0 \\
\forall i \in \Psi, \, k \in \Phi
\end{cases}
\end{cases}
\end{cases}
\tag{3.70}
$$

where $\Psi = \{1, 2, \cdots, n\}$, $\Phi = \{1, 2, \cdots, m\}$.

Because the lower decision maker, that is, the water manager in subarea i, chooses the value of y_{ik} after the value of x_i determined by the upper decision maker–the regional water agency the values of y_{ik} are not determined yet when the agency chooses the values of x_i. Thus, it follows that the leader determines the values of x_i on the assumption that the lower decision maker will choose the rational reaction, that is, an optimal solution to the linear programming problem with the fixed parameters x_i.

3.6.2.2 Equivalent Crisp Model

As generally known, it is hard to obtain optimal results in decision making when involving roughness, so, the rough interval parameters need to be transformed into deterministic ones. We take the expected value operator to convert the uncertain model with rough interval parameters into the crisp one in this section.

By taking the expected value operator, model (3.70) can be rewritten as

$$
\begin{cases}
\max\limits_{x_i} E[V_0] = E\left[ew + \sum\limits_{i=1}^{n}\left[\sum\limits_{k=1}^{m}\left(b_{ik}y_{ik} - c_{ik}(\tilde{d}_{ik} - y_{ik}) + p'_k y_{ik} - p_w \lambda_{ik} y_{ik}\right)\right.\right. \\
\left.\left. + (x_i - \sum\limits_{k=1}^{m} y_{ik})p(r)\right]\right] \\
\text{s.t.}
\begin{cases}
\sum\limits_{i=1}^{n} x_i + w = Q \\
x_i \geq \theta_i \ (\theta_i > 0) \\
w \geq \eta \ (\eta > 0) \\
\max\limits_{y_{ik}} E[V_i] = \sum\limits_{k=1}^{m}\left(b_{ik}y_{ik} - c_{ik}(\tilde{d}_{ik} - y_{ik}) - p'_k y_{ik} - p_w \lambda_{ik} y_{ik}\right) + (x_i - \sum\limits_{k=1}^{m} y_{ik})p(r) \\
\text{s.t.}
\begin{cases}
\sum\limits_{i=1}^{n}\sum\limits_{k=1}^{m} y_{ik} + w \leq Q \\
T_{ik\min} \leq y_{ik} \leq T_{ik\max} \\
\sum\limits_{k=1}^{m} \lambda_{ik} y_{ik} \leq q_i \\
p(r) = \kappa - \delta\left(\sum\limits_{i=1}^{n}\left(x_i - \sum\limits_{k=1}^{m} y_{ik}\right)\right) \ (\kappa > 0, \delta > 0) \\
y_{ik} \geq 0
\end{cases} \\
\forall i \in \Psi, k \in \Phi
\end{cases}
\end{cases}
$$

(3.71)

where $\Psi = \{1, 2, \cdots, n\}$, $\Phi = \{1, 2, \cdots, m\}$.

Furthermore, model (3.71) is equivalent to

$$
\begin{cases}
\max\limits_{x_i} E[V_0] = ew + \sum\limits_{i=1}^{n}\left[\sum\limits_{k=1}^{m}\left(b_{ik}y_{ik} - c_{ik}\left(\frac{d_{ik,1} + \bar{d}_{ik,1} + d_{ik,3} + \bar{d}_{ik,4}}{4} - y_{ik}\right) + p'_k y_{ik} - p_w \lambda_{ik} y_{ik}\right)\right. \\
\left. + (x_i - \sum\limits_{k=1}^{m} y_{ik})p(r)\right] \\
\text{s.t.}
\begin{cases}
\sum\limits_{i=1}^{n} x_i + w = Q \\
x_i \geq \theta_i \ (\theta_i > 0) \\
w \geq \eta \ (\eta > 0) \\
\max\limits_{y_{ik}} E[V_i] = E\left[\sum\limits_{k=1}^{m}\left(b_{ik}y_{ik} - c_{ik}\left(\frac{d_{ik,1} + \bar{d}_{ik,1} + d_{ik,3} + \bar{d}_{ik,4}}{4} - y_{ik}\right) - p'_k y_{ik} - p_w \lambda_{ik} y_{ik}\right)\right. \\
\left. + (x_i - \sum\limits_{k=1}^{m} y_{ik})p(r)\right] \\
\text{s.t.}
\begin{cases}
\sum\limits_{i=1}^{n}\sum\limits_{k=1}^{m} y_{ik} + w \leq Q \\
T_{ik\min} \leq y_{ik} \leq T_{ik\max} \\
\sum\limits_{k=1}^{m} \lambda_{ik} y_{ik} \leq q_i \\
p(r) = \kappa - \delta\left(\sum\limits_{i=1}^{n}\left(x_i - \sum\limits_{k=1}^{m} y_{ik}\right)\right) \ (\kappa > 0, \delta > 0) \\
y_{ik} \geq 0
\end{cases} \\
\forall i \in \Psi, k \in \Phi
\end{cases}
\end{cases}
$$

(3.72)

where $\tilde{d}_{ik} = ([\underline{d}_{ik,1}, \bar{d}_{ik,1}], [\underline{d}_{ik,2}, \bar{d}_{ik,2}]), \Psi = \{1, 2, \cdots, n\}, \Phi = \{1, 2, \cdots, m\}$. Now we have transformed the bilevel rough decision-making problem into a deterministic one. Some usual techniques can be applied to solve it.

3.6.2.3 Solution Approach

In order to handle the proceeding crisp bilevel problem, the interactive programming technique proposed by M. Sakawa [259], M. Sakawa et al. [261], S. Mishra and A. Ghosh [215], and G. Wang et al. [293] is applied to solve the nonlinear bilevel programming when the decision maker cannot decide which items are more important than others. In the present section, we deal with bi-level multifollowers programming (BLMFP) and take problem (3.72) as an example to illustrate this method.

It is natural that we take the uncertain objective function to evaluate the DM's imprecise consideration. For the objective function of each level in problem (3.72), DM has fuzzy goals such as "the goal should be more than or equal to a certain value." Suppose that the DM employs the linear membership function of the fuzzy goal whose parameters are determined by H. Zimmermann [346]. Therefore, we can denote the maximum and minimum values of each objective function, respectively, as follows:

$$V_j^{\max} = \max_{\mathbf{x} \in X, \mathbf{y} \in Y} E[V_j(\mathbf{x}, \mathbf{y})] \quad \forall j \in \Omega,$$

$$V_j^{\min} = \min_{\mathbf{x}^{k0} \in X, \mathbf{y}^{k0} \in Y} E[V_j(\mathbf{x}^{k0}, \mathbf{y}^{k0})] \quad \forall j, k \in \Omega,$$

where $j \neq k$, $\Omega = \{0, 1, \cdots, n\}$,

$$\mathbf{x} = (x_1, x_2, \cdots, x_n), \quad \mathbf{y} = \begin{bmatrix} y_{11}, y_{12}, \cdots, y_{1m} \\ y_{21}, y_{22}, \cdots, y_{2m} \\ \cdots, \cdots, \cdots \\ y_{n1}, y_{n2}, \cdots, y_{nm} \end{bmatrix}$$

where $(\mathbf{x}, \mathbf{y})$ is a feasible solution maximizing $E[V_j(\mathbf{x}, \mathbf{y})]$, and $(\mathbf{x}^{k0}, \mathbf{y}^{k0})$ is a feasible solution maximizing $E[V_k(\mathbf{x}, \mathbf{y})]$, which can be obtained by using standard linear programming techniques.

The function $\mu_i(E[V_i](\mathbf{x}, \mathbf{y}))$ strictly vary between $V_i^{\max}$ and $V_i^{\min}$. For the sake of simplicity, we can take the linear function to characterize the goal at each level. They can be defined as follows (see Figure 3.4):

$$\mu_i(E[V_i(\mathbf{x}, \mathbf{y})]) = \begin{cases} 1, & \text{if } E[V_i(\mathbf{x}, \mathbf{y})] > V_i^{\max} \\ \frac{E[V_i(\mathbf{x}, \mathbf{y})] - V_i^{\min}}{V_i^{\max} - V_i^{\min}}, & \text{if } V_i^{\min} < E[V_i(\mathbf{x}, \mathbf{y})] \leq V_i^{\max} \\ 0, & \text{if } V_i(\mathbf{x}, \mathbf{y}) \leq V_i^{\min} \end{cases} \quad (3.73)$$

for $i = 0, 1, \cdots, n$.

After eliciting the membership functions, the leader specifies the minimum satisfactory level $\lambda_0 \in [0, 1]$. To obtain the overall satisfactory optimal solution to both

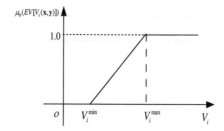

FIGURE 3.4   Linear function of $E[V_i]$.

levels, the upper decision maker needs to compromise with the lower decision maker to find a solution, which can be generated by solving the following problem:

$$\begin{cases} \min t \\ \text{s.t.} \begin{cases} \mu_0(E[V_0(\mathbf{x},\mathbf{y})]) \geq \lambda_0 \\ \lambda_i - \mu_i(E[V_i(\mathbf{x},\mathbf{y})]) \leq t \\ (\mathbf{x},\mathbf{y}) \in S \end{cases} \end{cases} \qquad (3.74)$$

for $i = 1, 2, \cdots, n$, $\lambda_i \in [0,1]$ is the minimal acceptable reference level specified by the ith follower for its membership function. By the definition of the membership function (3.73), problem (3.72) can be written as the following equivalent programming problem:

$$\begin{cases} \min t \\ \text{s.t.} \begin{cases} \frac{E[V_0(\mathbf{x},\mathbf{y})] - V_0^{\min}}{V_0^{\max} - V_0^{\min}} \geq \lambda_0 \\ \lambda_i - \frac{E[V_i(\mathbf{x},\mathbf{y})] - V_i^{\min}}{V_i^{\max} - V_i^{\min}} \leq t \\ \sum\limits_{i=1}^{n} x_i + w = Q \\ x_i \geq \theta_i \ (\theta_i > 0) \\ w \geq \eta \ (\eta > 0) \\ \sum\limits_{i=1}^{n} \sum\limits_{k=1}^{m} y_{ik} + w \leq Q \\ T_{ik\min} \leq y_{ik} \leq T_{ik\max} \\ \sum\limits_{k=1}^{m} \lambda_{ik} y_{ik} \leq q_i \\ p(r) = \kappa - \delta \left(\sum\limits_{i=1}^{n} \left(x_i - \sum\limits_{k=1}^{m} y_{ik} \right) \right) \ (\kappa > 0, \delta > 0) \\ y_{ik} \geq 0 \\ \forall i \in \Psi, k \in \Phi \end{cases} \end{cases} \qquad (3.75)$$

where $\Psi = \{1, 2, \cdots, n\}$, $\Phi = \{1, 2, \cdots, m\}$.

By solving problem (3.75), we can obtain the overall satisfactory solution to both levels.

THEOREM 3.6

If problem (3.75) has optimal solution $(\mathbf{x}^*, \mathbf{y}^*, t^*)$, *and* $t^* = 0$, *then* $(\mathbf{x}^*, \mathbf{y}^*, t^*)$ *is a Pareto-optimal solution to (3.72).*

PROOF Consider $(\mathbf{x}, \mathbf{y}, t)$ as deciding variables. Let $\mathbf{X}^* = (\mathbf{x}^*, \mathbf{y}^*, t^*)$ be an optimal solution to problem (3.75) and $t^* = 0$. Apparently, X^* is subject to all constraints. Assume that $X^* = (\mathbf{x}^*, \mathbf{y}^*, t^*)$ is not an Pareto optimal solution to (3.72). According to the definition of the Pareto optimal solution, there exists a solution $(\mathbf{x}', \mathbf{y}') \neq (\mathbf{x}^*, \mathbf{y}^*)$ such that $E[V_i(\mathbf{x}', \mathbf{y}')] \geq E[V_i(\mathbf{x}^*, \mathbf{y}^*)]$ and at least one inequality holds strictly, then

$$
\frac{E[V_i(\mathbf{x}',\mathbf{y}')]-V_i^{\min}}{V_i^{\max}-V_i^{\min}} = \mu_i(E[V_i(\mathbf{x}',\mathbf{y}')]) \geq \mu_i(E[V_i(\mathbf{x}^*,\mathbf{y}^*)])
$$
$$
= \frac{E[V_i(\mathbf{x}^*,\mathbf{y}^*)]-V_i^{\min}}{V_i^{\max}-V_i^{\min}} \tag{3.76}
$$

for $i = 1, 2, \cdots, n$.

Since $\mathbf{X}^* = (\mathbf{x}^*, \mathbf{y}^*, t^*)$ is an optimal solution to problem (3.72), then $\mathbf{X}^* = (\mathbf{x}^*, \mathbf{y}^*, t^*)$ satisfy $\lambda_i - \mu_i(E[V_i(\mathbf{x}^*, \mathbf{y}^*)]) \leq t^*$. Moreover,

$$
\lambda_i - \mu_i(E[V_i(\mathbf{x}',\mathbf{y}')]) \leq \lambda_i - \mu_i(E[V_i(\mathbf{x}^*,\mathbf{y}^*)]) \leq t^* = 0 \tag{3.77}
$$

for $i = 1, 2, \cdots, n$.

$(\mathbf{x}', \mathbf{y}', t')$ is the feasible solution to problem (3.75) by choosing $t' = \max\{\lambda_i - \mu_i(E[V_i(\mathbf{x}', \mathbf{y}')])\}$, $i = 1, 2 \cdots, n$. By (3.74) and (3.76), we know that the optimal solution t' for variables $\mathbf{x}'$, $\mathbf{y}'$ must be subject to $t' \leq t^*$. It shows a conflict that $X^* = (\mathbf{x}^*, \mathbf{y}^*, t^*)$ is the optimal solution too problem (3.75). That is, $(\mathbf{x}^*, \mathbf{y}^*)$ is a Pareto optimal solution to (3.72).

After proving the above theorem, we know that if problem (3.75) has an optimal solution, the bilevel programming must have optimal solutions.

To take account of the overall satisfactory balance between the leader and the followers for the decision makers, subjectively specify the membership function and the minimum satisfactory degrees. The following ratio of satisfactory degree of the upper level to that of the lower level is useful:

$$
\delta_i = \frac{\mu_i(E[V_i(\mathbf{x},\mathbf{y})])}{\mu_0(E[V_0(\mathbf{x},\mathbf{y})])} \tag{3.78}
$$

for $i = 1, 2, \cdots, n$.

Next, the balance of satisfactory degrees between both levels is given as follows:

1. The lower bound of δ_i:

$$
\delta_l = \frac{\min\limits_{i=1,\cdots,n} \mu_i(E[V_i(\mathbf{x},\mathbf{y})])}{\mu_0(E[V_0(\mathbf{x},\mathbf{y})])}.
$$

2. The upper bound of δ_i:

$$\delta_h = \frac{\max\limits_{i=1,\cdots,n} \mu_i(E[V_i(\mathbf{x},\mathbf{y})])}{\mu_0(E[V_0(\mathbf{x},\mathbf{y})])}.$$

If δ_i are not in the internal $[\Delta_l, \Delta_u]$ that is specified by the leader, the leader and the followers need to update their minimum satisfactory levels to reinforce their decision-making power according to the following procedure:

1. If $\delta_l > \Delta_u$, the leader needs to increase its satisfactory level λ_0 to reinforce his or her decision-making power.

2. If $\delta_u < \Delta_l$, the leader needs to reduce its satisfactory level λ_0 to reinforce the followers' decision-making power.

3. Otherwise, the ith follower increases its minimum satisfactory level, while the jth follower decreases its minimum satisfactory level, where $i \in I = \{i | \delta_i < \Delta_l\}$, $j \in J = \{j | \delta_j > \Delta_u\}$.

3.6.2.4 Computational Results and Analysis

The irrigation district of Gan-Fu plain is a large one that is located at the end of the Fu-He valley, and is also the largest irrigation area south of the Yangtze. Its designed irrigation area is 82,000 *ha*. At present, irrigation projects in the district mainly supply water for industrial, living, and irrigation uses, and also supplies to waterpower and environmental users. They do benefit to environment and economy. According to administrative division, the Gan-Fu plain irrigation area is divided into four subareas, which are recorded as u_1, u_2, u_3, u_4 in the present part. As the Gan-Fu plain irrigation area's water shortage is seasonal, we analyze the optimal allocation of water resources at a water shortage time. We consider the demand for water in industry, for resident living, and agricultural irrigation, which can be denoted as v_1, v_2, and v_3, respectively. The flow network and water users in the Gan-Fu plain irrigation district are shown in Figure 3.5.

Here, based on the water allocation system statistical data, we know that the available water for distribution is 440.82×10^6 m^3, the price function of water is $P(r) = 6 - 1.2r$, the cost of treating wastewater is 0.8 CNY/m^3. The ecoefficiency is 25.8 CNY/m^3 and the minimum ecological water requirement is 3.50×10^6 m^3. So, considering the rough factors in the water allocation system, the rough interval is characterized by $([a,b],[c,d])$, and the other parametric values are presented in Table 3.2–Table 3.4.

Based on the description and data presented above, we can use model (3.75) to deal with this case, and then get a satisfactory solution.

First, we transform the rough model to deterministic model based on the proposed method in the proceeding section (see Table 3.5). Then we plug the data into model (3.75) and solve the model using the proposed solution approach. The procedure of interactive fuzzy programming can be arrived at as follows:

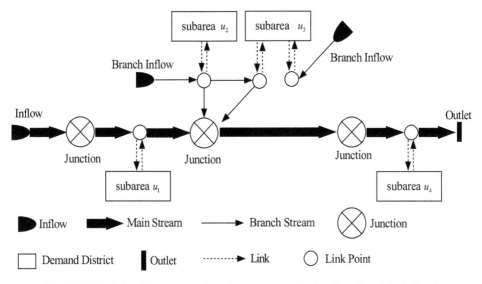

FIGURE 3.5 Flow network and water users in the Gan-Fu plain irrigation district.

TABLE 3.2 The parameters of rough intervals $\tilde{d}_{ik}$ $(i = 1, 2, 3, 4; k = 1, 2, 3)$

	u_1	u_2
v_1	([3.4,3.6],[3.05,3.90])	0
v_2	([3.18,3.46],[3.08,3.65])	0
v_3	([20.29,25.68],[18.15,28.62])	([40.28,56.38],[38.10,60.08])

	u_1	u_2
v_1	0	([2.12,2.26],[2,2.35])
v_2	([1.26,1.32],[1.18,1.37])	([4.02,4.36],[3.81,4.56])
v_3	([56.24,66.68],[44.51,70.19])	([260.66,366.98],[244.09,384.90)])

TABLE 3.3 The other parameters of water users in the water allocation system

Parameter	u_1			u_2			u_3			u_4		
	v_1	v_2	v_3	v_1	v_2	v_3	v_1	v_2	v_3	v_1	v_2	v_3
b_{ik}	55.5	71.4	23.5	0	0	26.4	0	68.2	25.8	58.2	67.8	24.3
c_{ik}	11.5	12.8	3.4	0	0	2.7	0	12.2	3.1	10.6	13.4	3.8
λ_{ik}	70%	34%	5.2%	0	0	6.7%	0	41%	4.5%	71%	37%	5.3%
$T_{ik\min}$	3.00	2.75	15.50	0	0	36.80	0	1.05	43.00	1.85	3.60	235.50
$T_{ik\max}$	3.42	3.30	20.50	0	0	48.20	0	1.22	56.80	2.18	4.15	308.50
p'_k	1.2	1	0.2	1.2	1	0.2	1.2	1	0.2	1.2	1	0.2

TABLE 3.4   The other parameters of water districts in the water allocation system

Parameter	u_1	u_2	u_3	u_4
θ_i	25.50	42.80	51.30	305.60
q_i	4.35	3.5	3.05	18.95
r_i	0.85	$\tilde{0}.9$	0.88	0.95
σ_i	0.05	0.08	0.03	0.01

TABLE 3.5   The expected value of rough intervals $\tilde{d}_{ik}$ $(i = 1,2,3,4; k = 1,2,3)$

Value	u_1	u_2	u_3	u_4
v_1	$E(\tilde{d}_{11})$	$E(\tilde{d}_{21})$	$E(\tilde{d}_{31})$	$E(\tilde{d}_{41})$
	3.4431	0	0	2.1869
v_2	$E(\tilde{d}_{12})$	$E(\tilde{d}_{22})$	$E(\tilde{d}_{32})$	$E(\tilde{d}_{42})$
	3.3374	0	1.2420	4.1921
v_3	$E(\tilde{d}_{13})$	$E(\tilde{d}_{23})$	$E(\tilde{d}_{33})$	$E(\tilde{d}_{43})$
	20.8829	49.0941	57.9352	314.4974

1. To identify the membership functions of the fuzzy goals for the objective functions, we solve five individual single objective problems for $i = 0,1,2,3,4$. The results are shown in Table 3.6.

2. By using the Zimmermann method ([346]) proposed in the previous section, we can get the maximum and minimum values, respectively, of each objective functions as follows:

$$\begin{cases} V_0^{\max} = 11515.33 \\ V_0^{\min} = 915.084 \end{cases} \quad \begin{cases} V_1^{\max} = 1007.696 \\ V_1^{\min} = -454.756 \end{cases} \quad \begin{cases} V_2^{\max} = 1762.440 \\ V_2^{\min} = 45.651 \end{cases}$$

$$\begin{cases} V_3^{\max} = 2191.410 \\ V_3^{\min} = -780.103 \end{cases} \quad \begin{cases} V_4^{\max} = 7860.982 \\ V_4^{\min} = -790.443 \end{cases} \tag{3.79}$$

3. By using the group decision-making approach, we can obtain the lower decision makers' degree of optimism, that is, the values of probability level σ_i and

TABLE 3.6   The results of solving five single objective problems (million CNY)

Results	$E[V_0]$	$E[V_1]$	$E[V_2]$	$E[V_3]$	$E[V_4]$
$\max E[V_0]$	11515.33	843.530	1225.442	1453.051	7542.453
$\max E[V_1]$	915.084	1007.696	328.820	−780.103	−790.443
$\max E[V_2]$	1312.928	−90.988	1762.440	108.843	−720.985
$\max E[V_3]$	1891.931	−454.756	45.651	2191.410	−145.264
$\max E[V_4]$	10306.23	513.173	722.516	931.642	7860.982

TABLE 3.7 Sensitivity analysis $(i = 1,2,3,4)$

Subarea	$\lambda_0 = 0.90$				$\lambda_0 = 0.91$			
	u_1	u_2	u_3	u_4	u_1	u_2	u_3	u_4
w		3.5				3.5		
x_i	26.4061	42.8000	51.7205	316.3934	27.2260	42.8000	55.9722	311.3218
y_{i1}	3.4200	0	0	2.1800	3.4200	0	0	2.1800
y_{i2}	3.3000	0	1.2200	4.1500	3.3	0	1.22	4.1500
y_{i3}	15.5000	48.2000	56.6622	280.7274	15.5000	48.2	56.6622	282.1750
$E[V_0]$		10455.30				10561.30		
$E[V_i]$	671.475	1367.026	1707.950	6426.617	663.4837	1358.365	1618.099	6639.261
$\mu_i(E[V_i])$	0.7700	0.7700	0.8373	0.8342	0.7646	0.7646	0.8071	0.8588
δ_i	0.8556	0.8556	0.9303	0.9269	0.8402	0.8402	0.8869	0.9437
Subarea	$\lambda_0 = 0.92$				$\lambda_0 = 0.93$			
	u_1	u_2	u_3	u_4	u_1	u_2	u_3	u_4
w		3.5				3.5		
x_i	28.2708	42.8000	55.3944	310.8548	29.6542	42.8000	54.6274	310.2384
y_{i1}	3.4200	0	0	2.1800	3.4200	0	0	2.1800
y_{i2}	3.3000	0	1.2200	4.1500	3.3000	0	1.2200	4.1500
y_{i3}	15.5000	48.2000	56.6622	283.6967	15.5000	48.2000	56.6622	285.3051
$E[V_0]$		10667.31				10774.54		
$E[V_i]$	655.083	1348.505	1624.312	6756.706	646.2056	1338.082	1630.908	6875.999
$\mu_i(E[V_i])$	0.7589	0.7589	0.8092	0.8725	0.7528	0.7528	0.8114	0.8885
δ_i	0.8249	0.8249	0.8796	0.9484	0.8095	0.8095	0.8725	0.9554
Subarea	$\lambda_0 = 0.94$				$\lambda_0 = 0.95$			
	u_1	u_2	u_3	u_4	u_1	u_2	u_3	u_4
w		3.5				3.5		
x_i	31.5848	42.8000	53.5556	309.3797	34.4914	42.8000	51.9861	308.0425
y_{i1}	3.4200	0	0	2.1800	3.4200	0	0	2.1800
y_{i2}	3.3000	0	1.2200	4.1500	3.3000	0	1.2200	4.1500
y_{i3}	15.5000	48.2000	56.6622	287.0170	15.5000	48.2000	56.6622	288.8553
$E[V_0]$		10879.31				10985.32		
$E[V_i]$	626.6076	1326.990	1637.946	6993.591	626.6076	1315.077	1023.778	1710.514
$\mu_i(E[V_i])$	0.7464	0.7464	0.8137	0.8997	0.7394	0.7394	0.8161	0.9136
δ_i	0.7940	0.7940	0.8656	0.9573	0.7783	0.7783	0.8591	0.9617

The unit of $E[V_i]$ $(i = 0,1,2,3,4)$ is million CNY.

possibility level r_i (see Table 3.4). Then determine the upper bound of the ratio of satisfactory degree to be 1.00, and the lower bound of the ratio of satisfactory degree to be 0.85, that is, $[\Delta_l, \Delta_u] = [0.85, 1]$. The upper decision maker sets his or her minimum satisfactory degrees at 0.90, and the lower decision makers set his or her minimum satisfactory degrees at 0.765, 0.765, 0.765, 0.765 respectively. The results can be seen in Table 3.7.

However, under a different decision-making environment and different conditions, the upper and lower decision makers may alter the minimal satisfactory degrees, and then the optimal solution will alter accordingly. In order to know how the optimal solutions change with the predetermined minimum satisfactory degrees, we do a sensitivity analysis. Alter the minimum satisfactory degrees of the upper decision maker by increasing it from 0.90 to 0.95, and record the changes of the satisfactory degrees for the lower decision makers. The Pareto-optimal solution to the problem is calculated in Table 3.7. Figures 3.6 and 3.7 show that the relationship between the satisfactory degrees and their ratios of the lower decision makers, and the satisfactory degrees of the upper decision maker λ_0. It shows that when the satisfactory degrees

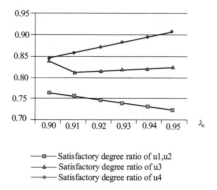

FIGURE 3.6 Satisfactory degrees of the lower decision makers and λ_0.

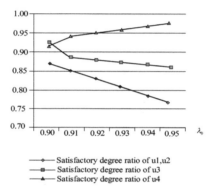

FIGURE 3.7 Satisfactory degrees of the upper decision makers and λ_0.

of the upper decision maker increases, the satisfactory degrees of the lower decision makers change in different directions and so do their ratios as well, which explains that there exists mutual influence between the upper and lower level of the model, which is consistent with the practical situation.

Chapter 4

Random Multiple Objective Rough Decision Making

Random decision-making problems are those based on probability theory. Probability theory is a mathematical branch dealing with random phenomena, which originated in the middle of the 17th century with Pascal, Fermat, and Huygens. In this chapter we first recapitulate the probability theory. Based on the expected value operator and chance operator of random variable, three classes of models are presented:

1. *Random expected value decision-making model.* Usually, decision makers find it difficult to make decisions when they encounter random parameters. A clear criteria must be introduced to help the decision. The expected value operator of random variables is introduced and a crisp equivalent model is deduced when the distribution is clear.

2. *Random chance-constrained decision-making model.* Sometimes, decision makers do not strictly require the objective value to be the maximal benefit but only need to obtain the maximum benefit under a predetermined confidence level. Then the chance-constrained model is proposed, and the crisp equivalent model is deduced when the distribution is clear.

3. *Random dependent-chance decision-making model.* When decision makers predetermine an objective value and require the maximal probability that objective values exceed the predetermined one.

By applying rough approximation to the above three classes of models, we will obtain the random expected value model (Ra-EVRM), the random chance-constrained rough model (Ra-CCRM), and the random dependent-chance rough model (Ra-DCRM). As an application, a large-scale hydropower construction project in the southwest region of China is considered to illustrate the effectiveness of the results in this chapter.

4.1 Resource-Constrained Project Scheduling Problem

In recent years, project scheduling has attracted growing attention both from the fields of theory and practice, which means that many well-known optimization problems are special cases of the more general project scheduling. Also, project scheduling applications can be found in research and development projects, construction engineering, software development, and so on. Moreover, because capacities have been cut in order to cope with lean management concepts, this is also very important for make-to-order companies (see P. Brucker et al. [40]). Project scheduling problems consist of resources, activities, performance measures, and precedence constrains (see R. Slowinski et al. [275]). The problems are known as resource-constrained project scheduling problems (rc-PSP) when the capacity of resources is limited. The activities of a project in the rc-PSP mode must be scheduled to minimize its project duration (i.e., make-span) subject to precedence relations between the activities and the limited resource availabilities, and it is known to be NP-hard (J. Blazewicz et al. [34]). There have been many project optimization scheduling models in project management theory, commonly referred to as rc-PSP, which have been reviewed by P. Brucker et al. [40], E. Demeulemeester et al. [74] and R. Kolisch and R. Padman [167].

However, in many real-life projects, such as in process industries and civil engineering, it is possible to perform the individual activities in alternative ways (modes). For example, we consider two modes of the activity "drilled drain hole (including flexible drain)," which belongs to the project "drilling grouting." If the activity is performed in the first mode, 3000 skilled workers and 1000 unskilled workers are needed, and the processing time equals 34 months. If it is performed in the second mode, we need 2000 skilled workers, and its duration equals 37 months. The modes in rc-PSP/mM differ in processing time and time lags to other activities, and resource requirements. They reflect time-resource tradeoffs and resource-resource tradeoffs (see [102] and [228]).

However, when developing the most appropriate schedule for a project, uncertainty is unavoidable for the decision manager in practice, such as processing time, quantities of resources, the due date of the project, the maximum limited resources, etc. J. Yagi [323] reformulates the concept of time to match dynamic scheduling through a nondeterministic approach. H. Aytug et al. [11] indicates that works in a deterministic environment often can be engineered to work under at least certain random conditions in a system. Then S. Vonder et al. [291], W. Herroelen [131], and J. Bidot et al. [30] consider the objective function for the case of project scheduling with random activity durations. The literature above proves the existence of randomness in rc-PSP/mM. However, no one has considered the roughness in rc-PSP/mM, which also exists. For instance, let us consider the activity "drilled drain hole (including flexible drain)" belonging to our example project "drilling grouting", because the dates collected from a practical system often contains noise, which means the dates are imperfect and incomplete. In order to deal with the dates appropriately, we use the random variables to describe the phenomenon. In addition, applying rough approximation techniques to random model makes the decision process more flexible. Hence, there is a strong motivation for further research in project scheduling. Meanwhile, multiple management objectives are always under consideration by the decision manager in practice. Unfortunately, rc-PSP/mM with multiobjective under rough random environment (rc-PSP/mM/Ro-Ra), which should be focused on more specially, has not received deserving attention.

As a generalization of the classical job shop scheduling problem, rc-PSP belongs to the class of NP-hard optimization problems, shown in J. Blazewicz et al. [34].

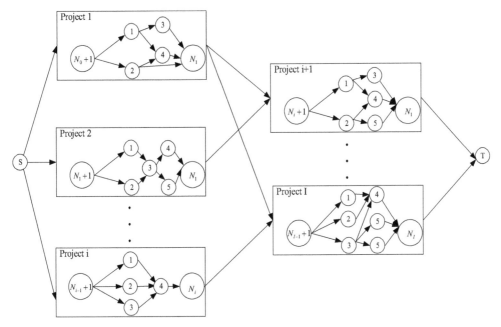

FIGURE 4.1 Precedence relations for rc-mPSP.

Therefore, as a generalized sequencing of rc-PSP, the rc-PSP/mM is also NP-hard. J. Brucker et al. [40] denoted the rc-PSP/mM as MPS–prec–C_{max}. Moreover, if there is more than one nonrenewable resource (e.g., energy, money), the problem of finding a feasible solution to the rc-PSP/mM is NP-complete (see R. Kolisch and A. Drexl [166]). As shown by A. Sprecher and A. Drexl [279], exact methods are unable to find optimal solutions to projects with more than 20 activities and three modes of peractivity when they are highly resource constrained. In practice, hence, heuristic algorithms to generate near-optimal schedules for large and highly constrained projects are needed. Several heuristic procedures have been proposed to solve rc-PSP/mM. F. Boctor presented heuristics for multimode problems without nonrenewable resources in [36, 37] and A. Drexl and J. Grunewald proposed a regret-based biased random sampling approach in [83]. A single-pass approach, a multipass approach, and a simulated annealing algorithm for multiobjective project scheduling were described by R. Slowinski et al. [275], and a local search procedure for nonpreemptive rc-PSP/mM was presented by R. Kolisch and A. Drexl [166]; meanwhile, a general class of nonpreemptive multimode time resource trade-off was presented by M. Mori and C. Tseng [219]. In 1999, L. Ozdamar [232] proposed a genetic algorithm based on a priority rule encoding. H. Zhang et al. [342] discussed particle swarm optimization for solving rc-PSP/mM.

A rc-PSP is illustrated in Figure 4.1. J. Xu and Z. Zhang [316] developed a mathematical model of the rc-mPSP for a large-scale water conservancy and hydropower construction project as follows:

$$
\begin{cases}
\min[tF, pT] = [\sum\limits_{i=1}^{I} t_{iJ}^{F}, \sum\limits_{i=1}^{I} c_{i}^{TP}(t_{ij}^{F} - t_{i}^{D})] \\
\text{s.t.} \begin{cases}
t_{(i-1)J}^{S} \leq t_{iJ}^{S} - t_{(i-1)J}, \forall i \\
t_{(i-1)J}^{S} + t_{(i-1)J} \leq t_{ij}^{S}, \forall i, \forall j \\
\sum\limits_{i \in Sp} \sum\limits_{j \in Sp} r_{ijk} \leq R_{k}, \forall k \\
t_{ij}^{S} \geq 0, \forall i, \forall j \\
R_{k} \geq 0, \forall k
\end{cases}
\end{cases}
\tag{4.1}
$$

The constraints in formulas (4.1) are used to impose precedence relations between activities, limit the resource demand imposed by the activities being processed at time p to the available capacity, and represent the usual integral constraint. More details about formulas (4.1) can be found in [316]. If the parameters in this model are random variable, the expected value model, chance-constrained model and dependent-chance model can be used to deal with them. As the rough approximation techniques are adopted to treat these model, we can obtain the random expected value rough model (Ra-EVRM), random chance-constrained rough model (Ra-CCRM), and random dependent-chance rough model (Ra-DCRM).

4.2   Random Variables

More than 50 years ago, stochastic programming was set up independently by E. Beale [25], G. Dantzig [71], A. Charnes and W. Cooper [52] and others who observed that, for many linear programs to be solved, the values of the presumably known coefficients were not available. They suggested replacing the deterministic view by a stochastic one, assuming that these unknown coefficients or parameters are random and their probability distribution P is known and independent of the decision variables. Dupačová [93] made a brief review of the history and achievements of stochastic programming.

    The prototype stochastic program aims at the selection of the "best possible" decision that fulfills given "hard" constraints, say $x \in X$, accepting that the outcome of this decision is influenced by the realization of a random event x. The realization of x is not known at the time of decision; however, to get the decision, one uses the knowledge of the probability distribution P of x. The random outcome of a decision $x \in X$ is quantified as $f_0(x, \omega)$. If the set of possible realizations of ω is finite, say $\{\omega^1, \omega^2, \cdots, \omega^S\}$, methods of multiobjective programming suggest choosing a solution that is efficient with respect to the objective functions $f_0(\cdot, \omega^s), s = 1, 2, \cdots, S$. Such efficient solutions can be obtained, for example, by the minimization (or maximization) of a weighted sum of $f_0(x, \omega^s), s = 1, 2, \cdots, S$. In our stochastic setting, the weights equal the known probabilities ω^s of the atoms or scenarios of the probability distribution P and the problem to be solved is

$$\min_{x \in X} p^s f_0(x, \omega^s).$$

Relaxation of hard constraints by the requirement that the constraints are fulfilled with a prescribed probability provides stochastic programs with probabilistic or chance constraints.

While simplifications cannot be avoided, multimodeling can help to discover model misidentifications. Complete knowledge of the underlying probability distribution cannot be expected and, in addition, approximations are needed to get numerically tractable problems. At the same time, optimal solutions to the approximate stochastic program should not be used without any further analysis in place of the sought solution of the "true" problem.

The results of theoretical analysis and software development for various types of stochastic programming models were influenced and supported by developments in optimization, probability and statistics, and computer technologies, with the progress recorded step by step in textbooks and monographs [32, 103, 154, 155, 248], in surveys, for example, [298], in special conference volumes, for example, [77], published dissertations, and in numerous focused issues of journals (see the preface of [300] for an extensive list of references). In the 1980s, a special care was devoted to software development and has resulted in an IIASA volume [299], in a recommended input format [31], in several monographs, for example, [132, 143, 210], software packages, and test batteries.

The first applications appeared in the 1950s in [55, 106]. They were based on simple types of stochastic programming those such as models with individual probabilistic constraints and stochastic linear programs with simple recourse. Moreover, special assumptions about the probability distribution P were exploited.

From the modeling point of view, stochastic vehicle routing, stochastic networks, and stochastic facility locations problems have been mostly treated as a natural extension of the stochastic transportation problem with simple recourse, whereas individual probabilistic constraints have appeared in the context of the stochastic nutrition models and in water resources management models.

Another important extension, to multistage stochastic programs, has aimed at a more realistic treatment of the dynamic or sequential structure of real-life decision problems. Several essential contributions in this direction have appeared already in [77, 247].

At present, the most popular seem to be financial applications of stochastic programming. The list of further favorable application areas contains, for instance, planning and allocation of resources (including water), energy production and transmission, production planning and optimization of technological processes, logistics problems (including aircraft allocation and yield management), and telecommunications. The achievements may be summarized as follows:

1. There are standardized types of stochastic programming models (e.g., two-stage and multistage stochastic programs with recourse, models with individual and joint probabilistic constraints, integer stochastic programs) with links to statis-

tics and probability, to parametric and multiobjective programming, to stochastic dynamic programming, and stochastic control with relevant software systems available or in progress.

2. There exist successful large-scale real-life applications. It became clear that their success is conditioned by a close collaboration with the users and that one can benefit from team work. Areas of further prospective applications have been delineated.

3. Also, the tradition of the triennial International Conferences on Stochastic Programming and numerous workshops; existence of focused groups, such as the Committee of Stochastic Programming (COSP) established in 1981 within the Mathematical Programming Society; or the WG 7.7 of IFIP; the periodically updated stochastic programming bibliography [48] and the stochastic programming electronic preprint series (SPEPS) belong among the evident achievements of the field.

The present boom in large-scale real-life applications has brought new challenging questions. An important task is an adequate reflection of the dynamic aspects, including the further development of tractable numerical approaches. Additional problems are related with the fact that the probability distribution P is rarely known completely and/or that it has to be approximated for reasons of numerical tractability so that one mostly solves an approximate stochastic program instead of the underlying true decision problem. The task is to generate the required input, that is, to approximate P bearing in mind the required type of the problem (see e.g., [91]). Moreover, without additional analysis, the obtained output (the optimal value and optimal solutions of the approximate stochastic program) should not be used to replace the sought solution of the true problem (see [89, 90]) for a discussion of suitable output analysis methods. These methods have to be tailored to the structure of the problem, and they should also reflect the source, character and precision of the input data.

As in current software systems at present, the methods of output analysis mainly address the two-stage (multiperiod) stochastic programs. The reason is that the structure of multistage problems is much more involved, and one cannot rely on intuitive straightforward generalizations. At the same time, validation experiments, as in [340], provide evidence that even three-stage stochastic programs may outperform significantly the existing static models. Hence, extensive all-round research in multistage stochastic programming is an important complex task of the day.

Next we recall some basic properties about random variables. Here we just research two special types of random variables that frequently appear in applications. One is discrete, and the other is continuous.

4.2.1   Discrete Random Variables

Next we recall the relative concepts of discrete random variables.

DEFINITION 4.1   *(K. Krickeberg [172]) Let ξ be a random variable on the probability space $(\Omega, \mathscr{A}, Pr)$. If $\Omega = \{\omega_1, \omega_2, \cdots\}$ is a set combined with* *finite or infinite discrete elements, where* $Pr\{\omega = \omega_i\} = p_i$ *and* $\sum_{i=1}^{\infty} p_i = 1$, *then* ξ *is called the discrete random variable.*

From the above definition, we know that a discrete random variable ξ is a mapping from the discrete probability space Ω to the real space **R**.

Example 4.1
Let $\Omega = \{1,2,3,4\}$ be the probability space and $Pr\{\omega_i = i\} = 0.25$, $i = 1, \cdots, 4$.
If $\xi(\omega_i) = 1/\omega_i$, then ξ is a discrete random variable.
   Intuitively, we want to know what the distribution of a discrete random variable is. The following equation is usually used to describe the distribution:

$$\begin{pmatrix} \xi(\omega_1) & \xi(\omega_2) & \cdots & \xi(\omega_n) \\ p_1 & p_2 & \cdots & p_n \end{pmatrix}.$$

In the following part, three special discrete random variables will be introduced.
   As we know, in the trial of tossing a coin, the probabilities of the front and the back are all 0.5. Then we denote that $\omega_1 = $ "Front" and $\omega_1 = $ "Back" and let ξ be a mapping from $\{\omega_1, \omega_2\}$ to $\{0,1\}$ satisfying $\xi(\omega_1) = 1$ and $\xi(\omega_2) = 0$.

DEFINITION 4.2   *(K. Krickeberg [172]) Suppose that n independent trials, each of which results in a "success" with probability p, are to be performed. If ξ represents the number of successes that occur in the n trials, then ξ is said to be a binomial random variable with parameters (n,p). Its probability mass function is given by*

$$Pr\{\xi = i\} = \binom{n}{i} p^i (1-p)^{n-i}, i = 0,1,\cdots,n \qquad (4.2)$$

where

$$\binom{n}{i} = \frac{n!}{i!(n-i)!}$$

is the binomial coefficient equal to the number of different subsets of i elements that can be chosen from a set of n elements. Obviously, in this example, Ω has n elements combined with the natural number $i = 1,2,\cdots,n$.

Since we assume that all trials are independent with each other, then the probability of any particular sequence of outcomes results in i successes and $n - i$ failures. Furthermore, it can be seen that Equation (4.2) is valid since there are $\binom{n}{i}$ different sequences of the n outcomes that result in i successes and $n - i$ failures, which can be seen by noting that there are $\binom{n}{i}$ different choices of the i trials that result in successes.

DEFINITION 4.3   *(K. Krickeberg [172]) A binomial random variable*
$(1, p)$ *is called a Bernoulli random variable.*

Since a binomial (n, p) random variable ξ represents the number of successes in
n independent trials, each of which results in a success with probability p, we can
represent it as follows:

$$\xi = \sum_{i=1}^{n} \xi_i \tag{4.3}$$

where

$$\xi_i = \begin{cases} 1, & \text{if the } i\text{th trial is a success} \\ 0, & \text{otherwise} \end{cases}$$

The following recursive formula expressing p_{i+1} in terms of p_i is useful when
computing the binomial probabilities:

$$\begin{aligned} p_{i+1} &= \frac{n!}{(n-i-1)!(i+1)!} p^{i+1} (1-p)^{n-i-1} \\ &= \frac{n!(n-i)}{(n-i)!i!(i+1)} p^i (1-p)^{n-i} \frac{p}{1-p} \\ &= \frac{n-i}{i+1} \frac{p}{1-p} p_i. \end{aligned}$$

DEFINITION 4.4   *(K. Krickeberg [172]) A random variable ξ that takes*
one of the values $0, 1, 2, \cdots$ is said to be a Poisson random variable with
parameter λ $(\lambda > 0)$ if its probability mass function is given by

$$p_i = Pr\{\xi = i\} = e^{-\lambda} \frac{\lambda^i}{i!}, \quad i = 1, 2, \cdots \tag{4.4}$$

The symbol e, defined by $e = \lim_{n \to \infty} (1 + 1/n)^n$, is a famous constant in mathematics
that is roughly equal to 2.7183.

Poisson random variables have a wide range of applications. One reason for this
is that such random variables may be used to approximate the distribution of the
number of successes in a large number of trials (which are either independent or at
most "weakly dependent") when each trial has a small probability of being a success.
To see why this is so, suppose that ξ is a binomial random variable with parameters
(n, p), and so represents the number of successes in n independent trials when each
trial is a success with probability p, and let $\lambda = np$. Then

$$\begin{aligned} Pr\{\xi = i\} &= \frac{n!}{(n-i)!i!} p^i (1-p)^{n-i} \\ &= \frac{n!}{(n-i)!i!} \left(\frac{\lambda}{n}\right)^i \left(1 - \frac{\lambda}{n}\right)^{n-i} \\ &= \frac{n(n-1)\cdots(n-i+1)}{n^i} \cdot \frac{\lambda^i}{i!} \cdot \frac{(1-\lambda/n)^n}{(1-\lambda/n)^i}. \end{aligned}$$

Now, for n large and p small,

$$\lim_{n\to\infty} \left(1 - \frac{\lambda}{n}\right)^n \to e^{-\lambda}, \lim_{n\to\infty} \frac{n(n-1)\cdots(n-i+1)}{n^i} \to 1, \lim_{n\to\infty} \left(1 - \frac{\lambda}{n}\right)^i \to 1.$$

Hence, for n large and p small,

$$Pr\{\xi = i\} \approx e^{-\lambda}\frac{\lambda^i}{i!}.$$

To compute the Poisson probabilities, we make use of the following recursive formula:

$$\frac{p_{i+1}}{p_i} = \frac{\frac{e^{-\lambda}\lambda^{(i+1)}}{(i+1)!}}{\frac{e^{-\lambda}\lambda^i}{i!}} = \frac{\lambda}{i+1},$$

or equivalently,

$$p_{i+1} = \frac{\lambda}{i+1}p_i, \ i \geq 0.$$

Consider independent trials, each of which is a success with probability p. If ξ represents the number of the first trial that is a success, then

$$Pr\{\xi = n\} = p(1-p)^{n-1} \tag{4.5}$$

which is easily obtained by noting that, in order for the first success to occur on the nth trial, the first $n-1$ must all be failures and the nth a success. Equation (4.5) is said to be a geometric random variable with parameter p.

If we let ξ denote the number of trials needed to amass a total of r successes when each trial is independently a success with probability p, then ξ is said to be a negative binomial, sometimes called a Pascal, random variable with parameters p and r. The probability mass function of such a random variable is given by

$$Pr\{\xi = n\} = \binom{n-1}{r-1}p^r(1-p)^{n-r}, \ n \geq r \tag{4.6}$$

To see why Equation (4.6) is valid, note that, in order for it to take exactly n trials to amass r successes, the first $n-1$ trials must result in exactly $r-1$ successes, and the probability of this is $\binom{n-1}{r-1}p^r(1-p)^{n-r}$, and then the nth trial must be a success, and the probability of this is p.

Consider an urn containing $N+M$ balls, of which N are light colored and M are dark colored. If a sample of size n is randomly chosen (in the sense that each of the $\binom{N+M}{n}$ subsets of size n is equally likely to be chosen), then ξ, the number of light colored balls selected, has probability mass function,

$$Pr\{\xi = i\} = \frac{\binom{N}{i}\binom{M}{n-i}}{\binom{N+M}{n}}.$$

A random variable ξ whose probability mass function is given by the preceding equation is called a hypergeometric random variable.

4.2.2 Continuous Random Variables

In this part we consider certain types of continuous random variables.

DEFINITION 4.5 *(K. Krickeberg [172]) A random variable ξ is said to be uniformly distributed over the interval (a,b), $a < b$, if its probability density function is given by*

$$f(x) = \begin{cases} \frac{1}{b-a}, & \text{if } a < x < b \\ 0, & \text{otherwise} \end{cases}$$

In other words, ξ is uniformly distributed over (a,b) if it puts all its mass on that interval and it is equally likely to be "near" any point on that interval.

The distribution function of ξ is given, for $a < x < b$, by

$$F(x) = Pr\{\xi \le x\} = \int_a^x (b-a)^{-1}dx = \frac{x-a}{b-a}.$$

DEFINITION 4.6 *(K. Krickeberg [172]) A random variable ξ is said to be normally distributed with mean μ and variance σ^2 if its probability density function is given by*

$$f(x) = \frac{1}{\sqrt{2\pi}\sigma}e^{-(x-\mu)^2/2\sigma^2}, -\infty < x < \infty.$$

The normal density is a bell-shaped curve that is symmetric about μ.

An important fact about normal random variables is that if ξ is normal with mean μ and variance σ^2, then for any constants a and b, $a\xi + b$ is normally distributed with mean $a\mu + b$ and variance $a^2\sigma^2$. It follows from this that if ξ is normal with mean μ and variance σ^2, then

$$\zeta = \frac{\xi - \mu}{\sigma}$$

is normal with mean 0 and variance 1. Such a random variable ζ is said to have a standard (or unit) normal distribution. Let Φ denote the distribution function of a standard normal random variable, that is

$$\Phi(x) = \frac{1}{\sqrt{2\pi}} \int_{-\infty}^x e^{-x^2/2}dx, -\infty < x < \infty.$$

The result that $\zeta = (\xi - \mu)/\sigma$ has a standard normal distribution when ξ is normal with mean μ and variance σ^2 is quite useful because it allows us to evaluate all probabilities concerning ξ in terms of Φ. For example, the distribution function of ξ can be expressed as

$$
\begin{aligned}
F(x) &= Pr\{\xi \le x\} \\
&= Pr\left\{\frac{\xi-\mu}{\sigma} \le \frac{x-\mu}{\sigma}\right\} \\
&= Pr\left\{\zeta \le \frac{x-\mu}{\sigma}\right\} \\
&= \Phi\left(\frac{x-\mu}{\sigma}\right).
\end{aligned}
$$

The value of $\Phi(x)$ can be determined either by looking it up in a table or by writing a computer program to approximate it. For a in the interval $(0,1)$, let ζ_a be such that

$$Pr\{\zeta > \zeta_a\} = 1 - \Phi(\zeta_a) = a.$$

That is, a standard normal will exceed ζ_a with probability a. The value of ζ_a can be obtained from a table of the values of Φ. For example, since

$$\Phi(1.64) = 0.95, \quad \Phi(1.96) = 0.975, \quad \Phi(2.33) = 0.99,$$

we see that

$$\zeta_{0.05} = 1.64, \quad \zeta_{0.025} = 1.96, \quad \zeta_{0.01} = 2.33 \tag{4.7}$$

The wide applicability of normal random variables results from one of the most important theorems of probability theory –the central limit theorem–which asserts that the sum of a large number of independent random variables has approximately a normal distribution.

DEFINITION 4.7   *(K. Krickeberg [172]) A continuous random variable having probability density function,*

$$
f(x) = \begin{cases} \lambda e^{-\lambda x}, & \text{if } 0 \le x < \infty \\ 0, & \text{otherwise} \end{cases}
$$

for some $\lambda > 0$ is said to be an exponential random variable with parameter λ.

Its cumulative distribution is given by

$$F(x) = \int_0^x \lambda e^{-\lambda x} dx = 1 - e^{-\lambda x}, \quad 0 < x < \infty.$$

The key property of exponential random variables is that they possess the "memoryless property," where we say that the nonnegative random variable ξ is memoryless if

$$Pr\{\xi > s+t | \xi > s\} = Pr\{\xi > t\}, \text{ for all } s,t \ge 0 \tag{4.8}$$

To understand why the foregoing is called the memoryless property, imagine that ξ represents the lifetime of some unit, and consider the probability that a unit of age s will survive an additional time t. Since this will occur if the lifetime of the unit exceeds $t + s$ given that it is still alive at time s, we see that

$$Pr\{\text{additional life of an item of age } s \text{ exceeds } t\} = Pr\{\xi > s + t | \xi > s\}.$$

Thus, Equation (4.8) is a statement of fact that the distribution of the remaining life of an item of age s does not depend on s. That is, it is not necessary to remember the age of the unit to know the distribution of its remaining life.

Equation (4.8) is equivalent to

$$Pr\{\xi > s + t\} = Pr\{\xi > s\}Pr\{\xi > t\}.$$

As the above equation is satisfied whenever ξ is an exponential random variable–since, in this case, $Pr\{\xi > x\} = e^{-\lambda x}$, we see that exponential random variables are memoryless (and indeed it is not difficult to show that they are the only memoryless random variables).

Another useful property of exponential random variables is that they remain exponential when multiplied by a positive constant. To see this, suppose that ξ is exponential with parameter λ, and let c be a positive number. Then

$$Pr\{c\xi \leq x\} = Pr\left\{\xi \leq \frac{x}{c}\right\} = 1 - e^{-\lambda x/c},$$

which shows that $c\xi$ is exponential with parameter λ/c.

Let $\xi_1, \xi_2, \cdots, \xi_n$ be independent exponential random variables with respective rates $\lambda_1, \lambda_2, \cdots, \lambda_n$. A useful result is that $\min\{\xi_1, \xi_2, \cdots, \xi_n\}$ is exponential with rate $\sum_i \lambda_i$ and is independent of which one of the ξ_i is the smallest. To verify this, let $M = \min\{\xi_1, \xi_2, \cdots, \xi_n\}$. Then

$$\begin{aligned}
Pr\left\{\xi_j = \min_i \xi_i | M > t\right\} &= Pr\{\xi_j - t = \min_i(\xi_i - t) | M > t\} \\
&= Pr\{\xi_j - t = \min_i(\xi_i - t) | \xi_i > t, i = 1, 2, \cdots, n\} \\
&= Pr\{\xi_j = \min_i \xi_i\}.
\end{aligned}$$

The final equality follows because, by the lack of memory property of exponential random variables, given that ξ_i exceeds t, the amount by which it exceeds it is exponential with rate λ_i. Consequently, the conditional distribution of $\xi_1 - t, \cdots, \xi_n - t$, given that all ξ_i exceed t, is the same as the unconditional distribution of $\xi_1, \cdots, \xi_n$. Thus, M is independent of which of the ξ_i is the smallest.

The result that the distribution of M is exponential with rate $\sum_i \lambda_i$ follows from

$$Pr\{M > t\} = Pr\{\xi_i > t, i = 1, 2, \cdots, n\} = \prod_{i=1}^{n} Pr\{\xi_i > t\} = e^{-\sum_{i=1}^{n} \lambda_i t}.$$

The probability that ξ_j is the smallest is obtained from

$$
\begin{aligned}
Pr\{\xi_j = M\} &= \int Pr\{\xi_j = M | \xi_j = t\} \lambda_j e^{-\lambda_j t} dt \\
&= \int Pr\{\xi_j > t, i \neq j | \xi_j = t\} \lambda_j e^{-\lambda_j t} dt \\
&= \int Pr\{\xi_j > t, i \neq j\} \lambda_j e^{-\lambda_j t} dt \\
&= \int \left(\prod_{i \neq j} e^{-\lambda_i t} \right) e^{-\lambda_j t} dt \\
&= \lambda_j \int e^{-\sum_i \lambda_i t} dt \\
&= \frac{\lambda_j}{\sum_i \lambda_i}.
\end{aligned}
$$

4.3   Ra-EVRM

Consider following multiple objective model

$$
\begin{cases}
\min[f_1(\mathbf{x}), f_2(\mathbf{x}), \cdots, f_m(\mathbf{x})] \\
\text{s.t.} \begin{cases} g_r(\mathbf{x}, \boldsymbol{\xi}) \leq 0, r = 1, 2, \cdots, p \\ \mathbf{x} \in X \end{cases}
\end{cases}
\tag{4.9}
$$

where $\mathbf{x}$ is a n-dimensional decision vector, $\boldsymbol{\xi} = (\xi_1, \xi_2, \cdots, \xi_m)$ is a random vector, $f_i(\mathbf{x}, \boldsymbol{\xi})$ are objective functions, $i = 1, 2, \cdots, m$, $g_r(\mathbf{x}, \boldsymbol{\xi})$ are constraint functions,$r = 1, 2, \cdots, p$.

Because of the existence of random vector $\boldsymbol{\xi}$, problem (4.9) is not well defined. That is, the meaning of maximizing $f_i(\mathbf{x}, \boldsymbol{\xi}), i = 1, 2, \cdots, m$ is not clear, and constraints $g_r(\mathbf{x}, \boldsymbol{\xi}), r = 1, 2, \cdots, p$ do not define a deterministic feasible set. In the following, we use the random expected value model to deal with the meaningless model. In addition, if we applied rough approximation technique to treat the random expected value model, we can obtain the random expected value rough model (Ra-EVRM).

4.3.1   General Model for Ra-EVRM

Now let us recall the well-known newsboy problem in which a boy operating a news stall has to determine the number x of newspapers to order in advance from the publisher at a cost of c per one newspaper every day. It is known that the selling price is a per one newspaper. However, if the newspapers are not sold at the end of the day, then the newspapers have a small value of b per one newspaper at the recycling center. Assuming that the demand for newspapers is denoted by ξ in a day, then the number of newspapers at the end of the day is clearly $x - \xi$ if $x > \xi$ or 0 if $x < \xi$. Thus the profit of the newsboy should be

$$
f(x, \xi) = \begin{cases} (a - c)x, & \text{if } x \leq \xi \\ (b - c)x + (a - b)\xi, & \text{if } x > \xi \end{cases}
\tag{4.10}
$$

In practice, the demand ξ for newspapers is usually a stochastic variable, so is the profit function $f(x,\xi)$. Since we cannot predict how profitable the decision of ordering x newspapers will actually be, a natural idea is to employ the expected profit, shown as follows:

$$E[f(x,\xi)] = \int_0^x [(b-c)x + (a-b)r]dF(r) + \int_x^{+\infty} (a-c)x dF(r) \qquad (4.11)$$

where E denotes the expected value operator and $F(\cdot)$ is the distribution function of demand ξ. The newsboy problem is related to determining the optimal integer number x of newspapers such that the expected profit $E[f(x,\xi)]$ achieves the maximal value, that is,

$$\begin{cases} \max E[f(x,\xi)] \\ \text{s.t. } x \geq 0, \text{ integers} \end{cases} \qquad (4.12)$$

This is a typical example of an expected value model.

Then we should first give the basic definition of the expected value. For the discrete random variable, we can define its expected value as follows.

DEFINITION 4.8   *Let ξ be a discrete random variable on the probability $(\Omega, \mathscr{A}, Pr)$ as*

$$\xi(\omega) = \begin{cases} x_1 & \text{if } \omega = \omega_1 \\ x_2 & \text{if } \omega = \omega_2 \\ \cdots \cdots \end{cases} \qquad (4.13)$$

where the probability of $\omega = \omega_i (i = 1,2,\cdots)$ is p_i. If the series $\sum_{\omega \in \Omega} \xi(\omega_i) Pr\{\omega = \omega_i\}$ is absolutely convergent, then it is said to be the expected value of ξ, denoted by $E[\xi]$.

For the continuous random variable, its expected value can be defined as follows.

DEFINITION 4.9   *Let ξ be a random variable on the probability space $(\Omega, \mathscr{A}, Pr)$. Then the expected value of ξ is defined by*

$$E[\xi] = \int_0^{+\infty} Pr\{\xi \geq r\}dr - \int_{-\infty}^0 Pr\{\xi \leq r\}dr \qquad (4.14)$$

There is another equivalent definition by the density function.

DEFINITION 4.10   *The expected value of a random variable ξ with probability density function $f(x)$ is*

$$E[\xi] = \int_{-\infty}^{+\infty} x f(x) dx \qquad (4.15)$$

Expected value, average, and mean are the same thing, but median is entirely different. The median is defined below, but only to make the distinction clear. After this, we will not make further use of the median.

DEFINITION 4.11   *The median of a random variable ξ is the unique value r in the range of ξ such that $Pr\{\xi < r\} \leq 1/2$ and $Pr\{\xi > r\} < 1/2$.*

For example, with an ordinary die, the median thrown value is 4, which is not the same as the mean 3.5. The median and the mean can be very far apart. For example, consider a $2n$-side die, with n 0s and 100s. The mean is 50, and the median is 100.

To deeply understand random variables, the variance of a random variable is given as follows.

DEFINITION 4.12   *(K. Krickeberg [172]) The variance of a random variable ξ is defined by*

$$V[\xi] = E[(\xi - E[\xi])^2] \tag{4.16}$$

The following properties of the expected value and variance of a random variable are very useful in the case of the decision-making problems with random parameters.

LEMMA 4.1
(K. Krickeberg [172]) Let ξ and η be random variables with finite expected values. Then for any numbers a and b, we have

$$E[a\xi + b\eta] = aE[\xi] + bE[\eta] \tag{4.17}$$

LEMMA 4.2
(K. Krickeberg [172]) For two independent random variables ξ and η, we have

$$E[\xi\eta] = E[\xi]E[\eta] \tag{4.18}$$

LEMMA 4.3
(K. Krickeberg [172]) For the random variable ξ, we have

$$V[\xi] = E[\xi^2] - (E[\xi])^2 \tag{4.19}$$

and for $a, b \in \mathbf{R}$, we have

$$V[a\xi + b] = a^2 V[\xi] \tag{4.20}$$

Let us consider the typical single objective with random parameters,

$$\begin{cases} \max f(\mathbf{x}, \boldsymbol{\xi}) \\ \text{s.t.} \begin{cases} g_j(\mathbf{x}, \boldsymbol{\xi}) \leq 0, j = 1, 2, \cdots, p \\ \mathbf{x} \in X \end{cases} \end{cases} \tag{4.21}$$

where $f(\mathbf{x}, \boldsymbol{\xi})$ and $g_j(\mathbf{x}, \boldsymbol{\xi}), j = 1, 2 \cdots, p$ are continuous functions in X, and $\boldsymbol{\xi} = (\xi_1, \xi_2, \cdots, \xi_n)$ is a random vector on the probability space $(\Omega, \mathscr{A}, Pr)$. Then it follows from the expected operator that

$$\begin{cases} \max E[f(\mathbf{x}, \boldsymbol{\xi})] \\ \text{s.t.} \begin{cases} E[g_j(\mathbf{x}, \boldsymbol{\xi})] \leq 0, j = 1, 2, \cdots, p \\ \mathbf{x} \in X \end{cases} \end{cases} \qquad (4.22)$$

After being dealt with by the expected value operator, the problem (4.21) has been converted into a certain programming, and then decision makers can easily obtain the optimal solution. However, whether problem (4.22) has optimal solutions is a spot that decision makers pay more attention to, and then its convexity is the focus. We will discuss in the following part.

DEFINITION 4.13 $\mathbf{x}$ *is said to be a* feasible solution *to problem (4.22) if and only if* $E[g_j(\mathbf{x}, \boldsymbol{\xi})] \leq 0, j = 1, 2, \cdots, p$.

DEFINITION 4.14   *Let* $\mathbf{x}^*$ *be feasible solution to problem (4.22). Then* $\mathbf{x}^*$ *is said to be an optimal solution to problem (4.22) if and only if* $E[f(\mathbf{x}^*, \boldsymbol{\xi})] \geq E[f(\mathbf{x}, \boldsymbol{\xi})]$ *for any feasible solution* $\mathbf{x}$.

A mathematical programming model is called convex if both the objective function and the feasible set are convex. For the expected value model (4.22), we have the following result on convexity.

THEOREM 4.1
If the functions $f(\mathbf{x}, \boldsymbol{\xi})$ *and* $g_j(\mathbf{x}, \boldsymbol{\xi})$ *are convex in X for each* $\boldsymbol{\xi}, j = 1, 2, \cdots, p$, *then the expected value model (4.22) is a convex programming.*

PROOF   For each $\boldsymbol{\xi}$, since the function $f(\mathbf{x}, \boldsymbol{\xi})$ is convex in X, we have

$$f(\lambda \mathbf{x}_1 + (1 - \lambda)\mathbf{x}_2, \boldsymbol{\xi}) \leq \lambda f(\mathbf{x}_1, \boldsymbol{\xi}) + (1 - \lambda)f(\mathbf{x}_2, \boldsymbol{\xi})$$

for any given solutions $\mathbf{x}_1$, $\mathbf{x}_2$ and any scalar $\lambda \in [0, 1]$. It follows from the expected value operator that

$$E[f(\lambda \mathbf{x}_1 + (1 - \lambda)\mathbf{x}_2, \boldsymbol{\xi})] \leq \lambda E[f(\mathbf{x}_1, \boldsymbol{\xi})] + (1 - \lambda)E[f(\mathbf{x}_2, \boldsymbol{\xi})],$$

which proves the convexity of the objective function $E[f(\mathbf{x}, \boldsymbol{\xi})]$ in X.

Let us prove the convexity of the feasible set by verifying that $\lambda \mathbf{x}_1 + (1 - \lambda)\mathbf{x}_2$ is feasible for any feasible solutions $\mathbf{x}_1$ and $\mathbf{x}_2$ constrained by $E[g_j(\mathbf{x}, \boldsymbol{\xi})] \leq 0$, $j = 1, 2, \cdots, p$ and any scalar $\lambda \in [0, 1]$. By the convexity of the functions $g_j(\mathbf{x}, \boldsymbol{\xi})$, $j = 1, 2, \cdots, p$, we know that

$$g_j(\lambda \mathbf{x}_1 + (1 - \lambda)\mathbf{x}_2, \boldsymbol{\xi}) \leq \lambda g_j(\mathbf{x}_1, \boldsymbol{\xi}) + (1 - \lambda)g_j(\mathbf{x}_2, \boldsymbol{\xi}),$$

which yields that

$$E[g_j(\lambda \mathbf{x}_1 + (1-\lambda)\mathbf{x}_2, \boldsymbol{\xi})] \le \lambda E[g_j(\mathbf{x}_1, \boldsymbol{\xi})] + (1-\lambda)E[g_j(\mathbf{x}_2, \boldsymbol{\xi})]$$

for $j = 1, 2, \cdots, p$. It follows that $\lambda \mathbf{x}_1 + (1-\lambda)\mathbf{x}_2$ is a feasible solution. Hence the feasible set is convex. This completes the proof.

Now let us consider the case of multiple objectives. If the DM wants to maximize the expected value of these objective functions in (4.9) and do not strictly require that the constraints holds, then the expected value model of problem ((4.9)) is obtained as follows:

$$\begin{cases} \max\{E[f_1(\mathbf{x}, \boldsymbol{\xi})], E[f_2(\mathbf{x}, \boldsymbol{\xi})], \cdots, E[f_m(\mathbf{x}, \boldsymbol{\xi})]\} \\ \text{s.t.} \begin{cases} g_r(\mathbf{x}, \boldsymbol{\xi}) \le 0, r = 1, 2, \cdots, p \\ \mathbf{x} \in X \end{cases} \end{cases} \tag{4.23}$$

As we know, the feasible set $I = \{\mathbf{x} \in X | g_r(\mathbf{x}, \boldsymbol{\xi}) \le 0, r = 1, 2, \cdots, p, \mathbf{x} \in X\}$ is not crisp because of the existence of the random vector $\boldsymbol{\xi}$. Then we use the rough set to deal with the feasible set I under the probabilistic environment. Since Pawlak proposed the concept of rough set, it has been rapidly developed and applied to many fields by many scholars. Here we apply the technique of rough -approximation to deal with random decisionmaking problems. According to R. Slowiński and D. Vanderpooten [277], we can extend it to the rough approximation under the similarity relationship.

First, we define the following binary relationship R_{h_r} for the constraints $g_r(\mathbf{x}, \boldsymbol{\xi}) \le 0, r = 1, 2, \cdots, p$ as follows:

$$\mathbf{x}R_{h_r}\mathbf{y} \Leftrightarrow E[|g_r(\mathbf{x}, \boldsymbol{\xi}) - g_r(\mathbf{x}, \boldsymbol{\xi})|] \le h_r \tag{4.24}$$

where $\mathbf{x}, \mathbf{y} \in X$, and $h_r(> 0)$ is the deviation that DM permits. In fact, for any $\mathbf{x} \in X$, we have that

$$E[|g_r(\mathbf{x}, \boldsymbol{\xi}) - g_r(\mathbf{x}, \boldsymbol{\xi})|] = 0 \le h_r \tag{4.25}$$

must hold. So R_{h_r} has reflexivity. Then R_{h_r} is a similarity relationship. For the converted feasible set, it requires two steps to solve the decision-making problem. The first step is to require that the lower and upper approximation sets are close enough to the initial feasible set. This means that we must guarantee the accuracy of the approximation to extend at some level. The second step is that we directly solve the decision-making problem in the lower and upper approximation set. After ensuring that the required accuracy is achieved, we can only solve the decision-making problem in the upper approximation. Then we can get the EVM based on rough approximation of problem (4.23) as

$$\begin{cases} \max\{E[f_1(\mathbf{x}, \boldsymbol{\xi})], E[f_2(\mathbf{x}, \boldsymbol{\xi})], \cdots, E[f_m(\mathbf{x}, \boldsymbol{\xi})]\} \\ \text{s.t.} \begin{cases} \alpha(I_r) \ge \theta_r \\ \mathbf{x} \in \bar{I}_r \\ r = 1, 2, \cdots, p \end{cases} \end{cases} \tag{4.26}$$

where $\alpha(I_r)$ expresses the accuracy of the approximation. Model (4.26) is called Ra-EVRM.

By the first constraint in problem (4.26), we can fix h_r by ensuring that the accuracy of rough approximation is more than θ_r; furthermore, the similarity relationship is fixed. In fact, if $\theta_r = 1, h_r = 0$ must hold according to the definition of the accuracy. It is obvious that h_r is gradually decreasing if θ_r increases. Hence, we obtain the deviation by decreasing h_r such that $\alpha(I_r) \geq \theta_r$. It means that DM can expand the feasible space by decreasing the accuracy θ_r. For example, for a manager of the supermarket, in the next holiday, DM only knows that the products will be sold very much but do not know the exact number, then DM can increasing the order number by decrease the accuracy θ_r. As the maximum h_r satisfying $\alpha(I_r) \geq \theta_r$ is fixed, the constraint $\mathbf{x} \in \bar{I}_r$ is crisp.

DEFINITION 4.15   *Let $\mathbf{x}^*$ be a feasible solution to problem (4.23). Then $\mathbf{x}^*$ is said to be an efficient solution if and only if that there does not exist feasible solutions $\mathbf{x}$ such that $E[f_i(\mathbf{x}, \boldsymbol{\xi})] \geq E[f_i(\mathbf{x}^*, \boldsymbol{\xi})], i = 1, 2, \cdots, m$ and there is at least one $i_0 \in \{1, 2, \cdots, m\}$ such that $E[f_{i_0}(\mathbf{x}, \boldsymbol{\xi})] > E[f_{i_0}(\mathbf{x}^*, \boldsymbol{\xi})]$.*

We can also formulate a random decision system as an expected value goal decision-making (EVGDM) model according to the priority structure and target levels set by the decision maker:

$$\begin{cases} \min \sum\limits_{j=1}^{l} P_j \sum\limits_{i=1}^{m} (u_{ij}d_i^+ + v_{ij}d_i^-) \\ \text{s.t.} \begin{cases} E[f_i(\mathbf{x}, \boldsymbol{\xi})] + d_i^- - d_i^+ = b_i, \ i = 1, 2, \cdots, m \\ \alpha(I_r) \geq \theta_r \\ x \in \bar{I}_r \\ r = 1, 2, \cdots, p \end{cases} \end{cases}$$

where P_j is the preemptive priority factor that expresses the relative importance of various goals, $P_j \gg P_{j+1}$, for all j; u_{ij} is the weighting factor corresponding to positive deviation for goal i with priority j assigned; v_{ij} is the weighting factor corresponding to negative deviation for goal i with priority j assigned; d_i^+ is the positive deviation from the target of goal i, defined as

$$d_i^+ = [E[f_i(x, \boldsymbol{\xi})] - b_i] \vee 0,$$

and d_i^- is the negative deviation from the target of goal i, defined as

$$d_i^- = [b_i - E[f_i(x, \boldsymbol{\xi})]] \vee 0,$$

where f_i is a function in goal constraints, g_j is a function in real constraints, b_i is the target value according to goal i, l is the number of priorities, m is the number of goal constraints, and p is the number of real constraints.

4.3.2   Ra-LEVRM and ε-Constraint Method

In order to solve the multiple objective decision-making problem (4.23), we must compute the crisp expected value of ξ. However, as we know, this process is a hard work. Next, we will focus on the linear random multiple objective decision making problem as follows:

$$\begin{cases} \max[\tilde{\mathbf{c}}_1^T\mathbf{x}, \tilde{\mathbf{c}}_2^T\mathbf{x}, \cdots, \tilde{\mathbf{c}}_m^T\mathbf{x}] \\ \text{s.t.} \begin{cases} \tilde{\mathbf{a}}_r^T\mathbf{x} \le \tilde{b}_r, r = 1, 2, \cdots, p \\ \mathbf{x} \ge \mathbf{0} \end{cases} \end{cases} \quad (4.27)$$

where $\tilde{\mathbf{c}}_i = (\tilde{c}_{i1}, \tilde{c}_{i1}, \cdots, \tilde{c}_{in})^T, \tilde{\mathbf{a}}_r = (\tilde{a}_{r1}, \tilde{a}_{r1}, \cdots, \tilde{a}_{rn})^T$ are random vectors, $\tilde{b}_r$ are random variables, $i = 1, 2, \cdots, m; r = 1, 2, \cdots, p$.

By using random expected value operator and rough approximation techniques, we have the following random linear expected value rough model (Ra-LEVRM), rewritten as

$$\begin{cases} \max[E[\tilde{\mathbf{c}}_1^T\mathbf{x}], E[\tilde{\mathbf{c}}_2^T\mathbf{x}], \cdots, E[\tilde{\mathbf{c}}_m^T\mathbf{x}]] \\ \text{s.t.} \begin{cases} \alpha(I_r) \ge \theta_r \\ x \in \bar{I}_r \\ r = 1, 2, \cdots, p \end{cases} \end{cases} \quad (4.28)$$

where $I_r = \{\mathbf{x} | \tilde{\mathbf{a}}_r^T\mathbf{x} \le \tilde{b}_r, \mathbf{x} \ge \mathbf{0}\}$, and $\theta_r \in [0, 1]$ for $r = 1, 2, \cdots, p$.

4.3.2.1   Crisp Equivalent Model

For a special case, we have a theorem and then get the crisp equivalent model for (4.28).

THEOREM 4.2
Let random vector $\tilde{\mathbf{c}}_i = (\tilde{c}_{i1}, \tilde{c}_{i2}, \cdots, \tilde{c}_{in})^T$ and $\tilde{\mathbf{a}}_r = (\tilde{a}_{r1}, \tilde{a}_{r2}, \cdots, \tilde{a}_{rn})^T$, and let $\tilde{c}_{ij}, \tilde{a}_{rj}$ and $\tilde{b}_{rj}$ be independently random variables for any $i = 1, 2, \cdots, m; j = 1, 2, \cdots, n$ and $r = 1, 2, \cdots, p$. Then problem (4.28) is equivalent to the following model under the similarity relationship R_{h_r}:

$$\begin{cases} \max[E[\tilde{\mathbf{c}}_1^T]\mathbf{x}, E[\tilde{\mathbf{c}}_2^T]\mathbf{x}, \cdots, E[\tilde{\mathbf{c}}_m^T]\mathbf{x}] \\ \text{s.t.} \begin{cases} E[\tilde{\mathbf{a}}_r^T\mathbf{x}] \le E[\tilde{b}_r] + (h_r)_{\sup}, r = 1, 2, \cdots, p \\ \mathbf{x} \ge \mathbf{0} \end{cases} \end{cases} \quad (4.29)$$

where R_{h_r} is defined by

$$\mathbf{x}R_{h_r}\mathbf{y} \Leftrightarrow E[|\tilde{\mathbf{a}}_r^T\mathbf{x} - \tilde{\mathbf{a}}_r^T\mathbf{y}|] \le h_r$$

and $(h_r)_{\sup}$ is the supremum value h_r such that

$$\int\int\cdots\int_{L_r} dx_1 dx_2 \cdots dx_n - \theta_r \int\int\cdots\int_{\bar{I}_r} dx_1 dx_2 \cdots dx_n \ge 0$$

holds.

PROOF   Denote $I_r = \{x | \tilde{\mathbf{a}}_r^T x \le \bar{b}_r, x \ge 0\}$. Then we have its upper and lower approximation under the similarity relationship R_{h_r} as follows:

$$\underline{I}_r = \{x | E[\tilde{\mathbf{a}}_r^T x] \le E[\tilde{b}_r] - h_r, x \ge 0\},$$

$$\bar{I}_r = \{x | E[\tilde{\mathbf{a}}_r^T x] \le E[\tilde{b}_r] + h_r, x \ge 0\}.$$

As we know, $|A|$ expresses the cardinality of the set A in the finite universe. For the infinite universe, we use it to express the Lebesgue measure. Then we have

$$|\underline{I}_r| = \int \int \cdots \int_{\underline{I}_r} dx_1 dx_2 \cdots dx_n$$

and

$$|\bar{I}_r| = \int \int \cdots \int_{\bar{I}_r} dx_1 dx_2 \cdots dx_n.$$

It is should be noted that there may be not less than one h_r such the approximation accuracy $\alpha(I_r)$ exceeds θ_r. However, there is always a supremum $(h_r)_{\sup}$ such that $\alpha(I_r) \ge \theta_r$. It is obvious that $(h_r)_{\sup}$ is a monotone decreasing function with respect to approximation accountancy θ_r. It follows that problem (4.28) can be rewritten as

$$\begin{cases} \max \left[E[\tilde{\mathbf{c}}_1^T] x, E[\tilde{\mathbf{c}}_2^T] x, \cdots, E[\tilde{\mathbf{c}}_m^T] x \right] \\ \text{s.t.} \begin{cases} E[\tilde{\mathbf{a}}_r^T x] \le E[\tilde{b}_r] + (h_r)_{\sup}, r = 1, 2, \cdots, p \\ x \ge 0 \end{cases} \end{cases}$$

where $(h_r)_{\sup}$ is the maximum value h_r such that

$$\int \int \cdots \int_{\underline{I}_r} dx_1 dx_2 \cdots dx_n - \theta_r \int \int \cdots \int_{\bar{I}_r} dx_1 dx_2 \cdots dx_n \ge 0$$

holds. This completes the proof.

COROLLARY 4.1

If $\tilde{\mathbf{c}}_i = (\tilde{c}_{i1}(\omega), \tilde{c}_{i2}(\omega), \cdots, \tilde{c}_{in}(\omega))^T$ are characterized by $\tilde{c}_{i1}(\omega) \sim \mathscr{U}(\underline{c}_{ij}, \bar{c}_{ij})$, $\tilde{\mathbf{a}}_r = (\tilde{a}_{r1}(\omega), \tilde{a}_{r2}(\omega), \cdots, \tilde{a}_{rn}(\omega))^T$, $\tilde{a}_{rj}(\omega) \sim \mathscr{U}(\underline{a}_{rj}, \bar{a}_{rj})$, $\tilde{b}_r \sim \mathscr{U}(\underline{b}_r, \bar{b}_r)$, $i = 1, 2, \cdots, m; r = 1, 2, \cdots, p; j = 1, 2, \cdots, n$, then model (4.29) can be written as

$$\begin{cases} \max \left[\frac{1}{2} \sum_{j=1}^n (\underline{c}_{1j} + \bar{c}_{1j}) x_j, \frac{1}{2} \sum_{j=1}^n (\underline{c}_{2j} + \bar{c}_{2j}) x_j, \cdots, \frac{1}{2} \sum_{j=1}^n (\underline{c}_{mj} + \bar{c}_{mj}) x_j \right] \\ \text{s.t.} \begin{cases} \sum_{j=1}^n (\underline{a}_{rj} + \bar{a}_{rj}) x_j \le \sum_{j=1}^n (\underline{b}_r + \bar{b}_r) + 2(h_r)_{\sup}, r = 1, 2, \cdots, p \\ x \ge 0 \end{cases} \end{cases} \quad (4.30)$$

where $\mathscr{U}$ denotes the uniform distribution, $\underline{I}_r = \{x | \sum_{j=1}^n (\underline{a}_{rj} + \bar{a}_{rj}) x_j \le \underline{b}_r - 2h_r, x \ge 0\}$ and $\bar{I}_r = \{x | \sum_{j=1}^n (\bar{a}_{rj} + \underline{a}_{rj}) x_j \le \bar{b}_r + 2(h_r)_{\sup}, x \ge 0\}$ and $(h_r)_{\sup}$ is the supre-

mum value h_r such that following inequality holds:

$$\int\int\cdots\int_{L_r} dx_1 dx_2\cdots dx_n - \theta_r \int\int\cdots\int_{\bar{I}_r} dx_1 dx_2\cdots dx_n \geq 0.$$

COROLLARY 4.2

If $\tilde{\mathbf{c}}_i = (\tilde{c}_{i1}, \tilde{c}_{i2}, \cdots, \tilde{c}_{in})^T$ are normally distributed with mean vectors $\boldsymbol{\mu}_i^c = (\mu_{i1}^c, \mu_{i2}^c, \cdots, \mu_{in}^c)^T$ and positive definite covariance matrix $\mathbf{V}_i^c$ on the probability space $(\Omega, \mathscr{A}, Pr)$, written as $\tilde{c}_i \sim \mathscr{N}(\boldsymbol{\mu}_i^c, \mathbf{V}_i^c)(i=1,2,\cdots,m)$, and random vector $\bar{a}_r \sim \mathscr{N}(\boldsymbol{\mu}_r^a, \mathbf{V}_r^a)$, $\bar{b}_r \sim \mathscr{N}(\mu_r^b, (\sigma_r^b)^2)$ $(r=1,2,\cdots,p)$, then model (4.29) can be written as

$$\begin{cases} \max \left[\boldsymbol{\mu}_1^{cT}\mathbf{x}, \boldsymbol{\mu}_2^{cT}\mathbf{x}, \cdots, \boldsymbol{\mu}_m^{cT}\mathbf{x}\right] \\ \text{s.t.} \begin{cases} \boldsymbol{\mu}_r^{aT}\mathbf{x} \leq \mu_r^b + (h_r)_{\sup}, r=1,2,\cdots,p \\ \mathbf{x} \geq \mathbf{0} \end{cases} \end{cases} \tag{4.31}$$

where $\underline{I}_r = \{\mathbf{x}|\mu_r^{aT}\mathbf{x} \leq \mu_r^b - (h_r)_{\sup}, \mathbf{x} \geq \mathbf{0}\}, \bar{I}_r = \{\mathbf{x}|\mu_r^{aT}x \leq \mu_r^b + h_r, \mathbf{x} \geq \mathbf{0}\}$ and $(h_r)_{\sup}$ is the supremum value h_r such that

$$\int\int\cdots\int_{\underline{I}_r} dx_1 dx_2\cdots dx_n - \theta_r \int\int\cdots\int_{\bar{I}_r} dx_1 dx_2\cdots dx_n \geq 0$$

holds.

COROLLARY 4.3

If $\tilde{\mathbf{c}}_i = (\tilde{c}_{i1}, \tilde{c}_{i2}(\omega), \cdots, \tilde{c}_{in})^T$ are exponentially distributed on the probability space $(\Omega, \mathscr{A}, Pr)$, written as $\tilde{c}_{ij} \sim \exp(\lambda_{ij}^c)(i=1,2,\cdots,m; j=1,2,\cdots,n)$, and random variable $\tilde{a}_{rj} \sim \exp(\lambda_{rj}^a)$, $\tilde{b}_r \sim \exp(\lambda_r^b)(r=1,2,\cdots,p; j=1,2,\cdots,n)$, then model (4.29) can be written as

$$\begin{cases} \max \left[\lambda_1^{cT}\mathbf{x}, \lambda_2^{cT}\mathbf{x}, \cdots, \lambda_m^{cT}\mathbf{x}\right] \\ \text{s.t.} \begin{cases} \lambda_r^{aT}\mathbf{x} \leq \frac{1}{\lambda_r^b} + (h_r)_{\sup}, r=1,2,\cdots,p \\ \mathbf{x} \in X \end{cases} \end{cases} \tag{4.32}$$

where $\lambda_i^c = (\frac{1}{\lambda_{i1}^c}, \frac{1}{\lambda_{i2}^c}, \cdots, \frac{1}{\lambda_{in}^c})^T$ and $\lambda_i^u = (\frac{1}{\lambda_{r1}^a}, \frac{1}{\lambda_{r2}^a}, \cdots, \frac{1}{\lambda_{rn}^e})^T, \underline{I}_r = \{\mathbf{x}|\lambda_r^{aT}\mathbf{x} \leq \frac{1}{\lambda_r^b} - (h_r)_{\sup}, \mathbf{x} \geq \mathbf{0}\}, \bar{I}_r = \{x|\lambda_r^{aT}\mathbf{x} \leq \frac{1}{\lambda_r^b} + (h_r)_{\sup}, \mathbf{x} \geq \mathbf{0}\}$ and $(h_r)_{\sup}$ is the supremum value h_r such that

$$\int\int\cdots\int_{\underline{I}_r} dx_1 dx_2\cdots dx_n - \theta_r \int\int\cdots\int_{\bar{I}_r} dx_1 dx_2\cdots dx_n \geq 0$$

holds.

Readers can get similar results when random parameters are subject to other distribution. If there are more than two different distributions in the same problem, readers can deal with it by the expected value operator and convert it into the crisp one. If the expected model has been converted into the crisp one described in the previous section, the ε-constraint method is applied to obtain a satisfactory solution. If not, the random simulation-based genetic algorithm are used to obtain an approximate solution. We discuss them in the following part.

4.3.2.2 ε-Constraint Method

The ε-constraint method (also be called the reference objective method) was presented by Y. Haimes et al. ([50]). The main idea of the ε-constraint method is to select one primary objective $f_{i_0}(x)$ to be minimization (if the problem is to minimize the objectives) and other objectives are converted into constraints. If the multiobjective problem is

$$(P) \min_{x \in X}[f_1(\mathbf{x}), f_2(\mathbf{x}), \cdots, f_m(\mathbf{x})],$$

then the ε-constraint method for the problem may be denoted as follows:

$$(P_{i_0}(\varepsilon)) \quad \begin{cases} \min f_{i_0}(\mathbf{x}) \\ \text{s.t.} \begin{cases} f_i(\mathbf{x}) \leq \varepsilon_i, \ i = 1, 2, \cdots, m, i \neq i_0 \\ \mathbf{x} \in X \end{cases} \end{cases}$$

where parameters $\varepsilon_i(i = 1, 2, \cdots, m, i \neq i_0)$ are predetermined by DM, which represent the DM's tolerant threshold for objective i.

REMARK 4.1 Consider the problem the maximization of multiobjectives

$$(P') \max_{x \in X}[f_1(\mathbf{x}), f_2(\mathbf{x}), \cdots, f_m(\mathbf{x})].$$

Then the ε-constraint method for the problem may be denoted as follows:

$$(P'_{i_0}(\varepsilon)) \quad \begin{cases} \max f_{i_0}(\mathbf{x}) \\ \text{s.t.} \begin{cases} f_i(\mathbf{x}) \geq \varepsilon_i, \ i = 1, 2, \cdots, m, i \neq i_0 \\ \mathbf{x} \in X \end{cases} \end{cases}$$

The relationship between the efficient solution to problem (P) and the optimal solution to problem $(P_{i_0}(\varepsilon))$ is given by the following Lemma.

LEMMA 4.4
(V. Chakong and Y. Haimes [50]) 1. $\mathbf{x}^*$ is an efficient solution to problem (P) if and only if $\mathbf{x}^*$ solves $P_i(\varepsilon^*)$ for every $i = 1, 2, \cdots, m$.
2. If $\mathbf{x}^*$ solves $P_i(\varepsilon^*)$ for some i, and the solution is unique, then $\mathbf{x}^*$ is an efficient solution to problem (P).

LEMMA 4.5
(J. Li [185]) Assume that x^* is an efficient solution of (P). Then there exists $\varepsilon_i(i = 1, 2, \cdots, m, i \neq i_0)$ such that x^* is a optimal solution of $(P_{i_0}(\varepsilon))$.

For the ε-constraint method, there are two advantages as follows:

1.Guaranteeing the i_0th optimization and considering other objectives simultaneously, which is preferred by the DM in practice decision making.

2. The trade-off rate can be obtained by the Kuhn–Tuker operator of the point x^*, which help DM find a more preferred decision.

We have noted that how to decide the proper value of ε? If the value of every ε_i is too small, it is possible that problem $(P_{Pi_0}(\varepsilon))$ has no feasible solution. Conversely, If the value of ε_i is too large, the loss of the other objective f_i may be great. Usually, the analyst can provide $f_i^0 = \min\limits_{\mathbf{x} \in X} f_i(\mathbf{x})(i = 1, 2, \cdots, m)$ and objective values $[f_1(\mathbf{x}), f_2(\mathbf{x}), \cdots, f_m(\mathbf{x})]$ of some feasible solution x. Then, the DM determine the values of ε_i according to the experience and some specific requirements. For more details see [50].

4.3.2.3   Numerical Example

Example 4.2
Consider the problem

$$\begin{cases} \max f_1 = \xi_1 x_1 + \xi_2 x_2 + \xi_3 x_3 \\ \max f_2 = \xi_4 x_1 + \xi_5 x_2 + \xi_6 x_3 \\ \text{s.t.} \begin{cases} \xi_7 x_1 + \xi_8 x_2 + \xi_9 x_3 \leq \xi_{10} \\ x_1 + x_2 + x_3 \leq 50 \\ x_1 + x_2 + x_3 \geq 20 \\ x_j \geq 0, j = 1, 2, 3 \end{cases} \end{cases} \tag{4.33}$$

where $\xi_j (j = 1, 2, 3)$ are independently random variables defined as

$$\xi_1 \sim \mathscr{N}(5,5), \xi_2 \sim \mathscr{N}(8,6), \xi_3 \sim \mathscr{N}(12,5), \xi_4 \sim \mathscr{N}(18,6), \xi_5 \sim \mathscr{N}(15,2),$$

$$\xi_6 \sim \mathscr{N}(28,3), \xi_7 \sim \mathscr{N}(4,4), \xi_8 \sim \mathscr{N}(3,6), \xi_9 \sim \mathscr{N}(5,2), \xi_{10} \sim \mathscr{N}(100,1).$$

If the DM requires that the approximation accuracy exceeds 0.9 under the probabilistic rough set, then we get the following expected value model:

$$\begin{cases} \max E[\xi_1 x_1 + \xi_2 x_2 + \xi_3 x_3] \\ \max E[\xi_4 x_1 + \xi_5 x_2 + \xi_6 x_3] \\ \text{s.t.} \begin{cases} \alpha(I) \geq 0.9 \\ x \in \bar{I} \end{cases} \end{cases} \tag{4.34}$$

where I expresses the feasible set. We get the supremum deviation $h = 7.5$ such that $\alpha(I) = \frac{\int_I dx_1 dx_2 dx_3}{\int_{\bar{I}} dx_1 dx_2 dx_3} \geq 0.9$, where $\underline{I} = \{(x_1, x_2, x_3) | E[\xi_7 x_1 + \xi_8 x_2 + \xi_9 x_3] \leq E[\xi_{10}] - 1.5, x_1 + x_2 + x_3 \leq 50, x_1 + x_2 + x_3 \geq 20, x_1, x_2, x_3 \geq 0\}$, and $\bar{I} = \{(x_1, x_2, x_3) \mid E[\xi_7 x_1 + \xi_8 x_2 + \xi_9 x_3] \leq E[\xi_{10}] + 1.5, x_1 + x_2 + x_3 \leq 50, x_1 + x_2 + x_3 \geq 20, x_1, x_2, x_3 \geq 0\}$. According to Corollary 4.2, problem (4.34) is equivalent to

$$\begin{cases} \max F_1(x) = 5x_1 + 8x_2 + 12x_3 \\ \max F_2(x) = 18x_1 + 15x_2 + 28x_3 \\ \text{s.t.} \begin{cases} 4x_1 + 3x_2 + 5x_3 \leq 107.5 \\ x_1 + x_2 + x_3 \leq 50 \\ x_1 + x_2 + x_3 \geq 20 \\ x_j \geq 0, j = 1,2,3 \end{cases} \end{cases} \tag{4.35}$$

Let $F_1(\mathbf{x})$ be the reference constraint and $F_2(\mathbf{x})$ be the objective functions. Then compute $\max_{\mathbf{x}} F_1(\mathbf{x})$, and we have $\varepsilon_0 = 286.7$. Let $F_2(\mathbf{x})$ be the objective function and construct the following single objective problem:

$$\begin{cases} \max F_2(\mathbf{x}) = 18x_1 + 15x_2 + 28x_3 \\ \text{s.t.} \begin{cases} 5x_1 + 8x_2 + 12x_3 \leq 286.7 \\ 4x_1 + 3x_2 + 5x_3 \leq 107.5 \\ x_1 + x_2 + x_3 \leq 50 \\ x_1 + x_2 + x_3 \geq 20 \\ x_j \geq 0, i = 1,2,3 \end{cases} \end{cases} \tag{4.36}$$

Then we obtain the optimal solution $x^* = (0,0,21.5)$ and $(F_1^*, F_2^*) = (258,602)$.

4.3.3   Ra-NLEVRM and Random Simulation-Based PSO

In the case of some complex problems, it is usually difficult to convert them into crisp ones and obtain their expected values. For example, let us consider the problem: $\max_{x \in X} E\left[\sqrt{x_1 \xi_1^2 + (x_2 + \xi_2)^2}\right]$, where ξ_1 is a uniformly distributed random variable and ξ_2 is a normally distributed random variable. As we know, it is almost impossible to convert it into a crisp one. Thus, an intelligent algorithm should be provided to solve it. The technique of random simulation-based PSO is a useful and efficient tool when dealing with them. Let us consider the following random nonlinear expected value rough model (Ra-NLEVRM):

$$\begin{cases} \max[E[f_1(\mathbf{x},\boldsymbol{\xi})], E[f_2(\mathbf{x},\boldsymbol{\xi})], \cdots, E[f_m(\mathbf{x},\boldsymbol{\xi})]] \\ \text{s.t.} \begin{cases} \alpha(I_r) \geq \eta_r \\ \mathbf{x} \in \bar{I}_r \\ r = 1,2,\cdots,p \end{cases} \end{cases}$$

where $f_i(\mathbf{x},\boldsymbol{\xi})$ or $g_j(\mathbf{x},\boldsymbol{\xi})$ or both of them are nonlinear with respect to $\mathbf{x}$ and $\boldsymbol{\xi}$, $i = 1,2,\cdots,m; j = 1,2,\cdots,p$. $\boldsymbol{\xi} = (\xi_1,\xi_2,\cdots,\xi_n)$ is a random vector on probability space $(\Omega,\mathscr{A},Pr)$, $I_r = \{\mathbf{x} \in X | g_r(\mathbf{x},\boldsymbol{\xi}) \leq 0\}$, and the similarity relationship is defined as R_{h_r} as

$$\mathbf{x}R_{h_r}\mathbf{y} \Leftrightarrow E[|g_r(\mathbf{x},\boldsymbol{\xi}) - g_r(\mathbf{x},\boldsymbol{\xi})|] \leq h_r,$$

where $\mathbf{x},\mathbf{y} \in X$, h_r is the deviation that the decision maker permits. Because of the existence of the nonlinear functions, we cannot usually convert it into the crisp one. Then we have to apply the technique of random simulation (Monte Carlo Simulation) to compute its expected value.

4.3.3.1 Random Simulation 1 for Expected Value

Let ξ be an n-dimensional random vector defined on the probability space $(\Omega, \mathscr{A}, Pr)$ (equivalently, it is characterized by a probability distribution $F(\cdot)$), and $f : \mathbf{R}^n \to \mathbf{R}$ a measurable function. Then $f(\xi)$ is a random variable. In order to calculate the expected value $E[f(\xi)]$, we generate ω_k from Ω according to the probability measure Pr, and write $\xi_k = \xi(\omega_k)$ for $k = 1, 2, \cdots, N$. Equivalently, we generate random vectors $\xi_k, k = 1, 2, \cdots, N$ according to the probability distribution $F(\cdot)$. It follows from the strong law of large numbers that

$$\frac{\sum\limits_{k=1}^{N} f(\xi_k)}{N} \to E[f(\xi)] \qquad (4.37)$$

as $N \to \infty$. Therefore, the value $E[f(\xi)]$ can be estimated by $\frac{1}{N} \sum\limits_{k=1}^{N} f(\xi_k)$ provided that N is sufficiently large.

The process of random simulation 1 for expected value can be summarized as follows:

Step1. Set $L = 0$.
Step2. Generate ω from Ω according to the probability measure Pr.
Step3. $L \leftarrow L + f(\xi(\omega))$.
Step4. Repeat the second and third steps N times.
Step5. $E[f(\xi)] = L/N$.

Example 4.3
Let ξ_1 be an exponentially distributed variable $exp(2)$, ξ_2 a normally distributed variable $\mathscr{N}(4, 1)$, and ξ_3 a uniformly distributed variable $\mathscr{U}(4, 8)$. A run of random simulation with 10000 cycles shows that $E[\sqrt{\xi_1^3 + \xi_2^2 + \xi_3}] = 6.8861$.

4.3.3.2 PSO

Population-based random local search techniques are a relatively new paradigm in the field of optimization. There are several nature inspired techniques belonging to this family that use metaphors as guides in solving problems. The most famous members are genetic algorithms that use the metaphor of genetic and evolutionary principles of fitness selection for reproduction to search solution spaces. In a similar fashion, the collective behavior of insect colonies, bird flocks, fish schools, and other animal societies are the motivation for Swarm Intelligence that is an attempt to design algorithms or distributed problem-solving devices inspired by the collective behavior of social insect colonies and other animal societies [38]. The Particle Swarm Optimization (PSO) method is a relatively new member of the Swarm Intelligence field for solving optimization problems.

PSO was proposed by J. Kennedy and R. Eberhard [162] and is one of the latest evolutionary optimization techniques for optimizing continuous nonlinear functions.

Its biological inspiration is based on the metaphor of social interaction and communication in a flock of birds or school of fishes. In these groups, there is a leader who guides the movement of the whole swarm. The movement of every individual is based on the leader and on his or her own knowledge. Since PSO is population-based and evolutionary in nature, the individuals (i.e., particles) in a PSO algorithm tend to follow the leader of the group, that is, the one with the best performance. In general, it can be said that the model that PSO inspired assumes that the behavior of every particle is a compromise between its individual memory and a collective memory.

In the PSO algorithm, a solution to a specific problem is represented by an n-dimensional position of a particle. A swarm of fixed number of particles is generated and each particle is initialized with a random position in a multidimensional search space. Each particle files through the multidimensional search space with a velocity. In each step of the iteration, the velocity of each particle is adjusted based on three components. The first component is the current velocity of the particle, which represents the inertia term or momentum of the particle, that is, the tendency to continue to move in the same direction. The second component is based on the position corresponding to the best solution, and is usually referred to as the personal best. The third component is based on the position corresponding to the best solution achieved so far by all the particles, that is, the global best. Once the velocity of each particle is updated, the particles are then moved to the new positions. The cycle repeats until the stopping criterion is met. The specific expressions used in the original PSO algorithm is discussed bellow.

The PSO algorithm consists of a population of particle initialized with random position and velocity. This population of particles is usually called a swarm. In the iteration step, each particle is first evaluated to find individual objective function value. For each particle, if a position is reached that has a better objective function than the previous best solution, the personal best position is updated. Also, if an objective function is found that is better than the previous best objective function of the swarm, the global best position is updated. The velocity is then updated on the particle's personal best position and the global best position found so far by the swarm. Every particle is then moved from the current position to the new position based on its velocity. The process repeats until the stopping criterion is met.

In PSO, a swarm of L particles served as searching agents for a specific problem solution. A particle's position (Θ_l), which consists of H dimensions, represents (directly of indirectly) a solution to the problem. The ability of a particle to search for a solution is represented by its velocity vector (Ω), which drives the movement of the particle. In each PSO iteration, every particle moves from one position to the next based on its velocity. By moving from one position to the next, a particle is reaching different prospective solutions to the problem. The basic particle movement equation if presented below:

$$\theta_{lh}(t+1) = \theta_{lh}(t) + \omega_{lh}(t+1) \tag{4.38}$$

where $\theta_{lh}(t+1)$ expresses the position of the lth particle at the hth dimension in the $(t+1)$th iteration, $\theta_{lh}(t)$ is the position of the lth particle at the hth dimension in the

tth iteration,and $\omega_{lh}(t+1)$ is the velocity of the lth particle at the hth dimension in the $(t+1)$th iteration.

PSO also imitated the swarm's cognitive and social behavior as local and global search abilities. In the basic version of PSO, the particle's personal best position (Ψ_l) and the global best position (Ψ_g) are always updated and maintained. The personal best position of a particle, which expresses cognitive or self-learning behavior, is defined as the position that gives the best objective function among the positions that have been visited by that particle. Once a particle reaches a position that has a better objective function than the previous best objective function for this particle, that is, $Z(\Theta_l) < Z(\Psi_l)$, the personal best position is updated, The global best position, which expresses the social behavior, is the position that gives the best objective function among the positions that have been visited by all particles in the swarm. Once a particle reaches a position that has a better objective function than the previous best objective function for whole swarm, that is, $Z(\Psi_l) < Z(\Psi_g)$, the global best position is also updated.

The personal best and global best position are used as the basis to update the velocity of the particle. In each iteration step, the velocity Ω is updated based on three terms: inertia, cognitive learning, and social learning terms.

The inertia term forces the particle to move in the same direction as in previous iteration. This term is calculated as a product of current velocity with the inertia weight (w).

The cognitive term forces the particle to go back to its personal best position. This term is calculated as a product of a random number (u), personal best acceleration constant (c_p), and the difference between the personal best position Ψ_l and current position Θ_l.

The social term forces the particle to move toward the global best position. This term is calculated as a product of random number (u), global best acceleration constant (c_g), and the difference between global best position Ψ_g and current position Θ_l. Specifically, the equation for updated velocity is expressed as follows:

$$\omega_{lh}(t+1) = w\omega_{lh}(t) + c_p u(\psi_{lh} - \theta_{lh}(t)) + c_g u(\psi_{gh} - \theta_{lh}(t)) \qquad (4.39)$$

where $\omega_{lh}(t+1)$ is the velocity of the lth particle at the hth dimension in the tth iteration; ψ_{lh} is the personal best position of the lth particle at the hth dimension in the tth iteration; ψ_{gh} is the global best position at the hth dimension in the tth iteration.

In the velocity-updating formula, random numbers are incorporated in order to randomize particle movement. Hence, two different particles may move to different position in the subsequent iteration even though they have similar personal best, and global best.

The notations used in the algorithm are as follows:

t	iteration index, $t = 1, 2 \cdots, T$
l	particle index, $l = 1, 2, \cdots, L$
h	dimension index, $h = 1, 2, \cdots, H$
u	uniform random number in the interval $[0, 1]$
$w(t)$	inertia weight in the tth iteration
$\omega_{lh}(t+1)$	velocity of the lth particle at the hth dimension in the tth iteration
$\theta_{lh}(t)$	position of the lth particle at the hth dimension in the tth iteration
ψ_{lh}	personal best position of the lth particle at the hth dimension in the tth iteration
ψ_{gh}	global best position at the hth dimension in the tth iteration
c_p	personal best position acceleration constant
c_g	global best position acceleration constant
θ^{max}	maximum position value
θ^{min}	minimum position value
Θ_i	vector position of the lth particle, $[\theta_{l1}, \theta_{l2}, \cdots, \theta_{lH}]$
Ω_l	vector velocity of the lth particle, $[\omega_{l1}, \omega_{l2}, \cdots, \omega_{lH}]$
Ψ_l	vector personal best position of the lth particle, $[\psi_{l1}, \psi_{l2}, \cdots, \psi_{lH}]$
Ψ_g	vector global best position, $[\psi_{l1}, \psi_{l2}, \cdots, \psi_{lH}]$
R_l	the lth set of solution
$Z(\Theta_l)$	fitness value of Θ_l

The procedure of PSO algorithm is summarized as follows (see Figure 4.2):

Step 1. Initialize L particle as a swarm. Generate the lth particle with random position Θ_l in the range $[\theta^{min}, \theta^{max}]$, velocity $\Omega_l = 0$, and personal best $\Psi_l = \Omega_l$ for $l = 1, 2, \cdots, L$. Set iteration $t = 1$.

Step 2. Decode particles into solutions. For $l = 1, 2, \cdots, L$, decode $\Theta_l(t)$ to a solution R_l. (This step is only needed if the particles are not directly representing the solutions.)

Step 3. Evaluate the particles. For $l = 1, 2, \cdots, L$, compute the performance measurement of R_l, and set this as the fitness value of Θ_l, represent by $Z(\Theta_l)$.

Step 4. Update pbest: For $l = 1, 2, \cdots, L$, update $\Psi_l = \Theta_l$, if $Z(\Theta_l) < Z(\Psi_l)$.

Step 5. Update pbest: For $l = 1, 2, \cdots, L$, update $\Psi_g = \Psi_l$, if $Z(\Psi_l) < Z(\Psi_g)$.

Step 6. Update the velocity and the position of each lth particle:

$$w(t) = w(T) + \frac{t - T}{1 - T}[w(1) - w(T)] \tag{4.40}$$

$$\omega_{lh}(t+1) = w\omega_{lh}(t) + c_p u(\psi_{lh} - \theta_{lh}(t)) + c_g u(\psi_{gh} - \theta_{lh}(t)) \tag{4.41}$$

$$\theta_{lh}(t+1) = \theta_{lh}(t) + \omega_{lh}(t+1) \tag{4.42}$$

If $\theta_{lh}(t+1) > \theta^{max}$, then

$$\theta_{lh}(t+1) = \theta^{max} \tag{4.43}$$

$$\omega_{lh}(t+1) = 0 \tag{4.44}$$

If $\theta_{lh}(t+1) < \theta^{max}$, then

$$\theta_{lh}(t+1) = \theta^{min} \tag{4.45}$$

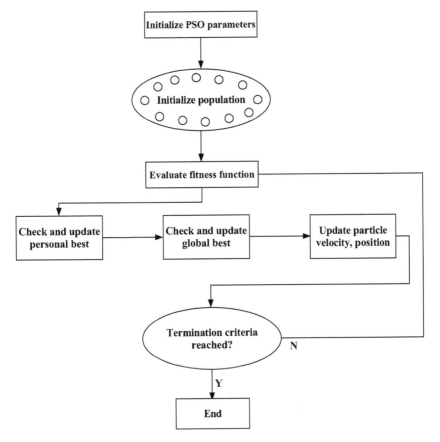

FIGURE 4.2 Flowchart of PSO.

$$\omega_{lh}(t+1) = 0 \tag{4.46}$$

Step 7. If the stopping criteria is met, that is $t = T$, stop; otherwise, $t = t + 1$ and return to Step 2.

The value $\theta^{\max}$ and $\theta^{\min}$ in equations (4.43) and (4.45) are the upper and lower bounds on the position of particles.

Let us discuss possible qualifications and effects of each parameter on the performance of PSO. The parameters consist of the population size (L), two acceleration constants (c_p and c_g), and the inertia weight (w).

Population size L represents the number of particles in the system. It is one important parameter of PSO, because it affects the fitness value and computation time. Furthermore, increasing the size of the population always increases computation time, but might not improve the fitness value. Generally speaking, too small a population size can lead to poor convergence, while too large a population size can yield good convergence at the expense of long running time.

The constants c_p and c_g are the acceleration constants of the personal best position and the global best position, respectively. Each acceleration constant controls the maximum distance that a particle is allowed to move from the current position to each best position. The new velocity can be viewed as a vector that combines the current velocity and the vectors of the best positions. Each best positions's vector consists of the direction that is pointed from the particle's current position to the best position, and the magnitude of the movement can be between 0 and the acceleration constant of the best position multiplied by the distance between the best position and the current position.

The new velocity is produced from the combination of vectors. One of these vectors is the current velocity. Inertia weight is a weight to control the magnitude of the current velocity on updating the new velocity. For $w = c$, it means that this vector has the same direction of the current velocity, and the parameters to control the search behavior.

Some PSO algorithms are implemented with a bound on velocity. For each dimension, the magnitude of a velocity cannot be greater than V_{max}. This parameter is one of the parameters to control the search behavior of the swarm. The smaller value of this parameter makes the particles in the population less aggressive in the search.

In the PSO particle movement mechanism, it is also common to limit the search space of particle location, that is, the position value of the particle dimension is bounded in the interval $[\theta^{min}, \theta^{max}]$. The use of position boundary θ^{max} is to force each particle to move within the feasible region to avoid solution divergence. Hence, the position value of certain particle dimensions is being set at the minimum or maximum value whenever it moves beyond the boundary. In addition, the velocity of the corresponding dimension is reset to zero to avoid further movement beyond the boundary.

Multiobjective optimization problems represent an important class of real-world problems. Typically, such problems involve trade-offs. For example, a car manufacturer may wish to maximize its profit, but meanwhile also to minimize its production cost. These objectives are typically conflicting with each other. For example, a higher profit could increase the production cost. Generally, there is no single optimal solution. Often, the manufacturer needs to consider many possible "trade-off" solutions before choosing the one that suits its need. The curve or surface (for more than two objectives) describing the optimal trade-off solutions between objectives is known as the Pareto front. A multiobjective optimization algorithm is required to find solutions as close as possible to the Pareto front while maintaining a good solution diversity along the Pareto front.

To apply PSO to multiobjective optimization problems, several issues have to be taken into consideration:

1. How to choose a leader for each particle? The PSO needs to favor nondominated particles over dominated ones, and drive the population toward different parts of the Pareto front, not just towards a single point. This requires that particles be allocated to different leaders.

2. How to identify nondominated particles with respect to all particles' current positions and personal best positions? And how to retain these solutions during

the search process? One strategy is to combine all particles' personal best positions and current positions and then extract the nondominated solutions from the combined population.

3. How to maintain particle diversity so that a set of well-distributed solutions can be found along the Pareto front? Some classic niching methods (e.g., crowding or sharing) can be adopted for this purpose.

The main difference between single objective PSO and MOPSO is how to choose the global best. An *lbest* PSO was used, and p_g was chosen from a local neighborhood using a ring topology. All personal best positions were kept in an archive. At each particle update, the current position is compared with solutions in this archive to see if the current position represents a nondominated solution. The archive is updated at each iteration to ensure that it contains only nondominated solutions.

Interestingly, it was not until 2002 that the next publication on PSO for multiobjective optimization appeared. M. Coello and J. Lechuga [64] proposed MOPSO (Multiobjective PSO), which uses an external archive to store nondominated solutions. The diversity of solutions is maintained by keeping only one solution within each hypercube, which is predefined by a user in the objective space. K. Parsopoulos and M. Vrahatis [237] adopted a more traditional weighted-sum approach. However, by using gradually changing weights, their approach was able to find a diverse set of solutions along the Pareto front.

With the aim of increasing the efficiency of extracting nondominated solutions from a swarm, X. Li [189] proposed NSPSO (Nondominated Sorting PSO), which follows the principal idea of the well-known NSGA II algorithm [73]. In NSPSO, instead of comparing solely a particle's personal best with its potential offspring, all particles' personal best positions and offspring are first combined to form a temporary population. After this, domination comparisons for all individuals in this temporary population are carried out. This approach will ensure more non-dominated solutions can be discovered through the domination comparison operations than the above-mentioned multiobjective PSO algorithms.

Many more multiobjective PSO variants have been proposed in recent years. A survey conducted by M. Sierra and C. Coello [254] in 2006 shows that there are currently 25 different PSO algorithms for handling multiobjective optimization problems. Interested readers should refer to X. Hu and R. Eberhart [138], S. Ho et al. [133], M. Gill et al. [116], Z. Xiao et al. [307], and Y. Feng et al. [105] for more information on these different approaches.

The principles that govern PSO algorithm can be stated as follows:

Step 1. Every particle k, a potential solution generated by the random simulation, in the swarm begins with a randomized position and randomized velocity. The position and velocity of particle k in the n-dimensional search space are represented by the vectors $X_k = (x_{k1}, x_{k2}, \cdots, x_{kn})$ and $V_k = (v_{k1}, v_{k2}, \cdots, v_{kn})$, respectively, where $x_{kd}(d = 1, \cdots, n)$ represents the location, and $v_{kd}(d = 1, ..., n)$ represents the flying velocity of particle k in the dth dimension of the search space.

Step 2. Every particle k knows its position and the value of the objective function for that position. It also remembers at which position $P_k^t = (P_{k1}^t, P_{k2}^t, \cdots, P_{kn}^t)$ it has achieved its highest performance.

Step 3. Every particle can generate a neighborhood from every position. Hence, it is also a member of some neighborhood of particles, and remembers which particle (given by the index g) has achieved the best overall position in that neighborhood. This neighborhood can either be a subset of the particles (local neighborhood) or all the particles (global neighborhood).

Step 4. In each iteration t, the behavior of a particle is a compromise among three possible alternatives: following its current pattern of exploration; going back toward its best previous position; going back toward the best historic value of all particles. This compromise is executed by the following equations at the current iteration of the algorithm:

$$\begin{cases} v_{kd}^{t+1} = wv_{kd}^{t} + c_1r_1(p_{kd}^{t} - x_{kd}^{t}) + c_2r_2(p_{gd}^{t} - p_{gd}^{t}) \\ x_{kd}^{t+1} = x_{kd}^{t} + v_{kd}^{t+1} \end{cases}$$

where w, called the inertia weight, is a constant value chosen by the user to control the impact of the previous velocities on the current velocity; c_1 is the weight given to the attraction to the previous best location of the current particle and c_2 is the weight given to the attraction to the previous best location of the particle neighborhood; r_1 and r_2 are uniformly distributed random variables in $[0,1]$.

4.3.3.3   Numerical Example

Example 4.4
Consider the problem

$$\begin{cases} \max f_1(\mathbf{x}, \boldsymbol{\xi}) = 2\xi_1 x_1^2 + 3\xi_2 x_2 - \xi_3 x_3 + \sqrt{(1-\xi_4)^2 + (3-\xi_5)^2 + (2-\xi_6)^2} \\ \max f_2(\mathbf{x}, \boldsymbol{\xi}) = 5\xi_1 x_1 - 2\xi_2 x_2 + 2\xi_3 x_3 + \sqrt{(5-\xi_4)^2 + (2-\xi_5)^2 + (1-\xi_6)^2} \\ \text{s.t.} \begin{cases} 5x_1\xi_1^2 - 3x_2^2\xi_2 + 6\sqrt{x_3}\xi_3^3 \leq 50 \\ 4\sqrt{x_1} + 6x_2 - 4.5x_3 \leq 20 \\ x_1 + x_2 + x_3 \leq 15 \\ x_1, x_2, x_3 \geq 0 \end{cases} \end{cases} \tag{4.47}$$

$\xi_j (j = 1, 2, \cdots, 6)$ are random variables characterized as

$$\xi_1 \sim \mathcal{N}(4, 1), \xi_2 \sim \mathcal{N}(5, 1), \xi_3 \sim \mathcal{U}(1, 9),$$
$$\xi_4 \sim \mathcal{U}(2, 10), \xi_5 \sim \exp(2), \xi_6 \sim \exp(4).$$

From the mathematical view, problem (4.47) is not well defined because of the uncertain parameters. Then we apply the expected value technique to deal with this uncertain programming. By using random technique, we have

$$E[\sqrt{(1-\xi_4)^2 + (3-\xi_5)^2 + (2-\xi_6)^2}] = 6.76,$$

$$E[\sqrt{(5-\xi_4)^2 + (2-\xi_5)^2 + (1-\xi_6)^2}] = 4.86.$$

$E[\xi_1^2] = 16.95, E[\xi_1^3] = 29.83$. If the DM require that the approximation accuracy exceeds 0.95 under the probabilistic rough set-then we get the following expected value model:

$$\begin{cases} \max E[2\xi_1 x_1^2 + 3\xi_2 x_2 - \xi_3 x_3 + \sqrt{(1-\xi_4)^2 + (3-\xi_5)^2 + (2-\xi_6)^2}] \\ \max E[f_2(x,\xi) = 5\xi_1 x_1 - 2\xi_2 x_2 + 2\xi_3 x_3 + \sqrt{(5-\xi_4)^2 + (2-\xi_5)^2 + (1-\xi_6)^2}] \\ \text{s.t.} \begin{cases} \alpha(I) \geq 0.95 \\ x \in \bar{I} \end{cases} \end{cases}$$

(4.48)

where I expresses the feasible set. First, we get the maximum deviation $h = 2.6$ such that $\alpha(I) = \frac{\int_I dx_1 dx_2 dx_3}{\int_{\bar{I}} dx_1 dx_2 dx_3} \geq 0.9$, where $\underline{I} = \{(x_1, x_2, x_3) | E[5x_1\xi_1^2 - 3x_2^2\xi_2 + 6\sqrt{x_3}\xi_3^3] \leq 50 - 2.6, 4\sqrt{x_1} + 6x_2 - 4.5x_3 \leq 20, x_1 + x_2 + x_3 \leq 15x_1, x_2, x_3 \geq 0\}$ and $\bar{I} = \{(x_1, x_2, x_3) | E[5x_1\xi_1^2 - 3x_2^2\xi_2 + 6\sqrt{x_3}\xi_3^3] \leq 50 + 2.6, 4\sqrt{x_1} + 6x_2 - 4.5x_3 \leq 20, x_1 + x_2 + x_3 \leq 15x_1, x_2, x_3 \geq 0\}$. Then we have

$$\begin{cases} \max F_1(x) = 8x_1^2 + 15x_2 - 5x_3 + 6.76 \\ \max F_2(x) = 20x_1 - 10x_2 + 10x_3 + 4.86 \\ \text{s.t.} \begin{cases} 84.75x_1 - 15x_2^2 + 178.98\sqrt{x_3} \leq 52.6 \\ 4\sqrt{x_1} + 6x_2 - 4.5x_3 \leq 20 \\ x_1 + x_2 + x_3 \leq 15 \\ x_1, x_2, x_3 \geq 0 \end{cases} \end{cases}$$

(4.49)

Next we will use PSO to solve the above problem. The process of generating a new position for a selected individual in the swarm is depicted in the following equation:

$$v_{t+1}^i = wv_t^i + c_1 \cdot \text{rand}() \cdot (P_t^i - x_t^i) + c_2 \cdot \text{rand}() \cdot (G_t - x_t^i),$$

where v_t^i and x_t^i are the ith particle current velocity and position; P_t^i and G_t are the kth particle best position and the global best position visited so far, respectively; $w = 0.7298$ is the inertia weight c_1, c_2 are learning factors, and rand() is a random number in [0,1]. After the simulation with many cycles, we get the optimal solution under different weights as shown in Table 4.1. The Matlab® file is presented in A.4.

4.4 Ra-CCRM

In this section, we will discuss the random CCM based on rough approximation, which is called random chance-constrained rough model (Ra-CCRM). For the linear case, the crisp equivalent model will be obtained for some special cases and then is solved by two-stage method. For the nonlinear case, which is hard to be converted into crisp equivalent model, we will solve it by the random simulation-based PSO with selection operator. Numerical examples will illustrate the effectiveness of the results we discussed.

TABLE 4.1   The optimal solution by rough simulation-based PSO

c_1	c_2	x_1	x_2	x_3	$\bar{f}_1$	$\bar{f}_2$
2	2.05	2.49	6.18	5.25	122.81	45.36
2	2.10	2.30	6.35	5.93	114.68	46.66
2	2.50	2.27	6.31	6.46	110.33	51.76
2	3.00	2.36	6.30	6.81	111.77	57.16
3	2.05	2.39	6.20	5.37	118.61	44.36
3	2.10	2.40	6.29	5.88	117.9	48.76
3	2.50	2.21	6.33	6.42	108.68	49.96
3	3.00	2.47	6.18	5.36	121.47	46.06

4.4.1   General Model for Ra-CCRM

In practice, the goal of decision makers is to maximize the objective value on the condition of probability α, where α is predetermined confidence level. Next, we will introduce the concept of the chance measure of random variables. The chance measure of a random event is considered as the probability of the event $f(\mathbf{x}, \boldsymbol{\xi}) \geq \bar{f}$. Then the chance constraint is considered as $Pr\{f(\mathbf{x}, \boldsymbol{\xi}) \geq \bar{f}\} \geq \alpha$, where α is the predetermined confidence level. A natural idea is to provide a confidence level α at which it is desired that the random constraints hold. Let us still consider the following model:

$$\begin{cases} \max f(\mathbf{x}, \boldsymbol{\xi}) \\ \text{s.t.} \begin{cases} g_j(\mathbf{x}, \boldsymbol{\xi}) \leq 0, j = 1, 2, \cdots, p \\ \mathbf{x} \in X \end{cases} \end{cases}$$

where $f(\mathbf{x}, \boldsymbol{\xi})$, and $g_j(\mathbf{x}, \boldsymbol{\xi})$, $j = 1, 2 \cdots, p$ are continuous functions in X and $\boldsymbol{\xi} = (\xi_1, \xi_2, \cdots, \xi_n)$ is a random vector on probability space $(\Omega, \mathscr{A}, Pr)$. Based on the chance-constrained operator, the random chance-constrained model (CCM):

$$\begin{cases} \max \bar{f} \\ \text{s.t.} \begin{cases} Pr\{f(\mathbf{x}, \boldsymbol{\xi}) \geq \bar{f}\} \geq \beta \\ Pr\{g_j(\mathbf{x}, \boldsymbol{\xi}) \leq 0\} \geq \alpha_j, j = 1, 2, \cdots, p \\ \mathbf{x} \in X \end{cases} \end{cases} \tag{4.50}$$

where β and α_j are the predetermined confidence levels, and $\bar{f}$ is the critical value that needs to be determined.

If the objective is to be minimized (for example, the objective is a cost function), the CCM should be

$$\begin{cases} \min \bar{f} \\ \text{s.t.} \begin{cases} Pr\{f(\mathbf{x}, \boldsymbol{\xi}) \leq \bar{f}\} \geq \beta \\ Pr\{g_j(\mathbf{x}, \boldsymbol{\xi}) \leq 0\} \geq \alpha_j, j = 1, 2, \cdots, p \\ \mathbf{x} \in X \end{cases} \end{cases} \tag{4.51}$$

where β and α_j are the predetermined confidence levels.

Now we consider the case of multiple objectives. If the DM wants to maximize the values of these objective functions in (4.9) at some predetermined confidence level and do not strictly require that the constraints hold, then the problem (4.9) can be rewritten as follows :

$$\begin{cases} max \left[\bar{f}_1, \bar{f}_2, \cdots, \bar{f}_m\right] \\ \text{s.t.} \begin{cases} Pr\{f_i(\mathbf{x}, \boldsymbol{\xi}) \geq \bar{f}_i\} \geq \delta_i, i = 1, 2, \cdots, m \\ g_r(\mathbf{x}, \boldsymbol{\xi}) \leq 0, r = 1, 2, \cdots, p \\ \mathbf{x} \in X \end{cases} \end{cases} \quad (4.52)$$

where δ_i are predetermined by the DM, representing the predetermined confidence level. As we know, the feasible set $I = \{\mathbf{x} \in X | Pr\{f_i(\mathbf{x}, \boldsymbol{\xi}) \geq \bar{f}_i\} \geq \delta_i, i = 1, 2, \cdots, m, g_r(\mathbf{x}, \boldsymbol{\xi}) \leq 0, r = 1, 2, \cdots, p\}$ is not crisp because of the existence of the random vector $\boldsymbol{\xi}$. Here, we apply the technique of rough approximation to deal with the random programming problems.

First, we define the following binary relationship $R_{h_r}^{\eta_r}$ for the constraints $g_r(\mathbf{x}, \boldsymbol{\xi}) \leq 0$:

$$\mathbf{x} R_{h_r}^{\eta_r} \mathbf{y} \Leftrightarrow Pr[|g_r(\mathbf{x}, \boldsymbol{\xi}) - g_r(\mathbf{y}, \boldsymbol{\xi})| \leq h_r] \geq \eta_r, r = 1, 2, \cdots, p \quad (4.53)$$

where $\mathbf{x}, \mathbf{y} \in X$, $\eta_r (0 \leq \eta \leq 1)$ is the confidence level, and $h_r (> 0)$ is the deviation that the DM permits. In fact, for any $\mathbf{x} \in X$,

$$Pr[|g_r(\mathbf{x}, \boldsymbol{\xi}) - g_r(\mathbf{y}, \boldsymbol{\xi})| \leq h_r] = 1 \geq \eta_r$$

must hold, $r = 1, 2, \cdots, p$, so $R_{h_r}^{\eta_r}$ has reflexivity. Then $R_{h_r}^{\eta_r}$ is a similarity relationship. For the converted feasible set, it requires two steps to solve the programming problem. The first step is to require that the lower and upper approximation sets are close enough to the initial feasible set. This means that we must guarantee the accuracy of the approximation to reach at some level. The second step is that we directly solve the programming problem in the lower and upper approximation set, respectively. After ensuring that the required accuracy is achieved, we can only solve the programming problem in the upper approximation. Then we can get the chance-constrained model based on rough approximation of the problem (4.52) as

$$\begin{cases} max \left[\bar{f}_1, \bar{f}_2, \cdots, \bar{f}_m\right] \\ \text{s.t.} \begin{cases} \alpha(I_r) \geq \theta_r \\ \mathbf{x} \in \bar{I}_r \\ r = 1, 2, \cdots, p \end{cases} \end{cases} \quad (4.54)$$

where $\alpha(I_r)$ expresses the accuracy of the approximation, $I_r = \{\mathbf{x} \in X | g_r(\mathbf{x}, \boldsymbol{\xi}) \leq 0, Pr\{f_i(\mathbf{x}, \boldsymbol{\xi}) \geq \bar{f}_i\} \geq \delta_i, i = 1, 2, \cdots, m\}$. Model (4.54) is called random chance-constrained rough model (Ra-CCRM).

By the first constraint in problem (4.54), we can fix h_r by ensuring that the accuracy of the rough approximation is more than θ_r. Furthermore, the similarity relationship is fixed. In fact, if $\theta_r = 1, h_r = 0$ must hold according to the definition of the accuracy. It is obvious that h_r is gradually decreasing if $r\theta$ increases. Hence, we obtain the deviation by minimizing h_r such that $\alpha(I_r) \geq \theta_r$. It means that the DM can expand the feasible space by decreasing the accuracy, θ_r. For example, the DM, as manager of a supermarket, knows that more products will sell during the next holiday but not how much more, and therefore he or she can increase the order for items

by decreasing the accuracy θ_r. As the maximum h_r satisfying $\alpha(I_r) \geq \theta_r$ fixed, the constraint $x \in \bar{I}_r$ is crisp.

DEFINITION 4.16 $\mathbf{x}^*$ *is said to be a feasible solution at θ_r-accuracy levels of Problem (4.54) if and only if it satisfies $\alpha(I_r) \geq \theta_r, r = 1, 2, \cdots, p$.*

DEFINITION 4.17   *A feasible solution at θ_r-accuracy, $\mathbf{x}^*$ is said to be a δ_i-efficient solution to problem (4.54) if and only if there exists no other feasible solution at θ_r-accuracy levels such that $Pr\{f_i(\mathbf{x}, \boldsymbol{\xi}) \geq \bar{f}_i\} \geq \delta_i$ with $\bar{f}_i(\mathbf{x}) \geq \bar{f}_i(\mathbf{x}^*)$ for all i, and $\bar{f}_{i_0}(\mathbf{x}) \geq \bar{f}_{i_0}(x^*)$ for at least one $i_0 \in \{1, 2, \cdots, m\}$.*

4.4.2   Ra-LCCRM and the Two-Stage Method

In order to solve the multiple objective decision-making problem (4.54), we must compute the crisp expected value of $\boldsymbol{\xi}$. However, as we know, this process is hard work sometimes. Next, we will focus on the linear model.

By using the random chance-constrained operator and rough approximation technique, we have the following random linear chance-constrained rough model (Ra-LCCRM):

$$
\begin{cases}
\max \left[\bar{f}_1, \bar{f}_2, \cdots, \bar{f}_m\right] \\
\text{s.t.} \begin{cases} \alpha(I_r) \geq \theta_r \\ \mathbf{x} \in \bar{I}_r \\ r = 1, 2, \cdots, p \end{cases}
\end{cases}
\tag{4.55}
$$

where $\tilde{\mathbf{c}}_i = (\tilde{c}_{i1}, \tilde{c}_{i1}, \cdots, \tilde{c}_{in})^T, \tilde{\mathbf{a}}_r = (\tilde{a}_{r1}, \tilde{a}_{r1}, \cdots, \tilde{a}_{rn})^T$ are random vectors, $\tilde{b}_r$ are random variables, $i = 1, 2, \cdots, m; r = 1, 2, \cdots, p$, $I_r = \{\mathbf{x} | Pr\{\tilde{\mathbf{c}}_i^T \mathbf{x} \geq \bar{f}_i\} \geq \delta_i, i = 1, 2, \cdots, m; \tilde{\mathbf{a}}_r^T \mathbf{x} \leq \tilde{b}_r, \mathbf{x} \geq \mathbf{0}\}$ and $\theta_r \in [0, 1]$ for $r = 1, 2, \cdots, p$.

4.4.2.1   Crisp Equivalent Model

One way of solving an Ra-CCRM is to convert the objectives and constraints into their respective crisp equivalents and then solve them with traditional multiobjective decision-making methods. However, this process is usually hard work and only successful in some special cases. Next, we will consider a special case and present the result in this subsection.

THEOREM 4.3
Let the random vector $\boldsymbol{\xi}$ degenerate into a random variable ξ with distribution function Φ, and the function $g(\mathbf{x}, \boldsymbol{\xi})$ has the form $g(\mathbf{x}, \boldsymbol{\xi}) = h(\mathbf{x}) - \xi$. Then

$Pr\{g(\pmb{x},\xi)\leq 0\}\geq \alpha$ *if and only if* $h(\pmb{x})\leq K_{\alpha}$, *where* $K_{\alpha}=sup\{K|K=\Phi^{-1}(1-\alpha)\}$.

PROOF The assumption implies that $Pr\{g(\pmb{x},\xi)\leq 0\}\geq \alpha$ can be written in the following form:

$$Pr\{h(\pmb{x})\leq \xi\}\geq \alpha \qquad (4.56)$$

It is clear that, for each given confidence level $\alpha(0<\alpha<1)$, there exists a number K_{α} (may be multiple or ∞) such that

$$Pr\{K_{\alpha}\leq \xi\}=\alpha \qquad (4.57)$$

and the probability $Pr\{\alpha\leq \xi\}$ will increase if K_{α} is replaced with a smaller number. Hence, $Pr\{h(\mathbf{x})\leq \xi\}\geq \alpha$ if and only if $h(\mathbf{x})\leq K_{\alpha}$.

Notice that the equation $Pr\{\alpha\leq \xi\}=1-\Phi(K_{\alpha})$ always holds, and we have

$$K_{\alpha}=\Phi^{-1}(1-\alpha),$$

where Φ^{-1} is the inverse function of Φ. Sometimes the solution to (4.57) is not unique. Equivalently, the function Φ^{-1} is multivalued. For this case, we should choose it as the largest one, that is,

$$K_{\alpha}=sup\{K|K=\Phi^{-1}(1-\alpha)\}.$$

Thus the deterministic equivalent is $h(\pmb{x})\leq K_{\alpha}$. The theorem is proved.

THEOREM 4.4

Let the random vector $\pmb{\xi}=(a_1,a_2,\cdots,a_n,b)$, *and the function* $g(\mathbf{x},\pmb{\xi})$ *has the form* $g(\mathbf{x},\pmb{\xi})=a_1x_1+a_2x_2+\cdots+a_nx_n-b$. *If* a_i *and* b *are assumed to be independently normally distributed variables, then* $Pr\{g(\mathbf{x},\pmb{\xi})\leq 0\}\geq \alpha$ *if and only if*

$$\sum_{i=1}^{n}E[a_i]x_i+\Phi^{-1}(\alpha)\sqrt{\sum_{i=1}^{n}V[a_i]x_i^2+V[b]}\leq E[b] \qquad (4.58)$$

where Φ *is the standardized normal distribution.*

PROOF The chance constraint $Pr\{g(\mathbf{x},\pmb{\xi})\leq 0\}\geq \alpha$ can be written in the following form:

$$Pr\left\{\sum_{i=1}^{n}a_ix_i\leq b\right\}\geq \alpha \qquad (4.59)$$

Since a_i and b are assumed to be independently normally distributed variables, the function

$$y(\mathbf{x}) = \sum_{i=1}^{n} a_i x_i - b$$

is also normally distributed with the following expected value and variance:

$$E[y(\mathbf{x})] = \sum_{i=1}^{n} E[a_i] x_i - E[b],$$

$$V[y(\mathbf{x})] = \sum_{i=1}^{n} V[a_i] x_i^2 + V[b].$$

We note that

$$\frac{\sum\limits_{i=1}^{n} a_i x_i - b - \left(\sum\limits_{i=1}^{n} E[a_i] x_i - E[b] \right)}{\sqrt{\sum\limits_{i=1}^{n} \sum\limits_{i=1}^{n} V[a_i] x_i^2 + V[b]}}$$

must be standardized normally distributed. Since the inequality $\sum\limits_{i=1}^{n} a_i x_i \leq b$ is equivalent to

$$\frac{\sum\limits_{i=1}^{n} a_i x_i - b - \left(\sum\limits_{i=1}^{n} E[a_i] x_i - E[b] \right)}{\sqrt{\sum\limits_{i=1}^{n} \sum\limits_{i=1}^{n} V[a_i] x_i^2 + V[b]}} \leq - \frac{\sum\limits_{i=1}^{n} E[a_i] x_i - E[b]}{\sqrt{\sum\limits_{i=1}^{n} \sum\limits_{i=1}^{n} V[a_i] x_i^2 + V[b]}},$$

the chance constraint (4.59) is equivalent to

$$Pr \left\{ \eta \leq - \frac{\sum\limits_{i=1}^{n} E[a_i] x_i - E[b]}{\sqrt{\sum\limits_{i=1}^{n} \sum\limits_{i=1}^{n} V[a_i] x_i^2 + V[b]}} \right\} \geq \alpha \qquad (4.60)$$

where η is the standardized normally distributed variable. Then the chance constraint (4.60) holds if and only if

$$\Phi^{-1}(\alpha) \leq - \frac{\sum\limits_{i=1}^{n} E[a_i] x_i - E[b]}{\sqrt{\sum\limits_{i=1}^{n} \sum\limits_{i=1}^{n} V[a_i] x_i^2 + V[b]}} \alpha \qquad (4.61)$$

That is, the deterministic equivalent of chance constraint is (4.58). The theorem is proved.

If f_i, g_r have linear forms, $i = 1, 2, \cdots, m; r = 1, 2, \cdots, p$, model (4.52) can be rewritten as

$$
\begin{cases}
\max \left[\bar{f}_1, \bar{f}_2, \cdots, \bar{f}_m \right] \\
\text{s.t.} \begin{cases} Pr\{\tilde{\mathbf{c}}_i^T \mathbf{x} \geq \bar{f}_i\} \geq \delta_i, i = 1, 2, \cdots, m \\ \tilde{\mathbf{a}}_r^T x \leq \tilde{b}_r, r = 1, 2, \cdots, p \\ \mathbf{x} \geq \mathbf{0} \end{cases}
\end{cases} \tag{4.62}
$$

where $\tilde{\mathbf{c}}_i = (\tilde{c}_{i1}, \tilde{c}_{i1}, \cdots, \tilde{c}_{in})^T, \tilde{\mathbf{a}}_r = (\tilde{a}_{r1}, \tilde{a}_{r1}, \cdots, \tilde{a}_{rn})^T$ are random vectors, $\tilde{b}_r$ are random variables, $i = 1, 2, \cdots, m, r = 1, 2, \cdots, p$.

Next, let us consider the first case when the decision maker only needs to require the accuracy of the approximation to exceed some value.

THEOREM 4.5

Assume that $\tilde{\mathbf{c}}_i$ are normally distributed random vectors with mean vector μ_i^c and positive definite covariance matrix V_i^c, written as $\tilde{\mathbf{c}}_i \sim \mathcal{N}(\mu_i^c, V_i^c), i = 1, 2, \cdots, m$. Similarly, $\tilde{\mathbf{a}}_r \sim \mathcal{N}(\mu_r^a, V_r^a)$. $\tilde{b}_r$ are normally distributed random variables with mean vector μ_r^b and positive definite covariance $(\sigma_r^b)^2$, written as $\tilde{b}_r \sim \mathcal{N}(\mu_r^b, (\sigma_r^b)^2)$. Define the similarity relationship $R_{h_r}^{\eta_r}$ by

$$
\mathbf{x} R_{h_r}^{\eta_r} \mathbf{y} \Leftrightarrow Pr\{|\tilde{\mathbf{a}}_r^T \mathbf{x} - \tilde{\mathbf{a}}_r^T \mathbf{y}| \leq h_r\} \geq \eta_r,
$$

where $\mathbf{x}, \mathbf{y} \in X$, $\eta_r(\in [0,1])$ is the confidence level, and h_r is the deviation that the decision maker permits. Then we have the following equivalent model of Ra-CCRM of problem (4.62) based on the similarity relationship $R_{h_r}^{\eta_r}$:

$$
\begin{cases}
\max \left[F_1(\mathbf{x}), F_2(\mathbf{x}), \cdots, F_m(\mathbf{x}) \right] \\
\text{s.t.} \begin{cases} \alpha(I_r) \geq \theta_r \\ \mathbf{x} \in \bar{I}_r, r = 1, 2, \cdots, p \\ \mathbf{x} \geq \mathbf{0} \end{cases}
\end{cases} \tag{4.63}
$$

where $F_i(\mathbf{x}) = \Phi^{-1}(1 - \delta_i)\sqrt{\mathbf{x}^T V_i^c \mathbf{x}} + \mu_i^{cT} \mathbf{x}, 0 \leq \theta_r, \eta_r \leq 0, \Omega_r = \{\mathbf{x} \in X | \mu_r^{aT} \mathbf{x} - \mu_r^b + \Phi^{-1}(\eta_r)\sqrt{\mathbf{x}^T V_i^a \mathbf{x}} + h_r \leq 0\}, \Theta_r = \{\mathbf{x} \in X | \mu_r^{aT} \mathbf{x} - \mu_r^b + \Phi^{-1}(\eta_r)\sqrt{\mathbf{x}^T V_i^a \mathbf{x}} - h_r \leq 0\}$ and θ_r satisfyies

$$
\left(\int \cdots \int_{\Omega_r} 1 dx_1 dx_2 \cdots dx_m \right) - \theta_r \left(\int \cdots \int_{\Theta_r} 1 dx_1 dx_2 \cdots dx_m \right) \geq 0, r = 1, 2, \cdots, p.
$$

PROOF Since $\tilde{\mathbf{c}}_i \sim \mathcal{N}(\mu_i^c, V_i^c)$ is normally distributed with mean vector μ_i^c and positive definite covariance matrix V_i^c, it follows that $\tilde{\mathbf{c}}_i^T \mathbf{x} \sim \mathcal{N}(\mu_i^{cT} \mathbf{x}, \mathbf{x}^T V_i^c \mathbf{x})$

is also a random variable. Then we have

$$Pr\{\tilde{c}_i^T \mathbf{x} \geq \bar{f}_i\} \geq \delta_i$$

$$\Leftrightarrow \delta_i \leq Pr\left\{\frac{\tilde{c}_i^T \mathbf{x} - \mu_i^{cT} \mathbf{x}}{\sqrt{\mathbf{x}^T V_i^c \mathbf{x}}} \geq \frac{\bar{f}_i - \mu_i^{cT} \mathbf{x}}{\sqrt{\mathbf{x}^T V_i^c \mathbf{x}}}\right\}$$

$$\Leftrightarrow \delta_i \leq 1 - Pr\left\{\frac{\tilde{c}_i^T \mathbf{x} - \mu_i^{cT} \mathbf{x}}{\sqrt{\mathbf{x}^T V_i^c \mathbf{x}}} \leq \frac{\bar{f}_i - \mu_i^{cT} \mathbf{x}}{\sqrt{\mathbf{x}^T V_i^c \mathbf{x}}}\right\}$$

$$\Leftrightarrow \delta_i \leq 1 - \Phi\left(\frac{\bar{f}_i - \mu_i^{cT} \mathbf{x}}{\sqrt{\mathbf{x}^T V_i^c \mathbf{x}}}\right)$$

$$\Leftrightarrow \bar{f}_i \leq \Phi^{-1}(1 - \delta_i)\sqrt{\mathbf{x}^T V_i^c \mathbf{x}} + \mu_i^{cT} \mathbf{x},$$

where Φ is the standardized normal distribution.

Denote $I_r = \{\mathbf{x}|\tilde{a}_r^T \mathbf{x} \leq \tilde{b}_r, r = 1, 2, \cdots, p\}$, then we have its upper and lower approximation under the similarity relationship R as follows:

$$\underline{I}_r = \{\mathbf{x}|Pr\{\tilde{a}_r^T \mathbf{x} \leq \tilde{b}_r - h_r\} \geq \eta_r\},$$

$$\bar{I}_r = \{\mathbf{x}|Pr\{\tilde{a}_r^T \mathbf{x} \leq \tilde{b}_r + h_r\} \geq \eta_r\}.$$

Since $\tilde{a}_r \sim \mathcal{N}(\mu_r^a, V_r^a), \tilde{b}_r \sim \mathcal{N}(\mu_r^b, (\sigma_r^b)^2)$, it follows that $\tilde{a}_r^T \mathbf{x} + \tilde{b}_r \sim \mathcal{N}(\mu_r^{aT} \mathbf{x} - \mu_r^b, \mathbf{x}^T V_r^a \mathbf{x} + (\sigma_r^b)^2)$, then

$$Pr\{\tilde{a}_r^T \mathbf{x} \leq \tilde{b}_r - h_r\} \geq \eta_r$$

$$\Rightarrow Pr\left\{\frac{(\tilde{a}_r^T \mathbf{x} - \tilde{b}_r) - (\mu_r^{aT} \mathbf{x} - \mu_r^b)}{\sqrt{\mathbf{x}^T V_r^a \mathbf{x} + (\sigma_r^b)^2}} \leq \frac{-h_r - (\mu_r^{aT} \mathbf{x} - \mu_r^b)}{\sqrt{\mathbf{x}^T V_r^a \mathbf{x} + (\sigma_r^b)^2}}\right\} \geq \eta_r$$

$$\Rightarrow \mu_r^{aT} \mathbf{x} - \mu_r^b + \Phi^{-1}(\eta_r)\sqrt{\mathbf{x}^T V_r^a \mathbf{x}} + h_r \leq 0.$$

Similarly, we have the following equivalent formula:

$$Pr\{\tilde{a}_r^T \mathbf{x} \leq \tilde{b}_r + h_r\} \geq \eta_r$$
$$\Rightarrow \mu_r^{aT} \mathbf{x} - \mu_r^b + \Phi^{-1}(\eta_r)\sqrt{\mathbf{x}^T V_i^a \mathbf{x}} - h_r \leq 0.$$

Then the upper and lower approximation can be rewritten as

$$\underline{I}_r = \{\mathbf{x}|\mu_r^{aT} \mathbf{x} - \mu_r^b + \Phi^{-1}(\eta_r)\sqrt{\mathbf{x}^T V_i^a \mathbf{x}} + h_r \leq 0\}$$

and

$$\bar{I}_r = \{\mathbf{x}|\mu_r^{aT} \mathbf{x} - \mu_r^b + \Phi^{-1}(\eta_r)\sqrt{\mathbf{x}^T V_i^a \mathbf{x}} - h_r \leq 0\}$$

As we know, $|A|$ expresses the cardinality of the set A in the finite universe. For the infinite universe, we use it to express the Lebesgue measure. Then we have

$$|\underline{I}| = \int \cdots \int_{\Omega_r} 1 dx_1 dx_2 \cdots dx_m,$$

where $\Omega_r = \{\mathbf{x}|\mu_r^{aT} \mathbf{x} - \mu_r^b + \Phi^{-1}(\eta_r)\sqrt{\mathbf{x}^T V_i^a \mathbf{x}} + h_r \leq 0\}$ and

$$|\bar{I}| = \int \cdots \int_{\Theta_r} 1 dx_1 dx_2 \cdots dx_m,$$

where $\Theta_r = \{x | \mu_r^{aT} x - \mu_r^b + \Phi^{-1}(\eta_r) \sqrt{x^T V_i^a x} - h_r \leq 0\}$. Since $\bar{f}_i$ obtains its maximal value as $\bar{f}_i = \Phi^{-1}(1 - \delta_i) \sqrt{x^T V_i^c x} + \mu_i^{cT} x$. It follows that the Ra-CCRM of problem (4.62) based on the similarity relationship $R_{h_r}^{\eta_r}$ is equivalent to

$$\begin{cases} \max [F_1(\mathbf{x}), F_2(\mathbf{x}), \cdots, F_m(\mathbf{x})] \\ \text{s.t.} \begin{cases} \alpha(I_r) \geq \theta_r \\ \mathbf{x} \in \bar{I}_r, r = 1, 2, \cdots, p \\ \mathbf{x} \geq \mathbf{0} \end{cases} \end{cases}$$

where $F_i(\mathbf{x}) = \Phi^{-1}(1 - \delta_i) \sqrt{\mathbf{x}^T V_i^c \mathbf{x}} + \mu_i^{cT} \mathbf{x}, i = 1, 2 \cdots, m$. This completes the proof.

By the first constraint in problem (4.63), we can fix h_r by ensuring that the accuracy of rough approximation is more than θ_r. Furthermore, the similarity relationship is fixed.

Second, let us convert the second constraint in problem (4.63) into a crisp one under the similarity relationship $R_{h_r}^{\eta_r}$. We have the following equivalent model of problem (4.63) under the similarity relationship based on $R_{h_r}^{\eta_r}$:

$$\begin{cases} \max [F_1(\mathbf{x}), F_2(\mathbf{x}), \cdots, F_m(\mathbf{x})] \\ \text{s.t.} \begin{cases} \mu_r^{aT} \mathbf{x} - \mu_r^b + \Phi^{-1}(\eta_r) \sqrt{\mathbf{x}^T V_i^a \mathbf{x}} - h_r \leq 0, r = 1, 2, \cdots, m \\ \mathbf{x} \geq \mathbf{0} \end{cases} \end{cases} \qquad (4.64)$$

where $F_i(\mathbf{x}) = \Phi^{-1}(1 - \delta_i) \sqrt{\mathbf{x}^T V_i^c \mathbf{x}} + \mu_i^{cT} \mathbf{x}, i = 1, 2 \cdots, m$.

4.4.2.2 Two-Stage Method

In this section, we will use the two-stage method to seek the efficient solution to the crisp multiobjective decision-making problem

$$\begin{cases} \max [H_1(\mathbf{x}), H_2(\mathbf{x}), \cdots, H_m(\mathbf{x})] \\ \text{s.t. } \mathbf{x} \in X \end{cases} \qquad (4.65)$$

The two-stage method is proposed by R. Li [187] on the basis of the maximin method proposed by H. Zimmermann [347].

The first stage. Apply Zimmermann's minimum operator to obtain the maximal satisfying degree α^0 of the objective set and the related feasible solution x^0, that is,

$$\begin{cases} \max \alpha \\ \text{s.t.} \begin{cases} \mu_k(\mathbf{x}) = \dfrac{H_k(\mathbf{x}) - H_k'}{H_k^* - H_k'} \geq \alpha, k = 1, 2, \cdots, m \\ \mathbf{x} \in X \end{cases} \end{cases} \qquad (4.66)$$

Assume that the optimal solution to problem (4.66) is $(\mathbf{x}^0, \alpha^0)$, where α^0 is the optimal satisfying degree of the whole objective sets. If the optimal solution to problem (4.66) is unique, $\mathbf{x}^0$ is the efficient solution to problem (4.65). However, as we cannot usually know if the optimal solution to problem (4.66) is unique, then the efficiency of $\mathbf{x}^0$ must be checked by the following stage.

The second stage. Check the efficiency of $\mathbf{x}^0$ or seek the new efficient solution $\mathbf{x}^1$. Construct a new model whose objective function is to maximize the average satisfying degree of all objects subject to the additional constraint $\alpha_k \geq \alpha^0 (k = 1, 2, \cdots, m)$. Since the compensatory of the arithmetic mean operator, the solution obtained in the second stage is efficient. The existence of the constraint $\alpha_k \geq \alpha^0 (k = 1, 2, \cdots, m)$ guarantees the mutual equilibrium of every objective function.

$$
\begin{cases}
\max \frac{1}{m} \sum\limits_{k=1}^{m} \alpha_k \\
\text{s.t.} \begin{cases}
\alpha^0 \leq \alpha_k \leq \mu_k(x), k = 1, 2, \cdots, m \\
0 \leq \alpha_k \leq 1 \\
x \in X
\end{cases}
\end{cases}
\tag{4.67}
$$

Assume that the optimal solution to problem (4.67) is $\mathbf{x}^1$. It is easy to prove that $\mathbf{x}^1$ is also the solution to problem (4.66), thus we have $\mathbf{x}^1 = \mathbf{x}^0$ if the solution to problem (4.66) is unique. However, if the solution to problem (4.66) is not unique, $\mathbf{x}^0$ may be efficient solution or not, and we can guarantee that $\mathbf{x}^1$ is definitely efficient. Thus, in any case, the two-stage method can provide an efficient solution in the second stage.

4.4.2.3 Numerical Example

Example 4.5

Consider the problem

$$
\begin{cases}
\max[\bar{f}_1, \bar{f}_2] \\
\text{s.t.} \begin{cases}
Pr\{\tilde{\xi}_1 x_1 + \tilde{\xi}_2 x_2 + \tilde{\xi}_3 x_3 + \tilde{\xi}_4 x_4 \geq \bar{f}_1\} \geq 0.9 \\
Pr\{\tilde{\xi}_5 x_1 + \tilde{\xi}_6 x_2 + \tilde{\xi}_7 x_3 + \tilde{\xi}_8 x_4 \geq \bar{f}_2\} \geq 0.8 \\
2x_1 \tilde{\xi}_1 + 2x_2 \tilde{\xi}_2 + x_3 \tilde{\xi}_3 + x_4 \tilde{\xi}_4 \leq 100 \tilde{\xi}_9 \\
x_1 + x_2 + x_3 + x_4 \geq 30 \\
x_1 + x_2 + x_3 + x_4 \leq 100 \\
x_i \geq 5, i = 1, 2, 3, 4
\end{cases}
\end{cases}
\tag{4.68}
$$

where $\tilde{\xi}_i (i = 1, 2, \cdots, 8)$ are independent random variables defined as

$$
\begin{aligned}
&\xi_1 \sim \mathcal{N}(5,5), \quad \xi_2 \sim \mathcal{N}(8,6), \quad \xi_3 \sim \mathcal{N}(12,5), \\
&\xi_4 \sim \mathcal{N}(18,6), \quad \xi_5 \sim \mathcal{N}(15,2) \quad \xi_6 \sim \mathcal{N}(28,3), \\
&\xi_7 \sim \mathcal{N}(20,4), \quad \xi_8 \sim \mathcal{N}(20,3), \quad \xi_9 \sim \mathcal{N}(8,4).
\end{aligned}
$$

Since $\Phi^{-1}(1 - 0.9) = -1.28, \Phi^{-1}(1 - 0.8) = -0.84$, it follows from Theorem

4.5 that problem (4.68) is equivalent to

$$
\begin{cases}
\max F_1(x) = -1.28(\sqrt{5x_1^2 + 6x_2^2 + 5x_3^2 + 6x_4^2}) + (5x_1 + 8x_2 + 12x_3 + 18x_4) \\
\max F_2(x) = -0.84\sqrt{2x_1^2 + 3x_2^2 + 4x_3^2 + 3x_4^2} + (15x_1 + 28x_2 + 20x_3 + 20x_4) \\
\text{s.t.} \begin{cases}
\int\int \cdots \int_{\underline{I}} dx_1 dx_2 \cdots dx_n - \theta \int\int \cdots \int_{\bar{I}} dx_1 dx_2 \cdots dx_n \geq 0 \\
x \in \bar{I} \\
x_1 + x_2 + x_3 + x_4 \geq 30 \\
x_1 + x_2 + x_3 + x_4 \leq 100 \\
x_i \geq 5, i = 1,2,3,4
\end{cases}
\end{cases}
$$

(4.69)

Let $\eta = 0.9$, then $\underline{I} = \{x | 1.28\sqrt{20x_1^2 + 24x_2^2 + 5x_3^2 + 6x_4^2 + 40000 + 10x_1 + 24x_2 + 12x_3 + 18x_4 - 800 + h \leq 0}\}, \bar{I} = \{x | 1.28\sqrt{20x_1^2 + 24x_2^2 + 5x_3^2 + 6x_4^2 + 40000 + 10x_1 + 24x_2 + 12x_3 + 18x_4 - 800 - h \leq 0}\}$.

If the DM requires that the approximation accuracy exceed 0.9, we get the maximum deviation $h = 14.6$, such that $\alpha(I) \geq 0.9$. After that, it follows from (4.64) that (4.69) is equivalent to

$$
\begin{cases}
\max F_1(x) = -1.28(\sqrt{5x_1^2 + 6x_2^2 + 5x_3^2 + 6x_4^2}) + (5x_1 + 8x_2 + 12x_3 + 18x_4) \\
\max F_2(x) = -0.84\sqrt{2x_1^2 + 3x_2^2 + 4x_3^2 + 3x_4^2} + (15x_1 + 28x_2 + 20x_3 + 20x_4) \\
\text{s.t.} \begin{cases}
1.28\sqrt{20x_1^2 + 24x_2^2 + 5x_3^2 + 6x_4^2 + 40000 + 10x_1 + 24x_2 + 12x_3 + 18x_4} \\
\leq 814.6 \\
x_1 + x_2 + x_3 + x_4 \geq 30 \\
x_1 + x_2 + x_3 + x_4 \leq 100 \\
x_i \geq 5, i = 1,2,3,4
\end{cases}
\end{cases}
$$

Then we use the two-stage method to solve the above problem. Construct the membership function as

$$
\mu_i = \frac{F_i(x) - F_i^0}{F_i^1 - F_i^0}, i = 1,2,
$$

where $F_i^1 = \max_{x \in X} F_i(x), F_i^0 = \min_{x \in X} F_i(x)$. Then we get

$$
F_1^1 = 387.19, F_1^0 = 214.61, F_2^1 = 749.66, F_2^0 = 542.78.
$$

According to the two-stage method, the problem (4.70) can be written as

$$
\begin{cases}
\max \alpha \\
\text{s.t.}
\begin{cases}
\dfrac{-1.28\sqrt{5x_1^2+6x_2^2+5x_3^2+6x_4^2}+(5x_1+8x_2+12x_3+18x_4)-214.61}{172.58} \geq \alpha \\[2mm]
\dfrac{-0.84\sqrt{2x_1^2+3x_2^2+4x_3^2+3x_4^2}+(15x_1+28x_2+20x_3+20x_4)-542.78}{206.88} \geq \alpha \\[2mm]
1.28\sqrt{20x_1^2+24x_2^2+5x_3^2+6x_4^2+40000}+10x_1+24x_2+12x_3+18x_4 \\
\quad \leq 814.6 \\
x_1+x_2+x_3+x_4 \geq 30 \\
x_1+x_2+x_3+x_4 \leq 100 \\
x_i \geq 5, i=1,2,3,4
\end{cases}
\end{cases}
$$

(4.70)

Then we obtain the optimal solution to problem $\mathbf{x}^* = (5,5,20.8,7)$ and $\alpha = 0.932$. At the second stage, construct the following problem to check the efficiency of $\mathbf{x}^*$:

$$
\begin{cases}
\max \frac{1}{2}(\alpha_1 + \alpha_2) \\
\text{s.t.}
\begin{cases}
\dfrac{-1.28\sqrt{5x_1^2+6x_2^2+5x_3^2+6x_4^2}+(5x_1+8x_2+12x_3+18x_4)-214.61}{172.58} \geq \alpha_1 \\[2mm]
\dfrac{-0.84\sqrt{2x_1^2+3x_2^2+4x_3^2+3x_4^2}+(15x_1+28x_2+20x_3+20x_4)-542.78}{206.88} \geq \alpha_2 \\[2mm]
1.28\sqrt{20x_1^2+24x_2^2+5x_3^2+6x_4^2+40000}+10x_1+24x_2+12x_3+18x_4 \\
\quad \leq 814.6 \\
x_1+x_2+x_3+x_4 \geq 30 \\
x_1+x_2+x_3+x_4 \leq 100 \\
x_i \geq 5, i=1,2,3,4
\end{cases}
\end{cases}
$$

(4.71)

We get $\mathbf{x}'^* = (5,5,23.8,5)$.

4.4.3 Ra-NLCCRM and Random Simulation-Based APSO

As problem (4.54) has nonlinear forms, that is, there is one nonlinear function of f_i, g_r at least, which is called random nonlinear chance-constrained rough model (Ra-NLCCRM). It is usually impossible to be converted into crisp equivalent model. Random simulation-based adaptive particle swarm optimization (APSO) with selection operator, a solving method, is used to deal with it.

4.4.3.1 Random Simulation 2 for Critical Value

Suppose that ξ is an n-dimensional random vector defined on the probability space $(\Omega, \mathscr{A}, Pos)$, and $f : \mathbf{R}^n \to \mathbf{R}$ is a measurable function. The problem is to determine the maximal value $\bar{f}$ such that

$$
Pr\{f(\xi) \geq \bar{f}\} \geq \alpha
$$

(4.72)

where α is a predetermined confidence level with $0 < \alpha < 1$. We generate ω_k from Ω according to the probability measure Pr and write $\xi_k = \xi(\omega_k)$ for $k = 1, 2, \cdots, N$.

Now we define

$$h(\xi_k) = \begin{cases} 1, & \text{if } f(\xi_k) \geq \bar{f} \\ 0, & \text{otherwise} \end{cases} \tag{4.73}$$

for $k = 1, 2, \cdots, N$, which are a sequence of random variables, and $E[h(\xi_k)] = \alpha$ for all k. By the strong law of large numbers, we obtain

$$\frac{\sum_{k=1}^{N} h(\xi_k)}{N} \to \alpha$$

as $N \to \infty$. Note that the sum $\sum_{k=1}^{N} h(\xi_k)$ is just the number of ξ_k satisfying $f(\xi_k) \geq \bar{f}$ for $k = 1, 2, \cdots, N$. Thus the value $\bar{f}$ can be taken as the N'th largest element in the sequence $\{f(\xi_1), f(\xi_2), \cdots, f(\xi_N)\}$, where N' is the integer part of αN. The procedure of random simulation for CCM are summarized as follows:

Step 1. Set N' as the integer part of αN.
Step 2. Generate $\omega_1, \omega_2, \cdots, \omega_N$ from Ω according to the probability measure Pr.
Step 3. Return the N'th largest element in $\{f(\xi_1), f(\xi_2), \cdots, f(\xi_N)\}$.

Example 4.6
Let us employ the stochastic simulation to search for the maximal $\bar{f}$ such that

$$Pr\left\{\sqrt{\xi_1^2 + \xi_2^2 + \xi_3^2} \geq \bar{f}\right\} \geq 0.8$$

where $\xi_1 \sim exp(1)$ is an exponentially distributed variable, $\xi_2 \sim \mathcal{N}(3, 1)$ is a normally distributed variable, and $\xi_3 \sim \mathcal{U}(0, 1)$ is a uniformly distributed variable. A run of stochastic simulation with 1000 cycles shows that $\bar{f} = 2.0910$.

4.4.3.2   APSO

PSO is based on an analogy with the choreography of flight of a flock of birds. Developments and resources in the particle swarm algorithm are in [179, 181]. Although PSO has shown some important advances by providing high speed of convergence in specific problems, it does exhibit some shortages. It is found that PSO has a poor ability to search at a fine grain because it lacks a velocity control mechanism [9]. Some improvements in the PSO algorithm are proposed in [134]. Many approaches have been attempted to improve the performance of PSO by variable inertia weight. The inertia weight is critical to the performance of PSO, which balances global exploration and local exploitation abilities of the swarm. A big inertia weight facilitates exploration, but it takes the particle a long time to converge. Similar to other evolutionary algorithms, PSO method conducts its search using a population of particles corresponding to individuals. Each particle in the swarm represents a candidate solution to the problem. It starts with a random initialization of a population of individuals in the search space and works on the social behavior of the particles in the swarm, such as birds flocking, fish schooling, and the swarm theory. Therefore, it finds the global optimum by simply adjusting the trajectory of each individual toward its own best location and toward the best particle of the swarm at each generation of evolution. However, the trajectory of each individual in the search space is adjusted by dynamically altering the velocity of each particle according to its own flying experience and that of other particles in the search space. This population-based robust algorithm always ensures convergence to the global optimum solution as compared to a GA.

Many scholars ([216, 236, 296, 332, 334]) proposed the APSO algorithm to dynamically adjust the velocity of each particle to accelerate the convergence. This section mainly introduces the APSO algorithm proposed by B. Panigrahi [236], and interested readers can refer to the related literatures. In the simple PSO method, the inertia weight is made constant for all the particles in a single generation, but the most important parameter that moves the current position toward the optimum position is the inertia weight ω. In order to increase the search ability, the algorithm should be redefined in such manner that the movement of the swarm should be controlled by the objective function. In the proposed adaptive PSO, the particle position is adjusted such that the highly fitted particle (best particle) moves slowly when compared to the lowly fitted particle. This can be achieved by selecting different ω values for each particle according to its rank, between ω_{min} and ω_{max} as in the following form:

$$\omega_i = \omega_{min} + \frac{(\omega_{max} - \omega_{min}) \cdot \text{Rank}_i}{\text{Total population}} \tag{4.74}$$

Hence, it follows from Equation (4.74) that the best particle takes the first rank, and the inertia weight for that particle is set to the minimum value, while that for the lowest fitted particle takes the maximum inertia weight, which makes that particle move with a high velocity. The velocity of each particle is updated using Equation (4.75), and if any updated velocity goes beyond V_{max}, it is limited to V_{max} using Equation (4.76):

$$v_{ij}(t+1) = \omega_i v_{ij}(t) + c_1 r_1(p_{ij}(t) - x_{ij}(t)) + c_2 r_2(p_{gj}(t) - x_{gj}(t)) \tag{4.75}$$

$$v_{ij}(t+1) = \text{sign}(v_{ij}(t+1)) \cdot \min(v_{ij}(t+1), V_{jmax}) \tag{4.76}$$

where $j = 1, 2, \cdots, d$ and $i = 1, 2, \cdots, n$. The new particle position is obtained by using Equation (4.77), and if any particle position goes beyond the range specified, it is adjusted to its boundary using Equation (4.78):

$$x_{ij}(t+1) = x_{ij}(t) + v_{ij}(t+1) \tag{4.77}$$

$$\begin{cases} x_{ij}(t+1) = \min(x_{ij}(t), \text{range}_{jmax}) \\ x_{ij}(t+1) = \max(x_{ij}(t), \text{range}_{jmin}) \end{cases} \tag{4.78}$$

where $j = 1, 2, \cdots, d$ and $i = 1, 2, \cdots, n$. The concept of reinitialization is introduced in the proposed APSO algorithm after a specific number of generations if there is no improvement in the convergence of the algorithm. The population of the proposed

APSO at the end of the above-mentioned specific generation is reinitialized with new randomly generated individuals. The number of these new individuals is selected from the k least-fit individuals of the original population, where k is the percentage of the total population to be changed. This effect is favorable when the algorithm prematurely converges to a local optimum and further improvement is not noticeable. This reinitialization of the population is performed after checking the changes in the "Fbest" value in each and every specific number of generations. The procedure for the APSO algorithm can be summarized as follows:

Step 1. Get the input parameters like range [min max] for each of the variables, c_1, c_2, iteration counter=0, V_{max}, ω_{max} and ω_{min}.

Step 2. Initialize n number of population of particles of dimension d with random positions and velocities.

Step 3. Increment the iteration counter by one.

Step 4. Evaluate the fitness function of all particles in the population, find the particles best position Pbest for each particle, and update its objective value. Similarly, find the global best position (Gbest) among all the particles and update its objective value.

Step 5. If the stopping criterion is met, go to Step 11. Otherwise continue.

Step 6. Evaluate the inertia factor according to Equation (4.74), so that each particles movement is directly controlled by its fitness value.

Step 7. Update the velocity using Equation (4.75) and correct it using Equation (4.76).

Step 8. Update the position of each particle according to Equation (4.77), and if the new position goes out of range, set it to the boundary value using Equation (4.78).

Step 9. The elites are inserted in the first position of the new population in order to maintain the best particle found so far.

Step 10. For every five generations, this $F_{Best,new}$ value (at the end of these 5 generations) is compared with the $F_{Best,old}$ value (at the beginning of these five generations, if there is no noticeable change, then re-initialize $k\%$ of the population. Go to Step 3.

Step 11. Output the Gbest particle and its objective value.

4.4.3.3 Numerical Example

Example 4.7

Consider the problem

$$
\begin{cases}
\max[\bar{f}_1, \bar{f}_2] \\
\text{s.t.} \begin{cases}
Pr\{2\xi_1 x_1^2 + 3\xi_2 x_2 - \xi_3 x_3 \geq \bar{f}_1\} \geq 0.8 \\
Pr\{5\xi_1 x_1 - 2\xi_2 x_2 + 2\xi_3 x_3 \geq \bar{f}_2\} \geq 0.8 \\
5x_1 \xi_1^2 - 3x_2^2 \xi_2 + 6\sqrt{x_3}\xi_3^3 \leq 50 \\
4\sqrt{x_1} + 6x_2 - 4.5x_3 \leq 20 \\
x_1 + x_2 + x_3 \leq 15 \\
x_1, x_2, x_3 \geq 0
\end{cases}
\end{cases}
\tag{4.79}
$$

TABLE 4.2   The optimal solution by APSO

c_1	c_2	x_1	x_2	x_3	$\bar{f}_1$	$\bar{f}_2$
2	2.05	2.49	6.18	5.25	122.81	45.36
2	2.1	2.30	6.35	5.93	114.68	46.66
2	2.5	2.27	6.31	6.46	110.33	51.76
2	3	2.36	6.30	6.81	111.77	57.16
3	2.05	2.39	6.20	5.37	118.61	44.36
3	2.1	2.40	6.29	5.88	117.9	48.76
3	2.5	2.21	6.33	6.42	108.68	49.96
3	3	2.47	6.18	5.36	121.47	46.06

ξ_1, ξ_2, ξ_3 are random variables characterized as

$$\xi_1 \sim \mathcal{N}(4,1), \xi_2 \sim \mathcal{N}(5,1), \xi_3 \sim \exp(2).$$

Define the similarity relationship R_h^η as $xR_{h,r}^{\eta}y : Pr\{(x_1,x_2,x_3),(y_1,y_2,y_3) \in X | (5x_1\xi_1^2 - 3x_2^2\xi_2 + 6\sqrt{x_3}\xi_3^3) - (5y_1\xi_1^2 - 3y_2^2\xi_2 + 6\sqrt{y_3}\xi_3^3)| \le h\} \ge \eta$, where $X = \{x|4\sqrt{x_1} + 6x_2 - 4.5x_3 \le 20, x_1 + x_2 + x_3 \le 15, x_1,x_2,x_3 \ge 0\}$.

If the DM require that $\eta = 0.8$ and the approximation accuracy exceeds 0.9 based on rough approximation, then we get the following Ra-CCRM:

$$
\begin{cases}
\max[\bar{f}_1, \bar{f}_2] \\
\text{s.t.} \begin{cases}
Pr\{2\xi_1 x_1^2 + 3\xi_2 x_2 - \xi_3 x_3 \ge \bar{f}_1\} \ge 0.8 \\
Pr\{5\xi_1 x_1 - 2\xi_2 x_2 + 2\xi_3 x_3 \ge \bar{f}_2\} \ge 0.8 \\
\alpha(I) \ge 0.95 \\
x \in \bar{I}
\end{cases}
\end{cases}
\tag{4.80}
$$

where I expresses the feasible set. First, we get the maximum deviation $h = 2.6$ such that $\alpha(I) = \frac{\int_{\underline{I}} dx_1 dx_2 dx_3}{\int_{\bar{I}} dx_1 dx_2 dx_3} \ge 0.9$, where $\underline{I} = \{(x_1,x_2,x_3) | Pr\{5x_1\xi_1^2 - 3x_2^2\xi_2 + 6\sqrt{x_3}\xi_3^3 \le 50 - h\} \ge \eta\}\}$ and $\bar{I} = \{(x_1,x_2,x_3) | Pr\{5x_1\xi_1^2 - 3x_2^2\xi_2 + 6\sqrt{x_3}\xi_3^3 \le 50 + h\} \ge \eta\}$. Then we have

$$
\begin{cases}
\max[\bar{f}_1, \bar{f}_2] \\
\text{s.t.} \begin{cases}
Pr\{2\xi_1 x_1^2 + 3\xi_2 x_2 - \xi_3 x_3 \ge \bar{f}_1\} \ge 0.8 \\
Pr\{5\xi_1 x_1 - 2\xi_2 x_2 + 2\xi_3 x_3 \ge \bar{f}_2\} \ge 0.8 \\
5x_1\xi_1^2 - 3x_2^2\xi_2 + 6\sqrt{x_3}\xi_3^3 \le 52.6 \\
4\sqrt{x_1} + 6x_2 - 4.5x_3 \le 20 \\
x_1 + x_2 + x_3 \le 15 \\
x_1,x_2,x_3 \ge 0
\end{cases}
\end{cases}
$$

Next we will use the APSO to solve the above problem. The process of generating a new position for a selected individual in the swarm is depicted in the following equation:

$$v_{t+1}^i = wv_t^i + c_1 \cdot \text{rand}() \cdot (P_t^i - x_t^i) + c_2 \cdot \text{rand}() \cdot (G_t - x_t^i),$$

where v_t^i and x_t^i are the ith particle current velocity and position, P_t^i and G_t are the kth particle best position and the global best position visited so far, respectively, $w = 0.7298$ is the inertia weight, c_1, c_2 are learning factors, and rand() is a random number in $[0,1]$. After simulation with many cycles, we get the optimal solution under different weights as shown in Table 4.2.

4.5 Ra-DCRM

The third way to deal with model (4.9) is DCM. As the rough approximation techniques are used to tackle the random DCM, we can get the random dependent chance rough model (Ra-DCRM).

4.5.1 General Model for Ra-DCRM

If the DM wants to maximize the probability of the objective function values to be not less than the aspiration levels in (4.9) and do not strictly requires that the constraints holds, then the dependent-chance model of the problem (4.9) is obtained as follows:

$$\begin{cases} \max[Pr\{f_1(\mathbf{x},\boldsymbol{\xi}) \geq \bar{f}_1\}, Pr\{f_2(\mathbf{x},\boldsymbol{\xi}) \geq \bar{f}_2\}, \cdots, Pr\{f_m(\mathbf{x},\boldsymbol{\xi}) \geq \bar{f}_m\}] \\ \text{s.t.} \begin{cases} g_r(\mathbf{x},\boldsymbol{\xi}) \leq 0, r = 1, 2, \cdots, p \\ \mathbf{x} \in X \end{cases} \end{cases} \qquad (4.81)$$

As we know, the feasible set $I = \{\mathbf{x} \in X | g_r(\mathbf{x},\boldsymbol{\xi}) \leq 0, r = 1, 2, \cdots, p\}$ is not crisp because of the existence of the random vector $\boldsymbol{\xi}$. Then we use the rough set to deal with the feasible set I under the probabilistic environment. Since Pawlak proposed the concept of rough set, it has been rapidly developed and applied to many fields by many scholars. Here, we apply the technique of rough approximation to deal with random programming problems. According to R. Slowiński and D. Vanderpooten [277], we can extend it to the rough approximation under the similarity relationship.

First, we define the following binary relationship $R_{h_r}^{\eta_r}$ for the constraints $g_r(\mathbf{x},\boldsymbol{\xi}) \leq 0, r = 1, 2, \cdots, p$:

$$\mathbf{x} R_{h_r}^{\eta_r} \mathbf{y} \Leftrightarrow Pr\{|g_r(\mathbf{x},\boldsymbol{\xi}) - g_r(\mathbf{y},\boldsymbol{\xi})| \leq h_r\} \geq \eta_r,$$

where $\mathbf{x}, \mathbf{y} \in X, \eta_r (0 \leq \eta_r \leq 1)$, and $h_r(>0)$ is the deviation the DM permits. In fact, for any $\mathbf{x} \in X$, we have that

$$Pr\{|g_r(\mathbf{x},\boldsymbol{\xi}) - g_r(\mathbf{y},\boldsymbol{\xi})| \leq h_r\} = 1 \geq \eta_r$$

So, $R_{h_r}^{\eta}$ has reflexivity. Then $R_{h_r}^{\eta_r}$ is a similarity relationship. For the converted feasible set, it requires two steps to solve the programming problem. The first step is to require that the lower and upper approximation sets are close enough to the initial feasible set. This means that we must guarantee the accuracy of the approximation to extend at some level. The second step is that we directly solve the programming problem in the lower and upper approximation set. After ensuring that the required accuracy is achieved, we can only solve the programming problem in the upper approximation. Then we can get the DCM based on rough approximation of the problem (4.81) as follows:

$$
\begin{cases}
\max[Pr\{f_1(\mathbf{x},\boldsymbol{\xi}) \geq \bar{f}_1\}, Pr\{f_2(\mathbf{x},\boldsymbol{\xi}) \geq \bar{f}_2\}, \cdots, Pr\{f_m(\mathbf{x},\boldsymbol{\xi}) \geq \bar{f}_m\}] \\
\text{s.t.} \begin{cases}
\alpha(I_r) \geq \theta_r \\
\mathbf{x} \in \bar{I}_r \\
r = 1, 2, \cdots, p
\end{cases}
\end{cases}
\tag{4.82}
$$

where $\alpha(I_r)$ expresses the accuracy of the approximation, $I_r = \{\mathbf{x} | g_r(\mathbf{x},\boldsymbol{\xi}) \leq 0\}$.

By introducing variables δ_i, model (4.82) can also be rewritten as

$$
\begin{cases}
\max[\delta_1, \delta_2, \cdots, \delta_m] \\
\text{s.t.} \begin{cases}
Pr\{f_i(\mathbf{x},\boldsymbol{\xi}) \geq \bar{f}_i\} \geq \delta_i, i = 1, 2, \cdots, m \\
\alpha(I_r) \geq \theta_r \\
\mathbf{x} \in \bar{I}_r \\
r = 1, 2, \cdots, p
\end{cases}
\end{cases}
\tag{4.83}
$$

DEFINITION 4.18   *A feasible solution $\mathbf{x}^*$ to problem (4.83) is said to be an efficient solution if and only if there not exist feasible solutions $\mathbf{x}$ such that*

$$
Pr\{f_i(\mathbf{x},\boldsymbol{\xi}) \geq \bar{f}_i\} \geq Pr\{[f_i(\mathbf{x}^*,\boldsymbol{\xi}) \geq \bar{f}_i\}, i = 1, 2, \cdots, m
\tag{4.84}
$$

and there is at least one $j \in \{1, 2, \cdots, m\}$ such that

$$
Pr\{f_i(\mathbf{x},\boldsymbol{\xi}) \geq \bar{f}_i\} > Pr\{[f_i(\mathbf{x}^*,\boldsymbol{\xi}) \geq \bar{f}_i\}.
$$

4.5.2   Ra-LDCRM and the SWT Method

In order to solve the multiple objective decision-making problem (4.81), we must compute the crisp probability value of the random event. However, as we know, this process is hard work sometimes. Next, we will focus on the linear random multiple objective decision making problem as follows:

$$
\begin{cases}
\max[\tilde{\mathbf{c}}_1^T \mathbf{x}, \tilde{\mathbf{c}}_2^T \mathbf{x}, \cdots, \tilde{\mathbf{c}}_m^T \mathbf{x}] \\
\text{s.t.} \begin{cases}
\tilde{\mathbf{a}}_r^T \mathbf{x} \leq \tilde{b}_r, r = 1, 2, \cdots, p \\
\mathbf{x} \in X
\end{cases}
\end{cases}
\tag{4.85}
$$

where $\tilde{\mathbf{c}}_i = (\tilde{c}_{i1}, \tilde{c}_{i1}, \cdots, \tilde{c}_{in})^T, \tilde{\mathbf{a}}_r = (\tilde{a}_{r1}, \tilde{a}_{r1}, \cdots, \tilde{a}_{rn})^T$ are random vectors, and $\tilde{b}_r$ are random variables, $i = 1, 2, \cdots, m, r = 1, 2, \cdots, p$.

By using the random dependent-chance model and rough approximation, we have the following random linear dependent-chance rough model (Ra-LDCRM):

$$
\begin{cases}
\max[\delta_1, \delta_2, \cdots, \delta_m] \\
\text{s.t.} \begin{cases}
Pr\{\tilde{\mathbf{c}}_i^T \mathbf{x} \geq \bar{f}_i\} \geq \delta_i, i = 1, 2, \cdots, m \\
\alpha(I_r) \geq \theta_r \\
\mathbf{x} \in \bar{I}_r \\
r = 1, 2, \cdots, p
\end{cases}
\end{cases} \tag{4.86}
$$

where $\alpha(I_r)$ expresses the accuracy of the approximation, and $I_r = \{\mathbf{x} | \tilde{\mathbf{a}}_r^T \mathbf{x} \leq \tilde{b}_r, r = 1, 2, \cdots, p, \mathbf{x} \in X\}, \theta_r \in [0, 1]$.

4.5.2.1 Crisp Equivalent Model

First, let us consider the first case when the decision maker only needs to require the accuracy of the approximation to exceed some value.

THEOREM 4.6
Assume that $\tilde{\mathbf{c}}_i$ is normally distributed with mean vector μ_i^c and positive definite covariance matrix V_i^c, written as $\tilde{\mathbf{c}}_i \sim \mathcal{N}(\mu_i^c, V_i^c)$. Similarly, $\tilde{a}_r \sim \mathcal{N}(\mu_r^a, V_r^a)$. Define the similarity relationship $\mathbf{x} R_{h_r}^{\eta_r} \mathbf{y}$ as

$$
\mathbf{x} R_{h_r}^{\eta_r} \mathbf{y} \Leftrightarrow Pr\{|g_r(\mathbf{x}, \boldsymbol{\xi}) - g_r(\mathbf{y}, \boldsymbol{\xi})| \leq h_r\} \geq \eta_r \tag{4.87}
$$

where $\mathbf{x}, \mathbf{y} \in X$, $\eta_r (\in [0, 1])$ is the confidence level, and h_r is the deviation that the decision maker permits. Then we have the following equivalent model of problem (4.86) under the similarity relationship $\mathbf{x} R_{h_r}^{\eta_r} \mathbf{y}$:

$$
\begin{cases}
\max \left[\Phi\left(\dfrac{\bar{f}_1 - \mu_1^{cT} \mathbf{x}}{\sqrt{\mathbf{x}^T V_1^c \mathbf{x}}} \right), \Phi\left(\dfrac{\bar{f}_2 - \mu_2^{cT} \mathbf{x}}{\sqrt{\mathbf{x}^T V_2^c \mathbf{x}}} \right), \cdots, \Phi\left(\dfrac{\bar{f}_m - \mu_m^{cT} \mathbf{x}}{\sqrt{\mathbf{x}^T V_m^c \mathbf{x}}} \right) \right] \\
\text{s.t.} \begin{cases}
\alpha(I_r) \geq \theta_r \\
\mathbf{x} \in \bar{I}_r \\
r = 1, 2, \cdots, p
\end{cases}
\end{cases} \tag{4.88}
$$

where $0 \leq \theta_r, \eta_r \leq 0, \Omega_r = \{\mathbf{x} | \mu_r^{aT} \mathbf{x} + \Phi^{-1}(\eta_r)\sqrt{\mathbf{x}^T V_m^a \mathbf{x}} \leq b_r - h_r\}, \Theta_r = \{\mathbf{x} | \mu_r^{aT} \mathbf{x} + \Phi^{-1}(\eta_r)\sqrt{\mathbf{x}^T V_m^u \mathbf{x}} \leq b_r + h_r\}$, and θ_r satisfies

$$
\left(\int \cdots \int_{\Omega_r} 1 dx_1 dx_2 \cdots dx_m \right) - \theta_r \left(\int \cdots \int_{\Theta_r} 1 dx_1 dx_2 \cdots dx_m \right), r = 1, 2, \cdots, p
$$

PROOF   Since $\tilde{\mathbf{c}}_i \sim \mathcal{N}(\mu_i^c, V_i^c)$ is normally distributed with mean vector μ_i^c and positive definite covariance matrix V_i^c, it follows that $\tilde{\mathbf{c}}_i^T \mathbf{x} \sim \mathcal{N}(\mu_i^{cT} \mathbf{x}, \mathbf{x}^T V_i^c \mathbf{x})$ is also a random variable. Then we have

$$
Pr\{\tilde{\mathbf{c}}_i^T \mathbf{x} \geq \bar{f}_i\} = Pr\left\{ \frac{\mu_i^{cT} \mathbf{x} - \mu_1^{cT} \mathbf{x}}{\sqrt{\mathbf{x}^T V_i^c \mathbf{x}}} \leq \frac{\bar{f}_1 - \mu_1^{cT} \mathbf{x}}{\sqrt{\mathbf{x}^T V_i^c \mathbf{x}}} \right\} = \Phi\left(\frac{\bar{f}_1 - \mu_1^{cT} \mathbf{x}}{\sqrt{\mathbf{x}^T V_i^c \mathbf{x}}} \right),
$$

where Φ is the standard normally distributed function. Denote $I_r = \{\mathbf{x}|\tilde{a}_r^T\mathbf{x} \leq \tilde{b}_r, r = 1, 2, \cdots, p, \mathbf{x} \in X\}$, and then we have its upper and lower approximation under the similarity relationship $R_{hr}^{\eta_r}$ as

$$\underline{I}_r = \{\mathbf{x} \in X | Pr\{\tilde{a}_r^T\mathbf{x} \leq \tilde{b}_r - h_r\} \geq \eta_r\}$$

and

$$\bar{I}_r = \{\mathbf{x} \in X | Pr\{\tilde{a}_r^T\mathbf{x} \leq \tilde{b}_r + h_r\} \geq \eta_r\}.$$

Since $\tilde{\mathbf{a}}_r \sim \mathcal{N}(\mu_r^a, V_r^a)$ is normally distributed with mean vector μ_r^a and positive definite covariance matrix V_r^a, it follows that $\tilde{a}_r^T\mathbf{x} \sim \mathcal{N}(\mu_a^{aT}\mathbf{x}, \mathbf{x}^T V_r^a\mathbf{x})$, and then

$$Pr\{\tilde{a}_r^T\mathbf{x} \leq \tilde{b}_r - h_r\} \geq \eta_r$$

$$\Rightarrow Pr\left\{\frac{\mu_r^{aT}\mathbf{x} - \mu_1^{aT}\mathbf{x}}{\sqrt{\mathbf{x}^T V_r^a\mathbf{x}}} \leq \frac{\tilde{f}_1 - \mu_1^{aT}\mathbf{x}}{\sqrt{\mathbf{x}^T V_r^a\mathbf{x}}}\right\} \geq \eta_r$$

$$\Rightarrow \mu_r^{aT}\mathbf{x} + \Phi^{-1}(\eta_r)\sqrt{\mathbf{x}^T V_i^a\mathbf{x}} \leq b_r - h_r.$$

Similarly, we have the following equivalent formula:

$$Pr\{\tilde{a}_r^T\mathbf{x} \leq \tilde{b}_r + h_r\} \geq \eta_r$$
$$\Rightarrow \mu_r^{aT}\mathbf{x} + \Phi^{-1}(\eta_r)\sqrt{\mathbf{x}^T V_i^a\mathbf{x}} \leq b_r + h_r$$

Then the upper and lower approximation can be rewritten as

$$\underline{I}_r = \{\mathbf{x} \in X | \mu_r^{aT}\mathbf{x} + \Phi^{-1}(\eta_r)\sqrt{\mathbf{x}^T V_i^a\mathbf{x}} \leq b_r - h_r\}$$

and

$$\bar{I}_r = \{\mathbf{x} \in X | \mu_r^{aT}\mathbf{x} + \Phi^{-1}(\eta_r)\sqrt{\mathbf{x}^T V_i^a\mathbf{x}} \leq b_r + h_r\}.$$

As we know, $|A|$ expresses the cardinality of the set A in the finite universe. For the infinite universe, we use it to express the Lebesgue measure. Then we have

$$|\underline{I}| = \int \cdots \int_{\Omega_r} 1 dx_1 dx_2 \cdots dx_m,$$

where $\Omega_r = \{x | \mu_r^{aT}x + \Phi^{-1}(\eta_r)\sqrt{x^T V_r^a x} \leq b_r - h_r\}$ and

$$|\bar{I}| = \int \cdots \int_{\Theta_r} 1 dx_1 dx_2 \cdots dx_m,$$

where $\Theta_r = \{x | \mu_r^{aT}x + \Phi^{-1}(\eta_r)\sqrt{x^T V_r^a x} \leq b_r + h_r\}$. It follows from that problem (4.86) can be rewritten as

$$\begin{cases} \max \left[\Phi\left(\frac{\tilde{f}_1 - \mu_1^{cT}\mathbf{x}}{\sqrt{\mathbf{x}^T V_1^c\mathbf{x}}}\right), \Phi\left(\frac{\tilde{f}_2 - \mu_2^{cT}\mathbf{x}}{\sqrt{\mathbf{x}^T V_2^c\mathbf{x}}}\right), \cdots, \Phi\left(\frac{\tilde{f}_m - \mu_m^{cT}\mathbf{x}}{\sqrt{\mathbf{x}^T V_m^c\mathbf{x}}}\right)\right] \\ \text{s.t.} \begin{cases} \alpha(I_r) \geq \theta_r \\ \mathbf{x} \in \bar{I}_r \\ r = 1, 2, \cdots, p. \end{cases} \end{cases}$$

This completes the proof.

4.5.2.2 Surrogate Worth Trade-off Method

The surrogate worth trade-off method, which is called the SWT method for short, was proposed by Y. Haimes et al. [125] in 1975 to solve the multiobjective programming problem. It can be applied to continuous variables, objective functions, and constraints that can be differentiated twice.

In its original version, SWT is, in principle, noninteractive and assumes continuous variables and twice-differentiable objective functions, and constraints. It consists of four steps:

1. Generate a representative subset of efficient solutions.
2. Obtain relevant trade-off information for each generated solution.
3. Interact with DM to obtain information about preference expressed in terms of worth.
4. Retrieve the best-compromise solution from the information obtained.

The detailed steps of the SWT method can be listed as follows:

Step 1. *Generation of a representative subset of efficient solutions.* The ε-constraint method is recommended to obtain the representative subset of efficient solutions. Without loss of generality, we choose a reference objective H_1 and formulate the ε-constraint problem:

$$\begin{cases} \max H_1(\mathbf{x}) \\ \text{s.t.} \begin{cases} H_i(\mathbf{x}) \geq \varepsilon_i, i = 2,3,\cdots,m \\ \mathbf{x} \in X \end{cases} \end{cases} \tag{4.89}$$

Although there is no rule to specify which objective should be chosen as a reference, the most important objective is recommended. To guarantee that the ε-constraint problem has feasible solution, a reasonable ε_i should be selected, usually, in the range $[a_i, b_i]$, where $a_i = \min_{\mathbf{x} \in X} H_i(\mathbf{x})$ and $b_i = \max_{\mathbf{x} \in X} H_i(\mathbf{x})$.

Step 2. *Obtaining trade-off Information.* In the process of solving problem (4.89), the trade-off information can easily be obtained merely by observing the optimal Kuhn–Tucker multipliers corresponding to the ε-constraints. Let these multipliers be denoted by $\lambda_{1i}(\mathbf{x}(\varepsilon))$. If $\lambda_{1k}(\mathbf{x}(\varepsilon)) > 0(k = 1,2,\cdots,m)$, and then the efficient surface in the objective function space around the neighborhood of $\boldsymbol{H}^\varepsilon = (H_1(\mathbf{x}(\varepsilon)), H_2(\mathbf{x}(\varepsilon)), \cdots, H_m(\mathbf{x}(\varepsilon)))^T$ can be represented by $\boldsymbol{H}_1 = (H_1, H_2, \cdots, H_m)$ and

$$\lambda_{1k}(\boldsymbol{x}(\varepsilon)) = -\left.\frac{\partial \boldsymbol{H}_1}{\partial H_k}\right|_{\boldsymbol{H}} = \boldsymbol{H}^\varepsilon, \ k = 2,3,\cdots,m \tag{4.90}$$

Thus, each $\lambda_{1k}(\mathbf{x}(\varepsilon))$ represents the efficient partial trade-off rate between H_1 and H_k at $\boldsymbol{H}^\varepsilon$ when all other objective are held fixed at their respective values at $\mathbf{x}(\varepsilon)$. The adjective "efficient" is used to signify that, after the trade-off is made, the resulting point remains on the efficient surface. For details see [50].

Step 3. *Interacting with the decision maker to elicit Preference.* The DM is supplied with trade-off information from Step 2 and the levels of all criteria. He or she then expresses his or her ordinal preference on whether or not (and by how much) he or she would like to make such a trade-off at that level. Y. Haimes et al. [50] constructed the following surrogate worth function: the DM is asked, "How much would you like to improve H_1 by $\lambda_{1k}(x(\varepsilon))$ units per one-unit degradation of H_k while all other objective remain fixed at $H_l(x(\varepsilon)), l \neq 1, k$?" Indicate your preference on a scale of -10 to 10, where the values have the following meanings:

(1) $+10$ means you have the greatest desire to improve improve H_1 by $\lambda_{1k}(x(\varepsilon))$ units per one-unit degradation of H_k.

(2) 0 means you are indifferent about the trade-off.

(3) -10 means you have the greatest desire to degrade improve H_1 by $\lambda_{1k}(x(\varepsilon))$ units per one-unit improvement in H_k.

Values between -10 and 0, and 0 and 10 show proportional desire to make the trade."

The DM's response is recorded as $w_{1k}(\mathbf{x}(\varepsilon))$, called the *surrogate worth* of the trade-off between H_1 and H_k at the efficient solution $x(\varepsilon)$. At a particular efficient solution, there will be $m - 1$ questions to obtain $w_{1k}(\mathbf{x}(\varepsilon)), k = 2, 3, \cdots, m$.

Step 4. *Retrieving the best-compromise solution.* If there exists an efficient solution $\mathbf{x}(\varepsilon_0)$ such that

$$w_{1k}(\mathbf{x}(\varepsilon_0)) = 0, \ k = 2, 3, \cdots, m \tag{4.91}$$

the DM has obtained a best-compromise solution. Thus equation (4.91) is the best-compromise condition of $\mathbf{x}(\varepsilon_0)$. If there is such $\mathbf{x}(\varepsilon_0)$ in the representative set, then stop and output $\mathbf{x}(\varepsilon_0)$. Otherwise we use multiple regression to construct the surrogate worth function as follows:

$$\mathbf{x}_{1k} = w_{1k}(H_1, H_2, \cdots, H_m), \ k = 2, 3, \cdots, m.$$

Then the system of equations

$$w_{1k}(H_1, H_2, \cdots, H_m) = 0, \ k = 2, 3, \cdots, m,$$

is solved to determine $(H_2^*, \cdots, H_m^*)$. Let $\varepsilon_{0k} = H_k^*(k = 2, \cdots, m), \varepsilon_0 = (\varepsilon_{02}, \cdots, \varepsilon_{0m})^T$. The best-compromise solution $\mathbf{x}(\varepsilon_0)$ is then found by solving problem (4.89).

4.5.2.3 Numerical Example

Example 4.8
Consider the problem

$$\begin{cases} \max f_1(\mathbf{x}, \boldsymbol{\xi}) = \xi_1 x1 + \xi_2 x2 + \xi_3 x3 \\ \max f_2(\mathbf{x}, \boldsymbol{\xi}) = c_1 \xi_4 x1 + c_2 \xi_5 x2 + c_3 \xi_6 x3 \\ \text{s.t.} \begin{cases} \xi_7 x1 + \xi_8 x2 + \xi_9 x3 \leq 6 \\ x_1 - x_2 + x_3 \geq 3.5 \\ x_1 + 4x_2 + 2x_3 \leq 10 \\ x_1, x_2, x_3 \geq 0 \end{cases} \end{cases} \tag{4.92}$$

where $c = (c_1, c_2, c_3) = (1.2, -0.5, 1.3)$, and $\xi_i (i = 1, 2, \cdots, 9)$ are independently random variables defined as follows:

$$\begin{aligned}
\xi_1 &\sim \mathcal{N}(2,1), & \xi_2 &\sim \mathcal{N}(3,0.5), & \xi_1 &\sim \mathcal{N}(1,0.2), \\
\xi_4 &\sim \mathcal{N}(5,1), & \xi_5 &\sim \mathcal{N}(2,1), & \xi_6 &\sim \mathcal{N}(2,0.5), \\
\xi_7 &\sim \mathcal{N}(3,0.2), & \xi_8 &\sim \mathcal{N}(0,1), & \xi_1 &\sim \mathcal{N}(1,0.5).
\end{aligned}$$

Assume that the DM aims at obtaining the maximum probability based on the predetermined level value $\bar{f}_1$ and $\bar{f}_2$ and require that the approximation accuracy exceeds 0.9 under the probabilistic rough set with the deviation $h = 1.5$. Let $\bar{f}_1 = 14, \bar{f}_2 = 22$ and I express the feasible set, and then we get the following probability maximization model:

$$\begin{cases}
Pr\{\xi_1 x_1 + \xi_2 x_2 + \xi_3 x_3 \geq 7\} \\
Pr\{c_1 \xi_4 x_1 + c_2 \xi_5 x_2 + c_3 \xi_6 x_3 \geq 2\} \\
\text{s.t.} \begin{cases} \mathbf{x} \in \bar{I} \\ \alpha(I) \geq 0.9 \end{cases}
\end{cases} \tag{4.93}$$

First, we get the maximum deviation $h = 1.23$ such that $\alpha(I) = \frac{\int_\Omega dx_1 dx_2 dx_3}{\int_\Theta dx_1 dx_2 dx_3} \geq 0.9$, where $\Omega = \{(x_1, x_2, x_3) | 3x_1 + x_3 + 1.28 \sqrt{0.2x_1^2 + x_2^2 + 0.5x_3^2} \leq 6 - h, x_1 - x_2 + x_3 \geq 3.5, x_1 + 4x_2 + 2x_3 \leq 10, x_1, x_2, x_3 \geq 0\}$ and $\Theta = \{(x_1, x_2, x_3) | 3x_1 + x_3 + 1.28 \sqrt{0.2x_1^2 + x_2^2 + 0.5x_3^2} \leq 6 + h, x_1 - x_2 + x_3 \geq 3.5, x_1 + 4x_2 + 2x_3 \leq 10, x_1, x_2, x_3 \geq 0\}$.
Second, it follows from Theorem 4.6 that Equation (4.95) is equivalent to

$$\begin{cases}
\max F_1 = 1 - \Phi\left(\frac{7 - (2x_1 + 3x_2 + x_3)}{\sqrt{x_1^2 + 0.5x_2^2 + 0.2x_3^2}} \right) \\
\max F_2 = 1 - \Phi\left(\frac{2 - (1.2x_1 - x_2 + 2.6x_3)}{\sqrt{1.44x_1^2 + 0.25x_2^2 + 0.845x_3^2}} \right) \\
\text{s.t.} \begin{cases} 3x_1 + x_3 + 1.28 \sqrt{0.2x_1^2 + x_2^2 + 0.5x_3^2} \leq 7.23 \\ x_1 - x_2 + x_3 \geq 3.5 \\ x_1 + 4x_2 + 2x_3 \leq 10 \\ x_1, x_2, x_3 \geq 0 \end{cases}
\end{cases} \tag{4.94}$$

Next, we will use the surrogate worth trade-off method solve the above problem. First, let $F_1(x)$ be the reference objective and compute $\min_x F_2(x) = 0.985$ and $\max_x F_2(x) = 0.989$. Take $\varepsilon = 0.986 \in [0.985, 0.989]$ and construct the following $\varepsilon-$ problem:

$$\begin{cases}
\max F_1 = 1 - \Phi\left(\frac{7 - (2x_1 + 3x_2 + x_3)}{\sqrt{x_1^2 + 0.5x_2^2 + 0.2x_3^2}} \right) \\
\text{s.t.} \begin{cases} 1 - \Phi\left(\frac{2 - (1.2x_1 - x_2 + 2.6x_3)}{\sqrt{1.44x_1^2 + 0.25x_2^2 + 0.845x_3^2}} \right) \leq 0.986 \\ 3x_1 + x_3 + 1.28 \sqrt{0.2x_1^2 + x_2^2 + 0.5x_3^2} \leq 7.23 \\ x_1 - x_2 + x_3 \geq 3.5 \\ x_1 + 4x_2 + 2x_3 \leq 10 \\ x_1, x_2, x_3 \geq 0 \end{cases}
\end{cases} \tag{4.95}$$

TABLE 4.3   Computational results

$\varepsilon = F_2$	F_1	x_1	x_2	x_3	λ_{12}	w_{12}
0.984	0.084	0	0.186	4.684	0.192	-10
0.985	0.083	0	0.197	4.565	0.192	-8
0.986	0.082	0	0.203	4.257	0.192	-6
0.987	0.081	0	0.221	4.188	0.192	-4
0.988	0.080	0	0.235	3.885	0.192	-2
0.989	0.079	0	0.266	3.806	0.190	0
0.989	0.078	0	0.287	3.782	0.190	2
0.989	0.077	0	0.295	3.565	0.190	4
0.989	0.076	0	0.312	3.452	0.190	6
0.989	0.076	0	0.352	3.182	0.190	8
0.989	0.076	0	0.386	3.052	0.190	10

Second, we get the optimal solution $x(\varepsilon) = (0, 0.285, 3.785)$ and the Kuhn–Tucker multiplier $\lambda_{12}x(\varepsilon) = 0.182$ of the constraint $F_2 \leq 0.986$.

Third, to interact with the DM, we prepare Table 4.3, listing a few representative efficient solutions. We can now go to the DM to elicit w_{12} for each efficient point in Table 4.3.

4.5.3   Ra-NLDCRM and Random Simulation-Based Tribe-PSO

As problem (4.82) has nonlinear forms, that is, there is one nonlinear function of f_i, g_r at least, it is called the random nonlinear dependent-chance rough model (Ra-NLDCRM). It is usually impossible to be converted into the crisp equivalent model. Random simulation-based tribe particle swarm optimization (Tribe-PSO) are used to deal with it.

4.5.3.1   Random Simulation 3 for Probability

Let ξ be an n-dimensional random vector defined on the probability space $(\Omega, \mathscr{A}, Pr)$, and $f : \mathbf{R}^n \to \mathbf{R}$ a measurable function. In order to obtain the probability

$$L = Pr\{f(\xi) \leq 0\} \tag{4.96}$$

we generate ω_k from Ω according to the probability measure Pr, and write $\xi_k = \xi(\omega_k)$ for $k = 1, 2, \cdots, N$. Let N' denote the number of occasions on which $f(\xi_k) \leq 0$ for $k = 1, 2, \cdots, N$ (i.e., the number of random vectors satisfying the system of inequalities). Let us define

$$h(\xi_k) = \begin{cases} 1, & \text{if } f(\xi_k) \leq 0 \\ 0, & \text{otherwise} \end{cases}$$

Then we have $E[h(\xi_k)] = L$ for all k, and $N' = \sum_{k=1}^{N} h(\xi_k)$. It follows from the strong law of large numbers that

$$\frac{N'}{N} = \frac{\sum_{k=1}^{N} h(\xi_k)}{N}$$

converges to L. Thus the probability L can be estimated by N'/N provided that N is sufficiently large. The process of random simulation 3 for probability can be summarized as follows:

Step 1. Set $N' = 0$.
Step 2. Generate ω from Ω according to the probability measure Pr.
Step 3. If $f(\xi(\omega)) \leq 0$, then $N' + +$.
Step 4. Repeat the second and third steps N times.
Step 5. Return $L = N'/N$.

Example 4.9
Let $\xi_1 \sim exp(2)$ be an exponentially distributed variable, $\xi_2 \sim \mathcal{N}(4,1)$ a normally distributed variable, and $\xi_3 \sim \mathcal{U}(0,2)$ a uniformly distributed variable. A run of stochastic simulation with 1000 cycles shows that

$$Pr\left\{ \sqrt{\xi_1^2 + \xi_2^2 + \xi_3^2} \leq 8 \right\} = 0.9642.$$

4.5.3.2 Tribe-PSO

Since the PSO algorithm was proposed by J. Kennedy and R. Eberhard [162], many improvements have bee made by some scholars. Tribe-PSO is one of them, which was proposed by K. Chen et al. [58]. In classical PSO, a parameter that the user must specify is the sociometry (topology, type and quantity of social relationships) of the swarm. In order to create a more robust PSO variant, M. Clerc [63] proposed a parameter-free PSO method called Tribes, in which details of the topology, including the size of the population, evolve over time in response to performance feedback. Tribes has attracted attention from researchers in different application areas such as the optimization of milling operations [229], flow shop scheduling [230], and particle swarm optimization [58, 80]. Tribe-PSO is inspired by the concept of hierarchical fair competition (HFC) developed by J. Hu et al. [137]. The main principles of HFC are that the competition is allowed only among individuals with comparable fitness, and it should be organized into hierarchical levels. Such principles help the algorithms to prevent the optimization procedure from too early loss of diversity and help the population to escape from the prematurity. J. Hu et al. [137] have successfully introduced these principles into GA and other EAs and developed the AHFC model. Keeping diversity in the population is very important for complex optimization problems, such as flexible docking. Thus, K. Chen et al. [58] proposed Tribe-PSO, a hybrid PSO model based on HFC principles. In this model, particles are divided into two layers, and the whole procedure of convergence is divided into three phases. Particles on different layers or in different phases are strictly controlled in order to preserve population diversities.

A tribe is a subswarm formed by particles that have the property that all particles inform all others belonging to the tribe (a symmetrical clique in graph theoretical lan-

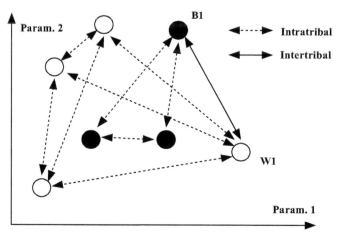

FIGURE 4.3   Tribal relationships.

guage). The concept is therefore related to the "cultural vicinity" (information neighborhood) and not to "spatial vicinity" (parameter space neighborhood). It should be noted that, due to the above definition, the set of informers of a particle (its so-called *i*-group) contains the whole of its tribe but is not limited to it. This is shown in Figure 4.3, where the *i*-group of particle B1 contains all particles of its tribe (black), and particle W1 belonging to the white tribe.

In the general PSO algorithm, K. Chen et al. [58] found that the part of $(pBest - p)$, and especially the part of $(pBest_{gBest} - p)$, still dominates the whole velocity function although these two vectors have random factors before each item. It results that the current position p is located in the neighborhood of $pBest$, while it is much far away from the position of $pBest_{gBest}$, and furthermore, the norm of the vector toward $pBest_{gBest}$ is much larger than that of the vector toward $pBest$. In this way, particles with worse fitness but promising diversities are strongly attracted by the best-particle of the swarm and get entrapped in the neighborhood of $gBest$ and lose their diversities. They also found that this kind of competition is inadequate and does harm to the healthy procedure of optimization. In order to prevent prematurity, we introduce the principle of HFC and propose the Tribe-PSO model.

In the following part, we will introduce the details of the Tribe-PSO model proposed by K. Chen et al. [58]. Tribe-PSO has two important concepts: layer, and phase. In a word, particles in the swarm are divided into two layers and the procedure of optimization is divided into three phases. Assume that there are totally $l \times m$ particles in the swarm. In the initiation step of Tribe-PSO, the swarm is divided into l sub-population, called tribes. Each tribe has the same structure of the basic PSO model: it has m particles, and the best particle from them is called $tBest$. Tribes form the basic layer, while the best particles from the l tribes form the upper layer in the two-layered structure of Tribe-PSO, which is shown in Figure 4.4. The convergence procedure of Tribe-PSO consists of three phases: *isolated phase, communing phase,*

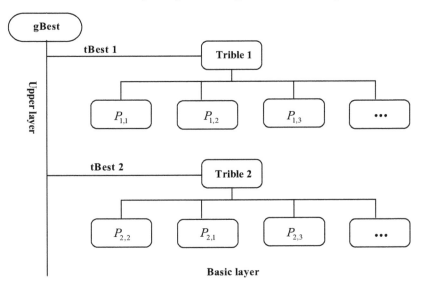

FIGURE 4.4 Two-layer structure of Tribe-PSO.

and *united phase*. Each phase occupies a portion of iterations. In the first phase, the tribes are isolated and work as *l* independent PSO models. No information is exchanged between each pair of tribes. The isolated phase ensures the tribes having enough time to develop before possible premature convergence. In the second phase, Tribe-PSO works in the standard two-layered model: tribe members *p* form the basic layer (the tribes) and the best particles from each tribe *tBest* form the upper layer. Information about searching experience is exchanged among certain basic tribe, and among the upper layer as well. The particle *tBest*, that is, the best particle from the tribe in the basic layer, serves as the information-exchanging agency between the tribe members *p* and the best particle of the whole swarm *gBest*. The communing phase leads the procedure to convergence in a more moderate way than basic PSO. Therefore the diversity of most particles is well preserved. In the last phase, all the tribes are united into one group. The model becomes a basic PSO model. The concept of *tBest* does not exist any longer. The united phase helps the swarm to converge as quickly as possible. In the Tribe-PSO model, particles in different layers or in different phases have different velocity functions. The velocity functions for a specific layer and phase are defined as follows:

Phase I (Isolated phase). There is no upper layer or *gBest* in this phase. All the *tBest* and *p* from different tribes have the same velocity function as Equation (4.97).

$$v = v \times w + 2 \times rand() \times (pBest - p) + 2 \times rand() \times (pBest_{tBest} - p) \quad (4.97)$$

Phase II (Communing phase). For tribe members, the velocity function is the same as they have in the first phase (4.97). For *tBest* and *gBest* particles, their velocity function is described in Equation (4.98). In this equation, *gBest* is regarded as the leader in the upper layer.

$$v = v \times w + 2 \times rand() \times (pBest - p) + 2 \times rand() \times (pBest_{gBest} - p) \quad (4.98)$$

Phase III (United phase). There is no *tBest* in this phase. All the tribes are united into one swarm. Thus, the velocity function for all the particles becomes the original one in the basic PSO (Equation (4.98)). Compared with basic PSO, Tribe-PSO has two more parameters. One is the ratio of tribe number l to tribe size m. The other is the partition of three phases. These two parameters have considerable effects on the global search ability and performance of the Tribe-PSO model. The discussion on the influence of the two parameters can be found in [58].

4.5.3.3   Numerical example

Example 4.10
Consider the problem

$$\begin{cases} \max Pr\{2\xi_1 x_1^2 + 3\xi_2 x_2 - \xi_3 x_3 \geq 120\} \\ \max Pr\{5\xi_1 x_1 - 2\xi_2 x_2 + 2\xi_3 x_3 \geq 50\} \\ \text{s.t.} \begin{cases} 5x_1\xi_1^2 - 3x_2^2\xi_2 + 6\sqrt{x_3}\xi_3^3 \leq 50 \\ 4\sqrt{x_1} + 6x_2 - 4.5x_3 \leq 20 \\ x_1 + x_2 + x_3 \leq 15 \\ x_1, x_2, x_3 \geq 0 \end{cases} \end{cases} \quad (4.99)$$

ξ_1, ξ_2, ξ_3 are random variables characterized as

$$\xi_1 \sim \mathcal{N}(4,1), \xi_2 \sim \mathcal{N}(5,1), \xi_3 \sim \exp(2).$$

Define the similarity relationship R_h^η as $\mathbf{x}R_{h_r}^{\eta_r}\mathbf{y} : Pr\{(x_1,x_2,x_3),(y_1,y_2,y_3) \in X | (5x_1\xi_1^2 - 3x_2^2\xi_2 + 6\sqrt{x_3}\xi_3^3) - (5y_1\xi_1^2 - 3y_2^2\xi_2 + 6\sqrt{y_3}\xi_3^3)| \leq h\} \geq \eta$ where $X = \{\mathbf{x}|4\sqrt{x_1} + 6x_2 - 4.5x_3 \leq 20, x_1 + x_2 + x_3 \leq 15, x_1, x_2, x_3 \geq 0\}$.

If the DM require that $\eta = 0.8$ and the approximation accuracy exceeds 0.95 based on rough approximation, then we get the following Ra-DCRM:

$$\begin{cases} \max Pr\{2\xi_1 x_1^2 + 3\xi_2 x_2 - \xi_3 x_3 \geq 120\} \\ \max Pr\{5\xi_1 x_1 - 2\xi_2 x_2 + 2\xi_3 x_3 \geq 50\} \\ \text{s.t.} \begin{cases} \alpha(I) \geq 0.95 \\ x \in \bar{I} \end{cases} \end{cases} \quad (4.100)$$

where I expresses the feasible set. First, by random simulations of DCM, we get the maximum deviation $h = 2.6$ such that

$$\alpha(I) = \frac{\oint_I dx_1 dx_2 dx_3}{\oint_{\bar{I}} dx_1 dx_2 dx_3} \geq 0.9,$$

where $\underline{I} = \{(x_1,x_2,x_3)|Pr\{5x_1\xi_1^2 - 3x_2^2\xi_2 + 6\sqrt{x_3}\xi_3^3 \le 50 - h \ge \eta\}\}$ and $\bar{I} = \{(x_1,x_2,$ $x_3)|Pr\{5x_1\xi_1^2 - 3x_2^2\xi_2 + 6\sqrt{x_3}\xi_3^3 \le 50 + h\} \ge \eta\}$. Then we have

$$
\begin{cases}
\max Pr\{2\xi_1 x_1^2 + 3\xi_2 x_2 - \xi_3 x_3 \ge 120\} \\
\max Pr\{5\xi_1 x_1 - 2\xi_2 x_2 + 2\xi_3 x_3 \ge 50\} \\
\text{s.t.} \begin{cases}
5x_1\xi_1^2 - 3x_2^2\xi_2 + 6\sqrt{x_3}\xi_3^3 \le 52.6 \\
4\sqrt{x_1} + 6x_2 - 4.5x_3 \le 20 \\
x_1 + x_2 + x_3 \le 15 \\
x_1,x_2,x_3 \ge 0
\end{cases}
\end{cases}
$$

After running the random simulation-based Tribe-PSO, we obtain

$$x^* = (2.41, 6.18, 5.36), \beta^* = [0.76, 0.68],$$

where $\beta^* = [Pr\{2\xi_1 x_1^* + 3\xi_2 x_2^* - \xi_3 x_3^* \ge 120\}, Pr\{5\xi_1 x_1^* - 2\xi_2 x_2^* + 2\xi_3 x_3^* \ge 50\}]$.

4.6 The rc-PSP/nM for Longtan Hydropower Station

In this section we focus on the resource-constrained project scheduling problems with multimode (rc-PSP/mM) and random parameters, applying them to a large-scale water conservancy and hydropower construction project in the southwest region of China. The objective functions are to minimize the total project time, minimize the total premature and deferred cost with random coefficients, and maximize the quality of the project. To realize multiple management objectives with random parameters in practice, we establish a random multiple objective mixed-integer linear decision making model that can describe the proposed problem appropriately. In order to deal with the abnormal situation, we apply the rough approximation technique to treat the random model. After stating the problem of the project and presenting the mathematical formulation model of rc-PSP/mM, we use the random expected value model and rough approximation to handle the random objective functions, and the random constrains for proposing an equivalent crisp model that can be solved by a computer program. As a NP-hard optimization problem, traditional optimization techniques could not cope with the rc-PSP/mM effectively according to its characteristics, so we present multiple objective particle swarm optimization (MOPSO) to solve the problem. The approach is applied to the drilling grouting in a large-scale water conservancy and hydropower construction project, and then we compare it with other approaches to rc-PSP/mM. The generated results validate the effectiveness of the proposed model and algorithm.

4.6.1 Problem Statement

Continuous surges in oil prices, as well as the combustion of fossil fuels such as coal, has exacerbated global warming. In this context, hydroelectric power has been favored, for example, the Merowe Dam in Pakistan, the Eight dam in Japan, the Jirau dam in Brazil, and the Nam Theun 2 Dam in Laos, etc. Some are planned to be constructed, and others have been constructed already. The economic situation of the world has seen an unusual turbulence since the financial crisis, and a series of measures to stabilize their economics have been adopted by the relevant countries and organizations, and thus the economic barycenter has shifted to the real economy. As an important link in the real economy, large-scale water conservancy and the hydropower construction projects play a very important role in flood control, power generation, environmental protection, employment, and so on.

The problem considered in this paper is from the Longtan hydropower station project, which is a large-scale water conservancy and hydropower construction project in the southwest region China. The Longtan hydropower station, located in Tiane county in Guangxi Zhuang Autonomous Region, is one of ten major indicative projects of the national strategies for the GreatWestern Development in China and the Power Transmission from West to East, and it is a control project in the cascade development of the Red River. The primary mission of the project with integrative efficiency is flood control and navigation. The level of the project is *I* and the type is the *I* large-scale. The layout is as follows: roller-compacted concrete gravity dam, flood building with seven outlets and two bottom outlets are arranged in the river bed, and power capacity of stream systems with nine installations are arranged on the left bank, with navigation structures arranged on the right bank, and finally a vertical ship lift is used for navigation. The project is designed 400 m in accordance with the normal pool level, and the installed plant capacity is 5400 MW. Tunnel diversion is applied in river diversion during construction, and two diversion openings are arranged on the left bank and right bank. The standard of diversion is ten-year flood, and the corresponding flow is 14700 m^3/s.

For the project, the GD Electric Power Design Institute was in charge of the exploration, design, and project scheduling for drilling grouting. Before planning the overall progress of the preparation, GD completed a preliminary drilling grouting project design, which is composed of a series of interrelated activities with different modes, each mode characterized by a certain known processing time and uncertain resource requirements. In the preliminary design report, the drilling grouting project duration is 5 years and 8 months (68 months). Experts recommend that the design of the construction organization and construction progress of the project are feasible, but in view of the current large-scale emergence of the national electricity supply and demand tension, the duration still needs to shortened and further studies in power generation. When the construction duration is optimized, it is necessary to handle the relationship of the construction progress, economic benefits, and quality assurance correctly. Therefore, we should utilize advanced technologies to optimize the construction process for shortening the construction duration ensuring quality while making safety a prerequisite, but the cost and economic benefits are important considerations.

Now, three objectives, that is, duration, cost, and quality, are taken into account synchronously. So, GD Electric Power Design Institute wishes to design a model of the resource-constrained project scheduling problems with multimode to determine not only the overall duration of the project but also its cost and quality. As we know, due to rc-PSP/mM of the project, it is hard to describe these problem parameters as crisp variables because there is no sufficient data to analyze in practice. For example, because the quantities of support hole plugging are uncertain, the duration of backfill grouting could be denoted as a random variable. The problem can be modeled as a random decision-making problem. There are a lot of techniques to deal with the random decision making problem such as the E-model (expectation optimization model), the V-model (variance minimization model), and the P-model (probability maximization model) proposed by A. Charnes and W. Cooper [53]. They all convert the model with random parameters into a crisp one. However, it may result in a loss of information when dealing with the foregoing realistic problem. For example, if we use the expected value operator to deal with the upper limit on the duration time, the exceeding condition of the upper limit is strictly forbidden. In fact, the system will be disabled when facing abnormal weather. A flexible constraint is more suited to realistic situations. The rough set is an efficient tool when dealing the indistinct information. Considering the decision manager's objectives, we minimize the total cost and overall duration of the project and maximize the quality of the project at the same time. Hence, we formulate a resource constrained project scheduling problem with multimode and random parameters based on rough approximation.

Rc-PSP/mM in this section could be stated as follows. A set of activities $\{1, 2, \cdots, I+1\}$ have to be scheduled under precedence and resource constraints, where activities 0 and $I+1$ are dummies, have no duration, just represent the initial and final activity (define the modes of activities 0 and $I+1$ both as 1). No activity may be started before all its predecessors are finished, and each activity can be executed in one of m_i modes. Resources involved in a project can be renewable (i.e., recoverable after serving an activity, such as equipment, manpower, etc.) or nonrenewable (i.e., limited in amount over project process and not recoverable, such as money, materials, etc.). There exists a set of renewable and nonrenewable resources types $\{1, 2, \cdots, K\}$ and $\{1, 2, \cdots, N\}$, respectively. The problem will be illustrated in Figure 4.1 in detail.

4.6.2 Modeling and Analysis

To model the rc-PSP/mM of drill grouting in Longtan large-scale water conservancy and hydropower construction project with a mixed rough and random environment, the assumptions are as follows:

1. A single project consists of a number of activities each with several known execution modes.

2. The start time of each activity is dependent upon the completion of some other activities (precedence constraints of activities). After completing a specific activity, the next activity must also be started in the project.

3. Activities cannot be interrupted, and each activity must be performed in a mode. When the activity is being executed, each activity mode combination requires a constant amount of one or more types of resources.

4. Mode switching is not allowed when activities are resumed after splitting.

5. Resources are available in certain limited quantities, but the consumption of each resource is a random variable.

4.6.2.1 Notations

In rc-PSP/mM, we consider a single project that consists of $i = 0, 1, \cdots, I+1$ activities with the precedence constraints between each pair of activities (e, i), where e immediately precedence i are taken into consideration. In addition, each activity i must be performed in one of m_i possible modes, where each activity mode combination has a fixed duration and requires one or more of types of renewable or nonrenewable resources when the activity is being executed. For start and end activities of the project, we have

$$p_{0,j} - p_{I+1,j} = 0, m_0 = m_{I+1} = 1,$$

where $j = 1, 2, \cdots, m_i$ (m_i is the number of possible modes of activity i).

In order to formulate the model, indices, variables, certain parameters, rough random interval and rough random coefficients, and decision variable are introduced as follows:

Indices
i activity index, $i = 0, 1, \cdots, e, \cdots, I+1$
j mode index, $j = 1, 2, \cdots, m_i$, (m_i is the number of possible modes of activity i)
k renewable resource type index, $k = 1, 2, \cdots, K$
n nonrenewable resource type index, $n = 1, 2, \cdots, N$
t period index, $t = t_i^{EF}, \cdots, t_i^{LF}$

Variables
t_i^{EF} early finish time of activity i
t_i^{LF} late finish time of activity i
T total project duration
C total project cost
Q total project quality

Certain parameters
$Pre(i)$ set of immediate predecessors of activity i
p_{ij} processing time of activity i of selected mode j
p_{ij}^S the earliest processing time of activity i of selected mode j
p_{ij}^L he latest processing time of activity i of selected mode j
q_i the quality of activity i
q_i^S the smallest acceptable quality of activity i
q_i^L the best quality of activity i
w_i the weight of the quality of activities i for the quality of the whole project
t_i^E expected finish time of activity i
r_k^M maximum-limited renewable resource k only available with the constant period availability
r_n^M maximum-limited nonrenewable resource k only available with the constant period availability

Random parameters

$\tilde{r}_{ijk}$ resource k required to execute activity i used in mode j
$\tilde{c}_i^{TP}$ random coefficient of the total penalty cost about activity i

Decision Variables

$$x_{ijt} = \begin{cases} 1 & \text{if activitiy } i \text{ executed in mode } j \text{ scheduled to be finished in time } t \\ 0 & \text{otherwise} \end{cases}$$

Note that x_{ijt} confirms the finish time of the current activity with the certain executed mode is scheduled in this certain time or not.

4.6.2.2   Modeling

Based on the assumptions and notations proceding, we propose a model of resource-constrained project-schedulingproblems with multi-mode and random parameters for drilling grouting in Longtan large-scale water conservancy and hydropower construction project. The goal of solving the rc-PSP/mM in this section is to find a mode combination for all activities as well as the resultant schedule (the order of activities) that leads to minimal project duration, minimizes total tardiness penalty for all activities, and maximizes the quality of the project. On the basis of the requirement of the manager's objectives for drilling grouting, and hydropower construction project, we propose a random decision-making model to tackle it. Furthermore, to make the decision more flexible, rough approximation is used to handle the random model.

Objective functions. As to a multiobjective rc-PSP/mM for drilling the grouting, the first objective is to minimize project duration. The first objective T seeks to minimize total project time, which is naturally considered to be the basic economic aim in project scheduling over time. We use the finish time of the last activity in the project, considering its entire possible executed mode to describe that as

$$T = \sum_{j=1}^{m_I} \sum_{t=t_I^{EF}}^{t_I^{LF}} t x_{Ijt} \tag{4.101}$$

Furthermore, the second objective is to measure and minimize the total cost. We minimize the total tardiness penalty of all activities, that is, C, which is the sum of penalty costs for all activities. Let $T = \sum_{j=1}^{m_I} \sum_{t=t_I^{EF}}^{t_I^{LF}} t x_{Ijt}$ be denoted as the actual finish time of activity i, and t_i^E be the expected finish time of activity i. Thus, we develop mathematical formulations of objectives as follows:

$$C = \left| \sum_{i=0}^{I+1} \tilde{c}_i^{TP} \left(\sum_{j=1}^{m_I} \sum_{t=t_I^{EF}}^{t_I^{LF}} t x_{Ijt} - t_i^E \right) \right| \tag{4.102}$$

where the cost coefficient c_i is a rough random interval, and $|\cdot|$, just keeping the value of C, is nonnegative.

Finally, as one of the key objectives in construction project management, the quality of the project is the precondition of the duration and cost. As we all know, different durations cause different quality of activities. In order to establish the mathematical model of quality evaluation, we assume that in the completion of the activities quality has a linear relationship with the duration of the activities.

Denote α_i as the slope of duration quality, and $\alpha_i = \frac{q_i^L - q_i^S}{p_{ij}^L - p_{ij}^S}$. Thus, we obtain that

$$q_i = q_i^S + \alpha_i(p_{ij} - p_{ij}^S).$$

Then, the whole quality of the project $Q = \sum_{i=0}^{I+1} w_i[q_i^S + \alpha_i(p_{ij} - p_{ij}^S)]$. So the third objective is stated in the following equation:

$$Q = \sum_{i=0}^{I+1} w_i[q_i^S + \alpha_i(p_{ij} - p_{ij}^S)] \tag{4.103}$$

where $\sum_{i=0}^{I+1} w_i = 1, w_i > 0$.

Precedence constraint. The constraints of rc-PSP/mM are divided into time constraints and resource constraints. A specific activity must be finished before changing to another activity when it is initiated, which means successive activities must be and only be scheduled after all the predecessors have been done. To ensuring that each activity is scheduled on of its modes and completed within $[t_i^{EF}, t_i^{LF}]$, we use that for each activity i considering its each immediate predecessors e, and there should be a relation as follows:

$$\sum_{j=1}^{m_e} \sum_{t=t_e^{EF}}^{t_e^{LF}} tx_{ejt} + \sum_{j=1}^{m_e} \sum_{t=t_e^{EF}}^{t_e^{LF}} p_{ij}x_{ijt} \leq \sum_{j=1}^{m_e} \sum_{t=t_e^{EF}}^{t_i^{LF}} tx_{ijt}, e \in Pre(i) \tag{4.104}$$

for $i = 0, 1, 2, \cdots, I+1$, where $Pre(i)$ is the set of the immediate predecessors of activity i, $\sum_{j=1}^{m_e} \sum_{t=t_e^{EF}}^{t_e^{LF}} tx_{ejt}$ is the actual finish time of all immediate predecessors e of activity i, $\sum_{j=1}^{m_e} \sum_{t=t_e^{EF}}^{t_e^{LF}} p_{ij}x_{ijt}$ is the process time of activity i, and $\sum_{j=1}^{m_e} \sum_{t=t_e^{EF}}^{t_i^{LF}} tx_{ijt}$ is the actual finish time of activity i.

Resources constraint. In project scheduling, it is also important to limit the total resource consumption of resources used by all activities in each period. In rc-PSP/mM, the amount of renewable resource k used by all activities could not exceed its limited quantity r_k^M in any time period. Also, the total resource consumption of nonrenewable resource n also limits the maximum available amount r_k^n. Thus, the constraints

$$\sum_{i=1}^{I} \sum_{j=1}^{m_i} \tilde{r}_{ijk} \sum_{s=t}^{t+p_{ij}+1} x_{ijt} \leq r_k^M \tag{4.105}$$

$$\sum_{i=1}^{I} \sum_{j=1}^{m_i} \tilde{r}_{ijn} \sum_{t=t_i^{EF}}^{t_i^{LF}} x_{ijt} \leq r_n^M \tag{4.106}$$

can be employed, where $k = 1, 2, \cdots, K$, $n = 1, 2, \cdots, N$ and $t = 1, 2, \cdots, T$.

Maturity constraint. As we all know, in order to schedule all activities in the project, the finish time of each activity must be in $[t_i^{EF}, t_i^{LF}]$. Therefore, we use the constraint

$$\sum_{j=1}^{m_i} \sum_{t=t_i^{EF}}^{t_i^{LF}} x_{ijt} = 1, i = 0, 1, \cdots, I+1 \tag{4.107}$$

to formulate the general feasibility of the problem.

Logical constraint. For describing some nonnegative variables and 0-1 variable in the model for practical sense, we can use the mathematical formulas as the time constraint and the resource constraint:

$$\begin{cases} p_{ij}^L \geq p_{ij} \geq p_{ij}^S > 0, q_i^L \geq q_i \geq q_j^S > 0 \\ t_{ij}^F \geq 0, t_{ij}^{EF} \geq 0, t_{ij}^{LF} \geq 0 \\ w_i > 0, \sum_{i=0}^{I+1} w_i = 1 \\ x_{ijt} = 1 \text{ or } 0 \end{cases} \tag{4.108}$$

where $k = 1, 2, \cdots, K$, $n = 1, 2, \cdots, N$ and $t = 1, 2, \cdots, T$.

From the above discussions, by the integration of equations. (4.101)–(4.108), the mathematical model of rc-PSP/mM for Longtan can be stated as follows:

$$\begin{cases} \min T = \sum_{j=1}^{m_I} \sum_{t=t_I^{EF}}^{t_I^{LF}} t x_{Ijt} \\[2mm] \min C = \left| \sum_{i=0}^{I+1} \tilde{c}_i^{TP} \left(\sum_{j=1}^{m_I} \sum_{t=t_I^{EF}}^{t_I^{LF}} t x_{Ijt} - t_i^E \right) \right| \\[2mm] \min Q = \sum_{i=0}^{I+1} w_i [q_i^S + \alpha_i (p_{ij} - p_{ij}^S)] \\[2mm] \text{s.t.} \begin{cases} \sum_{j=1}^{m_e} \sum_{t=t_e^{EF}}^{t_e^{LF}} t x_{ejt} + \sum_{j=1}^{m_e} \sum_{t=t_e^{EF}}^{t_e^{LF}} p_{ij} x_{ijt} \leq \sum_{j=1}^{m_e} \sum_{t=t_e^{EF}}^{t_e^{LF}} t x_{ijt} \\[2mm] \sum_{i=1}^{I} \sum_{j=1}^{m_i} \tilde{r}_{ijk} \sum_{s=t}^{t+p_{ij}+1} x_{ijt} \leq r_k^M \\[2mm] \sum_{i=1}^{I} \sum_{j=1}^{m_i} \tilde{r}_{iin} \sum_{t=t_i^{EF}}^{t_i^{LF}} x_{iit} \leq r_n^M \\[2mm] \sum_{j=1}^{m_i} \sum_{t=t_i^{EF}}^{t_i^{LF}} x_{ijt} = 1 \\[2mm] p_{ij}^L \geq p_{ij} \geq p_{ij}^S > 0 \\[2mm] q_i^L \geq q_i \geq q_j^S > 0 \\[2mm] t_{ij}^F \geq 0 \\[2mm] t_{ij}^{EF} \geq 0 \\[2mm] t_{ij}^{LF} \geq 0 \\[2mm] w_i > 0, \sum_{i=0}^{I+1} w_i = 1 \\[2mm] x_{ijt} = 1 \text{ or } 0 \end{cases} \end{cases} \tag{4.109}$$

where $e \in Pre(i), i = 0, 1, 2, \cdots, I+1; \ j = 1, 2, \cdots, m_i; \ k = 1, 2, \cdots, K; n = 1, 2, \cdots, N$ and $t = 1, 2, \cdots, T$.

Inconsistencies exist in three objectives of formula (4.109), and they all needed to be optimized. The constraints in (4.109) are used to impose precedence relations between activities, limit the resource demand imposed by the activities being processed at time t to the available capacity, and to represent the usual integrality constraint, respectively.

It is obvious that model (4.109) is meaningless mathematically for the existence of the random variable. If we use the random expected value operator to deal with it, we have

$$
\begin{cases}
\min T = \sum\limits_{j=1}^{m_I} \sum\limits_{t=t_I^{EF}}^{t_I^{LF}} tx_{Ijt} \\[2ex]
\min E[C] = E\left[\left| \sum\limits_{i=0}^{I+1} \tilde{c}_i^{TP} \left(\sum\limits_{j=1}^{m_I} \sum\limits_{t=t_I^{EF}}^{t_I^{LF}} tx_{Ijt} - t_i^E \right) \right| \right] \\[2ex]
\min Q = \sum\limits_{i=0}^{I+1} w_i [q_i^S + \alpha_i (p_{ij} - p_{ij}^S)] \\[2ex]
\text{s.t.} \begin{cases}
\sum\limits_{j=1}^{m_e} \sum\limits_{t=t_e^{EF}}^{t_e^{LF}} tx_{ejt} + \sum\limits_{j=1}^{m_e} \sum\limits_{t=t_e^{EF}}^{t_e^{LF}} p_{ij} x_{ijt} \leq \sum\limits_{j=1}^{m_e} \sum\limits_{t=t_e^{EF}}^{t_e^{LF}} tx_{ijt} \\[2ex]
\sum\limits_{i=1}^{I} \sum\limits_{j=1}^{m_i} \tilde{r}_{ijk} \sum\limits_{s=t}^{t+p_{ij}+1} x_{ijt} \leq r_k^M \\[2ex]
\sum\limits_{i=1}^{I} \sum\limits_{j=1}^{m_i} \tilde{r}_{ijn} \sum\limits_{t=t_i^{EF}}^{t_i^{LF}} x_{ijt} \leq r_n^M \\[2ex]
\sum\limits_{j=1}^{m_i} \sum\limits_{t=t_i^{EF}}^{t_i^{LF}} x_{ijt} = 1 \\[2ex]
p_{ij}^L \geq p_{ij} \geq p_{ij}^S > 0 \\[1ex]
q_i^L \geq q_i \geq q_j^S > 0 \\[1ex]
t_{ij}^F \geq 0 \\[1ex]
t_{ij}^{EF} \geq 0 \\[1ex]
t_{ij}^{LF} \geq 0 \\[1ex]
w_i > 0, \sum\limits_{i=0}^{I+1} w_i = 1 \\[1ex]
x_{ijt} = 1 \text{ or } 0
\end{cases}
\end{cases}
\tag{4.110}
$$

However, we still cannot easily find the optimal solution to problem (4.110) because of the stochastic constraint and need to convert it into a crisp one. In what follows, we will make use of the rough approximation proposed in the above section to deal with the constraint.

First, we define the following binary relationship for the random constraints as

$$
xR_{h_1} y \Leftrightarrow E\left[\left| \sum_{i=1}^{I} \sum_{j=1}^{m_i} \tilde{r}_{ijk} \sum_{s=t}^{t+p_{ij}+1} x_{ijt} - \sum_{i=1}^{I} \sum_{j=1}^{m_i} \tilde{r}_{ijk} \sum_{s=t}^{t+p_{ij}+1} y_{ijt} \right| \right] \leq h_1
$$

$$xR_{h_2}y \Leftrightarrow E\left[\left|\sum_{i=1}^{I}\sum_{j=1}^{m_i}\tilde{r}_{ijn}\sum_{t=t_i^{EF}}^{t_i^{LF}}x_{ijt} - \sum_{i=1}^{I}\sum_{j=1}^{m_i}\tilde{r}_{ijn}\sum_{t=t_i^{EF}}^{t_i^{LF}}y_{ijt}\right|\right] \le h_2$$

If the DM can give the approximation accuracy θ_1, θ_2 for the two random constraints, we have the following rough model for (4.110):

$$\begin{cases} \min T = \sum_{j=1}^{m_I}\sum_{t=t_I^{EF}}^{t_I^{LF}}tx_{Ijt} \\[2mm] \min E[C] = E\left[\left|\sum_{i=0}^{I+1}\tilde{c}_i^{TP}\left(\sum_{j=1}^{m_I}\sum_{t=t_I^{EF}}^{t_I^{LF}}tx_{Ijt} - t_i^E\right)\right|\right] \\[2mm] \min Q = \sum_{i=0}^{I+1}w_i[q_i^S + \alpha_i(p_{ij}-p_{ij}^S)] \\[2mm] \text{s.t.} \begin{cases} \sum_{j=1}^{m_e}\sum_{t=t_e^{EF}}^{t_e^{LF}}tx_{ejt} + \sum_{j=1}^{m_e}\sum_{t=t_e^{EF}}^{t_e^{LF}}p_{ij}x_{ijt} \le \sum_{j=1}^{m_e}\sum_{t=t_e^{EF}}^{t_e^{LF}}tx_{ijt} \\ \alpha(I_1) \ge \theta_1 \\ \alpha(I_2) \ge \theta_2 \\ x \in \bar{I}_1 \\ x \in \bar{I}_2 \\ \sum_{j=1}^{m_i}\sum_{t=t_i^{EF}}^{t_i^{LF}}x_{ijt} = 1 \\ p_{ij}^L \ge p_{ij} \ge p_{ij}^S > 0 \\ q_i^L \ge q_i \ge q_j^S > 0 \\ t_{ij}^F \ge 0 \\ t_{ij}^{EF} \ge 0 \\ t_{ij}^{LF} \ge 0 \\ w_i > 0, \sum_{i=0}^{I+1}w_i = 1 \\ x_{ijt} = 1 \text{ or } 0 \end{cases} \end{cases} \quad (4.111)$$

where

$$I_1 = \left\{x\Big|\sum_{i=1}^{I}\sum_{j=1}^{m_i}\tilde{r}_{ijk}\sum_{s=t}^{t+p_{ij}+1}x_{ijt} \le r_k^M\right\}, I_2 = \left\{x\Big|\sum_{i=1}^{I}\sum_{j=1}^{m_i}\tilde{r}_{ijn}\sum_{t=t_i^{EF}}^{t_i^{LF}}x_{ijt} \le r_n^M\right\}.$$

As the DM determines the approximation accuracy θ_1, θ_2, we can calculate the maximum h_1, h_2 such that $\alpha(I_1) \ge \theta_1, \alpha(I_2) \ge \theta_2$.

By the linearity of the random expected value operator, the expected value of the second objective function (4.102) can be transformed into

$$\begin{aligned} E[C, \tilde{r}_i^{TP}] &= \left[\sum_{i=0}^{I+1}\tilde{c}_i^{TP}\left(\sum_{j=1}^{m_I}\sum_{t=t_I^{EF}}^{t_I^{LF}}tx_{Ijt} - t_i^E\right)\right] \\ &= \sum_{i=0}^{I+1}\left(\sum_{j=1}^{m_I}\sum_{t=t_I^{EF}}^{t_I^{LF}}tx_{Ijt} - t_i^E\right)E[\tilde{r}_i^{TP}]. \end{aligned}$$

Meanwhile we use the expected value model to deal with the random coefficients in the resource constraints equations (4.105) and (4.106) as

$$E\left[\sum_{i=1}^{I}\sum_{j=1}^{m_i}\tilde{r}_{ijk}\sum_{s=t}^{t+p_{ij}+1}x_{ijt}-r_k^M\right]$$

$$=\sum_{i=1}^{I}\sum_{j=1}^{m_i}E\left[\tilde{r}_{ijk}\right]\sum_{s=t}^{t+p_{ij}+1}x_{ijt}-r_k^M$$

$$\leq 0$$

and

$$E\left[\sum_{i=1}^{I}\sum_{j=1}^{m_i}\tilde{r}_{ijn}\sum_{t=t_i^{EF}}^{t_i^{LF}}x_{ijt}-r_n^M\right]$$

$$=\sum_{i=1}^{I}\sum_{j=1}^{m_i}E\left[\tilde{r}_{ijk}\right]\sum_{t=t_i^{EF}}^{t_i^{LF}}x_{ijt}-r_n^M$$

$$\leq 0,$$

where $k=1,2,\cdots,K$; $n=1,2,\cdots,N$, and $t=1,2,\cdots,T$. Then, model (4.111) can be transformed into model (4.112) as

$$\begin{cases}
\min T=\sum_{j=1}^{m_I}\sum_{t=t_I^{EF}}^{t_I^{LF}}tx_{Ijt}\\[2mm]
\min C=\left|\sum_{i=0}^{I+1}\left(\sum_{j=1}^{m_I}\sum_{t=t_I^{EF}}^{t_I^{LF}}tx_{Ijt}-t_i^E\right)E[\tilde{c}_i^{TP}]\right|\\[2mm]
\min Q=\sum_{i=0}^{I+1}w_i[q_i^S+\alpha_i(p_{ij}-p_{ij}^S)]\\[2mm]
\text{s.t.}\begin{cases}
\sum_{j=1}^{m_e}\sum_{t=t_e^{EF}}^{t_e^{LF}}tx_{ejt}+\sum_{j=1}^{m_e}\sum_{t=t_e^{EF}}^{t_e^{LF}}p_{ij}x_{ijt}\leq\sum_{j=1}^{m_e}\sum_{t=t_e^{EF}}^{t_i^{LF}}tx_{ijt}\\[2mm]
\sum_{i=1}^{I}\sum_{j=1}^{m_i}E[\tilde{r}_{ijk}]\sum_{s=t}^{t+p_{ij}+1}x_{ijt}-r_k^M\leq 0\\[2mm]
\sum_{i=1}^{I}\sum_{j=1}^{m_i}E[\tilde{r}_{ijk}]\sum_{t=t_i^{EF}}^{t_i^{LF}}x_{ijt}-r_n^M\leq 0\\[2mm]
\sum_{j=1}^{m_i}\sum_{t=t_i^{EF}}^{t_i^{LF}}x_{ijt}=1\\[2mm]
p_{ij}^L\geq p_{ij}\geq p_{ij}^S>0\\
q_i^L\geq q_i\geq q_j^S>0\\
t_{ij}^F\geq 0\\
t_{ij}^{EF}\geq 0\\
t_{ij}^{LF}\geq 0\\
w_i>0,\sum_{i=0}^{I+1}w_i=1\\
x_{ijt}=1\text{ or }0
\end{cases}
\end{cases}\tag{4.112}$$

where $e \in Pre(i), i = 0, 1, 2, \cdots, I+1;\ j = 1, 2, \cdots, m_i;\ k = 1, 2, \cdots, K;\ n = 1, 2, \cdots, N$
and $t = 1, 2, \cdots, T$.

Now suppose that random variables $\tilde{c}_i^{TP}, \tilde{r}_{ijk}, \tilde{r}_{ijn}$ are defined as

$$\tilde{c}_i^{TP} \sim \mathscr{N}(\mu_{c_i}, \sigma_{c_i}^2), \tilde{r}_{ijk} \sim \mathscr{N}(\mu_{r_{ijk}}, \sigma_{r_{ijk}}^2), \tilde{r}_{ijn} \sim \mathscr{N}(\mu_{r_{ijn}}, \sigma_{r_{ijn}}^2).$$

It follows from Corollary 4.2 that (4.112) can be calculated as

$$
\begin{cases}
\min T = \sum\limits_{j=1}^{m_I} \sum\limits_{t=t_I^{EF}}^{t_I^{LF}} t x_{Ijt} \\[2mm]
\min C = \frac{1}{4} \left| \left(\sum\limits_{j=1}^{m_I} \sum\limits_{t=t_I^{EF}}^{t_I^{LF}} t x_{Ijt} - t_i^E \right) \mu_{c_i} \right] \\[2mm]
\min Q = \sum\limits_{i=0}^{I+1} w_i [q_i^S + \alpha_i(p_{ij} - p_{ij}^S)] \\[2mm]
\text{s.t.}
\begin{cases}
\sum\limits_{j=1}^{m_e} \sum\limits_{t=t_e^{EF}}^{t_e^{LF}} t x_{ejt} + \sum\limits_{j=1}^{m_e} \sum\limits_{t=t_e^{EF}}^{t_e^{LF}} p_{ij} x_{ijt} \leq \sum\limits_{j=1}^{m_e} \sum\limits_{t=t_e^{EF}}^{t_e^{LF}} t x_{ijt} \\[2mm]
\sum\limits_{i=1}^{I} \sum\limits_{j=1}^{m_i} \mu_{r_{ijk}} \sum\limits_{s=t}^{t+p_{ij}+1} x_{ijt} - r_k^M \leq h_1 \\[2mm]
\sum\limits_{i=1}^{I} \sum\limits_{j=1}^{m_i} \mu_{r_{ijn}} \sum\limits_{t=t_i^{EF}}^{t_i^{LF}} x_{ijt} - r_n^M \leq h_2 \\[2mm]
\sum\limits_{j=1}^{m_i} \sum\limits_{t=t_i^{EF}}^{t_i^{LF}} x_{ijt} = 1 \\[2mm]
p_{ij}^L \geq p_{ij} \geq p_{ij}^S > 0 \\[1mm]
q_i^L \geq q_i \geq q_j^S > 0 \\[1mm]
t_{ij}^F \geq 0 \\[1mm]
t_{ij}^{EF} \geq 0 \\[1mm]
t_{ij}^{LF} \geq 0 \\[1mm]
w_i > 0, \sum_{i=0}^{I+1} w_i = 1 \\[1mm]
x_{ijt} = 1 \text{ or } 0
\end{cases}
\end{cases}
\tag{4.113}
$$

where $e \in Pre(i),\ i = 0, 1, 2, \cdots, I+1;\ j = 1, 2, \cdots, m_i;\ k = 1, 2, \cdots, K;\ n = 1, 2, \cdots, N$
and $t = 1, 2, \cdots, T$.

Thus, we transform rc-PSP/mM for drilling grouting in the Longtan large-scale water conservancy and hydropower construction projects into deterministic model. We then put the certainty number into the multiobjective particle swarm optimization (MOPSO) algorithm, and model (4.113), which is equivalent to (4.112), can be solved by a computer program.

4.6.2.3 Solution Method

PSO has superior search performance in dealing with many of hard optimization problems with faster and more stable convergence rates compared with other population-based random optimization methods. Since PSO can be implemented easily and effectively, it has been rapidly applied in solving real-world optimization problems in recent years, such as D. Sha and C. Hsu [269], S. Ling et al. [194], A. Kashan and B. Karimi [159], Q. Pan et al. [235], etc. Researchers are also seeing PSO as a very strong competitor to other algorithms in solving multiobjective optimal (MOP) problems even though very few works have been reported. Moreover, PSO has rarely been applied to solve multiobjective rc-PSP/mM in construction. In this book, we take up the challenge to apply MOPSO for rc-PSP/mM. PSO imitates the animal social behavior of birds flocking to a desired position for certain objectives in a multidimensional space, and a group of birds searching for food in an area randomly. There is only one piece of food (called efficient solution) in the area being searched, and all birds do not know where the food is, but they know how far the food is in each iteration. Like an evolutionary or metaheuristic algorithm that evolves to find the global optimum of a real-valued function (called fitness function), PSO is based on a set of potential solutions defined in a given space (called search space). It conducts searches using fixed-number populations (called swarm) of individuals (called particles) that are updated from iteration to iteration. An n-dimensional position of a particle (called solution), initialized with a random position in a multidimensional search space, represents a solution to the problem, and it resembles the chromosome of a genetic algorithm (J. Robinson et al. [256]). The particles, which are characterized by their positions and velocities, (see [48, 162]), fly through the problem space by following the current optimum particles. Unlike other population-based algorithms, the velocity and position of each particle are dynamically adjusted according to the flying experiences or discoveries of its own and those of its companions. Meanwhile, the particle-updating mechanism is easy to implement. There are many variants of PSO, and in 1955 J. Kennedy and R. Eberhart [162] proposed as follows to update the position and velocity of each particle:

$$v_{ld}(\tau+1) = w(\tau)v_{ld}(\tau) + c_p r_1 [p_{ld}^{best}(\tau) - p_{ld}(\tau)] + c_g r_2 [p_{gd}^{best}(\tau) - p_{ld}(\tau)],$$

$$p_{ld}(\tau) = p_{ld}(\tau) + v_{ld}(\tau+1),$$

where $v_{ld}(\tau+1)$ is the velocity of lth particle at the dth dimension in the τth iteration, w is an inertia weight, $p_{ld}(\tau)$ is the position of lth particle at the dth dimension, r_1 and r_2 are random numbers in the range [0, 1], c_p and c_g are personal and global best position acceleration constant, respectively, and, $p_{ld}^{best}(\tau)$ and $p_{gd}^{best}(\tau)$ are personal and global best positions, respectively, of the lth particle at the dth dimension.

Due to the fast convergence, PSO has been proved to be especially suitable for multiobjective optimization. Therefore, a number of proposals have been suggested to extend PSO to handle multiobjective problems in the last few years. Since an efficient-based selection scheme combined with an adaptive grid, which is adopted both to store the nondominated solutions found during the search and to distribute them uniformly along the efficient frontier, was employed by C. Coello [65], there have been some studies reported in the literature that extend PSO to multiobjective problems, such as [140, 306] and so on. The MOPSO approach uses the concept of efficient dominance to determine the flight direction of a particle and it maintains previously found nondominated vectors in a global repository that is later used by other particles to guide their own flight. In this study, we will use a modified version of the MOPSO proposed by C. Coello [65] with an application to rc-PSP/mM for drilling scheduling in the Longtan large-scale water conservancy and hydropower construction projects.

Notations

τ	iteration index, $\tau = 1, 2, \cdots, T$
l	article index, $l = 1, 2, \cdots, L$
d	dimension index, $d = 1, 2, \cdots, m_i(t_i^{LF} - t_i^{EF} + 1)$
i	index of activity, $i = 1, 2, \cdots, I + 1$
j	index of mode, $j = 1, 2, \cdots, m_i$
r_1, r_2	uniform distributed random number within [0, 1]
$w(\tau)$	inertia weight in the τth iteration
$v_{id}^l(\tau)$	velocity of the ith activity of the lth particle at the dth dimension in the τth iteration
l^{th}	particle at the dth dimension in the τth iteration
$p_{id}^l(\tau)$	position of the ith activity of the lth particle at the dth dimension in the τth iteration
$p_{id}^{best}(\tau)$	personal best position of the ith activity of the lth particle at the dth dimension
c_p	personal best position acceleration constant
$V_i^l(\tau)$	vector velocity of the ith activity of the lth particle in the τth iteration, $V_i^l(\tau) = [v_{i1}^l(\tau), v_{i2}^l(\tau), \cdots, v_{im_i(t_i^{LF} - t_i^{EF} + 1)}^l(\tau)]$
ϑ	initial velocity of the ith activity at the dth dimension, that is, $V_i^l(1) = [\vartheta_{i1}, \vartheta_{i2}, \cdots, \vartheta_{im_i(t_i^{LF} - t_i^{EF} + 1)}]$
$P_i^l(\tau)$	vector position of the ith activity of the lth particle in the τth iteration, $P_i^l(\tau) = [p_{i1}^l(\tau), p_{i2}^l(\tau), \cdots, p_{im_i(t_i^{LF} - t_i^{EF} + 1)}^l(\tau)]$
$P_i^{l,best}$	vector personal best position of the ith activity of the lth particle, $P_i^{l,best} = [p_{i1}^{l,best}, p_{i2}^{l,best}, \cdots, p_{im_i(t_i^{LF} - t_i^{EF} + 1)}^{l,best}]$
R_i^l	the lth set of solution of the ith activity
$Fitness_i(P_i^l(\tau))$	fitness value of $P_i^l(\tau)$
REP	the positions of the particles that represent nondominated vectors in the repository

Framework of MOPSO for rc-PSP/mM. In the MOPSO algorithm, the performances of different particles are always compared in terms of their dominance relations. The main characteristic of this algorithm is the application of an external repository to store nondominated solutions. All the particles of this swarm are compared to each other, and the nondominated particles are stored in the repository after an initial population is generated. The positions of the particles will be subsequently updated using the following:

$$v_{ld}(\tau + 1) = w v_{ld}(\tau) + c_p r_1 [p_{ld}^{best}(\tau) - p_{ld}(\tau)] + c_g r_2 [REP_h(\tau) - p_{ld}(\tau)] \quad (4.114)$$

From Equation (4.114), we know that there is no such thing as the best position $p_{gd}^{best}(\tau)$ as in the standard PSO in MOPSO. Instead of $p_{gd}^{best}(\tau)$, there are several equally good nondominated solutions stored in the external repository. To update the velocity of each particle using Equation (4.114), the algorithm has to select one of the positions stored in the repository. The selections method is that nondominated solutions located in regions less densely populated in the objective space are given priority for being selected, leading to a better distribution of points in the efficient front. The approach followed un this study simply calculates differently from the adaptive grid . The density of points around each solution is stored in the repository and performs a roulette wheel selection such that the probability of choosing one point is inversely related to its associated density in the objective space. In every iteration τ, the new positions of all particles are compared among themselves, and the nondominated ones are then compared with all solutions stored in the repository. After adding new nondominated solutions and eliminating old solutions that are now dominated, the repository is then updated. Since the size of the repository is limited, whenever it gets full and a new nondominated solution is found, then this new solution takes the place of another nondominated solution in the repository, which is selected using a similar procedure based on density as described above randomly. However, the nondominated solution is assigned higher probabilities of being selected to solutions located in denser regions of the objective space. MOPSO in this book handles constraints simply and efficiently. When comparing two different solutions with at least one infeasible, the algorithm does the following: one feasible solution dominates the infeasible one; an infeasible solution with smaller violation of the constraints dominates another infeasible one. MOPSO for rc-PSP/mM will stop until the maximum number of iterations is reached.

The particle-represented solution should be checked and adjusted for infeasibility due to violating nonrenewable resource constraints and then be transformed to a schedule though a serial generation scheme based on precedence and resource constraints. The best position vector of particle l, P_{ld}^{best}, is initially set equal to the initial position of particle l. In subsequent iterations, the best position is updated in the following way:

If the current $p_{ld}^{best}(\tau)$ dominates the new position $p_{ld}^{best}(\tau+1)$, then $p_{ld}^{best}(\tau+1) = p_{ld}^{best}(\tau)$.

If the new position $p_{ld}(\tau+1)$ dominates $p_{ld}(\tau)$, then $p_{ld}^{best}(\tau+1) = p_{ld}(\tau+1)$.

If no one dominates the other, then one of them is randomly selected to be the $p_{ld}^{best}(\tau+1)$.

Now we will present the framework of MOPSO to solve rc-PSP/mM in drilling grouting project as follows:

1. Initialize L particles as a swarm. Set iteration $\tau = 1$. For $l = 1, 2, \cdots, L$, generate the position of the ith activity of the lth particle with integer random position:
$$P_i^l(\tau) = [p_{i1t_i^{EF}}^l(\tau), p_{i1(t_i^{EF}+1)}^l(\tau), \cdots, p_{i1t_i^{LF}}^l(\tau); p_{i2t_i^{EF}}^l(\tau), p_{i2(t_i^{EF}+1)}^l(\tau), \cdots, p_{i2t_i^{LF}}^l(\tau)$$
$$; \cdots ; p_{ijt_i^{EF}}^l(\tau), p_{ij(t_i^{EF}+1)}^l(\tau), \cdots, p_{ijt_i^{LF}}^l(\tau); \cdots ; p_{im_it_i^{EF}}^l(\tau), p_{im_i(t_i^{EF}+1)}^l(\tau), \cdots, p_{im_it_i^{LF}}^l$$
$(\tau)]$ and the value is 0 or 1. If $p_{ij(t_i^{EF}+t_{order})}^l(\tau) = 1$, then the ith activity starts at

$t_i^{EF} + t_{order}, t_{order} \in \{0, 1, \cdots, t_i^{LF} - t_i^{EF}\}$ executed by mode j.

2. Decode particles into solutions. For $l = 1, 2, \cdots, L$, decode $P_i^l(\tau)$ to a solution R_i^l as $x_{ijt}(\tau) = p_{ijt}^l(\tau)$.

. Check the feasibility of solutions. For $l = 1, 2, \cdots, L$, if the feasibility criterion is met by all particles, that is,

$$\sum_{j=1}^{m_e} \sum_{t=t_e^{EF}}^{t_e^{LF}} t p_{ejt}^l + \sum_{j=1}^{m_i} \sum_{t=t_i^{EF}}^{t_i^{LF}} p_{ij} p_{ijt}^l \leq \sum_{j=1}^{m_i} \sum_{t=t_i^{EF}}^{t_i^{LF}} t p_{ijt}^l \tag{4.115}$$

$$\sum_{i=1}^{I} \sum_{j=1}^{m_i} \mu_{r_{ijk}} \sum_{s=t}^{t+p_{ij}+1} p_{sjt}^l - r_k^M \leq h_1 \tag{4.116}$$

$$\sum_{i=1}^{I} \sum_{j=1}^{m_i} \mu_{r_{ijn}} \sum_{t=t_i^{EF}}^{t_i^{EF}} p_{ijt}^l - r_n^M \leq h_2 \tag{4.117}$$

then continue. Otherwise, return to Step 1.

4. Initialize the speed of each particle and personal best position. For $l = 1, 2, \cdots, L$, we have

$$V_i^L(1) = [\vartheta_{i1}, \vartheta_{i2}, \cdots, \vartheta_{im_i(t_i^{LF} - t_i^{EF} + 1)}] \quad \text{and} \quad P_{id}^{l,best} = P_i^l(1) \tag{4.118}$$

5. Evaluate each of the particles. For $l = 1, 2, \cdots, L$, L, compute the performance measurement of R_i^l, and set this as the fitness value of $P_i^l(\tau)$, represented by $Fitness_i(P_i^l(\tau))$.

6. Store the positions of the particles that represent nondominated vectors in the repository denoted as *REP*.

7. Generate hypercubes of the search space explored so far, and locate the particles using these hypercubes as a coordinate system where the coordinates of each particle are defined according to the values of its objective functions.

8. Initialize the memory of each particle, which is employed as a guide to travel through the search space and also stored in the repository. Update personal best position $P_{id}^{l,best} = P_i^l(1)$.

9. While maximum number of cycles has not been reached, do

(a) Compute the speed of each particle using the following formula:

$$v_{id}^l(\tau + 1) = w(\tau) v_{id}^l(\tau) + c_p r_1 [p_{id}l, best - p_{id}l(\tau)] + c_R r_2 [REP_{ih}(\tau) - p_{id}^l(\tau)]$$

where c_R is an acceleration constant, $w(\tau) = w(T) + \frac{\tau - T}{1 - T}[w(1) - w(T)]$, $REP_{ih}(\tau)$ is a solution selected from the repository in each iteration τ for activity i, and the index h is selected in the following way: those hypercubes containing more than one particle are assigned a fitness equal to the result of dividing any number in the number of particles they contain. This aims to decrease the fitness of those hypercubes that contain more particles, and it can be seen as a form of fitness sharing. Then, we apply roulette-wheel selection using these fitness values to select the hypercube from which we will take the corresponding particle. Once the hypercube has been selected, we select randomly a particle within such hypercube.

(b) Adding the speed produced from the previous step, compute the new positions of the particles

$$p_{id}^l(\tau+1) = p_{id}^l(\tau) + v_{id}^l(\tau+1).$$

(c) Maintain the particles within the search space in case they go beyond their boundaries.

(d) Evaluate every particle.

(e) Update the contents of REP_h and the geographical representation of the particles within the hypercubes. Insert all the currently nondominated locations into the repository and eliminate any dominated locations from the repository.

(f) Using Pareto dominance, if the position in memory is dominated by the current position, the position of the particle is updated by $P_{id}^{l,best} = P_i^l$; otherwise, the position in memory replaces the one in current; if neither of them is dominated by the other, select one of them randomly.

(g) Increment the loop counter.

10. End while.

11. If the stopping criterion is met, that is, $\tau = T$, go to Step 12. Otherwise, $\tau = \tau + 1$ and return to Step 9.

12. Decode REP_{ih} as the solution set.

Through the serial generation scheme, the schedule that satisfies the precedence and resource constraints are generated from the particle represented. The precedence constraints, that is Equation (4.115), can be easily satisfied when using the serial generation scheme. The renewable resource constraints, that is, Equation (4.116), is also satisfied easily because they can be recovered or released later after serving other activities ultimately. However, nonrenewable resource constraints, that is, Equation (4.117), may be infeasible. After serving some activities, the requirement of nonrenewable resources would be greater than their total quantities, that is,

$\sum_{i=1}^{I} \sum_{j=1}^{m_i} \mu_{r_{ijn}} \sum_{t=t_i^{EF}}^{t_i^{EF}} p_{ijt}^l \geq r_n^M + h_2$. Therefore, the checking and adjusting of particle-represented solutions so as to avoid nonrenewable resource infeasibility are necessary. The procedure to check and adjust the infeasible particle-represented solution is presented as follows (see Figure 4.5):

Step 1. Let *Inf* denote the infeasibility of nonrenewable resources, according to the particle-represented in the *l*th particle; compute

$$Inf = \begin{cases} 1, & \text{if } \sum_{i=1}^{I} \sum_{j=1}^{m_i} \mu_{r_{ijn}} \sum_{t=t_i^{EF}}^{t_i^{EF}} p_{ijt}^l \geq r_n^M + h_2 \\ 0, & \text{otherwise} \end{cases}$$

Step 2. If $Inf = 0$ for all of the nonrenewable resources, go to Step 6, otherwise, that is, $Inf = 1$ for some nonrenewable resources, go to Step 3.

Step 3. Select an activity $i(i = 1, 2, \cdots, N)$ with multiple execution modes, that is, $m_i > 1$.

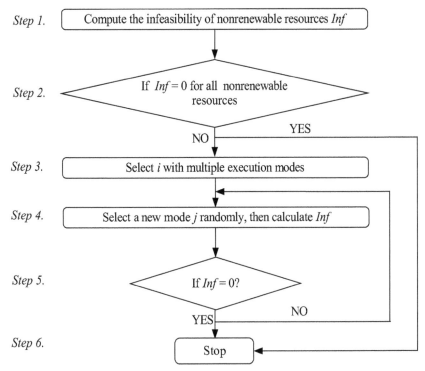

FIGURE 4.5 Overall procedure for checking and adjusting of infeasible particle-represented solutions.

Step 4. select a new mode $j' \neq j(\tau)$ and within $1,2,\cdots,m_i$ for activity i randomly, then compute *Inf* again.

Step 5. If *Inf=0*, replace j' with $j(\tau)$ and then go to Step 6; otherwise, repeat Step 4, until all modes of activity i have been iterated.

Step 6. Stop.

Solution representation is one of the key elements for effective implementation of MOPSO for rc-PSP/mM. An indirect representation, which consists of $m_i(t_i^{LF} - t_i^{EF} + 1)$ dimensional particle, is proposed here. The value of the particle in each dimension is encoded as 0 or 1, and the sum of the particles in all dimensions is 1. Meanwhile, the value of dth dimension represents the amount of the ith activity that starts at $t_i^{EF} + t_{order}, t_{order} \in \{0,1,\cdots,t_i^{LF} - t_i^{EF}\}$ executed by mode j. For $l = 1,2,\cdots,L$, decode P_i^l to a solution R_i^l as $x_{ijt}(\tau) = p_{ijt}^l(\tau)$. Mapping between one potential solution to rc-PSP/mMand particle representation is shown in Figure 4.6 for a more intuitive scene.

As mentioned above, this algorithm is designed for solving the problem and satisfying the demand of the decision manager. Since the objective functions of rc-PSP/mM in the drilling grouting project with a mixed rough and random environ-

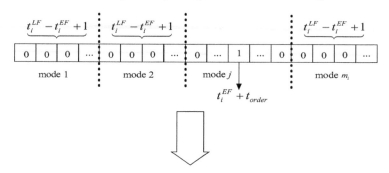

Activity i executed by mode j starts at $t_i^{EF} + t_{order}$

FIGURE 4.6 Particle-represented solution for rc-PSP/mM.

ment are minimal project duration, minimal total tardiness penalty for all activities and maximizing the quality of the project, the fitness value used to evaluate a particle is just the weighted value of the three objective functions. The weighted value, given by the decision manager in the preliminary design report, reflects the importance of the overall three objective functions i the project. Based on experiential value, $\lambda_{\text{minimal project duration}} = 0.45, \lambda_{\text{minimal project duration}} = 0.20$, and $\lambda_{\text{maximal project quality}} = 0.35$, respectively. Thus,

$$
Fitness_i(P_i^l(\tau)) = 0.45 \sum_{j=1}^{m_I} \sum_{t=t_I^{EF}}^{t_I^{LF}} tp_{Ijt}^l + 0.20 \left| \sum_{i=0}^{I+1} \sum_{j=1}^{m_i} \sum_{t=t_I^{EF}}^{t_I^{LF}} (tp_{ijt}^l - t_i^E)\mu_{c_1} \right|
$$
$$
+0.35 \left(-\sum_{i=0}^{I+1} \omega_i[q_i^s + \alpha_i(p_{ij} - p_{ij}^s)] \right)
$$

It is noted that we do not consider the units but are only taking into account their values in the above formula.

4.6.2.4 Case Study

In this section, an illustrative application of a working procedure in the project drilling grouting of Longtan large-scale water conservancy and hydropower construction project: illustrated to show the application of the proposed models and algorithms. Additionally, the same model and algorithm can be applied to other construction projects successfully. At present, 13 activities in the project drilling grouting are denoted as

 A_1 Rock multipoint bore-hole of deformation
 A_2 Osmometer hole drilling
 A_3 Bolt stress meter hole drilling
 A_4 Underground water-level observation hole drilling
 A_5 Drilled drain hole (including flexible drain)

TABLE 4.4 Executed modes, successors, and expected processing time

Activity	A_0	A_1	A_2	A_3	A_4
Mode	1	2	2	3	2
Successors	$A_1,A_2,A_3,A_4A_5,A_6,A_7$	A_{11},A_{12},A_{13}	A_{11},A_{12},A_{13}	A_{11},A_{12},A_{13}	A_{11},A_{12},A_{13}
Processing time (month)	0	48	48	48	48
Activity	A_5	A_6	A_7	A_8	A_9
Mode	3	2	3	2	2
Successors	A_8,A_9,A_{10}	A_8,A_9,A_{10}	$A11,A_{12},A_{13}$	A_{14}	A_{14}
Processing time (month)	36	24	30	30	30
Activity	A_{10}	A_{11}	A_{12}	A_{13}	A_{14}
Mode	2	3	2	2	1
Successors	A_{14}	A_{14}	A_{14}	A_{14}	A_{14}
Processing time (month)	20	18	18	18	0

Note: A_0 and A_{14}, which have no duration, are dummies, and just represent the initial and final activities respectively.

A_6 Inclinometer hole drilling

A_7 Embedded ϕ 50 type grouting PVC pipe

A_8 Consolidation grouting (including drilling)

A_9 Backfill grouting

A_{10} Consolidation grouting check hole drilling and water pressure test for check hole

A_{11} Closure grouting

A_{12} Embedded 25 type grouting steel pipe

A_{13} Contact grouting

Each activity has certain executed modes, successors, and an expected finish time according to the decision manager. The project uses months as a time unit (i.e., one month per unit) traditionally; at the same time, two dummy activities are used according to the convenience of the model. The corresponding date is shown in Table 4.4. According to the viewpoint of the decision-making manager, three types of resources, that is, manpower, equipment (renewable resources), and materials (nonrenewable resources) are considered in the scheduling. For calculating various kinds of resources expediently, we accord the measurement of all the resources to the consumption amount and unify the dimensionless unit into the cash value (10,000 CNY per unit). The detailed data of process time and resource consumption for each activity are shown in Table 4.5. The parameters of problem are set as follows: Population size $L = 50$, Iteration number $T = 100$, Acceleration constant $cp = 2.5$, and $cR = 3.0$, Inertia weight $w(1) = 0.9$ and $w(T) = 0.1$.

Each activity has certain maximal multiple unit requirements of three different resources:

(Manpower, Equipment, Materials)$= (r_1^M, r_2^M, r_M^3) = (750, 830, 75)$ (Units: 10,000 CNY).

The main certain variables are given below:

Activity	A_1	A_2	A_3	A_4	A_5	A_6	A_7	A_8	A_9	A_{10}	A_{11}	A_{12}	A_{13}
ω	0.115	0.115	0.115	0.115	0.086	0.058	0.072	0.072	0.072	0.048	0.044	0.044	0.044
α	0.58	0.60	0.60	0.56	0.45	0.30	0.39	0.41	0.40	0.25	0.23	0.20	0.22
q^S	2.31	2.31	2.31	2.31	1.73	1.15	1.44	1.44	1.44	0.96	0.87	0.87	0.87
t^E	48	48	48	48	36	24	30	30	30	20	18	18	18

TABLE 4.5 Project-related data set of drilling grouting project.

Activity	Mode	Processing time (month)	Resource consumptions (Units: 10,000 CNY)		
			Manpower	Equipment	Materials
A_0	1	Dummy activity			
A_1	1	46	$\tilde{r}_{111}$	$\tilde{r}_{112}$	$\tilde{r}_{113}$
	2	49	$\tilde{r}_{121}$	$\tilde{r}_{122}$	$\tilde{r}_{123}$
A_2	1	47	$\tilde{r}_{211}$	$\tilde{r}_{212}$	$\tilde{r}_{213}$
	2	49	$\tilde{r}_{221}$	$\tilde{r}_{222}$	$\tilde{r}_{223}$
A_3	1	48	$\tilde{r}_{311}$	$\tilde{r}_{312}$	$\tilde{r}_{313}$
	2	49	$\tilde{r}_{321}$	$\tilde{r}_{322}$	$\tilde{r}_{323}$
	3	49	$\tilde{r}_{331}$	$\tilde{r}_{332}$	$\tilde{r}_{333}$
A_4	1	47	$\tilde{r}_{411}$	$\tilde{r}_{412}$	$\tilde{r}_{413}$
	2	48	$\tilde{r}_{421}$	$\tilde{r}_{422}$	$\tilde{r}_{423}$
A_5	1	34	$\tilde{r}_{511}$	$\tilde{r}_{512}$	$\tilde{r}_{513}$
	2	37	$\tilde{r}_{521}$	$\tilde{r}_{522}$	$\tilde{r}_{523}$
A_6	1	23	$\tilde{r}_{611}$	$\tilde{r}_{612}$	$\tilde{r}_{613}$
	2	24	$\tilde{r}_{621}$	$\tilde{r}_{622}$	$\tilde{r}_{623}$
A_7	1	28	$\tilde{r}_{711}$	$\tilde{r}_{712}$	$\tilde{r}_{713}$
	2	29	$\tilde{r}_{721}$	$\tilde{r}_{722}$	$\tilde{r}_{723}$
	3	30	$\tilde{r}_{731}$	$\tilde{r}_{732}$	$\tilde{r}_{733}$
A_8	1	29	$\tilde{r}_{811}$	$\tilde{r}_{812}$	$\tilde{r}_{813}$
	2	31	$\tilde{r}_{821}$	$\tilde{r}_{822}$	$\tilde{r}_{823}$
A_9	1	30	$\tilde{r}_{911}$	$\tilde{r}_{912}$	$\tilde{r}_{913}$
	2	31	$\tilde{r}_{921}$	$\tilde{r}_{922}$	$\tilde{r}_{923}$
A_{10}	1	18	$\tilde{r}_{(10)11}$	$\tilde{r}_{(10)12}$	$\tilde{r}_{(10)13}$
	2	21	$\tilde{r}_{(10)21}$	$\tilde{r}_{(10)22}$	$\tilde{r}_{(10)23}$
A_{11}	1	18	$\tilde{r}_{(11)11}$	$\tilde{r}_{(11)12}$	$\tilde{r}_{(11)13}$
	2	17	$\tilde{r}_{(11)21}$	$\tilde{r}_{(11)22}$	$\tilde{r}_{(11)23}$
	3	19	$\tilde{r}_{(11)31}$	$\tilde{r}_{(11)32}$	$\tilde{r}_{(11)33}$
A_{12}	1	18	$\tilde{r}_{(12)11}$	$\tilde{r}_{(12)12}$	$\tilde{r}_{(12)13}$
	2	19	$\tilde{r}_{(12)21}$	$\tilde{r}_{(12)22}$	$\tilde{r}_{(12)23}$
A_{13}	1	18	$\tilde{r}_{(13)11}$	$\tilde{r}_{(13)12}$	$\tilde{r}_{(13)13}$
	2	16	$\tilde{r}_{(13)21}$	$\tilde{r}_{(13)22}$	$\tilde{r}_{(13)23}$
A_{14}	1	Dummy activity			

The random coefficients of cost are given as follows:

$$\tilde{c}_1^{TP} \sim \mathcal{N}(5.9,2),\ \tilde{c}_2^{TP} \sim \mathcal{N}(5.8,4),\ \tilde{c}_3^{TP} \sim \mathcal{N}(5.9,4),\ \tilde{c}_4^{TP} \sim \mathcal{N}(6.2,2),$$
$$\tilde{c}_5^{TP} \sim \mathcal{N}(4.5,2),\ \tilde{c}_6^{TP} \sim \mathcal{N}(2.9,6),\ \tilde{c}_7^{TP} \sim \mathcal{N}(3.6,1),\ \tilde{c}_8^{TP} \sim \mathcal{N}(3.8,1),$$
$$\tilde{c}_9^{TP} \sim \mathcal{N}(3.6,5),\ \tilde{c}_{10}^{TP} \sim \mathcal{N}(2.6,4),\ \tilde{c}_{11}^{TP} \sim \mathcal{N}(2.5,4),\ \tilde{c}_{12}^{TP} \sim \mathcal{N}(2.7,6),$$
$$\tilde{c}_{13}^{TP} \sim \mathcal{N}(2.5,3).$$

The value of related random variables in Table 4.5 are given as follows:

$$\tilde{r}_{111} \sim \mathcal{N}(67,2),\quad \tilde{r}_{121} \sim \mathcal{N}(66,8),\quad \tilde{r}_{211} \sim \mathcal{N}(65,2),\quad \tilde{r}_{221} \sim \mathcal{N}(63,5),$$
$$\tilde{r}_{311} \sim \mathcal{N}(66,2),\quad \tilde{r}_{321} \sim \mathcal{N}(65,9),\quad \tilde{r}_{331} \sim \mathcal{N}(65,2),\quad \tilde{r}_{411} \sim \mathcal{N}(65,4),$$
$$\tilde{r}_{421} \sim \mathcal{N}(66,4),\quad \tilde{r}_{511} \sim \mathcal{N}(51,6),\quad \tilde{r}_{521} \sim \mathcal{N}(50,2),\quad \tilde{r}_{611} \sim \mathcal{N}(34,6),$$
$$\tilde{r}_{621} \sim \mathcal{N}(33,4),\quad \tilde{r}_{711} \sim \mathcal{N}(43,6),\quad \tilde{r}_{721} \sim \mathcal{N}(43,4),\quad \tilde{r}_{731} \sim \mathcal{N}(42,1),$$
$$\tilde{r}_{811} \sim \mathcal{N}(69,3),\quad \tilde{r}_{821} \sim \mathcal{N}(68,4),\quad \tilde{r}_{911} \sim \mathcal{N}(70,3),\quad \tilde{r}_{921} \sim \mathcal{N}(68,4),$$
$$\tilde{r}_{(10)11} \sim \mathcal{N}(51,3),\ \tilde{r}_{(10)21} \sim \mathcal{N}(47,11),\ \tilde{r}_{(11)11} \sim \mathcal{N}(51,3),\ \tilde{r}_{(11)21} \sim \mathcal{N}(52,3),$$
$$\tilde{r}_{(11)31} \sim \mathcal{N}(50,5),\ \tilde{r}_{(12)11} \sim \mathcal{N}(52,2),\ \tilde{r}_{(12)21} \sim \mathcal{N}(51,7),\ \tilde{r}_{(13)11} \sim \mathcal{N}(51,3),$$
$$\tilde{r}_{(13)21} \sim \mathcal{N}(53,1).$$

The second renewable resource equipment:

$$\tilde{r}_{112} \sim \mathcal{N}(94,2),\quad \tilde{r}_{122} \sim \mathcal{N}(93,8),\quad \tilde{r}_{212} \sim \mathcal{N}(95,2),\quad \tilde{r}_{222} \sim \mathcal{N}(93,6),$$
$$\tilde{r}_{312} \sim \mathcal{N}(95,2),\quad \tilde{r}_{322} \sim \mathcal{N}(94,6),\quad \tilde{r}_{332} \sim \mathcal{N}(95,2),\quad \tilde{r}_{412} \sim \mathcal{N}(95,3),$$
$$\tilde{r}_{422} \sim \mathcal{N}(94,6),\quad \tilde{r}_{512} \sim \mathcal{N}(71,2),\quad \tilde{r}_{522} \sim \mathcal{N}(68,1),\quad \tilde{r}_{612} \sim \mathcal{N}(48,6),$$
$$\tilde{r}_{622} \sim \mathcal{N}(47,4),\quad \tilde{r}_{712} \sim \mathcal{N}(62,4),\quad \tilde{r}_{722} \sim \mathcal{N}(61,3),\quad \tilde{r}_{732} \sim \mathcal{N}(60,2),$$
$$\tilde{r}_{812} \sim \mathcal{N}(60,3),\quad \tilde{r}_{822} \sim \mathcal{N}(58,2),\quad \tilde{r}_{912} \sim \mathcal{N}(59,3),\quad \tilde{r}_{922} \sim \mathcal{N}(58,1),$$
$$\tilde{r}_{(10)12} \sim \mathcal{N}(40,3),\ \tilde{r}_{(10)22} \sim \mathcal{N}(40,1),\ \tilde{r}_{(11)12} \sim \mathcal{N}(35,2),\ \tilde{r}_{(11)22} \sim \mathcal{N}(35,1),$$
$$\tilde{r}_{(11)32} \sim \mathcal{N}(34,5),\ \tilde{r}_{(12)12} \sim \mathcal{N}(35,6),\ \tilde{r}_{(12)22} \sim \mathcal{N}(35,1),\ \tilde{r}_{(13)12} \sim \mathcal{N}(35,3),$$
$$\tilde{r}_{(13)22} \sim \mathcal{N}(34,2).$$

And the final nonrenewable resource materials:

$$\tilde{r}_{113} \sim \mathcal{N}(8.2,1),\quad \tilde{r}_{123} \sim \mathcal{N}(7.6,7),\quad \tilde{r}_{213} \sim \mathcal{N}(8.5,2),\quad \tilde{r}_{223} \sim \mathcal{N}(7.9,7),$$
$$\tilde{r}_{313} \sim \mathcal{N}(8.5,4),\quad \tilde{r}_{323} \sim \mathcal{N}(8.3,3),\quad \tilde{r}_{333} \sim \mathcal{N}(8.3,6),\quad \tilde{r}_{413} \sim \mathcal{N}(8.5,3),$$
$$\tilde{r}_{423} \sim \mathcal{N}(8.3,3),\quad \tilde{r}_{513} \sim \mathcal{N}(6.3,2),\quad \tilde{r}_{523} \sim \mathcal{N}(5.9,1),\quad \tilde{r}_{613} \sim \mathcal{N}(4.5,2),$$
$$\tilde{r}_{623} \sim \mathcal{N}(4.2,1),\quad \tilde{r}_{713} \sim \mathcal{N}(5.3,1),\quad \tilde{r}_{723} \sim \mathcal{N}(5.4,2),\quad \tilde{r}_{733} \sim \mathcal{N}(5.2,2),$$
$$\tilde{r}_{813} \sim \mathcal{N}(5.2,3),\quad \tilde{r}_{823} \sim \mathcal{N}(5,2),\quad \tilde{r}_{913} \sim \mathcal{N}(5.2,3),\quad \tilde{r}_{923} \sim \mathcal{N}(5.1,1),$$
$$\tilde{r}_{(10)13} \sim \mathcal{N}(3.4,8),\ \tilde{r}_{(10)23} \sim \mathcal{N}(3.3,9),\ \tilde{r}_{(11)13} \sim \mathcal{N}(3,4),\ \tilde{r}_{(11)23} \sim \mathcal{N}(3,1),$$
$$\tilde{r}_{(11)33} \sim \mathcal{N}(2.9,1),\ \tilde{r}_{(12)13} \sim \mathcal{N}(3.1,4),\ \tilde{r}_{(12)23} \sim \mathcal{N}(3,1),\ \tilde{r}_{(13)13} \sim \mathcal{N}(3,6),$$
$$\tilde{r}_{(13)23} \sim \mathcal{N}(3,7).$$

If the DM predetermines the accuracy of rough approximation $\theta_1 = \theta_2 = 0.8$, taking the data above into the proposed model (4.113), we use the proposed algorithm to solve it. After a run of the computer program, we obtained the following satisfactory solution, and the detailed results are shown in Table 4.6 including the dummy activities. Following the result in Table 4.6, the total duration was 5 years and 3 months (63 months). Compared with the initial design, we are short 5 months for the period of the total duration. In ensuring the quality of the project, total tardiness penalty cost is within an acceptable range. The result was very satisfactory for the decision manager.

TABLE 4.6   Optimal result of the project

The order of activities	A_0	A_4	A_2	A_1	A_3	A_5	A_6	A_7	A_9	A_8	A_{10}	A_{13}	A_{11}	A_{12}	A_{14}
Mode	1	2	1	1	1	1	2	3	1	1	1	2	2	1	1
Resource consumptions	Manpower: 723.69 Equipment: 810.66 Materials: 70.28														
Optimal total project duration	63 months														
Optimal total tardiness penalty cost	40.87 ten thousands CNY														
Optimal total project quality	2.64														
Optimal fitness value	29.4511														

Chapter 5

Fuzzy Multiple Objective Rough Decision Making

Fuzzy decision making problems are those based on fuzzy set theory. The fuzzy set theory is a mathematical branch dealing with fuzzy environment, presented by L. Zadeh [335]. In this chapter we first review the theory, and based on the expected value operator and chance operator of fuzzy variables, present three models.

1. *Fuzzy expected value decision-making model*. Usually, decision makers find it difficult to make a decision when they encounter the fuzzy parameters. A clear criteria must be brought forward to help them make the decision. The expected value operator of fuzzy variables is introduced, and a crisp equivalent model is deduced when the distribution is clear.

2. *Fuzzy chance-constrained decision-making model*. Sometimes, decision makers do not strictly require the objective value to be maximal benefit but only to obtain the maximum benefit under a predetermined confidence level. Then the chance-constrained model is proposed, and a crisp equivalent model is deduced when the distribution is clear.

3. *Fuzzy dependent-chance decision-making model*. Decision makers predetermine an objective value and require the maximal probability that objective values exceed the predetermined one.

By applying rough approximation to the above three classes of models, we will obtain the fuzzy expected value rough model (Fu-EVRM), the fuzzy chance-constrained rough model (Fu-CCRM) and the fuzzy dependent-chance rough model (Fu-DCRM).

5.1 Assignment Problem

The assignment problem, which is also called the allocation problem, is concerned with the allocation of materials from some supply points to some demand points through some intermediary points. The material allocation problem is basically taken up as an optimization problem with some constraints: it determines the allocation of a fixed amount of materials to a given number of activities in order to achieve the most effective results. It is one of the most important functions in finance, production, and operation management. It was studied by scholars, J. Xu et al. [320], J. Li and J. Xu [184], S. Jacobsen [152], E. Porteus and J. Yormark [245], S. Kaplan [156],

C. Tang [284], Q. Liu and J. Xu [198], and H. Luss [205].

However, in real-world allocation problems, the input data or parameters, such as demand, resources, costs and objective function are often imprecise or fuzzy because some information is incomplete or unobtainable. Conventional mathematical decision-making schemes clearly cannot solve all fuzzy decision-making problems. The current allocation model represents information in a fuzzy environment where the objective function and parameters are incompletely defined and cannot be accurately measured. In 1976, H. Zimmermann [345] first introduced fuzzy set theory into conventional linear decision making problems. That study considered linear decision-making problems with a fuzzy goal and fuzzy constraints. Following the fuzzy decision-making method proposed by Bellman and A. Zadeh [26] and using linear membership functions, that same study confirmed that there exists an equivalent linear decision-making problem. Thereafter, fuzzy linear decision making has been developed into a number of fuzzy optimization methods for solving the allocation problem.

For the earth-rock work allocation problem, several scholars did research work, where the objectives of the model are only to minimize the total cost. Actually, the earth-rock work allocation in a water conservancy and hydropower engineering project depends on the road condition. If the intensity is too high, the allocation scheme cannot be guaranteed, so minimizing the peak traffic intensity of the road is also to be considered as an objective. So, in this book, the source of the raw materials balance subsystem and the path transportation subsystem are brought into the large-scale systems model for the whole synthetic optimization. The allocation scheme provided through the joint optimization model is more appropriate.

In practice, we must consider the fuzzy allocation problem because some factors such as demand, supply, and even cost are usually vague. J. Xu and Zhou [318] considered a fuzzy earth-rock work allocation problem that is illustrated in Figure 5.1. C. Lin and M. Gen [193] presented a biobjective model for allocation problem as

$$
\begin{cases}
\max z_1(\mathbf{x}) = \sum_{i=1}^{n} \sum_{j=0}^{M} p_{ij} x_{ij} \\
\min z_2(\mathbf{x}) = \sum_{i=1}^{n} \sum_{j=0}^{M} c_{ij} x_{ij} \\
\text{s.t.} \begin{cases}
G_0(x) = \sum_{i=1}^{n} j, x_{ij} \le M \\
G_i(x) = \sum_{j=0}^{m} x_{ij} = 1, \quad \forall i \\
x_{ij} = 0 \text{ or } 1, \forall i, j
\end{cases}
\end{cases}
\tag{5.1}
$$

In addition, G. Pugh [249] researched the fuzzy allocation of manufacturing resources using the fuzzy logic to allocate the resources. However, if the parameters in the allocation problem are fuzzy variables, to our knowledge, there is little literature, so, in this section, we try to establish a multiobjective model with fuzzy parameters. Because of the fuzzy parameters, the feasible region and the objective values are changing, so we cannot solve the model. Thus, we need to use the expected value operator and the chance operator to deal with the model, and obtain

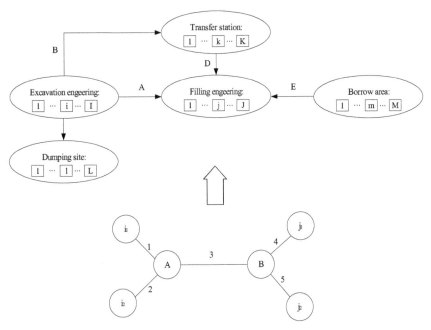

FIGURE 5.1   General structure of the earth-rock work allocation.

a fuzzy expectation multiobjective model with chance constraints. Since the feasible region is created by the chance constraints, it is a certain shadow of the original feasible region and can not reflect all of the cases of the original feasible region completely. So, in order to simulate the feasible region better and truly use rough approximation, we construct two approximation models by expanding and shrinking the chance-constrained feasible region.

In 1982, Z. Pawlak introduced rough set theory [241], which has emerged as another major mathematical tool for modeling the vagueness present in human classification mechanisms. This concept is fundamental to the examination of granularity in knowledge. It is a concept that has many applications in machine learning, pattern recognition, decision support systems, expert systems, data analysis, and data mining, among others.

The theory of fuzzy sets introduced by L. Zadeh [335] has provided a useful mathematical tool for describing the behavior of systems that are too complex or ill-defined to admit precise mathematical analysis by classical methods and tools. Extensive applications of the fuzzy set theory have been found in various fields. Fuzzy numbers which are a special kind of fuzzy sets defined by D. Dubois [87] and H. Zimmermann [347], are usually used to denote vague numbers. Since there are many vague factors in real life, fuzzy numbers are popular in decision making.

In 1988, D. Dubois and H. Prade [87] introduced a possibility space similar to the pattern space proposed by S. Nahmias [221] in 1978, and the fuzzy variable is first defined as a mapping from the possibility space to a real number by S. Nahmias, D. Dubois and H. Prade. After that, D. Dubois and H. Prade did much research work in possibility theory. They give the definition of the two basic measures of fuzzy variables, which are possibility and necessity, and present the relationship between the measures and the membership function proposed by L. Zadeh [335]. Actually, possibility may be understood as a measure of a fuzzy event with an optimistic attitude, and necessity could be understood as a measure of a fuzzy event with a pessimistic attitude. In practice, people's attitudes are usually different. So we need to combine these two basic measures to fit the varying attitudes. In this book we propose the measure *Me* which is the convex combination.

A fuzzy multiobjective model is one with fuzzy parameters. It is necessary for us to know that this model is a conceptual model rather than a mathematical model because we cannot maximize an uncertain quantity. There does not exist a natural ordership in an uncertain world. So, we need to transform the fuzzy multiobjective model into some approximate certain models to describe the uncertain model. In general, there are three kinds of models that could deal with the fuzzy multiobjective model, and are as follows (see [52, 53]):

1. *Expected value multiobjective model.* The model is used to optimize the expected value of the objective function subjected to the expected value of the constraints.

2. *Chance-constrained multiobjective model.* The model is used to optimize the critical value of the objective subjected to the chance constraints under some confidence levels.

3. *Dependent-chance constrained multiobjective model.* The model is used to optimize the measure of the objective function preceding a given value subjected to the chance constraints under some confidence levels.

However, in practical allocation problems, the many functional areas in an organization that yield an input to the aggregate plan normally have conflicting objectives. These objectives minimize costs/maximize profits, customer service, and so on. Moreover, the solution to fuzzy multiobjective optimization problems benefits from considering the imprecision of the decision maker's judgments. Especially, these conflicting objectives are required to be optimized simultaneously by the decision maker in the framework of fuzzy aspiration levels.

Therefore, the aim of this study is to develop a fuzzy multiobjective decision-making model for solving the allocation problem in a fuzzy environment. In order to handle it, we propose the expanded measure, and the corresponding expected value operator and chance operator. So, for a multiobjective decision-making model with fuzzy parameters. We use two operators to deal with it, we optimize the expected value of the objective functions subjected to the chance constraints, and we call this model the fuzzy expectation multiobjective model with chance constraints (ECM). In order to get better solutions to the original fuzzy multiobjective model, we use the knowledge of rough sets to divide the feasible region of ECM, and then we can obtain two approximation models, that is, the lower approximation model (LAM) and the upper approximation model (UAM). Actually, these two approximation models nicely connect with the possibility-chance constraints and the necessity-chance constraints. By solving these two models, we can get the optimal solution.

For general fuzzy multiple objective decision-making problems, the expected value operator and the chance operator convert them into crisp ones. As rough approximation techniques are adopted to treat these models, we can obtain fuzzy expected value rough model (Fu-EVRM), the fuzzy chance-constrained rough model (Fu-CCRM) and the fuzzy dependent-chance rough model (Fu-DCRM).

5.2   Fuzzy Variable

Let us introduce the basic knowledge regarding the fuzzy variable, including the definition, various measures, and the expected value operator of fuzzy variables.

5.2.1   Definition of Fuzzy Variable

Since its introduction in 1965 by L. Zadeh [335], the fuzzy set theory has been well developed and applied in a wide variety of real problems. The term fuzzy variable was first introduced by A. Kaufmann [160], then it appeared in S. Nahmias [221]. The possibility theory was proposed by L. Zadeh [337] and developed by many researchers such as D. Dubois and H. Prade [86].

In order to provide an axiomatic theory to describe fuzziness, S. Nahmias [221] suggested a theoretical framework. Let us give the definition of possibility space (also called pattern space by S. Nahmias).

DEFINITION 5.1   *(D. Dubois and H. Prade [84]) Let Θ be a nonempty set, and $P(\Theta)$ be the power set of Θ. For each $A \subseteq P(\Theta)$, there is a nonnegative number $Pos\{A\}$, called its possibility, such that*
   1. $Pos\{\emptyset\} = 0$;
   2. $Pos\{\Theta\} = 1$;
   3. $Pos\{\bigcup_k A_k\} = \sup_k Pos\{A_k\}$ for any arbitrary collection $\{A_k\}$ in $P(\Theta)$.
   The triplet $(\Theta, P(\Theta), Pos)$ is said to be a possibility space, and the function Pos is referred to as a possibility measure.

It is easy to obtain the following properties of *Pos* from the definition above.

LEMMA 5.1
Let Pos denote the possibility measure. Then Pos satisfies
   1. $0 \le Pos\{A\} \le 1, \forall A \in P(\Theta)$;
   2. $Pos\{A\} \le Pos\{B\}$, if $A \subseteq B$.

PROOF   1.   Since $\Theta = A \cup A^c$, we have $Pos\{A\} \vee Pos\{A^c\} = Pos\{\Theta\} = 1$, which implies that $Pos\{A\} \leq 1$. On the other hand, since $A = A \cup \phi$, we have $Pos\{A\} \vee 0 = Pos\{A\}$, which implies that $Pos\{A\} \geq 0$. It follows that $0 \leq Pos\{A\} \leq 1$ for any $A \subseteq P$. 2. Let $A \subseteq B$. Then there exists a set C such that $B = A \cup C$. Thus, we have $Pos\{A\} \vee Pos\{C\} = Pos\{B\}$, which gives that $Pos\{A\} \leq Pos\{B\}$.

Several researchers have defined fuzzy variable in different ways, such as A. Kaufman [160], L. Zadah [336, 337], and S. Nahmias [221]. In this book we use the following definition of fuzzy variable.

DEFINITION 5.2 *(S. Nahmias [221]) A fuzzy variable is defined as a function from the possibility space $(\Theta, P(\Theta), Pos)$ to the real line **R**.*

DEFINITION 5.3 *(D. Dubois and H. Prade [84]) Let ξ be a fuzzy variable on the possibility space $(\Theta, P(\Theta), Pos)$. Then its membership function $\mu : \mathbf{R} \mapsto [0,1]$ is derived from the possibility measure Pos by*

$$\mu(x) = Pos\{\theta \in \Theta | \xi(\theta) = x\} \tag{5.2}$$

REMARK 5.1 For any fuzzy variable ξ with membership function μ, we have $\sup_x \mu(x) = \sup_x Pos\{\theta \in \Theta | \xi(\theta) = x\} = Pos\{\Theta\} = 1$. That is, any fuzzy variables defined by Definition 5.3 are normalized.

REMARK 5.2 Let ξ be a fuzzy variable with membership function μ. Then ξ may be regarded as a function from the possibility space $(\Theta, P(\Theta), Pos)$ to **R**, provided that $Pos\{A\} = \sup\{\mu(\xi(\theta)) | \theta \in A\}$ for any $A \in P(\Theta)$.

THEOREM 5.1
(L. Zadeh [335]) Let $\tilde{a}_1, \tilde{a}_2, \cdots, \tilde{a}_n$ be fuzzy variables, and $f : \mathbf{R}^n \to \mathbf{R}$ be continuous functions. Then the membership function $\mu_{\tilde{a}}$ of $f(\tilde{a}_1, \tilde{a}_2, \cdots, \tilde{a}_n) \leq 0$ is derived from the membership functions $\mu_{\tilde{a}_1}, \mu_{\tilde{a}_2}, \cdots, \mu_{\tilde{a}_n}$ by

$$\mu_{\tilde{a}} = \sup_{x_1, x_2, \cdots, x_n \in R} \left\{ \min_{1 \leq i \leq n} \mu_{\tilde{a}_i}(x_i) | x = f(x_1, x_2, \cdots, x_n) \right\} \tag{5.3}$$

Now we introduce the LR fuzzy variable on which fuzzy arithmetics have good results.

DEFINITION 5.4 *(D. Dubois and H. Prade [85]) Let f be a function from real numbers set **R** to $[0,1]$. If f satisfies $(1)f(x) = f(-x), (2)f(0) = 1, (3)f(x)$ is decreasing on $[0, +\infty)$, then $f(x)$ is called the reference function of an LR fuzzy variable.*

The following reference functions are usually used in practical application:

1. $f(x) = \max\{0, 1 - |x|^p\}(p \geq 0)$;
2. $f(x) = \exp(-|x|^p)(p \geq 0)$;
3. $f(x) = \frac{1}{1+|x|^p}(p \geq 0)$;
4. $f(x) = \begin{cases} 1, & \text{if } x \in [-1,1] \\ 0, & \text{otherwise} \end{cases}$.

DEFINITION 5.5 *(D. Dubois and H. Prade [85]) Let $L(\cdot), R(\cdot)$ be two reference functions. If the membership function of fuzzy variable ξ has the following form:*

$$\mu_\xi(x) = \begin{cases} L(\frac{m-x}{\alpha}), & x \leq m, \alpha > 0 \\ R(\frac{x-m}{\beta}), & x \geq m, \beta > 0 \end{cases} \tag{5.4}$$

then ξ is called LR fuzzy variable; L,R are called left and right branch of ξ respectively; α, β are called left and right spread of ξ, respectively; and m is called the main value of ξ. Denote ξ by $(m, \alpha, \beta)_{LR}$. In addition, we assume that the LR fuzzy variable degenerates to a real number as $\alpha = \beta = 0$, that is, $(m, 0, 0)_{LR} = m$.

LEMMA 5.2

(D. Dubois and H. Prade [85]) Let $\xi_1 = (m_1, \alpha_1, \beta_1)_{LR}, \xi_2 = (m_2, \alpha_2, \beta_2)_{LR}$, and $k(\neq 0)$ be a real number. Then

1. $\xi_1 + \xi_2 = (m_1 + m_2, \alpha_1 + \alpha_2, \beta_1 + \beta_2)_{LR}$.
2. $\xi_1 - \xi_2 = (m_1 - m_2, \alpha_1 + \beta_2, \beta_1 + \alpha_2)_{LR}$.
3. $k\xi_1 = \begin{cases} (km_1, k\alpha_1, k\beta_1)_{LR} \\ (km_1, -k\alpha_1, -k\beta_1)_{LR}. \\ k > 0 \end{cases}$

Now let us introduce a kind of special *LR* fuzzy variables. By trapezoidal fuzzy variables we mean fuzzy variables fully determined by quadruples (r_1, r_2, r_3, r_4) of crisp numbers with $r_1 < r_2 \leq r_3 < r_4$, whose membership functions can be denoted by

$$\mu(x) = \begin{cases} \frac{x-r_1}{r_2-r_1}, & \text{if } r_1 \leq x \leq r_2 \\ 1, & \text{if } r_2 \leq x \leq r_3 \\ \frac{x-r_3}{r_3-r_4}, & \text{if } r_3 \leq x \leq r_4 \\ 0, & \text{otherwise} \end{cases}$$

The membership function curve of a trapezoidal fuzzy variable (r_1, r_2, r_3, r_4) is shown in Figure 5.2.

We note that the trapezoidal fuzzy variable is a triangular fuzzy variable if $r_2 = r_3$, denoted by a triple (r_1, r_2, r_3). That is , the membership function of the triangular

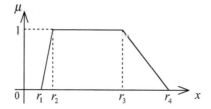

FIGURE 5.2 The membership function curve of trapezoidal fuzzy variable.

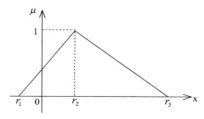

FIGURE 5.3 The membership function curve of a triangular fuzzy variable.

fuzzy variable (r_1, r_2, r_3) is

$$\mu(x) = \begin{cases} \frac{x-r_1}{r_2-r_1}, & \text{if } r_1 \le x \le r_2 \\ \frac{x-r_3}{r_2-r_3}, & \text{if } r_2 \le x \le r_3 \\ 0, & \text{otherwise} \end{cases}$$

The membership function curve of a triangular fuzzy variable (r_1, r_2, r_3) is shown in Figure 5.3.

From the fuzzy arithmetic, we can obtain the sum of trapezoidal fuzzy variables $\tilde{a} = (a_1, a_2, a_3, a_4)$ and $\tilde{b} = (b_1, b_2, b_3, b_4)$ as

$$\mu_{\tilde{a}+\tilde{b}}(z) = \sup\{\min\{\mu_{\tilde{a}}(x), \mu_{\tilde{b}}(y)\} | z = x+y\}$$
$$= \begin{cases} \frac{z-(a_1+b_1)}{(a_2+b_2)-(a_1+b_1)}, & \text{if } a_1+b_1 \le z \le a_2+b_2 \\ 1, & \text{if } a_2+b_2 \le z \le a_3+b_3 \\ \frac{z-(a_4+b_4)}{(a_3+b_3)-(a_4+b_4)}, & \text{if } a_3+b_3 \le z \le a_4+b_4 \\ 0, & \text{otherwise} \end{cases}$$

That is, the sum of two trapezoidal fuzzy variables is also a trapezoidal fuzzy variable, and

$$\tilde{a} + \tilde{b} = (a_1+b_1, a_2+b_2, a_3+b_3, a_4+b_4).$$

Next we consider the product of a trapezoidal fuzzy variable and a scalar number λ. We have

$$\mu_{\lambda \cdot \tilde{a}}(z) = \sup\{\mu_{\tilde{a}}(x) | z = \lambda x\},$$

which yields that

$$\lambda \cdot \tilde{a} = \begin{cases} (\lambda \tilde{a}_1, \lambda \tilde{a}_2, \lambda \tilde{a}_3, \lambda \tilde{a}_4), & \text{if } \lambda \geq 0 \\ (\lambda \tilde{a}_4, \lambda \tilde{a}_3, \lambda \tilde{a}_2, \lambda \tilde{a}_1), & \text{if } \lambda < 0 \end{cases}$$

That is, the product of a trapezoidal fuzzy variable and a scalar number is also a trapezoidal fuzzy variable. Thus, the weighted sum of trapezoidal fuzzy variables is also a trapezoidal fuzzy variable.

For example, we assume that $\tilde{a}_i$ are trapezoidal fuzzy variables $(a_{i1}, a_{i2}, a_{i3}, a_{i4})$, and λ_i are scalar numbers, $i = 1, 2, \cdots, n$, respectively. If we define

$$\lambda_i^+ = \begin{cases} \lambda_i, & \text{if } \lambda_i \geq 0 \\ 0, & \text{otherwise} \end{cases} \quad \text{and} \quad \lambda_i^- = \begin{cases} 0, & \text{if } \lambda_i \leq 0 \\ -\lambda_i, & \text{otherwise.} \end{cases}$$

for $i = 1, 2, \cdots, n$, then λ_i^+ and λ_i^- are all nonnegative and satisfy that $\lambda_i = \lambda_i^+ - \lambda_i^-$. By the sum and product operations of trapezoidal fuzzy variables, we can obtain

$$\tilde{a} = \sum_{i=1}^{n} \lambda_i \tilde{a}_i = \begin{pmatrix} \sum_{i=1}^{n} (\lambda_i^+ a_{i1} - \lambda_i^- a_{i4}) \\ \sum_{i=1}^{n} (\lambda_i^+ a_{i2} - \lambda_i^- a_{i3}) \\ \sum_{i=1}^{n} (\lambda_i^+ a_{i3} - \lambda_i^- a_{i2}) \\ \sum_{i=1}^{n} (\lambda_i^+ a_{i4} - \lambda_i^- a_{i1}) \end{pmatrix}^T.$$

As $\tilde{a}_i$ degenerates into triangular variables, that is, $\tilde{a}_i = (a_{i1}, a_{i2}, a_{i3})$, we have

$$\tilde{a} = \sum_{i=1}^{n} \lambda_i \tilde{a}_i = \begin{pmatrix} \sum_{i=1}^{n} (\lambda_i^+ a_{i1} - \lambda_i^- a_{i3}) \\ \sum_{i=1}^{n} (\lambda_i^+ a_{i2} - \lambda_i^- a_{i2}) \\ \sum_{i=1}^{n} (\lambda_i^+ a_{i3} - \lambda_i^- a_{i1}) \end{pmatrix}^T.$$

5.2.2 Possibility, Necessity, and General Fuzzy Measure

In order to measure the chances of occurrence of fuzzy events, three kinds of fuzzy measures of fuzzy events will be introduced in this section. If the decision making is performed optimistically by the decision maker, it is reasonable to use the possibility measure.

More generally, we give the following theorem on possibility of fuzzy events.

DEFINITION 5.6 *(D. Dubois and H. Prade [87]) Let $\tilde{a}_1, \tilde{a}_2, \cdots, \tilde{a}_n$ be fuzzy variables, and $f : \mathbf{R}^n \to \mathbf{R}$ be continuous functions. Then the possibility of the fuzzy event characterized by $f(\tilde{a}_1, \tilde{a}_2, \cdots, \tilde{a}_n) \leq 0$ is*

$$Pos\{f(\tilde{a}_1, \tilde{a}_2, \cdots, \tilde{a}_n) \leq 0\} = \sup_{x_1, x_2, \cdots, x_n} \left\{ \min_{1 \leq i \leq n} \mu_{\tilde{a}_i}(x_i) | f(x_1, x_2, \cdots, x_n) \leq 0 \right\} \quad (5.5)$$

Example 5.1

Let $\tilde{a}$ and $\tilde{b}$ be fuzzy variables on the possibility spaces $(\Theta_1, P(\Theta_1), Pos_1)$ and $(\Theta_2, P(\Theta_2), Pos_2)$, respectively. Then, $\tilde{a} \leq \tilde{b}$ is a fuzzy event defined on the product possibility space $(\Theta, P(\Theta), Pos)$, whose possibility is

$$Pos\{\tilde{a} \leq \tilde{b}\} = \sup_{x,y \in \mathbf{R}} \{\mu_{\tilde{a}}(x) \wedge \mu_{\tilde{b}}(y) | x \leq y\},$$

where the abbreviation *Pos* represents possibility. This means that the possibility of $\tilde{a} \leq \tilde{b}$ is the largest possibility, and that there exists at least one pair of values $x, y \in \mathbf{R}$ such that $x \leq y$, and the values of $\tilde{a}$ and $\tilde{b}$ are x and y, respectively. Similarly, the possibility of $\tilde{a} = \tilde{b}$ is given by $Pos\{\tilde{a} = \tilde{b}\} = \sup_{x \in \mathbf{R}}\{\mu_{\tilde{a}}(x) \wedge \mu_{\tilde{b}}(x)\}$. If the decision maker prefers a pessimistic decision in order to avoid risk, it may be approximate to replace the possibility measure *Pos* with the necessity measure.

A set function *Nec* defined on $P(\Theta)$ is said to be a necessity measure if it satisfies the following conditions:

1. $Nec\{\emptyset\} = 0$, and $Nec\{\Theta\} = 1$.
2. $Nec\{\bigcap_{i \in I} A_i\} = \inf_{i \in I}\{A_i\}$ for any subclass $\{A_i | i \in I\}$ of $P(\Theta)$, where I is an index set.

The necessity measure of a set A is defined as the impossibility of the opposite set A^c.

DEFINITION 5.7   *(D. Dubois and H. Prade [87]) Let* $(\Theta, P(\Theta), Pos)$ *be a possibility space, and A be a set in* $P(\Theta)$. *Then the necessity measure Nec of A is*

$$Nec\{A\} = 1 - Pos\{A^c\} \tag{5.6}$$

Thus the necessity measure is the dual of the possibility measure, that is, $Pos\{A\} + Nec\{A^c\} = 1$ *for any* $A \in P(\Theta)$.

LEMMA 5.3

(D. Dubois and H. Prade [87]) Let ξ_1 *and* ξ_2 *be two fuzzy variables. Then we have*

$$Pos\{\xi_1 \geq \xi_2\} = \sup\{\mu_{\xi_1}(u) \wedge \mu_{\xi_2}(v) | u > v\} \tag{5.7}$$

and

$$Pos\{\xi_1 > \xi_2\} = \sup\{\mu_{\xi_1}(u) \wedge \inf_v\{1 - \mu_{\xi_2}(v) | u \leq v\}\} \tag{5.8}$$

If the decision maker prefers a pessimistic decision in order to avoid risk, it may be approximate to replace the possibility measure with the necessity measure.

$$Nec\{\xi_1 \geq \xi_2\} = \inf\{(1 - \mu_{\xi_1}(u)) \vee \sup_v \mu_{\xi_2}(v) | u \geq v\} \tag{5.9}$$

and

$$Nec\{\xi_1 > \xi_2\} = \inf\{(1 - \mu_{\xi_1}(u)) \vee (1 - \mu_{\xi_2}(v)) | u \leq v\} \tag{5.10}$$

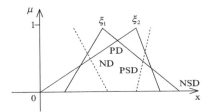

FIGURE 5.4 Conclusions of Lemma 5.3.

The conclusions of Lemma 5.3 can be shown by Figure 5.4.

By using the α-level sets of fuzzy variable ξ_1 and ξ_2, $[m_\alpha^L, m_\alpha^R]$ and $[n_\alpha^L, n_\alpha^R]$, Lemma 5.3 can be rewritten as

$$Pos\{\xi_1 \geq \xi_2\} \geq \alpha \Longleftrightarrow m_\alpha^R \geq n_\alpha^L \tag{5.11}$$

$$Pos\{\xi_1 > \xi_2\} \geq \alpha \Longleftrightarrow m_\alpha^R \geq n_{1-\alpha}^R \tag{5.12}$$

$$Nec\{\xi_1 \geq \xi_2\} \geq \alpha \Longleftrightarrow m_{1-\alpha}^R \geq n_\alpha^L \tag{5.13}$$

$$Nec\{\xi_1 > \xi_2\} \geq \alpha \Longleftrightarrow m_{1-\alpha}^L \geq n_{1-\alpha}^R \tag{5.14}$$

Sometimes the basic measures *Pos* and *Nec* proposed by D. Dubois and H. Prade are limited in dealing with the realistic uncertain decision making problem since these two measures reflect the attitudes of extremely optimism and pessimism, so we need a more general measure for the fuzzy decision-making problem. In fact, in the realistic uncertain decision-making process, the decision makers have different optimistic–pessimistic attitudes. They will decide the optimistic–pessimistic parameters according to their judgements, when they have good forecasts, they will be optimistic about a fuzzy event and vice versa. This optimistic-pessimistic parameter of a fuzzy event should be considered to avoid extreme attitudes. Therefore, we find it is necessary to design a more flexible measure to measure the fuzzy events in a decision-making problem by introducing the optimistic-pessimistic parameter, and this measure should be adjustable according to different attitudes of the decision makers. In addition, K. Iwamura and M. Horiike [151] defined a mathematical function by combining possibility and necessity from the mathematical point of view in the proceedings of the fifth international conference on information and management sciences. It is a nonrepresentational tool that can deal with the general fuzzy system. However, there is no realistic explanation in the definition, so as far as we know, there is not any promotion and application. Based on the above idea and discussion, we borrow the mathematical form by Iwamura, and introduce a fuzzy measure *Me*, which is suitable for the background of the decision making under the fuzzy environment and has practical significance.

DEFINITION 5.8   *(J. Xu and X. Zhou [318]) Let $(\Theta, P(\Theta), Pos)$ be a possibility space, and A be a set in $P(\Theta)$. Then the fuzzy measure of A is*

$$Me\{A\} = \lambda Pos\{A\} + (1 - \lambda)Nec\{A\} \qquad (5.15)$$

where λ is the optimistic and pessimistic index to determine the combined attitude of the decision maker.

REMARK 5.3   Note that the fuzzy measure Me is used to evaluate a degree that a fuzzy variable takes values in an interval with different optimistic-pessimistic attitudes. It is equal to the convex combination of Pos and Nec, that is, $Me\{A\} = \lambda Pos\{A\} + (1 - \lambda)Nec\{A\}$.

When $\lambda = 1$, we have $Me = Pos$, it means the decision maker is optimistic, it is the measure of the best case of that event, and it is the maximal chance that A holds;

When $\lambda = 0$, we have $Me = Nec$, which means the decision maker is pessimistic, it gives the measure of the worst case of that event, and it is the minimal chance that A holds;

When $\lambda = 0.5$, we have $Me = Cr$, where Cr is the credibility measure introduced by B. Liu [196]. It is a special case of Me, it means the decision maker takes compromise attitude. Cr is self dual, that is, $Cr\{A\} + Cr\{A^c\} = 1$ for any $A \in P(\Theta)$.

THEOREM 5.2

(J. Xu and X. Zhou [318]) Let $(\Theta, P(\Theta), Pos)$ be a possibility space, and A a set in $P(\Theta)$. Then

1. $Me\{\Theta\} = 1$, and $Me\{\Phi\} = 0$, where Θ is the collection and Φ is the empty set.

2. Since $\lambda(Pos\{A\} - Nec\{A\}) \geq 0$, so $Nec\{A\} \leq Me\{A\} \leq Pos\{A\}$ for any $A \in P(\Theta)$ and thus 0-1 boundedness holds for $Me\{\cdot\}$.

3. For any two sets that satisfy $A \subset B$, $Me\{A\} \leq Me\{B\}$, which means that monotonicity holds for $Me\{\cdot\}$.

4. For any $A \in P(\Theta)$, when the optimistic–pessimistic parameter $\lambda \geq 0.5$, we have $1 \leq Me\{A\} + Me\{A^c\} \leq 2$, and when $\lambda \leq 0.5$, we have $0 \leq Me\{A\} + Me\{A^c\} \leq 1$.

5. For any $A, B \in P(\Theta)$ and $\lambda \geq 0.5$, we have $Me\{A \cup B\} \leq Me\{A\} + Me\{B\}$, which means the restricted subadditivity holds for $Me\{\cdot\}$.

Example 5.2

Let $\Theta = \{\theta_1, \theta_2\}$, with $Pos\{\theta_1\} = 1.0$, and $Pos\{\theta_2\} = 0.7$. Then we have $Nec\{\theta_1\} = 1 - Pos\{\theta_2\} = 0.3, Nec\{\theta_2\} = 0$. So, based on the above results, we have

$$Me\{\theta_1\} = Nec\{\theta_1\} + \lambda(Pos\{\theta_1\} - Nec\{\theta_1\}) = 0.3 + 0.7\lambda,$$
$$Me\{\theta_2\} = 0.7\lambda,$$
$$Me\{\theta_1\} + Me\{\theta_2\} = 0.3 + 1.4\lambda.$$

If we want to make $Me\{\theta_1\} + Me\{\theta_2\} = 1$, we need to set the optimistic–pessimistic parameter $\lambda = 0.5$; If we set $\lambda = 0.2$, we have $Me\{\theta_1\} + Me\{\theta_2\} = 0.58$; If we set $\lambda = 0.6$, we have $Me\{\theta_1\} + Me\{\theta_2\} = 1.14$.

LEMMA 5.4
(J. Xu and X. Zhou [318]) Let $(\Theta, P(\Theta), Pos)$ be a possibility space, and A be a set in $P(\Theta)$. Then

$$Me^{\lambda_1}\{A\} \geq Me^{\lambda_2}\{A\} \tag{5.16}$$

holds for any $\lambda_1 \geq \lambda_2$, where $\lambda_1, \lambda_2 \in [0,1]$.

PROOF   According to Definition 5.8, we have that

$$Me^{\lambda_1}\{A\} - Me^{\lambda_2}\{A\}$$
$$= (\lambda_1 Pos\{A\} + (1-\lambda_1)Nec\{A\}) - (\lambda_2 Pos\{A\} + (1-\lambda_2)Nec\{A\})$$
$$= (\lambda_1 - \lambda_2)Pos\{A\} - (\lambda_1 - \lambda_2)Nec\{A\}$$
$$= (\lambda_1 - \lambda_2)(Pos\{A\} - Nec\{A\}).$$

Because $\lambda_1 - \lambda_2 \geq 0$. So, in order to prove $Me(\lambda_1)\{A\} - Me(\lambda_2)\{A\} \geq 0$, we just need to prove $Pos\{A\} \geq Nec\{A\}$.

If $Pos\{A\} = 1$, then it is obvious that $Pos\{A\} \geq Nec\{A\}$. Otherwise, we must have $Pos\{A^c\} = 1$, which implies that $Nec\{A\} = 1 - Pos\{A^c\} = 0$. Thus, $Pos\{A\} \geq Nec\{A\}$ always holds, so we have

$$(\lambda_1 - \lambda_2)(Pos\{A\} - Nec\{A\}) \geq 0 \Rightarrow Me^{\lambda_1}\{A\} \geq Me^{\lambda_2}\{A\}.$$

The theorem is proved.

COROLLARY 5.1
Let $(\Theta, P(\Theta), Pos)$ be a possibility space, and A a set in $P(\Theta)$. Then

$$Nec\{A\} \leq Me\{A\} \leq Pos\{A\} \tag{5.17}$$

PROOF   It follows from $Pos\{A\} = Me^1\{A\}$ and $Nec\{A\} = Me^0\{A\}$ that

$$Nec\{A\} \leq Me\{A\} \leq Pos\{A\}.$$

REMARK 5.4   A fuzzy event may fail even though its possibility achieves 1, and hold even though its necessity is 0. However, the fuzzy event must hold if its credibility is 1, and fail if its credibility is 0.

Example 5.3

By a triangular fuzzy variable we mean the fuzzy variable ξ fully determined by the triplet (r_1, r_2, r_3) of crisp numbers with $r_1 < r_2 < r_3$, whose membership function is given by

$$\mu(x) = \begin{cases} \frac{x-r_1}{r_2-r_1}, & \text{if } r_1 \le x \le r_2 \\ \frac{x-r_3}{r_2-r_3}, & \text{if } r_2 \le x \le r_3 \\ 0, & \text{otherwise} \end{cases}$$

From the definition of (5.6),(5.7), and (5.8), the possibility, necessity, and general measure of $\xi \le x$ are formulated as

$$Pos\{\xi \le x\} = \begin{cases} 0, & \text{if } x \le r_1 \\ \frac{x-r_1}{r_2-r_1}, & \text{if } r_1 \le x \le r_2 \\ 1, & \text{if } x \ge r_2 \end{cases}$$

$$Nec\{\xi \le x\} = \begin{cases} 0, & \text{if } x \le r_2 \\ \frac{x-r_2}{r_3-r_2}, & \text{if } r_2 \le x \le r_3 \\ 1, & \text{if } x \ge r_3 \end{cases}$$

$$Me\{\xi \le x\} = \begin{cases} 0, & \text{if } x \le r_1 \\ \lambda\frac{x-r_1}{r_2-r_1}, & \text{if } r_1 \le x \le r_2 \\ \lambda+(1-\lambda)\frac{x-r_2}{r_3-r_2}, & \text{if } r_2 \le x \le r_3 \\ 1, & \text{if } x \ge r_3 \end{cases}$$

$$Cr\{\xi \le x\} = \begin{cases} 0, & \text{if } x \le r_1 \\ \frac{x-r_1}{2(r_2-r_1)}, & \text{if } r_1 \le x \le r_2 \\ \frac{x-2r_2+r_3}{2(r_3-r_2)}, & \text{if } r_2 \le x \le r_3 \\ 1, & \text{if } x \ge r_3 \end{cases}$$

Example 5.4

By a trapezoidal fuzzy variable we mean the fuzzy variable ξ fully determined by quadruplet (r_1, r_2, r_3, r_4) of crisp numbers with $r_1 < r_2 < r_3 < r_4$,

$$\mu(x) = \begin{cases} \frac{x-r_1}{r_2-r_1}, & \text{if } r_1 \le x \le r_2 \\ 1, & \text{if } r_2 \le x \le r_3 \\ \frac{x-r_4}{r_3-r_4}, & \text{if } r_3 \le x \le r_4 \\ 0, & \text{otherwise} \end{cases}$$

From the definition of (5.6),(5.7) and (5.8), the possibility, necessity, and credibility of $\xi \leq x$ are formulated as

$$Pos\{\xi \leq x\} = \begin{cases} 1, & \text{if } x \geq r_2 \\ \frac{x-r_1}{r_2-r_1}, & \text{if } r_1 \leq x \leq r_2 \\ 0, & \text{otherwise} \end{cases}$$

$$Nec\{\xi \leq x\} = \begin{cases} 1, & \text{if } x \geq r_4, \\ \frac{x-r_3}{r_4-r_3}, & \text{if } r_3 \leq x \leq r_4, \\ 0, & \text{otherwise} \end{cases}$$

$$Me\{\xi \leq x\} = \begin{cases} 0, & \text{if } x \leq r_1, \\ \lambda \frac{x-r_1}{r_2-r_1}, & \text{if } r_1 \leq x \leq r_2 \\ \lambda, & \text{if } r_2 \leq x \leq r_3 \\ \lambda + (1-\lambda)\frac{x-r_3}{r_4-r_3}, & \text{if } r_3 \leq x \leq r_4 \\ 1, & \text{if } x \geq r_4 \end{cases}$$

$$Cr\{\xi \leq x\} = \begin{cases} 0, & \text{if } x \leq r_1 \\ \frac{x-r_1}{2(r_2-r_1)}, & \text{if } r_1 \leq x \leq r_2 \\ \frac{1}{2}, & \text{if } r_2 \leq x \leq r_3, \\ \frac{x-2r_3+r_4}{2(r_4-r_3)}, & \text{if } r_3 \leq x \leq r_4 \\ 1, & \text{otherwise} \end{cases}$$

LEMMA 5.5
(M. Lu [203]) Let $\xi = (r_1, r_2, r_3, r_4)$ be a trapezoidal fuzzy variable. Then for any given confidence level α with $0 < \alpha \leq 1$, we have
1. *$Pos\{\xi \leq 0\}$ if and only if $(1-\alpha)r_1 + \alpha r_2 \leq 0$.*
2. *$Nec\{\xi \leq 0\}$ if and only if $(1-\alpha)r_3 + \alpha r_4 \leq 0$.*
3. *When $\alpha \leq 0.5$, $Cr\{\xi \leq 0\}$ if and only if $(1-2\alpha)r_1 + 2\alpha \leq 0$.*
4. *When $\alpha > 0.5$, $Cr\{\xi \leq 0\}$ if and only if $(2-2\alpha)r_3 + (2\alpha-1)r_4 \leq 0$.*

5.2.3 Expected Value Operator of Fuzzy Variables

In order to measure the mean of a fuzzy variable, several researchers defined an expected value for fuzzy variables with different ways, such as D. Dubois and H. Prade [86], L. Campos and J. Verdegay [44], A. González [121], and R. Yager [321, 322]. In this book we define three kinds of expected value for a fuzzy variable based on three kinds fuzzy measures.

DEFINITION 5.9 *(J. Xu and X. Zhou [318]) Let ξ be a fuzzy variable on the possibility space $(\Theta, P(\Theta), Pos)$. The expected value of ξ is defined by*

$$E^{Me}[\xi] = \int_0^{+\infty} Me\{\xi \geq r\}dr - \int_{-\infty}^0 Me\{\xi \leq r\}dr \qquad (5.18)$$

where integrals are defined through the Lebesgue integral.

Similarly, we can define the expected value based on the *Pos*, *Nec*, and *Cr* measure, which are the special case of fuzzy measure *Me*.

$$
\begin{cases}
E^{Pos}[\xi] = \int_0^{+\infty} Pos\{\xi \geq r\}dr - \int_{-\infty}^0 Pos\{\xi \leq r\}dr \\
E^{Nec}[\xi] = \int_0^{+\infty} Nec\{\xi \geq r\}dr - \int_{-\infty}^0 Nec\{\xi \leq r\}dr \\
E^{Cr}[\xi] = \int_0^{+\infty} Cr\{\xi \geq r\}dr - \int_{-\infty}^0 Cr\{\xi \leq r\}dr
\end{cases}
\tag{5.19}
$$

Obviously, $E^{Me}[\xi] = \lambda E^{Pos}[\xi] + (1-\lambda)E^{Nec}[\xi]$, and $E^{Cr}[\xi] = \frac{1}{2}(E^{Pos}[\xi] + E^{Nec}[\xi])$. For different expected value operators, we have

THEOREM 5.3

Let ξ be a fuzzy variable on the possibility space $(\Theta, P(\Theta), Pos)$. Then

$$
E^{Me^{\lambda_1}}[\xi] \geq E^{Me^{\lambda_2}}[\xi]
\tag{5.20}
$$

holds for any $\lambda_1 \geq \lambda_2$, where $\lambda_1, \lambda_2 \in [0,1]$.

PROOF　It follows from the definition of E^{Me} that

$$E^{Me^{\lambda_1}}[\xi] - E^{Me^{\lambda_2}}[\xi]$$

$$= \left(\int_0^{+\infty} Me^{\lambda_1}\{\xi \geq r\}dr - \int_{-\infty}^0 Me^{\lambda_1}\{\xi \leq r\}dr \right)$$

$$- \left(\int_0^{+\infty} Me^{\lambda_2}\{\xi \geq r\}dr - \int_{-\infty}^0 Me^{\lambda_2}\{\xi \leq r\}dr \right)$$

$$= \int_0^{+\infty} (Me^{\lambda_1}\{\xi \geq r\} - Me^{\lambda_2}\{\xi \geq r\})dr - \int_{-\infty}^0 (Me^{\lambda_1}\{\xi \leq r\} - Me^{\lambda_2}\{\xi \leq r\})dr.$$

By Lemma 5.4, we have

$$Me^{\lambda_1}\{\xi \geq r\} - Me^{\lambda_2}\{\xi \geq r\} \geq 0$$

and

$$Me^{\lambda_1}\{\xi \leq r\} - Me^{\lambda_2}\{\xi \leq r\} \leq 0.$$

Then

$$\int_0^{+\infty} (Me^{\lambda_1}\{\xi \geq r\} - Me^{\lambda_2}\{\xi \geq r\})dr \geq 0$$

and

$$\int_{-\infty}^0 (Me^{\lambda_1}\{\xi \leq r\} - Me^{\lambda_2}\{\xi \leq r\})dr \leq 0.$$

Thus, $E^{Me^{\lambda_1}}[\xi] - E^{Me^{\lambda_2}}[\xi] \geq 0$, i.e. $E^{Me^{\lambda_1}}[\xi] \geq E^{Me^{\lambda_2}}[\xi]$.

COROLLARY 5.2

Let ξ be a fuzzy variable on the possibility space $(\Theta, P(\Theta), Pos)$. Then

$$
E^{Pos}[\xi] \geq E^{Me}[\xi] \geq E^{Nec}[\xi]
\tag{5.21}
$$

REMARK 5.5 The Definition 5.9 is not only applicable to the continuous fuzzy variables but also discrete fuzzy variables and functions of multiple fuzzy variables. If $E^{Me}[\xi]$ is defined as (5.18), then $E[\xi] = \lambda E^{Pos}[\xi] + (1-\lambda)E^{Nec}[\xi]$.

PROPOSITION 5.1

The expected value of triangular fuzzy variable $\xi = (r_1, r_2, r_3)$ is

$$E^{Me}[\xi] = \begin{cases} \frac{\lambda}{2}r_1 + \frac{1}{2}r_2 + \frac{(1-\lambda)}{2}r_3, & \text{if } r_3 \leq 0 \\ \frac{\lambda}{2}(r_1 + r_2) + \frac{\lambda r_3^2 - (1-\lambda)r_2^2}{2(r_3 - r_2)}, & \text{if } r_2 \leq 0 \leq r_3 \\ \frac{(1-\lambda)r_2^2 - \lambda r_1^2}{2(r_2 - r_1)} + \frac{\lambda}{2}(r_2 + r_3), & \text{if } r_1 \leq 0 \leq r_2 \\ \frac{(1-\lambda)}{2}r_1 + \frac{1}{2}r_2 + \frac{\lambda}{2}r_3, & \text{if } 0 \leq r_1 \end{cases} \tag{5.22}$$

where λ is the optimistic–pessimistic adjusting index.

PROOF Since there are four cases, let us discuss each case in turn based on Definition 5.9.

Case 1. $r_3 \leq 0$

$$\begin{aligned} E^{Me}[\xi] &= \int_0^{+\infty} Me\{\xi \geq r\}dr - \int_{-\infty}^0 Me\{\xi \leq r\}dr \\ &= -(\int_{-\infty}^{r_1} Me\{\xi \leq r\}dr + \int_{r_1}^{r_2} Me\{\xi \leq r\}dr + \int_{r_2}^{r_3} Me\{\xi \leq r\}dr \\ &\quad + \int_{r_3}^0 Me\{\xi \leq r\}dr) \\ &= -(\int_{r_1}^{r_2} \lambda \frac{r-r_1}{r_2-r_1}dr + \int_{r_2}^{r_3}(\lambda + (1-\lambda)\frac{r-r_2}{r_3-r_2})dr + \int_{r_3}^0 1 dr) \\ &= \frac{\lambda}{2}r_1 + \frac{1}{2}r_2 + \frac{(1-\lambda)}{2}r_3. \end{aligned}$$

Case 2. $r_2 \leq 0 \leq r_3$

$$\begin{aligned} E^{Me}[\xi] &= \int_0^{r_3} \lambda \frac{r_3-r}{r_3-r_2}dr - (\int_{r_1}^{r_2} \lambda \frac{r-r_1}{r_2-r_1}dr + \int_{r_2}^0(\lambda + (1-\lambda)\frac{r-r_2}{r_3-r_2})dr) \\ &= \frac{\lambda}{2}(r_1 + r_2) + \frac{\lambda r_3^2 - (1-\lambda)r_2^2}{2(r_3-r_2)}. \end{aligned}$$

Case 3. $r_1 \leq 0 \leq r_2$

$$\begin{aligned} E^{Me}[\xi] &= \int_0^{r_2}(\lambda + (1-\lambda)\frac{r_2-r}{r_2-r_1})dr + \int_{r_2}^{r_3} \lambda \frac{r_3-r}{r_3-r_2}dr - (\int_{r_1}^0 \lambda \frac{r-r_1}{r_2-r_1}dr) \\ &= \frac{(1-\lambda)r_2^2 - \lambda r_1^2}{2(r_2-r_1)} + \frac{\lambda}{2}(r_2 + r_3). \end{aligned}$$

Case 4. $0 \leq r_1$

$$\begin{aligned} E^{Me}[\xi] &= \int_0^{r_1} 1 dr + \int_{r_1}^{r_2}(\lambda + (1-\lambda)\frac{r_2-r}{r_2-r_1})dr + \int_{r_2}^{r_3} \lambda \frac{r_4-r}{r_4-r_3}dr + 0 \\ &= \frac{(1-\lambda)}{2}r_1 + \frac{1}{2}r_2 + \frac{\lambda}{2}r_3. \end{aligned}$$

The proof is complete.

REMARK 5.6 When $\lambda = 1/2$, the results of the four cases are consistent, that is,

$$E[(r_1, r_2, r_3)] = (r_1 + 2r_2 + r_3)/4.$$

PROPOSITION 5.2
The expected value of trapezoidal fuzzy variable $\xi = (r_1, r_2, r_3, r_4)$ *is*

$$
E^{Me}[\xi] = \begin{cases}
\frac{\lambda}{2}(r_1 + r_2) + \frac{1-\lambda}{2}(r_3 + r_4), & \text{if } r_4 \leq 0, \\
\frac{\lambda}{2}(r_1 + r_2) + \frac{\lambda r_4^2 - (1-\lambda)r_3^2}{2(r_4 - r_3)}, & \text{if } r_3 \leq 0 \leq r_4 \\
\frac{\lambda}{2}(r_1 + r_2 + r_3 + r_4), & \text{if } r_2 \leq 0 \leq r_3 \\
\frac{(1-\lambda)r_2^2 - \lambda r_1^2}{2(r_2 - r_1)} + \frac{\lambda}{2}(r_3 + r_4), & \text{if } r_1 \leq 0 \leq r_2 \\
\frac{1-\lambda}{2}(r_1 + r_2) + \frac{\lambda}{2}(r_3 + r_4), & \text{if } 0 \leq r_1
\end{cases}
\tag{5.23}
$$

where λ *is the optimistic–pessimistic adjusting index.*

PROOF   Since there are five cases, let us discuss each case in turn.
Case 1. $r_4 \leq 0$

$$
\begin{aligned}
E^{Pos}[\xi] &= \int_0^{+\infty} Pos\{\xi \geq r\}dr - \int_{-\infty}^0 Pos\{\xi \leq r\}dr \\
&= -(\int_{r_1}^{r_2} Pos\{\xi \leq r\}dr + \int_{r_2}^0 Pos\{\xi \leq r\}dr) \\
&= -(\int_{r_1}^{r_2} \frac{r - r_1}{r_2 - r_1}dr + \int_{r_2}^0 1dr) \\
&= \frac{r_1 + r_2}{2}, \\
E^{Nec}[\xi] &= \int_0^{+\infty} Nec\{\xi \geq r\}dr - \int_{-\infty}^0 Nec\{\xi \leq r\}dr \\
&= \int_{-\infty}^0 Pos\{\xi > r\}dr - \int_0^{+\infty} Pos\{\xi < r\}dr \\
&= \int_{-\infty}^{r_3} 1dr + \int_{r_3}^{r_4} \frac{r_4 - r}{r_4 - r_3}dr - \int_0^{+\infty} 1dr \\
&= \frac{r_3 + r_4}{2}.
\end{aligned}
$$

Then we have

$$
E^{Me}[\xi] = \lambda E^{Pos}[\xi] + (1 - \lambda)E^{Nec}[\xi] = \frac{\lambda}{2}(r_1 + r_2) + \frac{1 - \lambda}{2}(r_3 + r_4).
$$

Case 2. $r_3 \leq 0 \leq r_4$

$$
\begin{aligned}
E^{Pos}[\xi] &= \int_0^{r_4} \frac{r_4 - r}{r_4 - r_3}dr - (\int_{r_1}^{r_2} \frac{r - r_1}{r_2 - r_1}dr + \int_{r_2}^0 1dr) = \frac{r_4^2}{2(r_4 - r_3)} + \frac{r_1 + r_2}{2}, \\
E^{Nec}[\xi] &= \int_{-\infty}^{r_3} 1dr + \int_{r_3}^0 \frac{r_4 - r}{r_4 - r_3}dr - \int_0^{+\infty} 1dr = -\frac{r_3^2}{2(r_4 - r_3)}.
\end{aligned}
$$

Thus

$$
E^{Me}[\xi] = \frac{\lambda}{2}(r_1 + r_2) + \frac{\lambda r_4^2 - (1 - \lambda)r_3^2}{2(r_4 - r_3)}.
$$

Case 3. $r_2 \leq 0 \leq r_3$

$$
\begin{aligned}
E^{Pos}[\xi] &= (\int_0^{r_3} 1dr + \int_{r_3}^{r_4} \frac{r_4 - r}{r_4 - r_3}dr) - (\int_{r_1}^{r_2} \frac{r - r_1}{r_2 - r_1}dr + \int_{r_2}^0 1dr) = \frac{r_1 + r_2 + r_3 + r_4}{2}, \\
E^{Nec}[\xi] &= \int_{-\infty}^0 1dr - \int_0^{+\infty} 1dr = 0.
\end{aligned}
$$

It follows that

$$
E^{Me}[\xi] = \frac{\lambda}{2}(r_1 + r_2 + r_3 + r_4).
$$

Case 4. $r_1 \le 0 \le r_2$

$$E^{Pos}[\xi] = (\int_0^{r_3} 1 dr + \int_{r_3}^{r_4} \frac{r_4 - r}{r_4 - r_3} dr) - \int_{r_1}^0 \frac{r - r_1}{r_2 - r_1} dr = \frac{r_3 + r_4}{2} - \frac{r_1^2}{2(r_2 - r_1)},$$

$$E^{Nec}[\xi] = \int_{-\infty}^0 1 dr - (\int_0^{r_2} \frac{r - r_1}{r_2 - r_1} dr + \int_{r_2}^{+\infty} 1 dr) = \frac{r_2^2}{2(r_2 - r_1)}.$$

It implies that

$$E^{Me}[\xi] = \frac{(1 - \lambda)r_2^2 - \lambda r_1^2}{2(r_2 - r_1)} + \frac{\lambda}{2}(r_3 + r_4).$$

Case 5. $0 \le r_1$

$$E^{Pos}[\xi] = \int_0^{r_3} 1 dr + \int_{r_3}^{r_4} \frac{r_4 - r}{r_4 - r_3} dr = \frac{r_3 + r_4}{2},$$

$$E^{Nec}[\xi] = \int_{-\infty}^0 1 dr - (\int_{r_1}^{r_2} \frac{r - r_1}{r_2 - r_1} dr + \int_{r_2}^{+\infty} 1 dr) = \frac{r_2 + r_1}{2}.$$

Hence

$$E^{Me}[\xi] = \frac{1 - \lambda}{2}(r_1 + r_2) + \frac{\lambda}{2}(r_3 + r_4).$$

The proof is complete.

REMARK 5.7   When $\lambda = 1/2$, the results of the five cases are consistent, that is,

$$E[(r_1, r_2, r_3, r_4)] = (r_1 + r_2 + r_3 + r_4)/4.$$

PROPOSITION 5.3
Let ξ be a fuzzy variable with a membership function

$$\mu_\xi(x) = \begin{cases} 1, & \text{if } x \in [a, b] \\ 0, & \text{otherwise} \end{cases}$$

Then the expected value of ξ is

$$E^{Me}[\xi] = \begin{cases} (1 - \lambda)a + \lambda b, & \text{if } 0 \le a \le b \\ \lambda(a + b), & \text{if } a \le 0 \le b \\ \lambda a + (1 - \lambda)b, & \text{if } a \le b \le 0 \end{cases}$$

PROOF   First, we calculate $E^{Pos}[\xi]$ and $E^{Nec}[\xi]$. There are three cases. Let us discuss each case in turn.
Case 1. $0 \le a \le b$

$$\begin{aligned} E^{Pos}[\xi] &= \int_0^{+\infty} Pos\{\xi \ge r\} dr - \int_{-\infty}^0 Pos\{\xi \le r\} dr \\ &= \int_0^b Pos\{\xi \ge r\} dr \\ &= b, \\ E^{Nec}[\xi] &= \int_0^{+\infty} Nec\{\xi \ge r\} dr - \int_{-\infty}^0 Nec\{\xi \le r\} dr \\ &= \int_0^{+\infty} (1 - Pos\{\xi < r\}) dr - \int_{-\infty}^0 (1 - Pos\{\xi > r\}) dr \\ &= \int_{-\infty}^0 Pos\{\xi > r\} dr - \int_0^{+\infty} Pos\{\xi < r\} dr \\ &= \int_{-\infty}^0 1 dr - [\int_0^a Pos\{\xi < r\} dr + \int_a^b Pos\{\xi < r\} dr \\ &\quad + \int_b^{+\infty} Pos\{\xi < r\} dr] = a. \end{aligned}$$

It follows that

$$E^{Me}[\xi] = \lambda E^{Pos}[\xi] + (1 - \lambda)E^{Nec}[\xi]$$
$$= (1 - \lambda)a + \lambda b.$$

Case 2. $a \leq 0 \leq b$

$$
\begin{aligned}
E^{Pos}[\xi] &= \int_0^{+\infty} Pos\{\xi \geq r\}dr - \int_{-\infty}^0 Pos\{\xi \leq r\}dr \\
&= \int_0^b Pos\{\xi \geq r\}dr - \int_a^0 Pos\{\xi \leq r\}dr \\
&= b + a, \\
E^{Nec}[\xi] &= \int_0^{+\infty} Nec\{\xi \geq r\}dr - \int_{-\infty}^0 Nec\{\xi \leq r\}dr \\
&= \int_0^{+\infty}(1 - Pos\{\xi < r\})dr - \int_{-\infty}^0(1 - Pos\{\xi > r\})dr \\
&= \int_{-\infty}^0 Pos\{\xi > r\}dr - \int_0^{+\infty} Pos\{\xi < r\}dr \\
&= \int_{-\infty}^0 1dr - \int_0^{+\infty} 1dr \\
&= 0.
\end{aligned}
$$

Then we have

$$E^{Me}[\xi] = \lambda(a + b).$$

Case 3. $a \leq b \leq 0$

$$
\begin{aligned}
E^{Pos}[\xi] &= \int_0^{+\infty} Pos\{\xi \geq r\}dr - \int_{-\infty}^0 Pos\{\xi \leq r\}dr \\
&= -\int_a^0 Pos\{\xi \leq r\}dr \\
&= a, \\
E^{Nec}[\xi] &= \int_0^{+\infty} Nec\{\xi \geq r\}dr - \int_{-\infty}^0 Nec\{\xi \leq r\}dr \\
&= \int_0^{+\infty}(1 - Pos\{\xi < r\})dr - \int_{-\infty}^0(1 - Pos\{\xi > r\})dr \\
&= \int_{-\infty}^0 Pos\{\xi > r\}dr - \int_0^{+\infty} Pos\{\xi < r\}dr \\
&= \int_{-\infty}^b 1dr + \int_b^0 0dr - \int_0^{+\infty} 1dr \\
&= b.
\end{aligned}
$$

Therefore

$$E^{Me}[\xi] = \lambda a + (1 - \lambda)b.$$

Above all, for any cases we have

$$
E^{Me}[\xi] = \begin{cases}
(1 - \lambda)a + \lambda b, & \text{if } 0 \leq a \leq b \\
\lambda(a + b), & \text{if } a \leq 0 \leq b \\
\lambda a + (1 - \lambda)b, & \text{if } a \leq b \leq 0
\end{cases}
$$

This theorem is proved.

REMARK 5.8   If $\lambda = \frac{1}{2}$, that is, Me is actually Cr, then $E^{Cr}[\xi] = \frac{a+b}{2}$.

REMARK 5.9   If $r_2 = r_3$, that is, ξ degenerates to a triangular fuzzy variable, then

$$E^{Cr}[\xi] = \frac{1}{4}(r_1 + 2r_2 + r_4).$$

REMARK 5.10   If LR fuzzy number $\xi = (z, \alpha, \beta)_{LR}$ degenerates into a triangular fuzzy number, which means the reference function $L(x) = R(x) = 1 - x$, then

$$E^{Cr}[\xi] = z + \frac{1}{4}(\alpha + \beta).$$

PROPOSITION 5.4

Let ξ be a fuzzy variable with membership function

$$\mu_\xi(x) = e^{-k(x-a)^2} \quad (k > 0) \tag{5.24}$$

Then the expected value of ξ is

$$E^{Me}[\xi] = a - \int_0^a e^{-k(x-a)^2} dx \tag{5.25}$$

PROOF   Let us discuss each case in turn.

Case 1. $a \geq 0$

$$
\begin{aligned}
E^{Pos}[\xi] &= \int_0^{+\infty} Pos\{\xi \geq r\}dx - \int_{-\infty}^0 Pos\{\xi \leq x\}dx \\
&= \left(\int_0^a 1 dx + \int_a^{+\infty} e^{-k(x-a)^2} dx\right) - \int_{-\infty}^0 e^{-k(x-a)^2} dx \\
&= a - \int_0^a e^{-k(x-a)^2} dx, \\
E^{Nec}[\xi] &= \int_0^{+\infty} Nec\{\xi \geq x\}dx - \int_{-\infty}^0 Nec\{\xi \leq x\}dx \\
&= \int_{-\infty}^0 Pos\{\xi > x\}dx - \int_0^{+\infty} Pos\{\xi < x\}dx \\
&= \int_{-\infty}^0 1 dx - \left(\int_0^a e^{-k(x-a)^2} dx + \int_a^{+\infty} 1 dr\right) \\
&= a - \int_0^a e^{-k(x-a)^2} dx.
\end{aligned}
$$

It follows that

$$
\begin{aligned}
E^{Me}[\xi] &= \lambda E^{Pos}[\xi] + (1 - \lambda)E^{Nec}[\xi] \\
&= \lambda\left(a - \int_0^a e^{-k(x-a)^2} dx\right) + (1 - \lambda)\left(a - \int_0^a e^{-k(x-a)^2} dx\right) \\
&= a - \int_0^a e^{-k(x-a)^2} dx.
\end{aligned}
$$

Case 2. $a < 0$

$$
\begin{aligned}
E^{Pos}[\xi] &= \int_0^{+\infty} Pos\{\xi \geq r\}dx - \int_{-\infty}^0 Pos\{\xi \leq x\}dx \\
&= \int_0^{+\infty} e^{-k(x-a)^2} dx - \left(\int_{-\infty}^a e^{-k(x-a)^2} dx + \int_a^0 1 dx\right) \\
&= \int_a^0 e^{-k(x-a)^2} dx + a, \\
E^{Nec}[\xi] &= \int_0^{+\infty} Nec\{\xi \geq x\}dx - \int_{-\infty}^0 Nec\{\xi \leq x\}dx \\
&= \int_{-\infty}^0 Pos\{\xi > x\}dx - \int_0^{+\infty} Pos\{\xi < x\}dx \\
&= \left(\int_{-\infty}^a 1 dx + \int_a^0 e^{-k(x-a)^2} dx\right) - \int_0^{+\infty} 1 dr \\
&= a + \int_a^0 e^{-k(x-a)^2} dx.
\end{aligned}
$$

Then we have

$$
\begin{aligned}
E^{Me}[\xi] &= \lambda E^{Pos}[\xi] + (1 + \lambda)E^{Nec}[\xi] \\
&= \lambda\left(\int_a^0 e^{-k(x-a)^2} dx + a\right) + (1 + \lambda)\left(a + \int_a^0 e^{-k(x-a)^2} dx\right) \\
&= a - \int_0^a e^{-k(x-a)^2} dx.
\end{aligned}
$$

So we always have $E^{Me}[\xi] = a - \int_0^a e^{-k(x-a)^2} dx$. This theorem is proved.

REMARK 5.11 As is well known, it is hard to calculate integral like $\int_0^a e^{-k(x-a)^2} dx$. However, we can substitute $\sum_{i=0}^n \int_0^a \frac{[-k(x-a)^2]^i}{i!} dx$ for $\int_0^a e^{-k(x-a)^2} dx$ to obtain the approximate results.

The definition of expected value operator is not only applicable to the continuous case but also the discrete case.

PROPOSITION 5.5

Let ξ be a discrete fuzzy variable with membership function

$$\mu_\xi(x) = \begin{cases} \mu_1, & \text{if } x = a_1 \\ \mu_2, & \text{if } x = a_2 \\ \cdots \\ \mu_n, & \text{if } x = a_n \end{cases}$$

where $a_1 \leq a_2 \leq \cdots \leq a_n$. Then the expected value of ξ is

$$E^{Me}[\xi] = \sum_{i=1}^n w_i a_i \tag{5.26}$$

where the weights $w_i (i = 1, 2, \cdots, m)$ are given by

$$\begin{cases} w_1 = \lambda \mu_1 + (1 - \lambda)(\max_{1 \leq j \leq n} \mu_j - \max_{1 < j \leq n} \mu_j) \\ w_i = \lambda(\max_{1 \leq j \leq i} \mu_j - \max_{1 \leq j < n} \mu_j) + (1 - \lambda)(\max_{i \leq j \leq n} \mu_j - \max_{i < j \leq n} \mu_j), \; 2 \leq i \leq n-1 \\ w_n = \lambda \mu_m + (1 - \lambda)(\max_{1 \leq j \leq n} \mu_j - \max_{1 \leq j < n} \mu_j) \end{cases}$$

PROOF   It follows from the Definition of E^{Pos} that

$$E^{Pos}[\xi] = \int_0^{+\infty} Pos\{\xi \geq r\}dr - \int_{-\infty}^0 Pos\{\xi \leq r\}dr$$
$$= (\int_0^{x_{k+1}} Pos\{\xi \geq r\}dr + \int_{x_{k+1}}^{x_{k+2}} Pos\{\xi \geq r\}dr + \cdots$$
$$+ \int_{x_{n-1}}^{x_n} Pos\{\xi \geq r\}dr + \int_{x_n}^{+\infty} Pos\{\xi \geq r\}dr) - (\int_{-\infty}^{x_1} Pos\{\xi \leq r\}dr$$
$$+ \int_{x_1}^{x_2} Pos\{\xi \leq r\}dr + \cdots + \int_{x_{k-1}}^{x_k} Pos\{\xi \leq r\}dr + \int_{x_k}^0 Pos\{\xi \leq r\}dr)$$
$$= (\max_{k+1\leq i\leq n} \mu_i \cdot x_{k+1} + \max_{k+2\leq i\leq n} \mu_i \cdot (x_{k+2} - x_{k+1}) + \cdots$$
$$+ \mu_n \cdot (x_n - x_{n-1})) - (\mu_1 \cdot (x_2 - x_1) + \max_{1\leq i\leq 2} \mu_i \cdot (x_2 - x_1)$$
$$+ \cdots + \max_{1\leq i\leq k-1} \mu_i \cdot (x_k - x_{k-1}) + \max_{1\leq i\leq k} \mu_i \cdot (0 - x_k))$$
$$= \mu_1 \cdot x_1 + (\max_{1\leq i\leq 2} \mu_i - \mu_1)x_2 + \cdots + (\max_{1\leq i\leq k-1} \mu_i - \max_{1\leq i\leq k-2} \mu_i)x_{k-1}$$
$$+ (\max_{1\leq i\leq k} \mu_i - \max_{1\leq i\leq k-1} \mu_i)x_k + (\max_{k+1\leq i\leq n} \mu_i - \max_{k+2\leq i\leq n} \mu_i)x_{k+1} + \cdots$$
$$+ (\max_{n-1\leq i\leq n} \mu_i - \mu_n)x_{n-1} + \mu_n x_n$$
$$= \mu_1 \cdot x_1 + \sum_{j=2}^{k} (\max_{1\leq i\leq j} \mu_i - \max_{1\leq i\leq j-1} \mu_i)x_j + \sum_{j=k+1}^{n-1} (\max_{j\leq i\leq n} \mu_i - \max_{j+1\leq i\leq n} \mu_i)x_j + \mu_n \cdot x_n$$

If $\mu_1 \leq \mu_2 \leq \cdots \leq \mu_k$ and $\mu_{k+1} \geq \mu_{k+2} \geq \cdots \geq \mu_n$, then $E[\xi] = \sum_{i=1}^{n} \mu_i x_1$.

It follows from Definition of E^{Nec} that

$$E^{Nec}[\xi] = \int_0^{+\infty} Nec\{\xi \geq r\}dr - \int_{-\infty}^0 Nec\{\xi \leq r\}dr$$
$$= -\int_0^{+\infty} Pos\{\xi < r\}dr + \int_{-\infty}^0 Pos\{\xi > r\}dr$$
$$= -(\int_0^{x_{k+1}} Pos\{\xi < r\}dr + \int_{x_{k+1}}^{x_{k+2}} Pos\{\xi < r\}dr + \cdots$$
$$+ \int_{x_{n-1}}^{x_n} Pos\{\xi < r\}dr + \int_{x_n}^{+\infty} Pos\{\xi < r\}dr) + (\int_{-\infty}^{x_1} Pos\{\xi > r\}dr$$
$$+ \int_{x_1}^{x_2} Pos\{\xi > r\}dr + \cdots + \int_{x_{k-1}}^{x_k} Pos\{\xi > r\}dr + \int_{x_k}^0 Pos\{\xi > r\}dr)$$
$$= -(\max_{1\leq i\leq k} \mu_i \cdot x_{k+1} + \max_{1\leq i\leq k+1} \mu_i \cdot (x_{k+2} - x_{k+1}) + \cdots + \max_{1\leq i\leq n-1} \mu_i \cdot (x_n$$
$$- x_{n-1}) + \int_{x_n}^{+\infty} 1dr) + (\int_{-\infty}^{x_1} 1dr + \max_{2\leq i\leq n} \mu_i \cdot (x_2 - x_1) + \cdots$$
$$+ \max_{k\leq i\leq n} \mu_i \cdot (x_k - x_{k-1}) + \max_{k+1\leq i\leq n} \mu_i \cdot (0 - x_k))$$
$$= (-\max_{2\leq i\leq n} \mu_i) \cdot x_1 + (\max_{2\leq i\leq n} \mu_i - \max_{3\leq i\leq n} \mu_i) \cdot x_2 + \cdots + (\max_{k-1\leq i\leq n} \mu_i -$$
$$\max_{k\leq i\leq n} \mu_i) \cdot x_{k-1} + (\max_{k\leq i\leq n} \mu_i - \max_{k+1\leq i\leq n} \mu_i) \cdot x_k + (\max_{1\leq i\leq k+1} \mu_i -$$
$$\max_{1\leq i\leq k} \mu_i) \cdot x_{k+1} + (\max_{1\leq i\leq k+2} \mu_i - \max_{1\leq i\leq k+1} \mu_i) \cdot x_{k+2} + \cdots + (\max_{1\leq i\leq n-1} \mu_i -$$
$$\max_{1\leq i\leq n-2} \mu_i) \cdot x_{n-1} + (-\max_{1\leq i\leq n-1} \mu_i)x_n + (\int_{-\infty}^{x_1} 1dr - \int_{x_n}^{+\infty} 1dr).$$

Since

$$\int_{-\infty}^{x_1} 1dr - \int_{x_n}^{+\infty} 1dr = \int_{-\infty}^{x_1} 1dr + \int_{-x_1}^{x_n} 1dr - (\int_{-x_1}^{x_n} 1dr + \int_{x_n}^{+\infty} 1dr)$$
$$= (\int_{-\infty}^{x_1} 1dr - \int_{-x_1}^{+\infty} 1dr) + \int_{-x_1}^{x_n} 1dr$$
$$= x_n + x_1,$$

then we have

$$E^{Nec}(\xi) = \sum_{j=1}^{k}\left(\max_{j\leq i\leq n}\mu_i - \max_{j+1\leq i\leq n}\mu_i\right)x_j + \sum_{j=k+1}^{n}\left(\max_{1\leq i\leq j}\mu_i - \max_{1\leq i\leq j-1}\mu_i\right)x_j.$$

If $\mu_1 \geq \mu_2 \geq \cdots \geq \mu_k$ and $\mu_{k+1} \leq \mu_{k+2} \leq \cdots \leq \mu_n$, then $E^{Nec}[\xi] = \sum_{i=1}^{n}\mu_i x_1$.

Hence,

$$E^{Me}[\xi] = \lambda E^{Pos}[\xi] + (1-\lambda)E^{Nec}[\xi] = \sum_{i=1}^{n}w_i a_i.$$

where the weights $w_i, i = 1, 2, \cdots, m$ are given by

$$\begin{cases} w_1 = \lambda\mu_1 + (1-\lambda)\left(\max_{1\leq j\leq n}\mu_j - \max_{1<j\leq n}\mu_j\right) \\ w_i = \lambda\left(\max_{1\leq j\leq i}\mu_j - \max_{1\leq j<n}\mu_j\right) + (1-\lambda)\left(\max_{i\leq j\leq n}\mu_j - \max_{i<j\leq n}\mu_j\right), \ 2\leq i\leq n-1 \\ w_n = \lambda\mu_m + (1-\lambda)\left(\max_{1\leq j\leq n}\mu_j - \max_{1\leq j<n}\mu_j\right) \end{cases}$$

It is easy to verify that all $w_i \geq 0$ and $\sum_{i=1}^{n}w_i = \max_{1\leq n}\mu_i = 1$ since any fuzzy variables defined on a possibility space are normalized. This theorem is completed.

Based on the *Me* measure, we also have the linearity of fuzzy expected value operator given by the following theorem.

LEMMA 5.6
Let ξ and η be independent fuzzy variable with finite expected values. Then

$$E^{Me}[a\xi + b\eta] = aE^{Me}[\xi] + bE^{Me}[\eta] \tag{5.27}$$

holds for any numbers a and b.

PROOF   We prove this lemma by five steps.
Step 1. We first prove that $E^{Me}[\xi + b] = E^{Me}[\xi] + b$ for any real number b. If $b \geq 0$, we have

$$\begin{aligned} E^{Me}[\xi + b] &= \int_{0}^{+\infty}Me\{\xi + b \geq r\}dr - \int_{-\infty}^{0}Me\{\xi + b \leq r\}dr \\ &= \int_{0}^{+\infty}Me\{\xi \geq r-b\}dr - \int_{-\infty}^{0}Me\{\xi \leq r-b\}dr. \end{aligned}$$

We let $r' = r - b$, then the above equation is equal to the following equation:

$$\begin{aligned} &= \int_{-b}^{+\infty}Me\{\xi \geq r'\}dr' - \int_{-\infty}^{-b}Me\{\xi \leq r'\}dr' \\ &= \int_{-b}^{+\infty}Me\{\xi \geq r\}dr - \int_{-\infty}^{-b}Me\{\xi \leq r\}dr \\ &= \int_{0}^{+\infty}Me\{\xi \geq r\}dr - \int_{-\infty}^{0}Me\{\xi \leq r\}dr + \int_{-b}^{0}Me\{\xi \geq r\}dr + \int_{-b}^{0}Me\{\xi \leq r\}dr \\ &= E[\xi] + \int_{0}^{b}Me\{\xi \geq r-b\}dr + Me\{\xi \leq r-b\}dr \\ &= E[\xi] + b \end{aligned}$$

If $b < 0$, then

$$E^{Me}[\xi + b] = E^{Me}[\xi] - \int_b^0 Me\{\xi \geq r - b\}dr + Me\{\xi \leq r - b\}dr = E^{Me}[\xi] + b.$$

Step 2. We prove that $E^{Me}[a\xi] = aE^{Me}[\xi]$ for any real number a. If $a = 0$, then the equation $E^{Me}[a\xi] = aE^{Me}[\xi]$ holds trivially. Let $a > 0$. Then

$$
\begin{aligned}
E^{Me}[a\xi] &= \int_0^{+\infty} Me\{a\xi \geq r\}dr - \int_{-\infty}^0 Me\{a\xi \leq r\}dr \\
&= \int_0^{+\infty} Me\{\xi \geq \tfrac{r}{a}\}dr - \int_{-\infty}^0 Me\{\xi \leq \tfrac{r}{a}\}dr \\
&= a\int_0^{+\infty} Me\{\xi \geq \tfrac{r}{a}\}d(\tfrac{r}{a}) - a\int_{-\infty}^0 Me\{\xi \leq \tfrac{r}{a}\}d(\tfrac{r}{a}) \\
&= aE^{Me}[\xi].
\end{aligned}
$$

Meanwhile,

$$
\begin{aligned}
E^{Me}[a\xi] &= \int_0^{+\infty} Me\{a\xi \geq r\}dr - \int_{-\infty}^0 Me\{a\xi \leq r\}dr \\
&= \int_0^{+\infty} Me\{\xi \leq \tfrac{r}{a}\}dr - \int_{-\infty}^0 Me\{\xi \geq \tfrac{r}{a}\}dr \\
&= a\int_0^{+\infty} Me\{\xi \geq \tfrac{r}{a}\}d(\tfrac{r}{a}) - a\int_{-\infty}^0 Me\{\xi \leq \tfrac{r}{a}\}d(\tfrac{r}{a}) \\
&= aE^{Me}[\xi]
\end{aligned}
$$

for $a < 0$.

Step 3. When both ξ and η are simple fuzzy variables with the following membership functions, we prove that $E^{Me}[\xi + \eta] = E^{Me}[\xi] + E^{Me}[\eta]$.

$$
\mu_\xi(x) = \begin{cases} \mu_1, & \text{if } x = a_1 \\ \mu_2, & \text{if } x = a_2 \\ \cdots \\ \mu_m, & \text{if } x = a_m \end{cases}
\quad \text{and} \quad
\mu_\eta(x) = \begin{cases} \nu_1, & \text{if } x = b_1 \\ \nu_2, & \text{if } x = b_2 \\ \cdots \\ \nu_m, & \text{if } x = b_n \end{cases}
$$

Then $\xi + \eta$ is also a simple fuzzy variable taking values $a_i + b_j$ with membership degrees $\mu_i \wedge \nu_j$, $i = 1, 2, \cdots, m, j = 1, 2, \cdots, n$, respectively. Now according to Definition 5.9 and 5.8, we define

$$
\begin{aligned}
w_i = \lambda(&\max_{1 \leq k \leq m}\{\mu_k | a_k \leq a_i\} - \max_{1 \leq k \leq m}\{\mu_k | a_k < a_i\}) + (1 - \lambda)(\max_{1 \leq k \leq m}\{\mu_k | a_k \geq a_i\} \\
& - \max_{1 \leq k \leq m}\{\mu_k | a_k > a_i\}),
\end{aligned}
$$

$$
\begin{aligned}
w_j = \lambda(&\max_{1 \leq l \leq n}\{\nu_l | b_l \leq b_j\} - \max_{1 \leq l \leq n}\{\nu_l | b_l < b_j\}) + (1 - \lambda)(\max_{1 < l < n}\{\nu_l | b_l \geq b_j\} \\
& - \max_{1 \leq l \leq n}\{\nu_l | b_l > b_j\}),
\end{aligned}
$$

$$
\begin{aligned}
w_{ij} = \lambda(&\max_{1 \leq k \leq m, 1 \leq l \leq n}\{\mu_k \wedge \nu_l | a_k + b_l \leq a_i + b_j\} - \max_{1 \leq k \leq m, 1 \leq l \leq n}\{\mu_k \wedge \nu_l | a_k + b_l \\
& < a_i + b_j\}) + (1 - \lambda)(\max_{1 \leq k \leq m, 1 \leq l \leq n}\{\mu_k \wedge \nu_l | a_k + b_l \geq a_i + b_j\} - \max_{1 \leq k \leq m, 1 \leq l \leq n} \\
& \{\mu_k \wedge \nu_l | a_k + b_l > a_i + b_j\})
\end{aligned}
$$

for $i = 1, 2, \cdots, m$ and $j = 1, 2, \cdots, n$.

It is easy to verify that

$$w_i = \sum_{j=1}^n w_{ij}, \quad w_j = \sum_{i=1}^m w_{ij}$$

for $i = 1, 2, \cdots, m$ and $j = 1, 2, \cdots, n$.

If $\{a_i\}$, $\{b_j\}$ and $\{a_i + b_j\}$ are sequence consisting of distinct elements, then

$$E^{Me}[\xi] = \sum_{i=1}^{m} a_i w_i, \quad E^{Me}[\eta] = \sum_{j=1}^{n} b_j w_j, \quad E^{Me}[\xi + \eta] = \sum_{i=1}^{m} \sum_{j=1}^{n} (a_i + b_j) w_{ij}.$$

Thus $E^{Me}[\xi + \eta] = E^{Me}[\xi] + E^{Me}[\eta]$.

Step 4. We prove that $E^{Me}[\xi + \eta] = E^{Me}[\xi] + E^{Me}[\eta]$ when ξ and η are fuzzy variables such that

$$\begin{cases} \lim_{y \uparrow 0} Me\{\xi \le y\} \le \frac{1}{2} \le Me\{\xi \le 0\} \\ \lim_{y \uparrow 0} Me\{\eta \le y\} \le \frac{1}{2} \le Me\{\eta \le 0\} \end{cases} \tag{5.28}$$

We define simple fuzzy variables ξ_i as

$$Me\{\xi_i \le x\} = \begin{cases} \frac{k-1}{2^i}, & \frac{k-1}{2^i} \le Me\{\xi \le x\} < \frac{k}{2^i}, k = 1, 2, \cdots, 2^{i-1} \\ \frac{k}{2^i}, & \frac{k-1}{2^i} \le Me\{\xi \le x\} < \frac{k}{2^i}, k = 2^{i-1} + 1, \cdots, 2^i \\ 1, & Me\{\xi \le x\} = 1 \end{cases}$$

for $i = 1, 2, \cdots$.

Thus $\{\xi_i\}$ is a sequence of simple fuzzy variables satisfying

$$\begin{cases} Me\{\xi_i \le r\} \uparrow Me\{\xi \le r\}, & \text{if } r \le 0 \\ Me\{\xi_i \ge r\} \uparrow Me\{\xi \ge r\}, & \text{if } r \ge 0 \end{cases}$$

as $i \to \infty$.

Similarly, we can define simple fuzzy variables η_i, and it's clearly that $\{\xi_i + \eta_i\}$ is a sequence of simple fuzzy variables.

Furthermore, when $r \le 0$, it follows from (5.28) that

$$\begin{aligned} \lim_{i \to \infty} Me\{\xi_i + \eta_i \le r\} &= \lim_{i \to \infty} \sup_{x \le 0, y \le 0, x+y \le r} Me\{\xi_i \le x\} \wedge Me\{\eta_i \le y\} \\ &= \sup_{x \le 0, y \le 0, x+y \le r} \lim_{i \to \infty} Me\{\xi_i \le x\} \wedge Me\{\eta_i \le y\} \\ &= \sup_{x \le 0, y \le 0, x+y \le r} Me\{\xi \le x\} \wedge Me\{\eta \le y\} \\ &= Me\{\xi + \eta \le r\}. \end{aligned}$$

That is, $Me\{\xi_i + \eta_i \le r\} \uparrow Me\{\xi + \eta \le r\}$, if $r \le 0$. We can also prove that $Me\{\xi_i + \eta_i \ge r\} \uparrow Me\{\xi + \eta \ge r\}$, if $r \ge 0$.

Since the expected values $E^{Me}[\xi]$ and $E^{Me}[\eta]$ exist, then

$$\begin{aligned} E^{Me}[\xi_i] &= \int_0^{+\infty} Me\{\xi_i \ge r\} dr - \int_{-\infty}^0 Me\{\xi_i \le r\} dr \\ &\to \int_0^{+\infty} Me\{\xi \ge r\} dr - \int_{-\infty}^0 Me\{\xi \le r\} dr = E^{Me}[\xi] \end{aligned}$$

and

$$E^{Me}[\eta_i] \to E^{Me}[\eta], \quad E^{Me}[\xi_i + \eta_i] \to E^{Me}[\xi + \eta]$$

as $i \to \infty$.

It follows from Step 3 that $E^{Me}[\xi + \eta] = E^{Me}[\xi] + E^{Me}[\eta]$.

Step 5. When ξ and η are arbitrary fuzzy variables, we prove that $E^{Me}[\xi + \eta] = E^{Me}[\xi] + E^{Me}[\eta]$.

Since they have finite expected values, there exists two numbers o_1 and o_2 such that

$$\begin{cases} \lim_{y \uparrow 0} Me\{\xi + o_1 \le y\} \le \frac{1}{2} \le Me\{\xi + o_1 \le 0\} \\ \lim_{y \uparrow 0} Me\{\eta + o_2 \le y\} \le \frac{1}{2} \le Me\{\eta + o_2 \le 0\} \end{cases}$$

It follows from Step 1 and 4 that

$$\begin{aligned} E^{Me}[\xi + \eta] &= E^{Me}[(\xi + o_1) + (\eta + o_2) - o_1 - o_2] \\ &= E^{Me}[(\xi + o_1) + (\eta + o_2)] - o_1 - o_2 \\ &= E^{Me}[\xi + o_1] + E^{Me}[\eta + o_2] - o_1 - o_2 \\ &= E^{Me}[\xi] + o_1 + E^{Me}[\eta] + o_2 - o_1 - o_2 \\ &= E^{Me}[\xi] + E^{Me}[\eta]. \end{aligned}$$

Above all, $E^{Me}[\xi + \eta] = E^{Me}[\xi] + E^{Me}[\eta]$ is proved.

5.3 Fu-EVRM

We give the general form of the fuzzy multiple objective decision-making model as follows, and the discussion of this section is based on this model:

$$\begin{cases} \max \ [f_1(\mathbf{x}, \boldsymbol{\xi}), f_2(\mathbf{x}, \boldsymbol{\xi}), \cdots, f_m(\mathbf{x}, \boldsymbol{\xi})] \\ \text{s.t.} \begin{cases} g_r(\mathbf{x}, \boldsymbol{\xi}) \le 0, \ r = 1, 2, \cdots, p \\ \mathbf{x} \in X \end{cases} \end{cases} \tag{5.29}$$

where $\boldsymbol{\xi}$ are fuzzy vectors..

It is necessary for us to know that model (5.29) is a conceptual model rather than a mathematical model, because we cannot maximize an uncertain quantity. There does not exist a natural order in an uncertain world. So, we need to transform the fuzzy multiple objective model into some approximate certain models to describe the uncertain model. In general, there are three kinds of models that could deal with the fuzzy multiple objective model. The first type is the EVM. As we use rough approximation technique to deal with the EVM, we can obtain the fuzzy expected value rough model (Fu-EVRM).

5.3.1 General Model of Fu-EVRM

We use the expected value operator based on Me to deal with the objective functions and constraints, and we can obtain the following EVM:

$$\begin{cases} \max \ [E[f_1(\mathbf{x}, \boldsymbol{\xi})], E[f_2(\mathbf{x}, \boldsymbol{\xi})], \cdots, E[f_m(\mathbf{x}, \boldsymbol{\xi})]] \\ \text{s.t.} \begin{cases} E[g_r(\mathbf{x}, \boldsymbol{\xi})] \leq 0, \ r = 1, 2, \cdots, p \\ \mathbf{x} \in X \end{cases} \end{cases} \tag{5.30}$$

where E is the expected value operator for fuzzy variables.

For model (5.30), we let $D = \{\mathbf{x} | \mathbf{x} \in X, E^{Me}\{g_j(\mathbf{x}, \boldsymbol{\xi})\} \leq 0\}$, and we construct two sets P and N as follows:

$$P = \{\mathbf{x} | \mathbf{x} \in X, E^{Pos}\{g_j(\mathbf{x}, \boldsymbol{\xi})\} \leq 0\}, \ N = \{\mathbf{x} | \mathbf{x} \in X, E^{Nec}\{g_j(\mathbf{x}, \boldsymbol{\xi})\} \leq 0\} \tag{5.31}$$

THEOREM 5.4

For the feasible region D, and P, N defined by (5.31), we have

$$P \subseteq D \subseteq N.$$

PROOF   For any $\mathbf{x}_0 \in X$, it follows from Corollary 5.2 that

$$E^{Nec}[g_r(\mathbf{x}_0, \boldsymbol{\xi})] \leq E^{Me}[g_r(x_0, \boldsymbol{\xi})] \leq E^{Pos}[g_r(\mathbf{x}_0, \boldsymbol{\xi})] \tag{5.32}$$

If $\mathbf{x}_0 \in P$, then $E^{Pos}[g_r(\mathbf{x}_0, \boldsymbol{\xi})] \leq 0$. By equation (5.32), we have

$$E^{Me}[g_r(\mathbf{x}_0, \boldsymbol{\xi})] \leq 0.$$

Thus, $P \subseteq D$. Similarly, we can verify that $D \subseteq N$. The proof is complete.

Then we let $\underline{D} = P$ and $\overline{D} = N$, we use $\underline{D}$ and $\overline{D}$ to approximate D, it is obvious that $\underline{D} \subseteq D \subseteq \overline{D}$.

Then the model (5.30) can be transformed into two models:lower approximation model (LAM) and upper approximation model (UAM):

$$(LAM) \quad \begin{cases} \max \ [E^{Me}[f_1(\mathbf{x}, \boldsymbol{\xi})], E^{Me}[f_2(\mathbf{x}, \boldsymbol{\xi})], \cdots, E^{Me}[f_m(\mathbf{x}, \boldsymbol{\xi})]] \\ \text{s.t.} \begin{cases} E^{Pos}[g_r(\mathbf{x}, \boldsymbol{\xi})] \leq 0, r = 1, 2, \cdots, p \\ \mathbf{x} \in X \end{cases} \end{cases} \tag{5.33}$$

and

$$(UAM) \quad \begin{cases} \max \ [E^{Me}[f_1(\mathbf{x}, \boldsymbol{\xi})], E^{Me}[f_2(\mathbf{x}, \boldsymbol{\xi})], \cdots, E^{Me}[f_m(\mathbf{x}, \boldsymbol{\xi})]] \\ \text{s.t.} \begin{cases} E^{Nec}[g_r(\mathbf{x}, \boldsymbol{\xi})] \leq 0, r = 1, 2, \cdots, p \\ \mathbf{x} \in X \end{cases} \end{cases} \tag{5.34}$$

If "$\leq$" in model (5.30) are transformed into "$\geq$", then model (5.30) can be rewritten as

$$\begin{cases} \max \ [E[f_1(\mathbf{x}, \boldsymbol{\xi})], E[f_2(\mathbf{x}, \boldsymbol{\xi})], \cdots, E[f_m(\mathbf{x}, \boldsymbol{\xi})]] \\ \text{s.t.} \begin{cases} E[g_r(\mathbf{x}, \boldsymbol{\xi})] \geq 0, \ r = 1, 2, \cdots, p \\ \mathbf{x} \in X \end{cases} \end{cases} \tag{5.35}$$

For model (5.35), we let $D = \{x|x \in X, E^{Me}\{g_j(x, \xi)\} \geq 0\}$, and we construct two set P and N as follows:

$$P = \{\mathbf{x}|\mathbf{x} \in X, E^{Pos}\{g_j(\mathbf{x}, \boldsymbol{\xi})\} \geq 0\}, N = \{\mathbf{x}|\mathbf{x} \in X, E^{Nec}\{g_j(\mathbf{x}, \boldsymbol{\xi})\} \geq 0\} \quad (5.36)$$

THEOREM 5.5
For the feasible region D, and P,N defined in (5.36), we have

$$N \subseteq D \subseteq P.$$

PROOF   For any $x_0 \in X$, it follows from Corollary 5.2 that

$$E^{Nec}[g_r(\mathbf{x}_0, \boldsymbol{\xi})] \leq E^{Me}[g_r(\mathbf{x}_0, \boldsymbol{\xi})] \leq E^{Pos}[g_r(x_0, \xi)] \quad (5.37)$$

If $x_0 \in P$, then $E^{Nec}[g_r(\mathbf{x}_0, \boldsymbol{\xi})] \leq 0$. By equation (5.37), we have

$$E^{Me}[g_r(\mathbf{x}_0, \boldsymbol{\xi})] \leq 0.$$

Thus, $N \subseteq D$. Similarly, we can verify that $D \subseteq P$. The proof is complete.

Then we let $\underline{D} = N$ and $\overline{D} = P$, we use $\underline{D}$ and $\overline{D}$ to approximate D, and it is obvious that $\underline{D} \subseteq D \subseteq \overline{D}$.

Then model (5.35) can be transformed into two models: LAM and UAM, for short.

$$(LAM) \quad \begin{cases} \max \ [E^{Me}[f_1(\mathbf{x}, \boldsymbol{\xi})], E^{Me}[f_2(\mathbf{x}, \boldsymbol{\xi})], \cdots, E^{Me}[f_m(\mathbf{x}, \boldsymbol{\xi})]] \\ \text{s.t.} \begin{cases} E^{Nec}[g_r(\mathbf{x}, \boldsymbol{\xi})] \geq 0, r = 1, 2, \cdots, p \\ \mathbf{x} \in X \end{cases} \end{cases} \quad (5.38)$$

and

$$(UAM) \quad \begin{cases} \max \ [E^{Me}[f_1(\mathbf{x}, \boldsymbol{\xi})], E^{Me}[f_2(\mathbf{x}, \boldsymbol{\xi})], \cdots, E^{Me}[f_m(\mathbf{x}, \boldsymbol{\xi})]] \\ \text{s.t.} \begin{cases} E^{Pos}[g_r(\mathbf{x}, \boldsymbol{\xi})] \geq 0, r = 1, 2, \cdots, p \\ \mathbf{x} \in X \end{cases} \end{cases} \quad (5.39)$$

5.3.2   Fu-LEVRM and Satisfying Trade-off Method

Let us focus on the linear model, that is, the coefficients are triangular fuzzy variables.

$$\begin{cases} \max \ \left[E^{Me}\left[\sum_{j=1}^{n} \tilde{c}_{1j}x_j \right], E^{Me}\left[\sum_{j=1}^{n} \tilde{c}_{2j}x_j \right], \cdots, E^{Me}\left[\sum_{j=1}^{n} \tilde{c}_{mj}x_j \right] \right] \\ \text{s.t.} \begin{cases} E^{Me}\left[\sum_{j=1}^{n} \tilde{a}_{rj}x_j \right] \leq E^{Me}\left[\tilde{b}_r \right], r = 1, 2, \cdots, p \\ x_j \geq 0, j = 1, 2, \cdots, n \end{cases} \end{cases} \quad (5.40)$$

where $\tilde{c}_{ij} = (c_{ij} - \alpha^c_{ij}, c_{ij}, c_{ij} + \beta^c_{ij})$, $\tilde{a}_{rj} = (a_{rj} - \alpha^a_{rj}, a_{rj}, a_{rj} + \beta^a_{rj})$, and $\tilde{b}_r = (b_r - \alpha^b_r, b_r, b_r + \beta^b_r)$ are positive fuzzy variables.

According to LAM (5.33) and UAM (5.34), we present the linear lower approximation model (LLAM) and the linear upper approximation model (LUAM) of linear model (5.40):

$$
(LLAM) \quad \begin{cases} \max \left[E^{Me} \left[\sum_{j=1}^{n} \tilde{c}_{1j} x_j \right], E^{Me} \left[\sum_{j=1}^{n} \tilde{c}_{2j} x_j \right], \cdots, E^{Me} \left[\sum_{j=1}^{n} \tilde{c}_{mj} x_j \right] \right] \\ \text{s.t.} \begin{cases} E^{Pos} \left[\sum_{j=1}^{n} \tilde{a}_{rj} x_j \right] \geq E^{Pos} \left[\tilde{b}_r \right], r = 1, 2, \cdots, p \\ x_j \geq 0, j = 1, 2, \cdots, n \end{cases} \end{cases}
$$

$$(5.41)$$

and

$$
(LUAM) \quad \begin{cases} \max \left[E^{Me} \left[\sum_{j=1}^{n} \tilde{c}_{1j} x_j \right], E^{Me} \left[\sum_{j=1}^{n} \tilde{c}_{2j} x_j \right], \cdots, E^{Me} \left[\sum_{j=1}^{n} \tilde{c}_{mj} x_j \right] \right] \\ \text{s.t.} \begin{cases} E^{Nec} \left[\sum_{j=1}^{n} \tilde{a}_{rj} x_j \right] \geq E^{Nec} \left[\tilde{b}_r \right], r = 1, 2, \cdots, p \\ x_j \geq 0, j = 1, 2, \cdots, n \end{cases} \end{cases}
$$

$$(5.42)$$

which are collectively referred to as fuzzy linear expected value rough model (Fu-LEVRM).

5.3.2.1   Crisp Equivalent Model

THEOREM 5.6

Assume that $\tilde{c}_{ij} = (c_{ij} - \alpha^c_{ij}, c_{ij}, c_{ij} + \beta^c_{ij}), i = 1, 2, \ldots, m$ with $c_{ij} - \alpha^c_{ij} > 0$ are positive triangular fuzzy variables. Then, for a certain optimistic–pessimistic index λ, the expected value of the objective function can be computed as

$$
E \left[\sum_{j=1}^{n} \tilde{c}^T_{ij} x_j \right] = \sum_{j=1}^{n} \left(\frac{(1-\lambda)}{2} (c_{ij} - \alpha^c_{ij}) + \frac{c_{ij}}{2} + \frac{\lambda}{2} (c_{ij} + \beta^c_{ij}) \right) x_j.
$$

PROOF   Because $c_{ij} - \alpha^c_{ij} > 0$, we can easily derive the result from Lemma 5.6 and Case 4 in the proof of Proposition 5.1 .

THEOREM 5.7

Assume that $\tilde{a}_{rj} = (a_{rj} - \alpha^a_{rj}, a_{rj}, a_{rj} + \beta^a_{rj})$, and $\tilde{b}_r = (b_r - \alpha^b_r, b_r, b_r + \beta^b_r)$. Then

$$E^{Pos}\left[\sum_{j=1}^{n} \tilde{a}_{rj}x_j\right] \geq E^{Pos}\left[\tilde{b}_r\right] \quad \text{holds if and only if}$$

$$\begin{cases} \frac{r_1+r_2}{2} \leq 0, & \text{if } r_3 \leq 0 \\ \frac{r_1+r_2}{2} + \frac{r_3^2}{2(r_3-r_2)} \leq 0, & \text{if } r_2 \leq 0 \leq r_3 \\ \frac{r_2+r_3}{2} - \frac{r_1^2}{2(r_2-r_1)} \leq 0, & \text{if } r_1 \leq 0 \leq r_2 \\ \frac{r_2+r_3}{2} \leq 0, & \text{if } 0 \leq r_1 \end{cases}$$

where $r_1 = \sum_{j=1}^{n}(a_{rj} - \alpha_{rj}^a)x_j - b_r - \beta_r^b$, $r_2 = \sum_{j=1}^{n} a_{rj}x_j - b_r$, $r_3 = \sum_{j=1}^{n}(a_{rj}x_j + \beta_{rj}^a) - b_r + \alpha_r^b$.

PROOF It follows from the linearity of the expected value operator that

$$E^{Pos}\left[\sum_{j=1}^{n} \tilde{a}_{rj}x_j\right] \geq E^{Pos}\left[\tilde{b}_r\right]$$

$$\Leftrightarrow E^{Pos}\left[\sum_{j=1}^{n} \tilde{a}_{rj}x_j\right] - E^{Pos}\left[\tilde{b}_r\right] \geq 0$$

$$\Leftrightarrow E^{Pos}\left[\sum_{j=1}^{n} \tilde{a}_{rj}x_j - \tilde{b}_r\right] \geq 0.$$

By fuzzy arithmetic, $\sum_{j=1}^{n} \tilde{a}_{rj}x_j - \tilde{b}_r$ is still a triangular fuzzy variable with the form

$$\left(\sum_{j=1}^{n}(a_{rj} - \alpha_{rj}^a)x_j - b_r - \beta_r^b, \sum_{j=1}^{n} a_{rj}x_j - b_r, \sum_{j=1}^{n}(a_{rj} + \beta_{rj}^a)x_j - b_r + \alpha_r^b\right).$$

Let $r_1 = \sum_{j=1}^{n}(a_{rj} - \alpha_{rj}^a)x_j - b_r - \beta_r^b$, $r_2 = \sum_{j=1}^{n} a_{rj}x_j - b_r$, $r_3 = \sum_{j=1}^{n}(a_{rj} + \beta_{rj}^a)x_j - b_r + \alpha_r^b$. As we know, $E^{Pos} = E^{Me}$ if $\lambda = 1$. It follows that

$$\begin{cases} \frac{r_1+r_2}{2} \leq 0, & \text{if } r_3 \leq 0 \\ \frac{r_1+r_2}{2} + \frac{r_3^2}{2(r_3-r_2)} \leq 0, & \text{if } r_2 \leq 0 \leq r_3 \\ \frac{r_2+r_3}{2} - \frac{r_1^2}{2(r_2-r_1)} \leq 0, & \text{if } r_1 \leq 0 \leq r_2 \\ \frac{r_2+r_3}{2} \leq 0, & \text{if } 0 \leq r_1 \end{cases}$$

directly from Proposition 5.1. This proof is complete.

THEOREM 5.8
Assume that $\tilde{a}_{rj} = (a_{rj} - \alpha_{rj}^a, a_{rj}, a_{rj} + \beta_{rj}^a)$, *and* $\tilde{b}_r = (b_r - \alpha_r^b, b_r, b_r + \beta_r^b)$. *Then*

$$E^{Nec}\left[\sum_{j=1}^{n}\tilde{a}_{rj}x_j\right] \geq E^{Nec}[\tilde{b}_r] \text{ holds if and only if}$$

$$\begin{cases} \frac{r_2+r_3}{2} \leq 0, & \text{if } r_3 \leq 0 \\ -\frac{r_2^2}{2(r_3-r_2)} \leq 0, & \text{if } r_2 \leq 0 \leq r_3 \\ \frac{r_2^2}{2(r_2-r_1)} \leq 0, & \text{if } r_1 \leq 0 \leq r_2 \\ \frac{r_1+r_2}{2} \leq 0, & \text{if } 0 \leq r_1 \end{cases}$$

where $r_1 = \sum_{j=1}^{n}(a_{rj}-\alpha_{rj}^a)x_j - b_r - \beta_r^b$, $r_2 = \sum_{j=1}^{n}a_{rj}x_j - b_r$, and $r_3 = \sum_{j=1}^{n}(a_{rj}+\beta_{rj}^a)x_j - b_r + \alpha_r^b$.

PROOF　As we know $E^{Nec} = E^{Me}$ if $\lambda = 0$. Then we deduce

$$\begin{cases} \frac{r_2+r_3}{2} \leq 0, & \text{if } r_3 \leq 0 \\ -\frac{r_2^2}{2(r_3-r_2)} \leq 0, & \text{if } r_2 \leq 0 \leq r_3 \\ \frac{r_2^2}{2(r_2-r_1)} \leq 0, & \text{if } r_1 \leq 0 \leq r_2 \\ \frac{r_1+r_2}{2} \leq 0, & \text{if } 0 \leq r_1 \end{cases}$$

directly from Proposition 5.1. This proof is completed.

Based on Theorem 5.6, Theorem 5.7, and Theorem 5.8, there are four equivalents for (5.41) and (5.42), the linear lower and upper approximation models (E-LLAM, for short) and (E-LLAM) proposed as follows:

$$(E\text{-}LLAM)_1 \quad \begin{cases} \max[f_1(\mathbf{x}), f_2(\mathbf{x}), \cdots, f_m(\mathbf{x})] \\ \text{s.t.} \begin{cases} \frac{r_1+r_2}{2} \leq 0 \\ r_3 \leq 0 \\ x_j \geq 0, j = 1, 2, \cdots, n \end{cases} \end{cases} \quad (5.43)$$

$$(E\text{-}LUAM)_1 \quad \begin{cases} \max[f_1(\mathbf{x}), f_2(\mathbf{x}), \cdots, f_m(\mathbf{x})] \\ \text{s.t.} \begin{cases} \frac{r_2+r_3}{2} \leq 0 \\ r_3 \leq 0 \\ x_j \geq 0, j = 1, 2, \cdots, n \end{cases} \end{cases} \quad (5.44)$$

$$(E\text{-}LLAM)_2 \quad \begin{cases} \max[f_1(\mathbf{x}), f_2(\mathbf{x}), \cdots, f_m(\mathbf{x})] \\ \text{s.t.} \begin{cases} \frac{r_1+r_2}{2} + \frac{r_3^2}{2(r_3-r_2)} \leq 0 \\ r_2 \leq 0 \leq r_3, \\ x_j \geq 0, j = 1, 2, \cdots, n \end{cases} \end{cases} \quad (5.45)$$

$$(E\text{-}LUAM)_2 \quad \begin{cases} \max[f_1(\mathbf{x}), f_2(\mathbf{x}), \cdots, f_m(\mathbf{x})] \\ \text{s.t.} \begin{cases} -\frac{r_2^2}{2(r_3-r_2)} \leq 0 \\ r_2 \leq 0 \leq r_3 \\ x_j \geq 0, j = 1, 2, \cdots, n \end{cases} \end{cases} \quad (5.46)$$

$$(E\text{-}LLAM)_3 \quad \begin{cases} \max[f_1(\mathbf{x}), f_2(\mathbf{x}), \cdots, f_m(\mathbf{x})] \\ \text{s.t.} \begin{cases} \frac{r_2+r_3}{2} - \frac{r_1^2}{2(r_2-r_1)} \le 0 \\ r_1 \le 0 \le r_2 \\ x_j \ge 0, j = 1, 2, \cdots, n \end{cases} \end{cases} \quad (5.47)$$

$$(E\text{-}LUAM)_3 \quad \begin{cases} \max[f_1(\mathbf{x}), f_2(\mathbf{x}), \cdots, f_m(\mathbf{x})] \\ \text{s.t.} \begin{cases} \frac{r_2^2}{2(r_2-r_1)} \le 0 \\ r_1 \le 0 \le r_2 \\ x_j \ge 0, j = 1, 2, \cdots, n \end{cases} \end{cases} \quad (5.48)$$

$$(E\text{-}LLAM)_4 \quad \begin{cases} \max[f_1(\mathbf{x}), f_2(\mathbf{x}), \cdots, f_m(\mathbf{x})] \\ \text{s.t.} \begin{cases} \frac{r_2+r_3}{2} \le 0 \\ 0 \le r_1 \\ x_j \ge 0, j = 1, 2, \cdots, n \end{cases} \end{cases} \quad (5.49)$$

$$(E\text{-}LUAM)_4 \quad \begin{cases} \max[f_1(\mathbf{x}), f_2(\mathbf{x}), \cdots, f_m(\mathbf{x})] \\ \text{s.t.} \begin{cases} \frac{r_1+r_2}{2} \le 0 \\ 0 \le r_1 \\ x_j \ge 0, j = 1, 2, \cdots, n \end{cases} \end{cases} \quad (5.50)$$

where $f_i(\mathbf{x}) = \sum_{j=1}^{n} \left(\frac{(1-\lambda)}{2}(c_{ij} - \alpha_{ij}^c) + \frac{c_{ij}}{2} + \frac{\lambda}{2}(c_{ij} + \beta_{ij}^c) \right) x_j, r_1 = \sum_{j=1}^{n} (a_{rj} - \alpha_{rj}^a)x_j - b_r - \beta_r^b, r_2 = \sum_{j=1}^{n} a_{rj}x_j - b_r, r_3 = \sum_{j=1}^{n} (a_{rj} + \beta_{rj}^a)x_j - b_r + \alpha_r^b.$

So far, the linear fuzzy multiobjective model is transformed into the foregoing four model pairs (5.43)– (5.50), which are crisp and easy to solve. However, for the nonlinear models, it is very hard to convert the original model to equivalent models, so we use the hybrid algorithm in the next section to deal with it.

5.3.2.2 Satisfying Trade-off Method

The satisfying trade-off method for the multiobjective programming problems was proposed by Y. Sawaragi et al. [263]. It is an interactive method combining the satisfying level method with the ideal point method. This method can be applied to not only linear multiobjective but also nonlinear multiobjective programming.

Take the following problem as an example:

$$\begin{cases} \max[H_1(\mathbf{x}), H_2(\mathbf{x}), \cdots, H_m(\mathbf{x})] \\ \text{s.t. } \mathbf{x} \in X \end{cases} \quad (5.51)$$

At the begin, let us briefly introduce the simple satisfying level method. In some real decision-making problems, the DM usually provides a reference objective values $\bar{H} = (\bar{H}_1, \bar{H}_2, \cdots, \bar{H}_m)^T$. If the solution satisfies the reference value, take it. The simple satisfying level method can be summarized as follows:

Step 1. DM gives the reference objective values $\bar{H}$.

Step 2. Solve the following programming problem:

$$
\begin{cases}
\max \sum\limits_{i=1}^{m} H_i(\mathbf{x}) \\
\text{s.t.} \begin{cases} H_i(\mathbf{x}) \geq \bar{H}_i, \ i = 1,2,\cdots,m \\ \mathbf{x} \in X \end{cases}
\end{cases}
\tag{5.52}
$$

Step 3. If the problem (5.52) does not have the feasible solution, turn to Step 4. If the problem (5.52) has the optimal solution $\bar{\mathbf{x}}$, output $\bar{\mathbf{x}}$.

Step 4. DM regives the reference objective values $\bar{H}$, and turn to Step 2.

The satisfying trade-off method can be summarized as follows:

Step 1. Take the ideal point $\mathbf{H}^* = (H_1^*, H_2^*, \cdots, H_m^*)^T$ such that $H_i^* > \max_{\mathbf{x} \in X} f_i(\mathbf{x})$ $(i = 1, 2, \cdots, m)$.

Step 2. DM gives the objective level $\bar{\mathbf{H}}^k = (\bar{H}_1^k, \bar{H}_2^k, \cdots, \bar{H}_m^k)^T$ and $\bar{H}_i^k < \bar{H}_i^*(i = 1, 2, \cdots, m)$. Let $k = 1$.

Step 3. Solve the following problem to get the efficient solution:

$$
\min_{\mathbf{x} \in X} \max_{1 \leq i \leq m} w_i^k |H_i^* - H_i(\mathbf{x})|
\tag{5.53}
$$

where $w_i^k = \frac{1}{H_i^* - \bar{H}_i^k}$, $i = 1, 2, \cdots, m$,
or the equivalent problem:

$$
\begin{cases}
\min \lambda \\
\text{s.t.} \begin{cases} w_i^k(H_i^* - H_i(\mathbf{x})) \leq \lambda, \ i = 1,2,\cdots,m \\ \mathbf{x} \in X \end{cases}
\end{cases}
\tag{5.54}
$$

Suppose that the optimal solution is $\mathbf{x}^k$.

Step 4. According to the objective value $\mathbf{H}(\mathbf{x}^k) = (H_1(\mathbf{x}^k), H_2(\mathbf{x}^k), \cdots, H_m(\mathbf{x}^k))^T$, DM divide them into three classes: (1) which needs to improve, denote the related subscript set I_I^k, (2) which is permitted to release, denote the related subscript set I_R^k, (3) which is accepted, denote the related subscript set I_A^k. If $I_I^k = \Phi$, stop the iteration and output $\mathbf{x}^k$. Otherwise, DM gives the new reference objective values $\tilde{H}_i^k$, $i \in I_I^k \cup I_R^k$ and let $\tilde{H}_i^k = H_i(\mathbf{x}^k)$, $i \in I_A^k$.

Step 5. Let $u_i(i = 1, 2, \cdots, m)$ be the optimal Kuhn-Tucker operator of the first constraint. If there exists a minimal nonnegative number ε such that

$$
\sum_{i=1}^{m} u_i w_i^k(\tilde{H}_i^k - H_i(\mathbf{x}^k)) \geq -\varepsilon,
$$

then we deem that $\tilde{H}_i^k$ passes the check for feasibility. Let $\tilde{H}_{i+1} = \tilde{H}_i^k$ $(i = 1, 2, \cdots, m)$, turn to Step 3. Otherwise, $\tilde{H}_i^k$ isn't feasible. DM should re-give $\tilde{H}_i^k$, $i \in I_I^k \cup I_R^k$ and recheck it.

5.3.2.3 Numerical Example

Example 5.5

Consider the problem

$$\begin{cases} \max f_1(\mathbf{x}, \boldsymbol{\xi}) = \xi_1 x_1 + \xi_2 x_2 + \xi_3 x_3 + \xi_4 x_4 \\ \max f_2(\mathbf{x}, \boldsymbol{\xi}) = \xi_5 x_1 + \xi_6 x_2 + \xi_7 x_3 + \xi_8 x_4 \\ \text{s.t.} \begin{cases} \xi_5 x_1 + \xi_6 x_2 + \xi_7 x_3 + \xi_8 x_4 \le \xi_9 \\ 3x_1 + 5x_2 + 10x_3 + 15x_4 \le 100 \\ x_j \ge 0, j = 1,2,3,4 \end{cases} \end{cases} \quad (5.55)$$

wherein the coefficients are triangular fuzzy variables characterized as

$$\xi_1 = (3.5, 4, 4.5), \quad \xi_2 = (4.5, 5, 5.5), \ \xi_3 = (8.5, 9, 9.5),$$
$$\xi_4 = (10.5, 11, 11.5), \ \xi_5 = (6.5, 7, 7.5), \ \xi_6 = (4.5, 5, 5.5),$$
$$\xi_7 = (2.5, 3, 3.5), \quad \xi_8 = (1.5, 2, 2.5), \ \xi_9 = (110, 120, 130).$$

In order to solve it, we use the expected operator and rough approximation technique to deal with fuzzy objectives and fuzzy constraints, and then we can obtain the LAM and UAM as follows:

$$(LAM) \begin{cases} \max \ E^{Me}[\xi_1 x_1 + \xi_2 x_2 + \xi_3 x_3 + \xi_4 x_4] \\ \max \ E^{Me}[\xi_5 x_1 + \xi_6 x_2 + \xi_7 x_3 + \xi_8 x_4] \\ \text{s.t.} \begin{cases} E^{Pos}[\xi_5 x_1 + \xi_6 x_2 + \xi_7 x_3 + \xi_8 x_4] \le E^{Pos}[\xi_9] \\ 3x_1 + 5x_2 + 10x_3 + 15x_4 \le 100 \\ x_j \ge 0, j = 1,2,3,4 \end{cases} \end{cases} \quad (5.56)$$

and

$$(UAM) \begin{cases} \max \ E^{Me}[\xi_1 x_1 + \xi_2 x_2 + \xi_3 x_3 + \xi_4 x_4] \\ \max \ E^{Me}[\xi_5 x_1 + \xi_6 x_2 + \xi_7 x_3 + \xi_8 x_4] \\ \text{s.t.} \begin{cases} E^{Nec}[\xi_5 x_1 + \xi_6 x_2 + \xi_7 x_3 + \xi_8 x_4] \le E^{Nec}[\xi_9] \\ 3x_1 + 5x_2 + 10x_3 + 15x_4 \le 100 \\ x_j \ge 0, j - 1,2,3,4 \end{cases} \end{cases} \quad (5.57)$$

Here we suppose that $\lambda = 1/2$, and by Theorem 5.6, Theorem 5.7, and Theorem 5.8, we know that the problem has four equivalent model pairs as follows:

$$(E - LLAM)_1 \begin{cases} \max[4x_1 + 5x_2 + 9x_3 + 11x_4] \\ \max[7x_1 + 5x_2 + 3x_3 + 2x_4] \\ \text{s.t.} \begin{cases} \frac{r_1 + r_2}{2} \le 0 \\ r_3 \le 0 \\ 3x_1 + 5x_2 + 10x_3 + 15x_4 \le 100 \\ x_j \ge 0, j = 1,2,3,4 \end{cases} \end{cases} \quad (5.58)$$

$(E-LUAM)_1$
$$\begin{cases} \max[4x_1+5x_2+9x_3+11x_4] \\ \max[7x_1+5x_2+3x_3+2x_4] \\ \text{s.t.} \begin{cases} \frac{r_2+r_3}{2} \le 0 \\ r_3 \le 0, \\ 3x_1+5x_2+10x_3+15x_4 \le 100 \\ x_j \ge 0, j=1,2,3,4 \end{cases} \end{cases} \quad (5.59)$$

$(E-LLAM)_2$
$$\begin{cases} \max[4x_1+5x_2+9x_3+11x_4] \\ \max[7x_1+5x_2+3x_3+2x_4] \\ \text{s.t.} \begin{cases} \frac{r_1+r_2}{2} + \frac{r_3^2}{2(r_3-r_2)} \le 0 \\ r_2 \le 0 \le r_3 \\ 3x_1+5x_2+10x_3+15x_4 \le 100 \\ x_j \ge 0, j=1,2,3,4 \end{cases} \end{cases} \quad (5.60)$$

$(E-LUAM)_2$
$$\begin{cases} \max[4x_1+5x_2+9x_3+11x_4] \\ \max[7x_1+5x_2+3x_3+2x_4] \\ \text{s.t.} \begin{cases} -\frac{r_2^2}{2(r_3-r_2)} \le 0 \\ r_2 \le 0 \le r_3 \\ 3x_1+5x_2+10x_3+15x_4 \le 100 \\ x_j \ge 0, j=1,2,3,4 \end{cases} \end{cases} \quad (5.61)$$

$(E-LLAM)_3$
$$\begin{cases} \max[4x_1+5x_2+9x_3+11x_4] \\ \max[7x_1+5x_2+3x_3+2x_4] \\ \text{s.t.} \begin{cases} \frac{r_2+r_3}{2} - \frac{r_1^2}{2(r_2-r_1)} \le 0 \\ r_1 \le 0 \le r_2 \\ 3x_1+5x_2+10x_3+15x_4 \le 100 \\ x_j \ge 0, j=1,2,3,4 \end{cases} \end{cases} \quad (5.62)$$

$(E-LUAM)_3$
$$\begin{cases} \max[4x_1+5x_2+9x_3+11x_4] \\ \max[7x_1+5x_2+3x_3+2x_4] \\ \text{s.t.} \begin{cases} \frac{r_2^2}{2(r_2-r_1)} \le 0 \\ r_1 \le 0 \le r_2 \\ 3x_1+5x_2+10x_3+15x_4 \le 100 \\ x_j \ge 0, j=1,2,3,4 \end{cases} \end{cases} \quad (5.63)$$

$(E-LLAM)_4$
$$\begin{cases} \max[4x_1+5x_2+9x_3+11x_4] \\ \max[7x_1+5x_2+3x_3+2x_4] \\ \text{s.t.} \begin{cases} \frac{r_2+r_3}{2} \le 0 \\ 0 \le r_1 \\ 3x_1+5x_2+10x_3+15x_4 \le 100 \\ x_j \ge 0, j=1,2,3,4 \end{cases} \end{cases} \quad (5.64)$$

$(E-LUAM)_4$
$$\begin{cases} \max[f_1(x),f_2,\cdots,f_m] \\ \text{s.t.} \begin{cases} \frac{r_1+r_2}{2} \le 0 \\ 0 \le r_1 \\ 3x_1+5x_2+10x_3+15x_4 \le 100 \\ x_j \ge 0, j=1,2,3,4 \end{cases} \end{cases} \quad (5.65)$$

where $r_1 = 6.5x_1 + 4.5x_2 + 2.5x_3 + 1.5x_4 - 130, r_2 = 7x_1 + 5x_2 + 3x_3 + 2x_4 - 120, r_3 = 7.5x_1 + 5.5x_2 + 3.5x_3 + 2.5x_4 - 110$.

Now, we solve (E-LLAM)$_2$ to illustrate the satisfying trade-off method. By replacing r_1, r_2, and r_3 with $r_1 = 6.5x_1 + 4.5x_2 + 2.5x_3 + 1.5x_4 - 130, r_2 = 7x_1 + 5x_2 + 3x_3 + 2x_4 - 120$, and $r_3 = 7.5x_1 + 5.5x_2 + 3.5x_3 + 2.5x_4 - 110$ respectively, (E-LLAM)$_2$ can rewritten in the following form:

$$\text{(E-LUAM)}_2 \quad \begin{cases} \max F_1(\mathbf{x}) = 4x_1 + 5x_2 + 9x_3 + 11x_4 \\ \max F_2(\mathbf{x}) = 7x_1 + 5x_2 + 3x_3 + 2x_4 \\ \text{s.t.} \begin{cases} 14.5x_1 + 10.5x_2 + 6.5x_3 + 4.5x_4 - 230 \leq 0 \\ 7.5x_1 + 5.5x_2 + 3.5x_3 + 2.5x_4 - 110 \leq 0 \\ 3x_1 + 5x_2 + 10x_3 + 15x_4 \leq 100 \\ x_j \geq 0, j = 1,2,3,4 \end{cases} \end{cases} \quad (5.66)$$

For (5.66), first, calculate the problem $\max\limits_{x \in X} F_i(x), i = 1,2$, and we get $F_1^{\max} = 105.11$ and $F_2^{\max} = 102.67$. Then we set the ideal point $F^* = (F_1^*, F_2^*)^T = (110, 110)^T$. Second, let DM give the objective level $\tilde{F}^k = (\tilde{F}_1^k, \tilde{F}_2^k)^T = (108, 108)$ $(k = 1)$. Third, compute the weight coefficients by the following equation:

$$w_i^k = \frac{1}{F_i^* - \tilde{F}_1^k}, i = 1,2,$$

and we get $w_1^1 = 0.25$ and $w_2^1 = 0.25$. Solve the following problem:

$$\begin{cases} \max \lambda \\ \text{s.t.} \begin{cases} 0.25[110 - (4x_1 + 5x_2 + 9x_3 + 11x_4)] \leq \lambda \\ 0.25[110 - (7x_1 + 5x_2 + 3x_3 + 2x_4] \leq \lambda \\ 14.5x_1 + 10.5x_2 + 6.5x_3 + 4.5x_4 - 230 \leq 0 \\ 7.5x_1 + 5.5x_2 + 3.5x_3 + 2.5x_4 - 110 \leq 0 \\ 3x_1 + 5x_2 + 10x_3 + 15x_4 \leq 100 \\ x_j \geq 0, j = 1,2,3,4 \end{cases} \end{cases} \quad (5.67)$$

and we get $x^1 = (12.5, 0, 3.18, 2.05)^T$. Fourth, compute each objective value and we get $F_1(x_1) = 101.14$ and $F_2(x_1) = 101.14$. If the DM figures that the two objective values need not be changed, output $x^* = x^1$. The other models can be solved by a similar process.

5.3.3 Fu-NLEVRM and Fuzzy Simulation-Based SA

As problem (5.30) has nonlinear forms, that is at least one nonlinear function of f_i, g_r at least. By using rough approximation techniques, we can get the fuzzy nonlinear expected value rough model (Fu-NLEVRM). It is usually impossible to be converted into the crisp equivalent model. Therefore, uzzy simulation-based simulated annealing (SA), a solving method, is used to deal with it.

5.3.3.1 Fuzzy Simulation 1 for Expected Value

Let f be a real-valued function, and ξ_i be a fuzzy variable with membership functions μ_i, $i = 1, 2, \cdots, n$, respectively. We denote $\xi = (\xi_1, \xi_2, \cdots, \xi_n)$. Then $f(\xi)$ is also a fuzzy variable whose expected value is defined by

$$E^{Me}[f(\xi)] = \int_0^\infty Me\{f(\xi) \geq r\}dr - \int_{-\infty}^0 Me\{f(\xi) \leq r\}dr \qquad (5.68)$$

A fuzzy simulation will be designed to estimate $E^{Me}[f(\xi)]$. We randomly generate $u_{1j}, u_{2j}, \cdots, u_{nj}$ from the ε-level sets of $\xi_1, \xi_2, \cdots, \xi_n$, $j = 1, 2, \cdots, m$, respectively, where ε is a sufficiently small number. Let $u_j = (u_{1j}, u_{2j}, \cdots, u_{nj})$ and $\mu_j = \mu_1(u_{1j}) \wedge \mu_2(u_{2j}) \wedge \cdots \wedge \mu_n(u_{nj})$ for $j = 1, 2, \cdots, m$.

Then for any number $r \geq 0$, the fuzzy measure $Me\{f(\xi) \geq r\}$ can be estimated by

$$\begin{aligned} Me\{f(\xi) \geq r\} = \lambda \left(\max_{j=1,2,\cdots,m} \{\mu_j | f(u_j) \geq r\} \right) \\ + (1 - \lambda) \left(1 - \max_{j=1,2,\cdots,m} \{\mu_j | f(u_j) < r\} \right) \end{aligned} \qquad (5.69)$$

and for any number $r < 0$, $Me\{f(\xi) \leq r\}$ can be estimated by

$$\begin{aligned} Me\{f(\xi) \leq r\} = \lambda \left(\max_{j=1,2,\cdots,m} \{\mu_j | f(u_j) \leq r\} \right) \\ + (1 - \lambda) \left(1 - \max_{j=1,2,\cdots,m} \{\mu_j | f(u_j) > r\} \right) \end{aligned} \qquad (5.70)$$

provided that m is sufficiently large.

The processes are summarized as follows:

Step 1. Set $E = 0$.

Step 2. Randomly generate $u_{1j}, u_{2j}, \cdots, u_{nj}$ from the ε-level sets of $\xi_1, \xi_2, \cdots, \xi_n$, and denote $u_j = (u_{1j}, u_{2j}, \cdots, u_{nj})$, $j = 1, 2, \cdots, m$, respectively, where ε is a sufficiently small number.

Step 3. Set $a = f(u_1) \wedge f(u_2) \wedge \cdots \wedge f(u_m)$, $b = f(u_1) \vee f(u_2) \vee \cdots \vee f(u_m)$.

Step 4. Randomly generate r from $[a, b]$.

Step 5. If $r \geq 0$, then $E \leftarrow E + Me\{f(\xi) \geq r\}$.

Step 6. If $r < 0$, then $E \leftarrow E - Me\{f(\xi) \leq r\}$.

Step 7. Repeat the fourth to sixth steps for N times.

Step 8. $E[f(\xi)] = a \vee 0 + b \wedge 0 + E \cdot (b - a)/N$.

Example 5.6

We employ fuzzy simulation to calculate the expected value of $\xi_1 \xi_2 \xi_3 \xi_4$, where ξ_1, ξ_2, ξ_3, and ξ_4 are fuzzy variables defined as $\xi_1 = (1, 2, 3), \xi_2 = (2, 3, 4), \xi_3 = (3, 4, 5), \xi = (4, 5, 6)$. A run of fuzzy simulation with 1000 cycles shows that the expected value

$$E^{Me} = \begin{cases} 209.3, & \lambda = 0 \\ 222.0, & \lambda = 0.2 \\ 270.9, & \lambda = 0.5 \\ 326.6, & \lambda = 0.8 \\ 341.0 & \lambda = 1 \end{cases}$$

5.3.3.2 SA

The simulated annealing (SA) algorithm was proposed by S. Kirkpatrick et al. [165] for the problem of finding, numerically, a point of the global minimum of a function defined on a subset of an n-dimensional Euclidean space. The motivations for the methods lies in the physical process of annealing, in which a solid is heated to a liquid state and, when cooled sufficiently slowly, takes up the configuration with minimal inner energy. N. Metropolis et al. [213] described this process mathematically. Simulating annealing uses this mathematical description for the minimization of functions other than energy. The first results were published by V. Černý [47], S. Kirkpatrick et al. [165] and S. Geman et al. [112]. SA has been successfully used for extensive research (see [7, 82, 88, 92, 99, 164, 188, 233, 234, 305]).

Annealing, physically, refers to the process of heating up a solid to a high temperature, followed by slow cooling achieved by decreasing the temperature of the environment in steps. At each step the temperature is maintained constant for a period of time sufficient for the solid to reach thermal equilibrium. At equilibrium, the solid could have many configurations, each corresponding to different spins of the electrons and to a specific energy level. Simulated annealing is a computational stochastic technique for obtaining near-global optimum solutions to combinatorial and function optimization problems. The method is inspired by the thermodynamic process of cooling (annealing) of molten metals to attain the lowest free energy state. When molten metal is cooled slowly enough, it tends to solidify in a structure of minimum energy. This annealing process is mimicked by a search strategy. The key principle of the method is to allow occasional worsening moves so that these can eventually help locate the neighborhood to the true (global) minimum. The associated mechanism is given by the Boltzmann probability, namely,

$$\text{probability}(p) = \exp\left(\frac{-\triangle E}{K_B T}\right) \tag{5.71}$$

where $\triangle E$ is the change in the energy value from one point to the next, K_B the Boltzmann's constant, and T the temperature (control parameter). For the purpose of optimization the energy term, $\triangle E$ refers to the value of the objective function, and the temperature, T, is a control parameter that regulates the process of annealing. The consideration of such a probability distribution leads to the generation of a Markov chain of points in the problem domain. The acceptance criterion given by Equations. (5.71) is popularly referred to as the Metropolis criterion [213]. Another variant of this acceptance criterion (for both improving and deteriorating moves) has been proposed by R. Galuber [117] and can be written as

$$\text{probability}(p) = \frac{\exp(-\Delta E/T)}{1+\exp(-\Delta E/T)} \tag{5.72}$$

In simulated annealing search strategy, at the start, any move is accepted. This allows us to explore the solution space. Then, gradually, the temperature is reduced, which means that one becomes more and more selective in accepting the new solution. By the end, only the improving moves are accepted in practice. The temperature is systematically lowered using a problem-dependent schedule characterized by a set of decreasing temperatures. Next, we introduce the general framework for the SA algorithm.

The standard SA technique makes the analogy between the state of each molecule that determines the energy function and the value of each parameter that affects the objective functions. It then uses the statistical mechanics principle for energy minimization to minimize the objective function and optimize the parameter estimates. Starting with a high temperature, it randomly perturbs the parameter values and calculates the resulting objective function. The new state of objective function after perturbation is then accepted by a probability determined by the Metropolis criterion. The system temperature is then gradually reduced as the random perturbation proceeds, until the objective function reaches its global or near-global minimum. A typical SA algorithm is described as follows:

Step 1. Specify initial temperature $T_k = T_0$ for $k = 0$; randomly initialize the parameter set estimate $\theta^* = \theta_0$.

Step 2. Under kth temperature, if the inner loop break condition is met, go to Step 3; otherwise, for the $(j+1)$th perturbation, randomly produce a new parameter set θ_{j+1}, compute the change in the objective function $\Delta f = f(\theta^*) - f(\theta_{j+1})$. If $\Delta f \leq 0$, accept $\theta_{j+1}(\theta^* = \theta_j)$; if not, follow the Metropolis criterion to accept θ_{j+1} with a probability of $\min(1, e^{-\Delta f/T_k})$, and Step 2 continues.

Step 3. Reduce T_k to T_{k+1} following a specified cooling schedule. If the outer loop break condition is met, computation stops and the optimal parameter set is reached; if not, return to Step 2.

The steps outlined above consist of one inner loop (Step 2) and one outer loop (Step 3). The proceeding of SA are mainly controlled by

(a) the choice of T_0;

(b) the way a new perturbation is generated;

(c) the inner loop break conditions;

(d) the choice of cooling schedule;

(e) the outer loop break conditions.

5.3.3.3   Numerical Examples

Example 5.7
Consider the problem

$$\begin{cases} \max f_1(\mathbf{x}, \boldsymbol{\xi}) = \xi_1 x_1 + \xi_2^2 x_2 + \sqrt{\xi_3} x_3 \\ \max f_2(\mathbf{x}, \boldsymbol{\xi}) = \xi_4 x_1 + \xi_5 x_2 + \xi_6 x_3 \\ \text{s.t.} \begin{cases} \sqrt{\xi_1} x_1 + \xi_2 x_2 + \xi_3^2 \sqrt{x_3} \le \xi_7 \\ x_1 + x_2 + x_3 \le 100 \\ x_i \ge 10, i = 1, 2, 3 \end{cases} \end{cases} \quad (5.73)$$

wherein the coefficients are triangular fuzzy variables characterized as

$$\xi_1 = (3.5, 4, 4.5), \ \xi_2 = (4.5, 5, 5.5), \ \xi_3 = (8.5, 9, 9.5), \quad \xi_4 = (10.5, 11, 11.5),$$
$$\xi_5 = (2.5, 3, 3.5), \ \xi_6 = (4.5, 5, 5.5), \ \xi_7 = (150, 200, 250).$$

In order to solve it, we use the expected value operator and rough approximation technique to deal with fuzzy objectives and fuzzy constraints, and then we can obtain the LAM and UAM as follows:

$$(LAM) \quad \begin{cases} \max E^{Me}[\xi_1 \sqrt{x_1} + \xi_2^2 x_2 + \sqrt{\xi_3} x_3^2] \\ \max E^{Me}[\xi_4 x_1 + \xi_5 x_2 + \xi_6 x_3] \\ \text{s.t.} \begin{cases} E^{Pos}[\sqrt{\xi_1} x_1 + \xi_2 x_2 + \xi_3^2 \sqrt{x_3}] \le E^{Pos}[\xi_7] \\ x_1 + x_2 + x_3 \le 100 \\ x_i \ge 10, i = 1, 2, 3 \end{cases} \end{cases} \quad (5.74)$$

and

$$(UAM) \quad \begin{cases} \max E^{Me}[\xi_1 \sqrt{x_1} + \xi_2^2 x_2 + \sqrt{\xi_3} x_3^2] \\ \max E^{Me}[\xi_4 x_1 + \xi_5 x_2 + \xi_6 x_3] \\ \text{s.t.} \begin{cases} E^{Nec}[\sqrt{\xi_1} x_1 + \xi_2 x_2 + \xi_3^2 \sqrt{x_3}] \le E^{Nec}[\xi_7] \\ x_1 + x_2 + x_3 \le 100 \\ x_i \ge 10, i = 1, 2, 3 \end{cases} \end{cases} \quad (5.75)$$

Let $\lambda = 0.5$, and by fuzzy simulation we have

$$E^{Me}[\xi_1] - 4, \quad E^{Me}[\xi_2^2] - 22.8, \quad E^{Me}[\sqrt{\zeta_3}] = 3.0, \ E^{Me}[\zeta_4] = 11,$$
$$E^{Me}[\xi_5] = 7 \quad E^{Me}[\xi_6] = 5, \quad E^{Pos}[\sqrt{\xi_1}] = 2.0, \ E^{Nec}[\sqrt{\xi_1}] = 1.9,$$
$$E^{Pos}[\xi_2] = 5.2, \quad E^{Nec}[\xi_2] = 4.8, \quad E^{Pos}[\xi_3^2] = 79.0, \ E^{Nec}[\xi_3^2] = 74.6,$$
$$E^{Pos}[\xi_7] = 207.3, \ E^{Nec}[\xi_7] = 192.7.$$

Then (5.74) and (5.75) can be rewritten as

$$(LAM) \quad \begin{cases} \max F_1(x) = 4\sqrt{x_1} + 22.8 x_2 + 3 x_3^2 \\ \max F_2(x) = 11 x_1 + 3 x_2 + 5 x_3 \\ \text{s.t.} \begin{cases} 2 x_1 + 5.2 x_2 + 79\sqrt{x_3} \le 207.3 \\ x_1 + x_2 + x_3 \le 100 \\ x_i \ge 10, i = 1, 2, 3 \end{cases} \end{cases} \quad (5.76)$$

and

$$\text{(UAM)} \quad \begin{cases} \max F_1(x) = 4\sqrt{x_1} + 22.8x_2 + 3x_3^2 \\ \max F_2(x) = 11x_1 + 3x_2 + 5x_3 \\ \text{s.t.} \begin{cases} 1.9x_1 + 4.8x_2 + 74.6\sqrt{x_3} \geq 192.7 \\ x_1 + x_2 + x_3 \leq 100 \\ x_i \geq 10, i = 1, 2, 3 \end{cases} \end{cases} \quad (5.77)$$

After running SA, we get $x^* = (10.7, 16.5, 10.1)$ and $F^* = (693.0, 217.7)$ for (5.76) and $x^* = (10.0, 19.0, 757.6)$ and $F^* = (757.6, 218.6)$ for (5.77).

5.4   Fu-CCRM

The second way to deal with model (5.29) is CCM. As the rough approximation techniques are used to tackle the fuzzy CCM, we can get the fuzzy chance0constrained rough model (Fu-CCRM).

5.4.1   General Model of Fu-CCRM

The general fuzzy CCM is formulated as

$$\begin{cases} \max \ [\bar{f}_1, \bar{f}_2, \cdots, \bar{f}_m] \\ \text{s.t.} \begin{cases} Ch\{f_i(\mathbf{x}, \boldsymbol{\xi}) \geq \bar{f}_i\} \geq \delta_i, i = 1, 2, \cdots, m \\ Ch\{g_r(\mathbf{x}, \boldsymbol{\xi}) \leq 0\} \geq \theta_r, \ r = 1, 2, \cdots, p \\ \mathbf{x} \in X \end{cases} \end{cases} \quad (5.78)$$

where Ch represents fuzzy measures such as Pos, Nec, Me, and Cr; and δ_i, θ_r are the predetermined confidence levels.

For model (5.78), we let $D = \{\mathbf{x} | \mathbf{x} \in X, Me\{f_i(\mathbf{x}, \boldsymbol{\xi}) \geq \bar{f}_i\} \geq \delta_i, i = 1, 2, \cdots, m, Me\{g_r(\mathbf{x}, \boldsymbol{\xi}) \leq 0\} \geq \theta_r, r = 1, 2, \cdots, p\}$, and we construct two set L and U as

$$\begin{cases} L = \{\mathbf{x} | \mathbf{x} \in X, Nec\{f_i(\mathbf{x}, \boldsymbol{\xi}) \geq \bar{f}_i\} \geq \delta_i, i = 1, 2, \cdots, m, Nec\{g_r(\mathbf{x}, \boldsymbol{\xi}) \leq 0\} \geq \theta_r, \\ r = 1, 2, \cdots, p\} \\ U = \{\mathbf{x} | \mathbf{x} \in X, Pos\{f_i(\mathbf{x}, \boldsymbol{\xi}) \geq \bar{f}_i\} \geq \delta_i, i = 1, 2, \cdots, m, Pos\{g_r(\mathbf{x}, \boldsymbol{\xi}) \leq 0\} \geq \theta_r, \\ r = 1, 2, \cdots, p\} \end{cases}$$

$$(5.79)$$

THEOREM 5.9
For the feasible region D, and L, U defined by (5.79), we have

$$L \subseteq D \subseteq U.$$

PROOF   For any $\mathbf{x}_0 \in X$, If $\mathbf{x}_0 \in L$, that is, $Nec\{f_i(\mathbf{x}_0, \boldsymbol{\xi}) \geq \bar{f}_i\} \geq \delta_i, i = 1, 2, \cdots, m, Nec\{g_r(\mathbf{x}_0, \boldsymbol{\xi}) \leq 0\} \geq \theta_r, r = 1, 2, \cdots, p$. It follows from Corollary 5.1

that $Me\{f_i(\mathbf{x}_0,\boldsymbol{\xi}) \geq \bar{f}_i\} \geq Nec\{f_i(\mathbf{x}_0,\boldsymbol{\xi}) \geq \bar{f}_i\} \geq \delta_i, i = 1,2,\cdots,m, Me\{g_r(\mathbf{x}_0,\boldsymbol{\xi}) \leq 0\} \geq Nec\{g_r(\mathbf{x}_0,\boldsymbol{\xi}) \leq 0\} \geq \theta_r, r = 1,2,\cdots,p$, that is, $x_0 \in D$. Thus, $L \subseteq D$. In a similar way, we can prove $D \subseteq U$. The proof is complete.

Then we let $\underline{D} = L$ and $\overline{D} = U$, we use $\underline{D}$ and $\overline{D}$ to approximate D, it is obvious that $\underline{D} \subseteq D \subseteq \overline{D}$.

Then model (5.78) can transformed into two models: the lower approximation model (LAM) and the upper approximation model (UAM).

$$(LAM) \quad \begin{cases} \max\ [\bar{f}_1, \bar{f}_2, \cdots, \bar{f}_m] \\ \text{s.t.} \begin{cases} Pos\{f_i(\mathbf{x},\boldsymbol{\xi}) \geq \bar{f}_i\} \geq \delta_i, i = 1,2,\cdots,m \\ Pos\{g_r(\mathbf{x},\boldsymbol{\xi}) \leq 0\} \geq \theta_r, r = 1,2,\cdots,p \\ \mathbf{x} \in X \end{cases} \end{cases} \quad (5.80)$$

and

$$(UAM) \quad \begin{cases} \max\ [\bar{f}_1, \bar{f}_2, \cdots, \bar{f}_m] \\ \text{s.t.} \begin{cases} Nec\{f_i(\mathbf{x},\boldsymbol{\xi}) \geq \bar{f}_i\} \geq \delta_i, i = 1,2,\cdots,m \\ Nec\{g_r(\mathbf{x},\boldsymbol{\xi}) \leq 0\} \geq \theta_r, r = 1,2,\cdots,p \\ \mathbf{x} \in X \end{cases} \end{cases} \quad (5.81)$$

DEFINITION 5.10 $\mathbf{x}^*$ *is said to be a feasible solution at* θ_r-*necessity levels of problem (5.80) if and only if it satisfies*

$$Nec\{g_r(\mathbf{x},\boldsymbol{\xi}) \leq 0\} \geq \theta_r, r = 1,2,\cdots,p.$$

DEFINITION 5.11 *A feasible solution at* θ_r-*necessity levels,* $\mathbf{x}^*$, *is said to be a* δ_i-*efficient solution to problem (5.80) if and only if there exists no other feasible solution at* θ_r-*necessity levels* $\mathbf{x}$ *such that* $Nec\{f_i(\mathbf{x},\boldsymbol{\xi})\} \geq \delta_i$ *with* $f_i(\mathbf{x}) \geq \bar{f}_i(\mathbf{x}^*)$ *for all i and* $f_{i_0}(\mathbf{x}) > \bar{f}_{i_0}(\mathbf{x}^*)$ *for at least one* $i_0 \in \{1,2,\cdots,m\}$.

DEFINITION 5.12 *A feasible solution at* θ_r-*possibility levels,* $\mathbf{x}^*$, *is said to be a* δ_i-*efficient solution to problem (5.81) if and only if there exists no other feasible solution at* θ_r-*possibility levels* $\mathbf{x}$ *such that* $Pos\{f_i(\mathbf{x},\boldsymbol{\xi})\} \geq \delta_i$ *with* $f_i(\mathbf{x}) \geq \bar{f}_i(\mathbf{x}^*)$ *for all i and* $f_{i_0}(\mathbf{x}) > \bar{f}_{i_0}(\mathbf{x}^*)$ *for at least one* $i_0 \in \{1,2,\cdots,m\}$

If the objective is to minimize $\bar{f}_i$, we have the following model:

$$\begin{cases} \min[\bar{f}_1, \bar{f}_2, \cdots, \bar{f}_m] \\ \text{s.t.} \begin{cases} Ch\{f_i(\mathbf{x},\boldsymbol{\xi}) \geq \bar{f}_i\} \geq \delta_i, i = 1,2,\cdots,m \\ Ch\{g_r(\mathbf{x},\boldsymbol{\xi}) \leq 0\} \geq \theta_r, r = 1,2,\cdots,p \\ \mathbf{x} \in X \end{cases} \end{cases} \quad (5.82)$$

where *Ch* refers to fuzzy measure.

5.4.2 Fu-LCCRM and Step Method

We consider the linear decision-making with fuzzy parameters. Fuzzy constraints can no longer give a crisp feasible set. Naturally, one requires constraints to hold at a predetermined confidence level. The chance constraints with fuzzy parameters are expressed as follows:

$$Me \left\{ \sum_{j=1}^{n} \tilde{a}_{rj} x_j \leq \tilde{b}_r \right\} \geq \theta_r, \ r = 1, 2, \cdots, p \tag{5.83}$$

where $Me\{\cdot\}$ denotes the general measure of the fuzzy event $\{\cdot\}$.

If the decision maker hopes that the possibility of the objectives are not less than δ_i, and that the possibility of the constraints, being no greater than $\tilde{b}_r$, is not less than θ_r, then the chance-constrained linear decision-making model with fuzzy parameters is formulated as follows:

$$\begin{cases} \max \ [\bar{f}_1, \bar{f}_2, \cdots, \bar{f}_m] \\ \text{s.t.} \begin{cases} Me \left\{ \sum_{j=1}^{n} \tilde{c}_{ij} x_j \geq \bar{f}_i \right\} \geq \delta_i, \ i = 1, 2, \cdots, m \\ Me \left\{ \sum_{j=1}^{n} \tilde{a}_{rj} x_j \leq \tilde{b}_r \right\} \geq \theta_r, \ r = 1, 2, \cdots, p \\ x_j \geq 0, j = 1, 2, \cdots, n \end{cases} \end{cases} \tag{5.84}$$

where $\max \bar{f}_i$ is the δ_i-return defined as

$$\max \left\{ \bar{f}_i | Me \left\{ \sum_{j=1}^{n} \tilde{c}_{ij} x_j \geq \bar{f}_i \right\} \geq \delta_i \right\}.$$

According to LAM (5.80) and UAM (5.81), we present the linear lower approximation model (LLAM) and the linear upper approximation model (LUAM) of the linear model (5.84):

$$(LLAM) \begin{cases} \max \ [\bar{f}_1, \bar{f}_2, \cdots, \bar{f}_m] \\ \text{s.t.} \begin{cases} Nec \left\{ \sum_{j=1}^{n} \tilde{c}_{ij} x_j \geq \bar{f}_i \right\} \geq \delta_i, \ i = 1, 2, \cdots, m \\ Nec \left\{ \sum_{j=1}^{n} \tilde{a}_{rj} x_j \leq \tilde{b}_r \right\} \geq \theta_r, \ r = 1, 2, \cdots, p \\ x_j \geq 0, j = 1, 2, \cdots, n \end{cases} \end{cases} \tag{5.85}$$

and

$$(LLAM) \begin{cases} \max \ [\bar{f}_1, \bar{f}_2, \cdots, \bar{f}_m] \\ \text{s.t.} \begin{cases} Pos \left\{ \sum_{j=1}^{n} \tilde{c}_{ij} x_j \geq \bar{f}_i \right\} \geq \delta_i, \ i = 1, 2, \cdots, m \\ Pos \left\{ \sum_{j=1}^{n} \tilde{a}_{rj} x_j \leq \tilde{b}_r \right\} \geq \theta_r, \ r = 1, 2, \cdots, p \\ x_j \geq 0, j = 1, 2, \cdots, n \end{cases} \end{cases} \tag{5.86}$$

which are collectively referred to as fuzzy linear chance-constrained rough model (Fu-LCCRM).

5.4.2.1 Crisp Equivalent Model

The traditional mathematical solution methods require conversion of the chance constraints to their respective deterministic equivalents. However, this process is usually hard to perform and is only successful in some special cases. First we present a few useful results.

THEOREM 5.10

Let $\tilde{c}_{ij}$ be a LR fuzzy variable with membership function

$$\mu_{\tilde{c}_{ij}}(t) = \begin{cases} L\left(\frac{c_{ij}-t}{\alpha_{ij}^c}\right), & t \leq c_{ij}, \alpha_{ij}^c > 0 \\ R\left(\frac{t-c_{ij}}{\beta_{ij}^c}\right), & t \geq c_{ij}, \beta_{ij}^c > 0 \end{cases} \tag{5.87}$$

where the vector $(c_{ij})_{n\times 1} = (c_{i1}, c_{i2}, \cdots, c_{in})^{\mathrm{T}}$ is real number; α_{ij}^c and β_{ij}^c are the left and right spread of $\tilde{c}_{ij}$, $i = 1, 2, \cdots, m$, $j = 1, 2, \cdots, n$; the reference function $L, R : [0, 1] \to [0, 1]$ satisfies that $L(1) = R(1) = 0$, $L(0) = R(0) = 1$, and it is a monotone function. Then $\mathrm{Pos}\{\tilde{c}_i^{\mathrm{T}}x \geq \bar{f}_i\} \geq \delta_i$ is equivalent to

$$\bar{f}_i \leq c_i^T x + R^{-1}(\delta_i)\beta_i^{c\mathrm{T}}x, \quad i = 1, 2, \cdots, m \tag{5.88}$$

PROOF Because $\tilde{c}_{ij}$ is an LR fuzzy number, its membership function is $\mu_{\tilde{c}_{ij}}$. By extension principle [337], the membership function of fuzzy number $\tilde{c}_i^{\mathrm{T}}x$ is

$$\mu_{\tilde{c}_i^{\mathrm{T}}x}(r) = \begin{cases} L\left(\frac{c_i^{\mathrm{T}}x-r}{\alpha_i^{c\mathrm{T}}x}\right) & r \leq c_i^{\mathrm{T}}x \\ R\left(\frac{r-c_i^{\mathrm{T}}x}{\beta_i^{c\mathrm{T}}x}\right) & r \geq c_i^{\mathrm{T}}x \end{cases} \tag{5.89}$$

for $i = 1, 2, \cdots, m$. For convenience, we denote $\tilde{c}_{ij} = (c_{ij}, \alpha_{ij}^c, \beta_{ij}^c)_{\mathrm{LR}}$, $\tilde{c}_i^{\mathrm{T}}x = (c_i^{\mathrm{T}}x, \alpha_i^{c\mathrm{T}}x, \beta_i^{c\mathrm{T}}x)_{\mathrm{LR}}$.

According to Lemma 5.3, we can get

$$\mathrm{Pos}\{\tilde{c}_i^{\mathrm{T}}x \geq \bar{f}_i\} \geq \delta_i \Leftrightarrow c_i^{\mathrm{T}}x + R^{-1}(\delta_i)\beta_i^{c\mathrm{T}}x \geq \bar{f}_i, \quad i = 1, 2, \cdots, m.$$

The proof is completed.

REMARK 5.12 Especially when the reference function of the variable $\tilde{c}_{ij}$ is $L(x) = R(x) = 1 - x$ ($x \in [0, 1]$), then the LR fuzzy variable is specified as the triangular fuzzy variable, and $R^{-1}(\delta_i) = 1 - \delta_i$, so we have the following equivalent expressions:

$$\mathrm{Pos}\{\tilde{c}_i^{\mathrm{T}}x \geq \bar{f}_i\} \geq \delta_i \Leftrightarrow c_i^T x + (1 - \delta_i)\beta_i^{c\mathrm{T}}x \geq \bar{f}_i, \quad i = 1, 2, \cdots, m.$$

THEOREM 5.11

Let $\tilde{a}_{rj}, \tilde{b}_r$ be an LR fuzzy variables with membership functions

$$\mu_{\tilde{a}_{rj}}(t) = \begin{cases} L\left(\frac{a_{rj}-t}{\alpha_{rj}^a}\right), & t \le a_{rj}, \alpha_{rj}^a > 0 \\ R\left(\frac{t-a_{rj}}{\beta_{rj}^a}\right), & t \ge a_{rj}, \beta_{rj}^a > 0 \end{cases} \qquad (5.90)$$

$$\mu_{\tilde{b}_r}(t) = \begin{cases} L\left(\frac{b_r-t}{\alpha_r^b}\right), & t \le b_r, \alpha_r^b > 0 \\ R\left(\frac{t-b_r}{\beta_r^b}\right), & t \ge b_r, \beta_r^b > 0 \end{cases} \qquad (5.91)$$

where the vector $(a_{rj})_{n\times 1} = (a_{r1}, a_{r2}, \cdots, a_{rn})^\mathrm{T}$ is real number; α_{rj}^a and β_{rj}^a are the left and right spread of $\tilde{a}_{rj}$; α_r^b and β_r^b are the left and right spread of $\tilde{b}_r$, $r = 1, 2, \cdots, p$, $j = 1, 2, \cdots, n$; the reference function $L, R : [0,1] \to [0,1]$ satisfies that $L(1) = R(1) = 0$, $L(0) = R(0) = 1$, and it is monotone function. Suppose that a_{rj} and b_r are independent. Then $\mathrm{Pos}\{\tilde{a}_r^\mathrm{T} x \le \tilde{b}_r\} \ge \theta_r$ is equivalent to

$$b_r + R^{-1}(\theta_r)\beta_r^b \ge a_r^\mathrm{T} x - L^{-1}(\theta_r)\alpha_r^{a\mathrm{T}} x, \quad r = 1, 2, \cdots, p. \qquad (5.92)$$

PROOF   Because $\tilde{a}_{rj}$ is a LR fuzzy number, its membership function is $\mu_{\tilde{a}_{rj}}$. By the extension principle [337], the membership function of fuzzy number $\tilde{a}_r^\mathrm{T} x$ is

$$\mu_{\tilde{a}_r^\mathrm{T} x}(r) = \begin{cases} L\left(\frac{a_r^\mathrm{T} x - r}{\alpha_r^{a\mathrm{T}} x}\right), & r \le a_r^\mathrm{T} x \\ R\left(\frac{r - a_r^\mathrm{T} x}{\beta_r^{a\mathrm{T}} x}\right), & r \ge a_r^\mathrm{T} x \end{cases} \qquad (5.93)$$

for $r = 1, 2, \cdots, p$. And $\tilde{b}_r$ is also a LR fuzzy number with membership function $\mu_{\tilde{a}_{rj}}$. According to Lemma 5.3, we can get

$$\mathrm{Pos}\{\tilde{a}_r^\mathrm{T} x \le \tilde{b}_r\} \ge \theta_r \Leftrightarrow b_r + R^{-1}(\theta_r)\beta_r^b \ge a_r^\mathrm{T} x - L^{-1}(\theta_r)\alpha_r^{a\mathrm{T}} x, r = 1, 2, \cdots, p. \qquad (5.94)$$

The proof is completed.

REMARK 5.13   Especially when the reference function of the variables $\tilde{a}_{rj}, \tilde{b}_r$ are all $L(x) = R(x) = 1 - x$ ($x \in [0,1]$), then the LR fuzzy variables are specified as the triangular fuzzy variables, and $R^{-1}(\delta_i) = 1 - \delta_i$, so we have the following equivalent expressions:

$$\mathrm{Pos}\{\tilde{a}_r^\mathrm{T} x \le \tilde{b}_r\} \ge \theta_r \Leftrightarrow b_r + (1 - \theta_r)\beta_r^b \ge a_r^\mathrm{T} x - (1 - \theta_r)\alpha_r^{a\mathrm{T}} x, r = 1, 2, \cdots, p.$$

Then we give the fuzzy CCM crisp equivalent model based on *Nec* measure.

THEOREM 5.12

Assume that the fuzzy variables are the same as those in Theorem 5.10. Then $\mathrm{Nec}\{\tilde{c}_i^\mathrm{T} x \ge \bar{f}_i\} \ge \delta_i$ is equivalent to

$$\bar{f}_i \le c_i^\mathrm{T} x - L^{-1}(1 - \delta_i)\alpha_i^{c\mathrm{T}} x, \quad i = 1, 2, \cdots, m \qquad (5.95)$$

PROOF   The proof is similar to the proof of Theorem 5.10.

REMARK 5.14   Especially when the reference function of the variable $\tilde{c}_{ij}$ is $L(x) = R(x) = 1 - x$ ($x \in [0,1]$), then the LR fuzzy variable is specified as the triangular fuzzy variable, and $R^{-1}(\delta_i) = 1 - \delta_i$, so we have the following equivalent expressions:

$$\text{Nec}\{\tilde{c}_i^T x \geq \bar{f}_i\} \geq \delta_i \Leftrightarrow \bar{f}_i \leq c_i^T x - \delta_i \alpha_i^{cT} x, \quad i = 1, 2, \cdots, m.$$

THEOREM 5.13
Assume that the fuzzy variables $\tilde{a}_{rj}$ and $\tilde{b}_r$ are as the same as those in Theorem 5.11. Then $\text{Nec}\{\tilde{a}_r^T x \leq \tilde{b}_r\} \geq \theta_r$ is equivalent to

$$b_r - L^{-1}(1 - \theta_r)\alpha_r^b \geq a_r^T x + R^{-1}(\theta_r)\beta_r^{aT} x, \quad r = 1, 2, \cdots, p \qquad (5.96)$$

PROOF   The proof is similar as the proof of Theorem 5.11.

REMARK 5.15   Especially when the reference function of the variables $\tilde{a}_{rj}, \tilde{b}_r$ are all $L(x) = R(x) = 1 - x$ ($x \in [0,1]$), then the LR fuzzy variables are specified as the triangular fuzzy variables, and $R^{-1}(\delta_i) = 1 - \delta_i$, so we have the following equivalent expressions:

$$\text{Nec}\{\tilde{a}_r^T x \leq \tilde{b}_r\} \geq \theta_r \Leftrightarrow b_r - \theta_r \alpha_r^b \geq a_r^T x + (1 - \theta_r)\beta_r^{aT} x, r = 1, 2, \cdots, p \quad (5.97)$$

Then, according to the Theorems 5.10, 5.11, 5.12, and 5.13, the equivalent models for the linear lower approximation model (5.85) and the upper approximation model (5.86) are proposed as follows:

$$(E\text{-}LLAM) \begin{cases} \max[\bar{f}_1, \bar{f}_2, \cdots, \bar{f}_m] \\ \text{s.t.} \begin{cases} \bar{f}_i \leq c_i^T x - L^{-1}(1 - \delta_i)\alpha_i^{cT} x, \quad i = 1, 2, \cdots, m \\ b_r - L^{-1}(1 - \theta_r)\alpha_r^b - a_r^T x - R^{-1}(\theta_r)\beta_r^{aT} x \geq 0, r = 1, 2, \cdots, p \\ x \geq 0 \end{cases} \end{cases}$$

$$(5.98)$$

$$(E\text{-}LUAM) \begin{cases} \max[\bar{f}_1, \bar{f}_2, \cdots, \bar{f}_m] \\ \text{s.t.} \begin{cases} \bar{f}_i \leq c_i^T x + R^{-1}(\delta_i)\beta_i^{cT} x, \quad i = 1, 2, \cdots, m \\ b_r + R^{-1}(\theta_r)\beta_r^b - a_r^T x + L^{-1}(\theta_r)\alpha_r^{aT} x \geq 0, \ r = 1, 2, \cdots, p \\ x \geq 0 \end{cases} \end{cases}$$

$$(5.99)$$

If all the fuzzy variables are triangular fuzzy variables, then models (5.98) and (5.99) can be rewritten as

$$(E\text{-}LLAM) \begin{cases} \max[\bar{f}_1, \bar{f}_2, \cdots, \bar{f}_m] \\ \text{s.t.} \begin{cases} \bar{f}_i \leq c_i^T x - \delta_i \alpha_i^{cT} x, \quad i = 1, 2, \cdots, m, \\ b_r - \theta_r \alpha_r^b \geq a_r^T x + (1 - \theta_r)\beta_r^{aT} x, \quad r = 1, 2, \cdots, p \\ x \geq 0 \end{cases} \end{cases} \qquad (5.100)$$

$$(E\text{-}LUAM) \quad \begin{cases} \max[\bar{f}_1, \bar{f}_2, \cdots, \bar{f}_m] \\ \text{s.t.} \begin{cases} c_i^T x + (1-\delta_i)\beta_i^{cT} x \geq \bar{f}_i, & i = 1,2,\cdots,m \\ b_r + (1-\theta_r)\beta_r^b \geq a_r^T x - (1-\theta_r)\alpha_r^{aT} x, & r = 1,2,\cdots,p \\ x \geq 0 \end{cases} \end{cases}$$

$$(5.101)$$

5.4.2.2   The Step Method

In this section we use the step method, which is also called the STEM method and is the interactive programming method to deal with the multiobjective programming problem [27].

The STEM method is based on the norm ideal point method and its resolving process includes the analysis and decision stage. In the analysis stage, the analyzer resolves the problem by the norm ideal point method and provides the decision makers with the solutions and the related objective values and ideal objective values. At the decision stage, the DM gives the tolerance level of the satisfied object to the dissatisfied object to make its objective value better after comparing the objective values obtained in the analysis stage with the ideal point, and then provides the analyzer with the information to continue resolving the problem. Do it repeatedly and the DM will get the final satisfied solution.

Shimizu extended the STEM method to deal with the general nonlinear multiobjective programming problem. Interested readers can refer to this literature [263] and others [226, 281] regarding its further development.

Consider the following multiple objective decision-making problem:

$$\min_{x \in X} f(\mathbf{x}) = (f_1(\mathbf{x}), f_2(\mathbf{x}), \cdots, f_m(\mathbf{x})) \quad (5.102)$$

where $\mathbf{x} = (x_1, x_2, \cdots, x_n)$ and $X = \{\mathbf{x} \in \mathbf{R}^n | A\mathbf{x} = b, \mathbf{x} \geq 0\}$. Let $\mathbf{x}^i$ be the optimal solution to the problem $\min_{\mathbf{x} \in X} f_i(\mathbf{x})$, and compute each objective function $f_i(\mathbf{x})$ at $\mathbf{x}^k$, and then we get the m^2 objective function value

$$f_{ik} = f_i(\mathbf{x}^k), \quad i,k = 1,2,\cdots,m.$$

Denote $f_i^* = f_{ii} = f_i(\mathbf{x}^i)$, $\boldsymbol{f}^* = (f_1^*, f_2^*, \cdots, f_m^*)^T$, and f_i^* is a ideal point of problem (5.102). Compute the maximum value of the objective function $f_i(\mathbf{x})$ at every minimum point $\mathbf{x}^k$

$$f_i^{\max} = \max_{1 \leq k \leq m} f_{ik}, i = 1,2,\cdots,m.$$

To make it more clear, we list it in Table 5.1.

According to Table 5.1, we only look for solution $\boldsymbol{x}$ such that the distance between $f(\boldsymbol{x})$ and $\boldsymbol{f}^*$ is minimum, that is, such that each objective is close to the ideal point. Consider the problem

$$\min_{\mathbf{x} \in X} \max_{1 \leq i \leq m} w_i |f_i(\boldsymbol{x}) - f_i^*| = \min_{\mathbf{x} \in X} \max_{1 \leq i \leq m} w_i \left| \sum_{j=1}^{n} c_{ij} x_j - f_i^* \right| \quad (5.103)$$

TABLE 5.1   Payoff table

f	$\mathbf{x}^1$	$\cdots$	$\mathbf{x}^i$	$\cdots$	$\mathbf{x}^m$	max
f_1	$f_{11} = f_1^*$	$\cdots$	f_{1i}	$\cdots$	f_{1m}	$f_1^{\max}$
$\vdots$	$\vdots$	$\vdots$	$\vdots$	$\vdots$	$\vdots$	$\vdots$
f_i	f_{i1}	$\cdots$	$f_{ii} = f_i^*$	$\cdots$	f_{im}	$f_i^{\max}$
$\vdots$	$\vdots$	$\vdots$	$\vdots$	$\vdots$	$\vdots$	$\vdots$
f_m	f_{m1}	$\cdots$	f_{mi}	$\cdots$	$f_{mm} = f_m^*$	$f_m^{\max}$

where $\boldsymbol{w} = (w_1, w_2, \cdots, w_m)^T$ is the weight vector, and w_i is the ith weight, which can be decided as follows:

$$\alpha_i = \begin{cases} \frac{f_i^{\max} - f_i^*}{f_i^{\max}} \frac{1}{\|\boldsymbol{c}_i\|}, & f_i^{\max} > 0 \\ \frac{f_i^* - f_i^{\max}}{f_i^{\max}} \frac{1}{\|\boldsymbol{c}_i\|}, & f_i^{\max} \leq 0 \end{cases} \tag{5.104}$$

$$w_i = \alpha_i / \sum_{i=1}^m \alpha_i \tag{5.105}$$

for $i = 1, 2, \cdots, m$, where $\|\boldsymbol{c}_i\| = \sqrt{\sum_{j=1}^n c_{ij}^2}$. Then problem (5.102) is equivalent to

$$\begin{cases} \min \lambda \\ \text{s.t.} \begin{cases} w_i \left(\sum_{j=1}^n c_{ij} x_j - f_i^* \right) \leq \lambda, & i = 1, 2, \cdots, m \\ \lambda \geq 0 \\ \mathbf{x} \in X \end{cases} \end{cases} \tag{5.106}$$

Assume that the optimal solution to problem (5.106) is $(\tilde{\mathbf{x}}, \tilde{\lambda})^T$. It is obvious that $(\tilde{\mathbf{x}}, \tilde{\lambda})^T$ is a weak efficient solution to problem (5.102). In order to check if $\tilde{\mathbf{x}}$ is satisfied, the DM needs to compare $f_i(\tilde{\mathbf{x}})$ with the ideal objective value $f_i^*, i = 1, 2, \cdots, m$. If the DM has been satisfied with $f_s(\tilde{\mathbf{x}})$ but dissatisfied with $f_t(\tilde{\mathbf{x}})$, we add the following constraint in the next step in order to improve the objective value f_t:

$$f_t(\mathbf{x}) \leq f_t(\tilde{\mathbf{x}}).$$

For the satisfied object f_s, we add one tolerance level δ_s,

$$f_s(\boldsymbol{x}) \leq f_s(\tilde{\mathbf{x}}) + \delta_s.$$

Thus, in problem (5.106), we replace X with the following constraint set:

$$X^1 = \{\mathbf{x} \in X | f_s(\mathbf{x}) \leq f_s(\tilde{\mathbf{x}}) + \delta_s, f_t(\mathbf{x}) \leq f_t(\tilde{\mathbf{x}})\},$$

and delete the objective f_s (do it by letting $w_s = 0$), and then resolve the new problem to get better solutions.

The STEM method can be summarized as follows:

Step 1. Compute every single objective programming problem:

$$f_i(\mathbf{x}^i) = \min_{\mathbf{x} \in X} f_i(\mathbf{x}), \ i = 1, 2, \cdots, m.$$

If $\mathbf{x}^1 = \cdots = \mathbf{x}^m$, we obtain the optimal solution $\mathbf{x}^* = \mathbf{x}^1 = \cdots = \mathbf{x}^m$ and stop.

Step 2. Compute the objective value of $f_i(\mathbf{x})$ at every minimum point $\mathbf{x}^k$, then get m^2 objective values $f_{ik} = f_i(\mathbf{x}^k)(i, k = 1, 2, \cdots, m)$. List Table 5.1, and we have

$$f_i^* = f_{ii}, \ f_i^{\max} = \max_{1 \le k \le m} f_{ik}, \ i = 1, 2, \cdots, m.$$

Step 3. Give the initial constraint set and let $X^1 = X$.

Step 4. Compute the weight coefficients $w_1, w_2, \cdots, w_m$ by Equtions (5.104) and (5.105).

Step 5. Solve the auxiliary problem:

$$\begin{cases} \min \lambda \\ \text{s.t.} \begin{cases} w_i \left(\sum_{j=1}^{n} c_{ij} x_j - f_i^* \right) \le \lambda, \ i = 1, 2, \cdots, m \\ \lambda \ge 0 \\ \mathbf{x} \in X^k \end{cases} \end{cases} \quad (5.107)$$

Let the optimal of problem (5.107) be $(\mathbf{x}^k, \lambda^k)^T$.

Step 6. The DM compare the reference value $f_i(\mathbf{x}^k)(i = 1, 2, \cdots, m)$ with the ideal objective value f_i^*. (1) If the DM is satisfied with all objective values, output $\tilde{\mathbf{x}} = \mathbf{x}^k$. (2) If the DM is dissatisfied with all objective values, no satisfied solutions exist and therefore stop the process. (3) If the DM is satisfied with the object $f_{s_k}(1 \le s_k \le m, k < m)$, turn to **Step 7**.

Step 7. The DM gives the tolerance level $\delta_{s_k} > 0$ to the object f_{s_k} and constructs the new constraint set as follows:

$$X^{k+1} = \{\mathbf{x} \in X^k | f_{s_k}(\mathbf{x}) \le f_{s_k}(\mathbf{x}^k) + \delta_{s_k}, f_i(\mathbf{x}) \le f_i(\mathbf{x}^k), i \ne s_k\}.$$

Let $\delta_{s_k} = 0, k = k + 1$, and turn to **Step 4**.

5.4.2.3 Numerical Example

Example 5.8

Consider the problem

$$\begin{cases} \max \xi_1 x_1 + \xi_2 x_2 + \xi_3 x_3 + \xi_4 x_4 \\ \max \xi_5 x_1 + \xi_6 x_2 + \xi_7 x_3 + \xi_8 x_4 \\ \text{s.t.} \begin{cases} \xi_5 x_1 + \xi_6 x_2 + \xi_7 x_3 + \xi_8 x_4 \le \xi_9 \\ 3x_1 + 5x_2 + 2x_3 + 7x_4 \le 120 \\ x_j \ge 0, j = 1, 2, 3, 4 \end{cases} \end{cases} \quad (5.108)$$

wherein, the coefficients are triangular fuzzy variables:

$$\xi_1 = (3,4,5), \ \xi_2 = (5,6,7), \ \xi_3 = (8,9,10),$$
$$\xi_4 = (2,4,6), \ \xi_5 = (7,8,9), \ \xi_6 = (4,5,6),$$
$$\xi_7 = (2,3,4), \ \xi_8 = (1,2,3), \ \xi_9 = (120,160,200).$$

In order to solve it, we use the chance operator and rough approximation technique to deal with fuzzy objectives and fuzzy constraints, then we can obtain the LAM and UAM as follows:

$$(LAM) \begin{cases} \max[\bar{\bar{f}}_1, \bar{\bar{f}}_2] \\ \text{s.t.} \begin{cases} Nec\{\xi_1 x_1 + \xi_2 x_2 + \xi_3 x_3 + \xi_4 x_4 \geq \bar{\bar{f}}_1\} \geq \delta_1 \\ Nec\{\xi_5 x_1 + \xi_6 x_2 + \xi_7 x_3 + \xi_8 x_4 \geq \bar{\bar{f}}_2\} \geq \delta_2 \\ Nec\{\xi_5 x_1 + \xi_6 x_2 + \xi_7 x_3 + \xi_8 x_4 \leq \xi_9\} \geq \theta \\ 3x_1 + 5x_2 + 2x_3 + 7x_4 \leq 120 \\ x_j \geq 0, j = 1,2,3,4 \end{cases} \end{cases} \tag{5.109}$$

and

$$(UAM) \begin{cases} \max[\bar{\bar{f}}_1, \bar{\bar{f}}_2] \\ \text{s.t.} \begin{cases} Pos\{\xi_1 x_1 + \xi_2 x_2 + \xi_3 x_3 + \xi_4 x_4 \geq \bar{\bar{f}}_1\} \geq \delta_1 \\ Pos\{\xi_5 x_1 + \xi_6 x_2 + \xi_7 x_3 + \xi_8 x_4 \geq \bar{\bar{f}}_2\} \geq \delta_2 \\ Pos\{\xi_5 x_1 + \xi_6 x_2 + \xi_7 x_3 + \xi_8 x_4 \leq \xi_9\} \geq \theta \\ 3x_1 + 5x_2 + 2x_3 + 7x_4 \leq 120 \\ x_j \geq 0, j = 1,2,3,4 \end{cases} \end{cases} \tag{5.110}$$

Let confidence level $\delta_1 = \delta_2 = \theta = 0.8$. By models (5.100) and (5.101), we know that models (5.109) and (5.110) have equivalent models as follows:

$$(E\text{-}LLAM) \begin{cases} \max[\bar{\bar{f}}_1, \bar{\bar{f}}_2] \\ \text{s.t.} \begin{cases} 3.2x_1 + 5.2x_2 + 8.2x_3 + 2.4x_4 \geq \bar{\bar{f}}_1 \\ 7.2x_1 + 4.2x_2 + 2.2x_3 + 0.4x_4 \geq \bar{\bar{f}}_2 \\ 8.2x_1 + 5.2x_2 + 3.2x_3 + 2.2x_4 \leq 128 \\ 3x_1 + 5x_2 + 2x_3 + 7x_4 \leq 120 \\ x_j \geq 0, j = 1,2,3,4 \end{cases} \end{cases} \tag{5.111}$$

$$(E\text{-}LUAM) \begin{cases} \max[\bar{\bar{f}}_1, \bar{\bar{f}}_2] \\ \text{s.t.} \begin{cases} 4.2x_1 + 6.2x_2 + 9.2x_2 + 4.4x_4 \geq \bar{\bar{f}}_1 \\ 8.2x_1 + 5.2x_2 + 3.2x_3 + 2.2x_4 \geq \bar{\bar{f}}_2 \\ 7.8x_1 + 4.8x_2 + 2.8x_2 + 1.8x_4 \leq 168 \\ 3x_1 + 5x_2 + 2x_3 + 7x_4 \leq 120 \\ x_j \geq 0, j = 1,2,3,4 \end{cases} \end{cases} \tag{5.112}$$

Then we use the step method to solve the above problem. For (5.111), first, compute the optimal solution and value of each objective function as follows:

$$f_1^* = 328, \mathbf{x}^1 = (0,0,40,0), f_2^* = 112.4, \mathbf{x}^2 = (15.6,0,0,0).$$

TABLE 5.2   Payoff table
of Example 5.8

f	x^1	x^2	min
f_1	328	88	$f_1^{\min} = 88$
f_2	50	112.4	$f_2^{\min} = 64$

By the step method, we get the payoff table as shown in Table 5.2. Second, compute the weight coefficient by the following method:

$$\alpha_1 = \frac{f_1^* - f_1^{\min}}{f_1^{\min}} \frac{1}{\sqrt{3.2^2 + 5.2^2 + 8.2^2 + 2.4^2}} = 0.070,$$

$$\alpha_2 = \frac{f_2^* - f_2^{\min}}{f_2^*} \frac{1}{\sqrt{7.2^2 + 4.2^2 + 2.2^2 + 0.4^2}} = 0.064.$$

Then we get $w_1 = \alpha_1/(\alpha_1 + \alpha_2) = 0.522$ and $w_2 = \alpha_2/(\alpha_1 + \alpha_2) = 0.478$.
Problem (5.111) can be rewritten as

$$\begin{cases} \min \lambda \\ \text{s.t.} \begin{cases} 0.522[328 - (3.2x_1 + 5.2x_2 + 8.2x_3 + 2.4x_4)] \leq \lambda \\ 0.478[112.4 - (7.2x_1 + 4.2x_2 + 2.2x_3 + 0.4x_4)] \leq \lambda \\ 8.2x_1 + 5.2x_2 + 3.2x_3 + 2.2x_4 \leq 128 \\ 3x_1 + 5x_2 + 2x_3 + 7x_4 \leq 120 \\ x_j \geq 0, j = 1,2,3,4 \end{cases} \end{cases} \quad (5.113)$$

We obtain $x = (1.16, 0, 37.0, 0)$ and $f = (307.3, 89.8)$.

For (5.112), first, compute the optimal solution and value of each objective function as follows:

$$f_1^* = 552, \mathbf{x}^1 = (0,0,60,0), f_2^* = 192, \mathbf{x}^2 = (0,0,60,0).$$

Then $\mathbf{x}^* = (0,0,60,0)$ is the only optimal solution, and $f^* = (552.192)$ is the absolute optimal objective value. It is obvious that $(552, 192)$ is more efficient than $(307.3, 89.8)$.

5.4.3   Fu-NLCCRM and Fuzzy Simulation-Based ASA

As problem (5.78) has nonlinear forms, that is, there is at least one nonlinear function of f_i, g_r. By using the rough approximation technique, we can obtain the fuzzy nonlinear chance-constrained rough model (Fu-NLCCRM). It is usually impossible to be converted into the crisp equivalent model. Fuzzy simulation-based adaptive simulated annealing (ASA), a solving method, is used to deal with it.

5.4.3.1   Fuzzy Simulation 2 Critical Value

Suppose that f is a real-valued function, and ξ_i a fuzzy vector defined on the possibility space $(\Theta, P(\Theta), Pos)$. Let us find the maximal $\bar{f}$ such that the inequality

$$Me\{f(\xi) \geq \bar{f}\} \geq \alpha \quad (5.114)$$

holds. We randomly generate θ_k from Θ such that $Pos\{\theta_k\} \geq \varepsilon$ and write $v_k = Pos\{\theta_k\}, k = 1, 2, \cdots, N$, where ε is a sufficiently small number. For any number r, we have

$$L(r) = \lambda \max_{1 \leq k \leq N} \{v_k | f(\xi(\theta_k)) \geq r\} + (1 - \lambda) \min_{1 \leq k \leq N} \{1 - v_k | f(\xi(\theta_k)) \leq r\},$$

where λ is the optimism coefficient. It follows from monotonicity that we may employ bisection search to find the maximal value r such that $L(r) \geq \alpha$. This value is an estimation of $\bar{f}$. We summarize this process as follows:

Step 1. Generate θ_k from Θ such that $Pos\{\theta_k\} \geq \varepsilon$, where ε is a sufficiently small number.

Step 2. Find the maximal value r such that $L(r) \geq r$ holds.

Step 3. Return r.

Example 5.9

We assume that, ξ_1, ξ_2, and ξ_3 are fuzzy variables defined as $\xi_1 = (10, 11, 12), \xi_2 = (12, 13, 14), \xi_3 = (13, 14, 15)$. A run of fuzzy simulation with 1000 cycles shows that the maximal $\bar{f}$ satisfying

$$Me\{\xi_1\xi_2\xi_3 \geq \bar{f}\} \geq 0.8 = \begin{cases} 1575.3, & \lambda = 0 \\ 1577.0, & \lambda = 0.2 \\ 1581.9, & \lambda = 0.5 \\ 2161.6, & \lambda = 0.8 \\ 2275.5, & \lambda = 1 \end{cases}$$

5.4.3.2 ASA

Too often the management of complex systems is ill-served by not utilizing the best tools available. For example, requirements set by decision makers often are not formulated in the same language as the constructs formulated by powerful mathematical formalisms, and so the products of analyses are not properly or maximally utilized even if and when they come close to faithfully representing the powerful intuitions they are supposed to model. In turn, even powerful mathematical constructs are ill-served, especially when dealing with approximations to satisfy constraints of numerical algorithms familiar to particular analysts but which tend to destroy the power of the intuitive constructs developed by decision makers. In order to deal with fitting parameters or exploring sensitivities of variables, as models of systems have become more sophisticated in describing complex behavior, it has become increasingly important to retain and respect the nonlinearities inherent in these models, as they are indeed present in the complex systems they model. The adaptive simulated annealing (ASA) algorithm can help to handle these fits of nonlinear models of real-world data.

ASA, also known as very fast simulated reannealing, is a very efficient version of SA. ASA is a global optimization technique with certain advantages. It is versatile and needs very few parameters to tune. The distinct feature of this method is the temperature change mechanism, which is an important part of the transition probability equation. The conventional approach allows a higher chance of transition to a worse solution when beginning with a high temperature. By doing so, the search can move out of local optimization. However, as the search process develops, the continuously declining temperature will result in a reduced chance of uphill transition. Such an approach could be useful if local optimization is near the start point, but may not lead to a near-optimal solution if some local optimization is encountered at a relatively low temperature toward the end of the search. Therefore, an improved method should be considered to alleviate this difficulty. In the conventional method, the cooling schedule is usually monotonically nonincreasing. In ASA, an adaptive cooling schedule based on the profile of the search path to dynamically adjust the temperature is available. Such adjustments could be in any direction, including the possibility of reheating. Some scholars proposed the idea of reversing the temperature earlier. K. Dowsland [81] considered two functions together to control the temperature in the application to packing problem. The first function is a function that reduces the temperature, whereas the second function is used as a heating-up function that gradually increases the temperature if needed. In ASA, a single function is proposed to maintain the temperature above a minimum level. If there is any upward move, the heating process gradually takes place, but the cooling process may suddenly take place by the first downhill move. N. Azizi [12] proposed the following temperature control function:

$$T_i = T_{\min} + \lambda ln(1 + r_i) \qquad (5.115)$$

where $T_{\min}$ is the minimum value that the temperature can take, λ is a coefficient that controls the rate of temperature rise, and r_i is the number of consecutive upward moves at iteration i. The initial value of r_i is zero, and thus the initial temperature $T_0 = T_{\min}$. The purpose of the minimum temperature, $T_{\min}$, is two-fold. First, it prevents the probability function from becoming invalid when r_i is zero. Second, it determines the initial value of the temperature. $T_{\min}$ can take any value greater than zero. The parameter λ controls the rate of temperature rise. The greater the value of λ, the faster the temperature rise. The search spends less time looking for good solutions in its current neighborhood when a large value is assigned to λ. Similarly, by assigning a small value to the parameter λ, the search spends more time looking for better solutions in the neighborhood. Choosing a value for the parameter λ could be linked to computation time, which also depends on the size and complexity of the problem. We do not clarify other parts aspects of ASA here; readers can find more detailed analysis of the algorithm in [12, 59, 144–146].

Consider the following general optimization problem

$$\min_{\mathbf{x} \in X} f(\mathbf{x}) \qquad (5.116)$$

where $\mathbf{x} = [x_1, \cdots, x_n]^T$ is the n-dimensional decision vector to be optimized, X is the feasible set of $\mathbf{x}$. The cost function $f(\mathbf{x})$ can be multimodal and nonsmooth. ASA

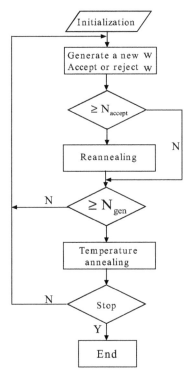

FIGURE 5.5 Flowchart of the adaptive simulated annealing.

is a global optimization scheme for solving such type of constrained optimization problems. Next, let us introduce how to use ASA to solve the above problem.

Although there are many possible realizations of the ASA, an implementation is illustrated in Figure 5.5, and this algorithm is detailed as follows:

Step 1. An initial $\mathbf{x} \in X$ is randomly generated, the initial temperature of the acceptance probability function, $T_{accept}(0)$, is set to $f(\mathbf{x})$, and the initial temperatures of the parameter generating probability functions, $T_{i,gen}(0), 1 < i < n$, are set to 1.0. A user-defined control parameter c in annealing is given, and the annealing times, k_i for $1 \le i \le n$ and k_a, are all set to 0.

Step 2. The algorithm generates a new point in the parameter space with

$$x_i^{new} = x_i^{old} + q_i M_i, \ 1 \le i \le n \tag{5.117}$$

where M_i is a positive number determined by the difference between the upper and lower boundary of x_i and

$$\mathbf{x}^{new} \in X \tag{5.118}$$

where q_i is calculated as

$$q_i = sgn\left(v_i - \frac{1}{2}\right) T_{i,gen}(k_i) \times \left(\left(1 + \frac{1}{T_{i,gen}(k_i)}\right)^{|2v_i - 1|} - 1\right) \tag{5.119}$$

and v_i is a uniformly distributed random variable in $[0,1]$. Notice that if a generated $\mathbf{x}^{new}$ is not in X, it is simply discarded, and a new point is tried again until $\mathbf{x}^{new} \in X$.

The value of the cost function $f(\mathbf{x}^{new})$ is then evaluated, and the acceptance probability function of $\mathbf{x}^{new}$ is given by

$$P_{accept} = \frac{1}{1 + exp((f(\mathbf{x}^{new}) - f(\mathbf{x}^{old}))/T_{accept}(K_a))} \tag{5.120}$$

A uniform random variable P_{unif} is generated in $[0,1]$. If $P_{unif} \leq P_{accept}$, $\mathbf{x}^{new}$ is accepted; otherwise it is rejected.

Step 3. After every N_{accept} acceptance points, reannealing takes place by first calculating the sensitives

$$s_i = \left| \frac{f(\mathbf{x}^{best} + e_i \delta) - f(\mathbf{x}^{best})}{\delta} \right|, \ 1 \leq i \leq n \tag{5.121}$$

where $\mathbf{x}^{best}$ is the best point found so far, δ is a small step size, the n-dimensional vector e_i has unit ith element, and the rest of the elements of e_i are all zeros. Let $s_{max} = max\{s_i, 1 \leq i \leq n\}$. Each parameter generating temperature $T_{i,gen}$ is scaled by a factor s_{max}/s_i and the annealing time k_i is reset:

$$T_{i,gen}(k_i) = \frac{s_{max}}{s_i} T_{i,gen}(k_i), \ k_i = \left(-\frac{1}{c} log\left(\frac{T_{i,gen}(k_i)}{T_{i,gen}(0)}\right)\right)^n.$$

Similarly, $T_{accept}(0)$ is reset to the value of the last accepted cost function $T_{accept}(k_a)$ is reset to $f(\mathbf{x}^{best})$, and the annealing time k_a is rescaled accordingly:

$$k_a = \left(-\frac{1}{c} log\left(\frac{T_{i,gen}(k_a)}{T_{i,gen}(0)}\right)\right)^n \tag{5.122}$$

Step 4. After every N_{gen} generated points, annealing takes place with

$$k_i = k_i + 1, \ T_{i,gen}(k_i) = T_{i,gen}(0) exp(-ck_i^{1/n})$$

and

$$k_a = k_a + 1, \ T_{accept}(k_a) = T_{accept}(0) exp(-ck_a^{1/n}).$$

Otherwise, go to **Step 5.**

Step 5. The algorithm is terminated if the parameters have remained unchanged for a few successive reannealing or a preset maximum number of cost function evaluations has been reached; otherwise, go to **Step 2.**

As in a standard SA algorithm, the ASA contains two loops. The inner loop ensures that the parameter space is searched sufficiently at a given temperature, which is necessary to guarantee that the algorithm finds a global optimum. The ASA also uses only the value of the cost function in the optimization process, and it is very simple to program.

Last, we discuss algorithm parameter tuning. For the foregoing ASA algorithm, most of the algorithm parameters are automatically set and "tuned", and the user only needs to assign a control parameter c and set two values N_{accept} and N_{genera}. Obviously, the optimal values of N_{accept} and N_{genera} are problem dependent, but our experience suggests that an adequate choice for N_{accept} is in the range of tens to hundreds, and an appropriate value for N_{gen} is in the range of hundreds to thousands. The annealing rate control parameter c can be determined form the chosen initial temperature, final temperature, and predetermined number of annealing steps. We have found that a choice of c in the range 1.0 to 10.0 is often adequate.

It should be emphasized that, as ASA has excellent self-adaptation ability, the performance of the algorithm is not critically influenced by the specific chosen values of c, N_{accept} and N_{gen}.

5.4.3.3 Numerical Example

Example 5.10

Consider the problem

$$\begin{cases} \max f_1(\mathbf{x}, \boldsymbol{\xi}) = \xi_1 x_1 + \xi_2^2 x_2 + \sqrt{\xi_3} x_3 \\ \max f_2(\mathbf{x}, \boldsymbol{\xi}) = \xi_4 x_1 + \xi_5 x_2 + \xi_6 x_3 \\ \text{s.t.} \begin{cases} \sqrt{\xi_1} x_1 + \xi_2 x_2 + \xi_3^2 \sqrt{x_3} \leq \xi_7 \\ x_1 + x_2 + x_3 \leq 100 \\ x_i \geq 10, i = 1, 2, 3 \end{cases} \end{cases} \tag{5.123}$$

wherein, the coefficients are triangular fuzzy variables characterized as

$$\xi_1 = (3.5, 4, 4.5), \ \xi_2 = (4.5, 5, 5.5), \ \xi_3 = (8.5, 9, 9.5), \qquad \xi_4 = (10.5, 11, 11.5),$$
$$\xi_5 = (2.5, 3, 3.5), \ \xi_6 = (4.5, 5, 5.5), \ \xi_7 - (150, 200, 250).$$

In order to solve it, we use the Fu-CCRM to deal with it, and then we can obtain the LAM and UAM as follows:

$$(LAM) \begin{cases} \max[\bar{f}_1, \bar{f}_2] \\ \text{s.t.} \begin{cases} Nec\{\xi_1 \sqrt{x_1} + \xi_2^2 x_2 + \sqrt{\xi_3} x_3^2 \geq \bar{f}_1\} \geq 0.8 \\ Nec\{\xi_4 x_1 + \xi_5 x_2 + \xi_6 x_3 \geq \bar{f}_2\} \geq 0.8 \\ Nec\{\sqrt{\xi_1} x_1 + \xi_2 x_2 + \xi_3^2 \sqrt{x_3} \leq \xi_7\} \geq 0.8 \\ x_1 + x_2 + x_3 \leq 100 \\ x_i \geq 10, i = 1, 2, 3 \end{cases} \end{cases} \tag{5.124}$$

and

$$(UAM) \quad \begin{cases} \max[\bar{f}_1, \bar{f}_2] \\ \text{s.t.} \begin{cases} Pos\{\xi_1\sqrt{x_1} + \xi_2^2 x_2 + \sqrt{\xi_3} x_3^2 \geq \bar{f}_1\} \geq 0.8 \\ Pos\{\xi_4 x_1 + \xi_5 x_2 + \xi_6 x_3 \geq \bar{f}_2\} \geq 0.8 \\ Pos\{\sqrt{\xi_1} x_1 + \xi_2 x_2 + \xi_3^2 \sqrt{x_3} \leq \xi_7\} \geq 0.8 \\ x_1 + x_2 + x_3 \leq 100 \\ x_i \geq 10, i = 1, 2, 3 \end{cases} \end{cases} \qquad (5.125)$$

Let $\lambda = 0.5$. Next we will use the random simulation-based simulated algorithm to solve the above problem. Set the initial temperature T_0, let the last temperature be 1, and the cooling method be 0.05% decrement once. The neighborhood can be constructed as follows:

$$x_1^1 = x_1^0 + rh, x_1^2 = x_1^0 + rh, x_1^3 = x_3^0 + rh,$$

where r is a random number in $(0,1)$ and h is the step length (here $h = 2.0$). After running the fuzzy simulation-based PSA, we obtain $\mathbf{x}^* = (10.6, 19.0, 16.9)$ and $F = (767.6, 238.6)$ for (5.125).

5.5 Fu-DCRM

The third way deal with model (5.29) is DCM. As rough approximation techniques are used to tackle the fuzzy DCM, we can get the fuzzy dependent-chance rough model (Fu-DCRM).

5.5.1 General Model of Fu-DCRM

The general fuzzy DCM is formulated as

$$\begin{cases} \max \ [Ch\{f_1(\mathbf{x}, \boldsymbol{\xi}) \geq \bar{f}_1\}, Ch\{f_2(\mathbf{x}, \boldsymbol{\xi}) \geq \bar{f}_2\}, \cdots, Ch\{f_m(\mathbf{x}, \boldsymbol{\xi}) \geq \bar{f}_m\}] \\ \text{s.t.} \begin{cases} Ch\{g_r(\mathbf{x}, \boldsymbol{\xi}) \leq 0\} \geq \theta_r, \ r = 1, 2, \cdots, p \\ \mathbf{x} \in X \end{cases} \end{cases} \qquad (5.126)$$

where Ch represents fuzzy measures such as Pos, Nec, Me, and Cr; $\bar{f}_i$ are aspiration level for the ith objectives; and θ_r are the predetermined confidence levels. Obviously, model (5.126) can be equivalent to the following models:

$$\begin{cases} \max \ [\delta_1, \delta_2, \cdots, \delta_m] \\ \text{s.t.} \begin{cases} Ch\{f_1(\mathbf{x}, \boldsymbol{\xi}) \geq \bar{f}_1\} \geq \delta_i, i = 1, 2, \cdots, m \\ Ch\{g_r(\mathbf{x}, \boldsymbol{\xi}) \leq 0\} \geq \theta_r, r = 1, 2, \cdots, p \\ \mathbf{x} \in X \end{cases} \end{cases} \qquad (5.127)$$

For model (5.127), we let $D = \{\mathbf{x}|\mathbf{x} \in X, Me\{f_1(x,\xi) \geq \bar{f}_1\} \geq \delta_i, i = 1,2,\cdots,m,$
$Me\{g_r(\mathbf{x},\xi) \leq 0\} \geq \theta_r, r = 1,2,\cdots,p\}$, we construct two sets L and U as follows:

$$\begin{cases} L = \{\mathbf{x}|\mathbf{x} \in X, Nec\{f_i(\mathbf{x},\xi) \geq \bar{f}_i\} \geq \delta_i, i = 1,2,\cdots,m, Nec\{g_r(\mathbf{x},\xi) \leq 0\} \geq \theta_r, \\ \quad r = 1,2,\cdots,p\} \\ U = \{\mathbf{x}|\mathbf{x} \in X, Pos\{f_i(\mathbf{x},\xi) \geq \bar{f}_i\} \geq \delta_i, i = 1,2,\cdots,m, Pos\{g_r(\mathbf{x},\xi) \leq 0\} \geq \theta_r, \\ \quad r = 1,2,\cdots,p\} \end{cases}$$

(5.128)

It follows from Theorem 5.9 that $L \subseteq D \subseteq U$.

By rough approximation techniques, model (5.127) can be transformed into two models, LAM and UAM:

$$(LAM) \quad \begin{cases} \max\ [\delta_1,\delta_2,\cdots,\delta_m] \\ \text{s.t.} \begin{cases} Nec\{f_1(\mathbf{x},\xi) \geq \bar{f}_1\} \geq \delta_i, i = 1,2,\cdots,m \\ Nec\{g_r(\mathbf{x},\xi) \leq 0\} \geq \theta_r, r = 1,2,\cdots,p \\ \mathbf{x} \in X \end{cases} \end{cases}$$

(5.129)

and

$$(UAM) \quad \begin{cases} \max\ [\delta_1,\delta_2,\cdots,\delta_m] \\ \text{s.t.} \begin{cases} Pos\{f_1(\mathbf{x},\xi) \geq \bar{f}_1\} \geq \delta_i, i = 1,2,\cdots,m \\ Pos\{g_r(\mathbf{x},\xi) \leq 0\} \geq \theta_r, r = 1,2,\cdots,p \\ \mathbf{x} \in X \end{cases} \end{cases}$$

(5.130)

DEFINITION 5.13 $\mathbf{x}^*$ *is said to be a feasible solution at θ_r-necessity levels to problem (5.129) if and only if it satisfies $Nec\{g_r(\mathbf{x},\xi) \leq 0\} \geq \theta_r, r = 1,2,\cdots,p$.*

DEFINITION 5.14 *A feasible solution at θ_r-necessity levels, $\mathbf{x}^*$, is said to be a efficient solution to problem (5.129) if and only if there exists no other feasible solution at θ_r-necessity levels x, such that $Nec\{f_i(\mathbf{x},\xi) \geq \bar{f}_i\} \geq Nec\{f_i(\mathbf{x}^*,\xi) \geq \bar{f}_i\}$ for all i and $Nec\{f_{i_0}(\mathbf{x},\xi) \geq \bar{f}_{i_0}\} \geq Nec\{f_{i_0}(\mathbf{x}^*,\xi) \geq \bar{f}_{i_0}\}$ for at least one $i_0 \in \{1,2,\cdots,m\}$.*

DEFINITION 5.15 $\mathbf{x}^*$ *is said to be a feasible solution at θ_r-possibility levels o problem (5.130) if and only if it satisfies $Pos\{g_r(\mathbf{x},\xi) \leq 0\} \geq \theta_r, r = 1,2,\cdots,p$.*

DEFINITION 5.16 *A feasible solution at θ_r-possibility levels, $\mathbf{x}^*$, is said to be an efficient solution to problem (5.130) if and only if there exists no other feasible solution at θ_r-possibility levels x, such that $Pos\{f_i(\mathbf{x},\xi) \geq \bar{f}_i\} \geq Pos\{f_i(\mathbf{x}^*,\xi) \geq \bar{f}_i\}$ for all i, and $Pos\{f_{i_0}(\mathbf{x},\xi) \geq \bar{f}_{i_0}\} \geq Pos\{f_{i_0}(\mathbf{x}^*,\xi) \geq \bar{f}_{i_0}\}$ for at least one $i_0 \in \{1,2,\cdots,m\}$.*

If the objective is to minimize δ_i, we have the following model:

$$\begin{cases} \min[\delta_1, \delta_2, \cdots, \delta_m] \\ \text{s.t.} \begin{cases} Ch\{f_i(\mathbf{x}, \boldsymbol{\xi}) \geq \bar{f_i}\} \leq \delta_i, i = 1, 2, \cdots, m \\ Ch\{g_r(\mathbf{x}, \boldsymbol{\xi}) \leq 0\} \geq \theta_r, r = 1, 2, \cdots, p \\ \mathbf{x} \in X \end{cases} \end{cases} \quad (5.131)$$

where Ch refers to the fuzzy measure.

5.5.2   Linear Fu-DCRM and Lexicographic Method

We consider linear decision making with fuzzy parameters. Fuzzy constraints can no longer give a crisp feasible set. Naturally, one requires constraints to hold at a predetermined confidence level. The chance constraints with fuzzy parameters are expressed as

$$Me\left\{ \sum_{j=1}^{n} \tilde{a}_{rj} x_j \leq \tilde{b}_r \right\} \geq \theta_r, \ r = 1, 2, \cdots, p \quad (5.132)$$

where $Me\{\cdot\}$ denotes the general measure of the fuzzy event $\{\cdot\}$.

If the DM wants to maximize $Me\left\{ \sum_{j=1}^{n} \tilde{c}_{ij} x_j \geq \bar{f_i} \right\}$ subject to chance constraints

$Me\left\{ \sum_{j=1}^{n} \tilde{a}_{rj} x_j \leq \tilde{b}_r \right\} \geq \theta_r$, then the dependent-chance linear decision-making model with fuzzy parameters is formulated as

$$\begin{cases} \max \ [\delta_1, \delta_2, \cdots, \delta_m] \\ \text{s.t.} \begin{cases} Me\left\{ \sum_{j=1}^{n} \tilde{c}_{ij} x_j \geq \bar{f_i} \right\} \geq \delta_i, \ i = 1, 2, \cdots, m \\ Me\left\{ \sum_{j=1}^{n} \tilde{a}_{rj} x_j \leq \tilde{b}_r \right\} \geq \theta_r, \ r = 1, 2, \cdots, p \\ x_j \geq 0, j = 1, 2, \cdots, n \end{cases} \end{cases} \quad (5.133)$$

where $\bar{f_i}$ are the aspiration levels, $i = 1, 2, \cdots, m$.

According to LAM (5.129) and UAM (5.130), we present the linear lower approximation model (LLAM) and the linear upper approximation model (LUAM) of the linear model (5.133):

$$(LLAM) \quad \begin{cases} \max \ [\delta_1, \delta_2, \cdots, \delta_m] \\ \text{s.t.} \begin{cases} Nec\left\{ \sum_{j=1}^{n} \tilde{c}_{ij} x_j \geq \bar{f_i} \right\} \geq \delta_i, \ i = 1, 2, \cdots, m \\ Nec\left\{ \sum_{j=1}^{n} \tilde{a}_{rj} x_j \leq \tilde{b}_r \right\} \geq \theta_r, \ r = 1, 2, \cdots, p \\ x_j \geq 0, j = 1, 2, \cdots, n \end{cases} \end{cases} \quad (5.134)$$

and

$$(LUAM) \quad \begin{cases} \max \ [\delta_1, \delta_2, \cdots, \delta_m] \\ \text{s.t.} \begin{cases} Pos\left\{ \sum_{j=1}^{n} \tilde{c}_{ij} x_j \geq \bar{f}_i \right\} \geq \delta_i, \ i = 1, 2, \cdots, m \\ Pos\left\{ \sum_{j=1}^{n} \tilde{a}_{rj} x_j \leq \bar{b}_r \right\} \geq \theta_r, \ r = 1, 2, \cdots, p \\ x_j \geq 0, j = 1, 2, \cdots, n \end{cases} \end{cases} \quad (5.135)$$

which are collectively referred to as the fuzzy dependent-chance rough model (Fu-LCCRM).

5.5.2.1 Crisp Equivalent Model

One way to solve problem (5.134) and (5.135) is to transform them into their crisp equivalent models and then solve them by traditional solution techniques. However, this process is usually hard to perform and only successful in some special cases. First we present some useful results.

THEOREM 5.14

Let $\tilde{c}_{ij}$ be a LR fuzzy variable with membership function

$$\mu_{\tilde{c}_{ij}}(t) = \begin{cases} L\left(\frac{c_{ij}-t}{\alpha_{ij}^c}\right), & t \leq c_{ij}, \alpha_{ij}^c > 0 \\ R\left(\frac{t-c_{ij}}{\beta_{ij}^c}\right), & t \geq c_{ij}, \beta_{ij}^c > 0 \end{cases} \quad (5.136)$$

where the vector $(c_{ij})_{n \times 1} = (c_{i1}, c_{i2}, \cdots, c_{in})^T$ is a real vector, α_{ij}^c and β_{ij}^c are the left and right spread of $\tilde{c}_{ij}$, $i = 1, 2, \cdots, m$, $j = 1, 2, \cdots, n$, the reference function $L, R : [0,1] \to [0,1]$ satisfies that $L(1) = R(1) = 0$, $L(0) = R(0) = 1$, and it is a monotone function. Then $Nec\{\tilde{c}_i^T x \geq f_i\} \geq \delta_i$ is equivalent to

$$L^{-1}(1 - \delta_i) \leq \frac{c_i^T x - f_i}{\alpha_i^{cT} x}, \quad i = 1, 2, \cdots, m \quad (5.137)$$

PROOF The proof is similar to the proof of Theorem 5.12.

Since we assume that the reference function $L(\cdot)$ is monotonically decreasing function, so max δ_i is equivalent to max $L^{-1}(1 - \delta_i)$. By Theorem 5.14 and Theorem 5.12, it is easy for us to derive the crisp equivalent model (5.138) of model (5.134) as

$$\begin{cases} \max \ \left[\frac{c_1^T x - f_1}{\alpha_1^{cT} x}, \frac{c_2^T x - f_2}{\alpha_2^{cT} x}, \cdots, \frac{c_m^T x - f_m}{\alpha_m^{cT} x}\right] \\ \text{s.t.} \begin{cases} b_r + R^{-1}(\theta_r)\beta_r^b - a_r^T x + L^{-1}(\theta_r)\alpha_r^{aT} x \geq 0, \ r = 1, 2, \cdots, p \\ x \geq 0 \end{cases} \end{cases} \quad (5.138)$$

REMARK 5.16   Especially when the reference functions of the variables $\tilde{a}_{rj}, \tilde{b}_r$ are all $L(x) = R(x) = 1 - x$ $(x \in [0,1])$, then the LR fuzzy variables are specified as the triangular fuzzy variables, and $R^{-1}(\delta_i) = 1 - \delta_i$, so model(5.138) can be rewritten as

$$
\begin{cases}
\max & \left[\dfrac{c_1^T x - f_1}{\alpha_1^{cT} x}, \dfrac{c_2^T x - f_2}{\alpha_2^{cT} x}, \cdots, \dfrac{c_m^T x - f_m}{\alpha_m^{cT} x} \right] \\
\text{s.t.} & \begin{cases} b_r + (1 - \theta_r)\beta_r^b \geq a_r^T x - (1 - \theta_r)\alpha_r^{aT} x, r = 1,2,\cdots,p \\ x \geq 0 \end{cases}
\end{cases}
\tag{5.139}
$$

THEOREM 5.15

Assume that $\tilde{c}_{ij}$ is a LR fuzzy variable, and the membership function of $\tilde{c}_{ij}$ is

$$
\mu_{\tilde{c}_{ij}}(t) = \begin{cases}
L\left(\dfrac{c_{ij} - t}{\alpha_{ij}^c} \right), & t \leq c_{ij}, \alpha_{ij}^c > 0 \\
R\left(\dfrac{t - c_{ij}}{\beta_{ij}^c} \right), & t \geq c_{ij}, \beta_{ij}^c > 0
\end{cases}
\tag{5.140}
$$

where the vector $(c_{ij})_{n \times 1} = (c_{i1}, c_{i2}, \cdots, c_{in})^T$ is a real vector, α_{ij}^c and β_{ij}^c are the left and right spread of $\tilde{c}_{ij}$, $i = 1,2,\cdots,m$, $j = 1,2,\cdots,n$, the reference function $L,R : [0,1] \to [0,1]$ satisfies that $L(1) = R(1) = 0$, $L(0) = R(0) = 1$, and it is a monotone function. Then $\text{Pos}\{\tilde{c}_i^T x \geq f_i\} \geq \delta_i$ is equivalent to

$$
R^{-1}(\delta_i) \geq \frac{f_i - c_i^T x}{\beta_i^{cT} x}, \quad i = 1,2,\cdots,m
\tag{5.141}
$$

PROOF   The proof is the same as that of Theorem 5.10.

Since we assume that the function $R(\cdot)$ is a monotonically decreasing function, so max δ_i is equivalent to min $R^{-1}(\delta_i)$. By Theorem 5.15 and Theorem 5.13, it is easy for us to derive the crisp equivalent model (5.142) of the model (5.135),

$$
\begin{cases}
\max & \left[\dfrac{c_1^T x - f_1}{\beta_1^{cT} x}, \dfrac{c_2^T x - f_2}{\beta_2^{cT} x}, \cdots, \dfrac{c_m^T x - f_m}{\beta_m^{cT} x} \right] \\
\text{s.t.} & \begin{cases} b_r + R^{-1}(\theta_r)\beta_r^b - a_r^T x + L^{-1}(\theta_r)\alpha_r^{aT} x \geq 0, r = 1,2,\cdots,p \\ x \geq 0 \end{cases}
\end{cases}
\tag{5.142}
$$

REMARK 5.17   Especially when the reference functions of the variables $\tilde{a}_{rj}, \tilde{b}_r$ are all $L(x) = R(x) = 1 - x$ $(x \in [0,1])$, then the LR fuzzy variables are specified as the triangular fuzzy variables, and $R^{-1}(\delta_i) = 1 - \delta_i$, so we have the following equivalent expressions:

$$
\begin{cases}
\max & \left[\dfrac{c_1^T x - f_1}{\beta_1^{cT} x}, \dfrac{c_2^T x - f_2}{\beta_2^{cT} x}, \cdots, \dfrac{c_m^T x - f_m}{\beta_m^{cT} x} \right] \\
\text{s.t.} & \begin{cases} b_r - \theta_r \alpha_r^b \geq a_r^T x + (1 - \theta_r)\beta_r^{aT} x, r = 1,2,\cdots,p \\ x \geq 0 \end{cases}
\end{cases}
\tag{5.143}
$$

5.5.2.2 Lexicographic Method

The basic idea of the lexicographic method is to rank the objective function by its importance to decision makers and then resolve the next objective function after resolving the foregoing one. We take the solution of the last programming problem as the final solution.

Consider the following multiobjective programming problem:

$$\begin{cases} \min[f_1(\mathbf{x}), f_2(\mathbf{x}), \cdots, f_m(\mathbf{x})] \\ \text{s.t. } \mathbf{x} \in X \end{cases} \tag{5.144}$$

Without loss of generality, assume the rank as $f_1(\mathbf{x}), f_2(\mathbf{x}), \cdots, f_m(\mathbf{x})$ according to different importance. Solve the following single objective problem in turn:

$$\begin{cases} \min f_i(\mathbf{x}) \\ \text{s.t. } \begin{cases} f_k(\mathbf{x}) = f_k(\mathbf{x}^k), k = 1, 2, \cdots, i-1 \\ \mathbf{x} \in X \end{cases} \end{cases} \tag{5.145}$$

where $i = 1, 2, \cdots, m$, X is the feasible area. Denote the feasible area of problem (5.145) as X^i.

THEOREM 5.16
Let $X \subset \mathbf{R}^n$, $\mathbf{f} : X \to \mathbf{R}^m$. If $\mathbf{x}^m$ be the optimal solution by the lexicographic method, then $\mathbf{x}^m$ is an efficient solution to problem (5.144).

PROOF If $\mathbf{x}^m$ is not an efficient solution to problem (5.144), there exists $\bar{\mathbf{x}} \in X$ such that $\mathbf{f}(\bar{\mathbf{x}}) \leq \mathbf{f}(\mathbf{x}^m)$. Since $f_1(\mathbf{x}^m) = f_1^* = f_1(\mathbf{x}^1)$, $f_1(\bar{\mathbf{x}}) < f_1(\mathbf{x}^m)$ cannot hold. It necessarily follows that $f_1(\bar{\mathbf{x}}) = f_1(\mathbf{x}^m)$.

If $f_k(\bar{\mathbf{x}}) = f_k(\mathbf{x}^m)(k = 1, 2, \cdots, i-1)$ but $f_i(\bar{\mathbf{x}}) < f_i(\mathbf{x}^m)$, it follows that $\bar{x}$ is a feasible solution to problem (5.145). Since $f_i(\bar{\mathbf{x}}) < f_i(\mathbf{x}^m) = f_i(\mathbf{x}^i)$, this results in the conflict that $\mathbf{x}^i$ is the optimal solution to problem (5.145). Thus, $f_k(\bar{\mathbf{x}}) = f_k(\mathbf{x}^m)(k = 1, 2, \cdots, i)$ necessarily holds. Then we can prove $f_k(\bar{\mathbf{x}}) = f_k(\mathbf{x}^m)(k = 1, 2, \cdots, m)$ by mathematical induction. This conflicts with $\mathbf{f}(\bar{\mathbf{x}}) \leq \mathbf{f}(\mathbf{x}^m)$. This completes the proof.

5.5.2.3 Numerical Example

Example 5.11
Consider the problem

$$\begin{cases} \max f_1(\mathbf{x}, \boldsymbol{\xi}) = \xi_1 x_1 + \xi_2 x_2 + \xi_3 x_3 + \xi_4 x_4 \\ \max f_2(\mathbf{x}, \boldsymbol{\xi}) = \xi_5 x_1 + \xi_6 x_2 + \xi_7 x_3 + \xi_8 x_4 \\ \text{s.t. } \begin{cases} \xi_5 x_1 + \xi_6 x_2 + \xi_7 x_3 + \xi_8 x_4 \leq \xi_9 \\ 3x_1 + 5x_2 + 2x_3 + 7x_4 \leq 120 \\ x_j \geq 0, j = 1, 2, 3, 4 \end{cases} \end{cases} \tag{5.146}$$

wherein the coefficients are triangular fuzzy variables characterized as

$$\tilde{\xi}_1 = (3,4,5),\ \tilde{\xi}_2 = (5,6,7),\ \tilde{\xi}_3 = (8,9,10),$$
$$\tilde{\xi}_4 = (2,4,6),\ \tilde{\xi}_5 = (7,8,9),\ \tilde{\xi}_6 = (4,5,6),$$
$$\tilde{\xi}_7 = (2,3,4),\ \tilde{\xi}_8 = (1,2,3),\ \tilde{\xi}_9 = (120,160,200).$$

In order to solve it, we use the DCM and rough approximation techniques, and then we can obtain the LAM and UAM as

$$(LAM) \begin{cases} \max Nec\{\tilde{\xi}_1 x_1 + \tilde{\xi}_2 x_2 + \tilde{\xi}_3 x_3 + \tilde{\xi}_4 x_4 \geq \bar{f}_1\} \\ \max Nec\{\tilde{\xi}_5 x_1 + \tilde{\xi}_6 x_2 + \tilde{\xi}_7 x_3 + \tilde{\xi}_8 x_4 \geq \bar{f}_2\} \\ \text{s.t.} \begin{cases} Nec\{\tilde{\xi}_5 x_1 + \tilde{\xi}_6 x_2 + \tilde{\xi}_7 x_3 + \tilde{\xi}_8 x_4 \leq \tilde{\xi}_9\} \geq \theta \\ 3x_1 + 5x_2 + 2x_3 + 7x_4 \leq 120 \\ x_j \geq 0, j = 1,2,3,4 \end{cases} \end{cases} \quad (5.147)$$

and

$$(UAM) \begin{cases} \max Pos\{\tilde{\xi}_1 x_1 + \tilde{\xi}_2 x_2 + \tilde{\xi}_3 x_3 + \tilde{\xi}_4 x_4 \geq \bar{f}_1\} \\ \max Pos\{\tilde{\xi}_5 x_1 + \tilde{\xi}_6 x_2 + \tilde{\xi}_7 x_3 + \tilde{\xi}_8 x_4 \geq \bar{f}_2\} \\ \text{s.t.} \begin{cases} Pos\{\tilde{\xi}_5 x_1 + \tilde{\xi}_6 x_2 + \tilde{\xi}_7 x_3 + \tilde{\xi}_8 x_4 \leq \tilde{\xi}_9\} \geq \theta \\ 3x_1 + 5x_2 + 2x_3 + 7x_4 \leq 120 \\ x_j \geq 0, j = 1,2,3,4 \end{cases} \end{cases} \quad (5.148)$$

or

$$(LAM) \begin{cases} \max[\delta_1, \delta_2] \\ \text{s.t.} \begin{cases} Nec\{\tilde{\xi}_1 x_1 + \tilde{\xi}_2 x_2 + \tilde{\xi}_3 x_3 + \tilde{\xi}_4 x_4 \geq \bar{f}_1\} \geq \delta_1 \\ Nec\{\tilde{\xi}_5 x_1 + \tilde{\xi}_6 x_2 + \tilde{\xi}_7 x_3 + \tilde{\xi}_8 x_4 \geq \bar{f}_2\} \geq \delta_2 \\ Nec\{\tilde{\xi}_5 x_1 + \tilde{\xi}_6 x_2 + \tilde{\xi}_7 x_3 + \tilde{\xi}_8 x_4 \leq \tilde{\xi}_9\} \geq \theta \\ 3x_1 + 5x_2 + 2x_3 + 7x_4 \leq 120 \\ x_j \geq 0, j = 1,2,3,4 \end{cases} \end{cases} \quad (5.149)$$

and

$$(UAM) \begin{cases} \max[\delta_1, \delta_2] \\ \text{s.t.} \begin{cases} Pos\{\tilde{\xi}_1 x_1 + \tilde{\xi}_2 x_2 + \tilde{\xi}_3 x_3 + \tilde{\xi}_4 x_4 \geq \bar{f}_1\} \geq \delta_1 \\ Pos\{\tilde{\xi}_5 x_1 + \tilde{\xi}_6 x_2 + \tilde{\xi}_7 x_3 + \tilde{\xi}_8 x_4 \geq \bar{f}_2\} \geq \delta_2 \\ Pos\{\tilde{\xi}_5 x_1 + \tilde{\xi}_6 x_2 + \tilde{\xi}_7 x_3 + \tilde{\xi}_8 x_4 \leq \tilde{\xi}_9\} \geq \theta \\ 3x_1 + 5x_2 + 2x_3 + 7x_4 \leq 120 \\ x_j \geq 0, j = 1,2,3,4 \end{cases} \end{cases} \quad (5.150)$$

Let confidence level $\theta = 0.8$. By models (5.139) and (5.143), we know that models (5.149) and (5.150) have equivalent models as

$$(E\text{-}LLAM) \begin{cases} \max F_1 = \frac{4x_1 + 6x_2 + 9x_3 + 4x_4 - f_1}{x_1 + x_2 + x_3 + 2x_4} \\ \max F_2 = \frac{8x_1 + 5x_2 + 3x_3 + 2x_4 - f_2}{x_1 + x_2 + x_3 + x_4} \\ \text{s.t.} \begin{cases} 8.2x_1 + 5.2x_2 + 3.2x_3 + 2.2x_4 \leq 128 \\ 3x_1 + 5x_2 + 2x_3 + 7x_4 \leq 120, \\ x_j \geq 0, j = 1,2,3,4 \end{cases} \end{cases} \quad (5.151)$$

$$(E\text{-}LUAM) \begin{cases} \max F_1 = \frac{4x_1+6x_2+9x_3+4x_4-f_1}{x_1+x_2+x_3+2x_4} \\ \max F_2 = \frac{8x_1+5x_2+3x_3+2x_4-f_2}{x_1+x_2+x_3+x_4} \\ \text{s.t.} \begin{cases} 7.8x_1+4.8x_2+2.8x_2+1.8x_4 \le 168 \\ 3x_1+5x_2+2x_3+7x_4 \le 120 \\ x_j \ge 0, j = 1,2,3,4 \end{cases} \end{cases} \tag{5.152}$$

We set the aspiration level $f_1 = 350, f_2 = 100$. Then we use the lexicographic method to solve the above problems. For problem (5.151), assume that the objective function $F_1(x)$ is more important than $F_2(x)$ to the DM. Let us first solve the problem

$$\begin{cases} \max F_1 = \frac{4x_1+6x_2+9x_3+4x_4-f_1}{x_1+x_2+x_3+2x_4} \\ \text{s.t.} \begin{cases} 8.2x_1+5.2x_2+3.2x_3+2.2x_4 \le 128 \\ 3x_1+5x_2+2x_3+7x_4 \le 120 \\ x_j \ge 0, j = 1,2,3,4 \end{cases} \end{cases} \tag{5.153}$$

Then we get the optimal solution $x^* = (0,0,40,0)$ and $F_1(x^*) = 0.25, F_2(x^*) = 0.5$. Second, construct the decision-making problem:

$$\begin{cases} \max F_2 = \frac{f_2-(8x_1+5x_2+3x_3+2x_4)}{x_1+x_2+x_3+x_4} \\ \text{s.t.} \begin{cases} \frac{4x_1+6x_2+9x_3+4x_4-f_1}{x_1+x_2+x_3+2x_4} = 0.25 \\ 8.2x_1+5.2x_2+3.2x_3+2.2x_4 \le 128 \\ 3x_1+5x_2+2x_3+7x_4 \le 120 \\ x_j \ge 0, j = 1,2,3,4 \end{cases} \end{cases} \tag{5.154}$$

Then we get the final optimal solution $x^* = (0,0,40,0)$ and $F_2(x^*) = 0.5$. Then $\delta_1^* = 0.75, \delta_2^* = 0.5$.

For problem (5.152), assume that the objective function $F_1(x)$ is more important than $F_2(x)$ to the DM. Let us first solve the problem:

$$\begin{cases} \max F_1 = \frac{4x_1+6x_2+9x_3+4x_4-f_1}{x_1+x_2+x_3+2x_4} \\ \text{s.t.} \begin{cases} 7.8x_1+4.8x_2+2.8x_2+1.8x_4 \le 168 \\ 3x_1+5x_2+2x_3+7x_4 \le 120 \\ x_j \ge 0, j = 1,2,3,4 \end{cases} \end{cases} \tag{5.155}$$

Then we get the optimal solution $x^* = (0,0,60,0)$ and $F_1(x^*) = 3.17, F_2(x^*) = 1.33$. Second, construct the decision-making problem:

$$\begin{cases} \max F_2 = \frac{f_2-(8x_1+5x_2+3x_3+2x_4)}{x_1+x_2+x_3+x_4} \\ \text{s.t.} \begin{cases} \frac{4x_1+6x_2+9x_3+4x_4-f_1}{x_1+x_2+x_3+2x_4} = 3.17 \\ 7.8x_1+4.8x_2+2.8x_2+1.8x_4 \le 168 \\ 3x_1+5x_2+2x_3+7x_4 \le 120 \\ x_j \ge 0, j = 1,2,3,4 \end{cases} \end{cases} \tag{5.156}$$

Then we get the final optimal solution $\mathbf{x}^* = (0,0,60,0)$ and $F_2(\mathbf{x}^*) = 1.13$. Then $\delta_1^* = 0.87, \delta_2^* = 0.67$.

5.5.3   Fu-NLDCRM and Fuzzy Simulation-Based PSA

As problem (5.126) has nonlinear forms, that is, there is at least one nonlinear function of f_i, g_r. By using the rough approximation technique, we can obtain the fuzzy nonlinear dependent-chance rough model (Fu-NLDCRM). It is usually impossible be to converted into the crisp equivalent model. Fuzzy simulation-based parallel simulated annealing (PSA) algorithm, a solving method, is used to deal with it.

5.5.3.1   Fuzzy Simulation 3 for *Me* measure

Suppose that $f : \mathfrak{R}^m \to \mathfrak{R}^n$ is a function, and $\xi = (\xi_1, \xi_2, \cdots, \xi_m)$ is a fuzzy vector on the possibility space $(\Theta, P(\Theta), Pos)$. We design a fuzzy simulation to compute the *Me* measure:

$$L = Me\{f(\xi) \le 0\} \qquad (5.157)$$

We randomly generate θ_k from Θ such that $Pos(\theta_k) \ge \varepsilon$ and write $v_k = Pos(\theta_k), k = 1, 2, \cdots, N$, respectively, where ε is a sufficiently small number. Equivalently, we randomly generate $u_{1k}, u_{2k}, \cdots, u_{nk}$ from the ε-level sets of $\xi_1, \xi_2, \cdots, \xi_n$, and write $v_k = \mu_1(u_{1k}) \wedge \mu_2(u_{2k}) \wedge \cdots \wedge \mu_n(u_{nk})$ for $k = 1, 2, \cdots, N$, where μ_i are membership functions of $\xi_i, i = 1, 2, \cdots, n$, respectively. Then the $Me\{f(\xi) \le 0\}$ can be estimated by the formula

$$L(r) = \lambda \max_{1 \le k \le N} \{v_k | f(\xi(\theta_k)) \ge r\} + (1 - \lambda) \min_{1 \le k \le N} \{1 - v_k | f(\xi(\theta_k)) \le r\}$$

where λ is the optimism coefficient. The processes of fuzzy simulation 3 can be summarized as follows:

Step 1. Generate θ_k from Θ such that $Pos\{\theta_k\} \ge \varepsilon$, where ε is a sufficiently small number.
Step 2. Set $v_k = Pos\{\theta_k\}, k = 1, 2, \cdots, N$.
Step 3. Return L via the estimation formula.

Example 5.12
We assume that ξ_1, ξ_2, and ξ_3 are fuzzy variables defined as $\xi_1 = (10, 11, 12), \xi_2 = (12, 13, 14), \xi_3 = (13, 14, 15)$. A run of fuzzy simulation with 1000 cycles shows that

$$Me\{\xi_1\xi_2\xi_3 \ge 1800\} \ge 0.8 = \begin{cases} 0.2176, & \lambda = 0 \\ 0.3284, & \lambda = 0.2 \\ 0.5882, & \lambda = 0.5 \\ 0.8150, & \lambda = 0.8 \\ 0.9668 & \lambda = 1 \end{cases}$$

5.5.3.2   PSA

PSA is a more efficient tool to deal with complex problems. Parallel processing appears to be the only viable way to substantially speed up the method and thus expand its applicability. For fast-tailored SAs, parallel implementations may also reduce the loss of robustness. In an effort to increase convergence speed, parallel SA optimization algorithms have been implemented with mixed results in various scientific fields [57]. Z. Czech and P. Czarnas [70] presented a parallel simulated annealing for the vehicle routing problem. Many scholars [14, 70, 123] introduced the PSA with the feature that multiple processing elements follow a single search path (Markov chain)(to be referred to as SMC PSA), but E. Aarts and J. Korst [1] described the idea of the multiple Markov chains PSA (to be referred to as MMC PSA), namely, the division algorithm, and S. Lee and K. Lee [177] proposed the MMC PSA and applied it to graph partition, and N. Li et al. [186] applied it to 3D engineering layout design. A. Bevilacqua [28] present a novel approach to the parallel implementation of SA on a symmetric multiprocessor system. In addition, we offer an adaptive method to dynamically change the program execution flow at run time so as to obtain the maximum benefit from these shared memory parallel architectures. J. Leite and B. Topping [180] considers the evaluation of parallel schemes for engineering problems where the solution spaces may be very complex and highly constrained, and function evaluations vary from medium to high cost. In addition, this study provides guidelines for the selection of appropriate schemes for engineering problems.

Although optimization performance can be greatly improved through parallelization of SA [78], there has few applications of this techniques in multiobjective programming problems. Since the "annealing community" has so far not achieved a common agreement with regard to a general approach for the serial SA, the best parallel scheme is still the object of current research.

In any case, a key issue in developing a PSA is to ensure that it maintains the same convergence property as the sequential version for which a formal proof to converge to a global minimum exists. For this purpose, it is easy to show that it is not necessary for a parallel implementation to generate exactly the same sequence of solutions as the sequential version. Frequently, it is enough to keep the same ratio between the number of accepted moves and the number of attempted moves (acceptance rate), averaged over a given temperature, as the correspondent sequential version.

Next, let us discuss the detailed algorithm of PSA. This section mainly refers to [28, 186, 211, 331], and readers can also consult the related literature. Let us first introduce the parallel Markov chain in [211]. Let X be a finite set and $U : X \to R^+$ be a nonconstant function to be minimized. We denote by as $X_{\min}$ the set of global minima of U. Suppose that we have $p > 1$ processors. The optimization algorithm which is the center of our interest, is described as follows. Choose any starting point $x_0 \in X$ and let each processor individually run a Metropolis Markov chain of fixed length $L > 1$ at inverse temperature $\beta(0)$ starting in x_0. After L transitions, the simulation is stopped, and only one state x_1 is selected from the p states according to a selection strategy. Again each processor individually simulates a Metropolis Markov chain of length L at an updated inverse temperature $\beta(1)$ starting in x_1. At the end of the simulation again, a state x_2 is selected from the p states, and the next simulation starts from x_2, etc. (see Figure 5.6). This algorithm is closely related to the so-called parallel chain algorithm. However, the main difference is that the number of parallel chains and the length of the Markov chains L is kept fixed in our model.

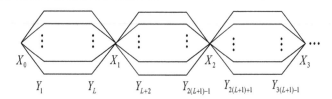

FIGURE 5.6   Illustration of the PSA algorithm's working method.

In the parallel chain algorithm, the length L is usually increased and the number of parallel Markov chains is decreased during the run of the algorithm so that the parallel chain algorithm asymptotically behaves like the so-called one-chain algorithm. Second, let us introduce the process of PSA. PSA starts with a high temperature. After generating an initial solution, it attempts to move randomly from the current solution to one of its neighborhood solutions. The changes in the objective function values Δf are computed (we usually consider a weighted sum of all objective functions in a multiobjective programming problem). If the new solution results in a better objective value, it is accepted. However, if the new solution yields a worse value, it can still be accepted according to the acceptance probability function $P(T_j)$. The PSA algorithm repeats this Metropolis algorithm L times at each temperature to reach thermal equilibrium, where L is a control parameter, usually called the Markov Chain Length. After some temperature annealing, solutions migrate from the neighboring processors at each interval with migration probability, P_m. Parameter T_j is gradually decreased by a cooling function as parallel SA proceeds until the terminating condition is met. Readers can refer to Figure 5.7 to know the workflow of PSA.

The PSA algorithms proposed can be summarized as follows:

Step 1. Initialize PSA parameters: starting temperature T_0, final temperature T_f, Markov chain length L, migration rate P_m, and migration interval M_i.

Step 2. Randomly generate an initial feasible **x**.

Step 3. Repeat the following L times:

1. Generate a neighborhood solution $\mathbf{x}_j^{new}$ through a random number generator.

2. Compute $\Delta f = f(\mathbf{x}_j^{new}) - f(\mathbf{x}_j^{old})$.

3. If $\Delta f \leq 0$ (when we minimize the objective function), set $\mathbf{x}_j^{old} = \mathbf{x}_j^{new}$.

4. If $\Delta f > 0$, generate a random number X in $(0,1)$, and compute $P(T_j)$. If $P(T_j) > X$, set $\mathbf{x}_j^{old} = \mathbf{x}_j^{new}$.

Step 4. The migration interval (M_i) is reached, then solutions migrate from the neighboring processors with migration rate P_m.

Step 5. If the terminating condition is met, stop; otherwise, let T_j decrease by the cooling schedule and go to **Step 3**.

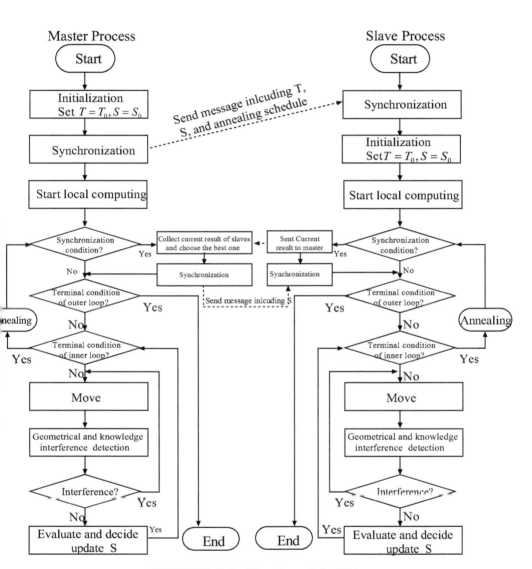

FIGURE 5.7 Flowchart of the PSA.

5.5.3.3　Numerical Number

Example 5.13

Consider the problem

$$
\begin{cases}
\max 2\xi_1^2 x_1 + \xi_2^2 x_2 + \sqrt{\xi_3} x_3 + \sqrt{\xi_4^2 x_4 + \xi_5^2 x_5 + \xi_6^2 x_6} \\
\max \xi_1 x_1 + \xi_2 x_2 + \xi_3 x_3 + \xi_4 x_4 + \xi_5 x_5 + \xi_6 x_6 \\
\text{s.t.} \begin{cases}
2x_1 + x_2 + x_3 + x_4 + x_5 + x_6 \le 260 \\
x_1 + x_2 + x_3 + x_4 + x_5 + x_6 \ge 150 \\
7x_1 + 4x_2 + 3x_3 + 5x_4 + 2x_5 + 6x_6 \le 1100 \\
2x_1 + 7x_2 + 3x_3 + 5x_4 + x_5 + 2x_6 \ge 450 \\
x_j \ge 10, j = 1,2,3,4,5,6
\end{cases}
\end{cases}
\tag{5.158}
$$

$\xi_j (j = 1,2,3,4,5,6)$ are fuzzy variables characterized as

$$\xi_1 = (1,3,7), \xi_2 = (1,2,3), \xi_3 = (4,5,6), \xi_4 = (6,7,8), \xi_5 = (7,8,9), \xi_6 = (3,5,9)$$

We set the aspiration level $f_1 = 600, f_2 = 800$. By using the DCM and rough approximation technique, we have

$$
(LAM) \quad
\begin{cases}
\max[\delta_1, \delta_2] \\
\text{s.t.} \begin{cases}
Nec\{2\xi_1^2 x_1 + \xi_2^2 x_2 + \sqrt{\xi_3} x_3 + \sqrt{\xi_4^2 x_4 + \xi_5^2 x_5 + \xi_6^2 x_6} \ge 600\} \ge \delta_1 \\
Nec\{\xi_1 x_1 + \xi_2 x_2 + \xi_3 x_3 + \xi_4 x_4 + \xi_5 x_5 + \xi_6 x_6 \ge 800\} \ge \delta_2 \\
2x_1 + x_2 + x_3 + x_4 + x_5 + x_6 \le 260 \\
x_1 + x_2 + x_3 + x_4 + x_5 + x_6 \ge 150 \\
7x_1 + 4x_2 + 3x_3 + 5x_4 + 2x_5 + 6x_6 \le 1100 \\
2x_1 + 7x_2 + 3x_3 + 5x_4 + x_5 + 2x_6 \ge 450 \\
x_j \ge 10, i = 1,2,3,4,5,6
\end{cases}
\end{cases}
\tag{5.159}
$$

and

$$
(UAM) \quad
\begin{cases}
\max[\delta_1, \delta_2] \\
\text{s.t.} \begin{cases}
Pos\{2\xi_1^2 x_1 + \xi_2^2 x_2 + \sqrt{\xi_3} x_3 + \sqrt{\xi_4^2 x_4 + \xi_5^2 x_5 + \xi_6^2 x_6} \ge 600\} \ge \delta_1 \\
Pos\{\xi_1 x_1 + \xi_2 x_2 + \xi_3 x_3 + \xi_4 x_4 + \xi_5 x_5 + \xi_6 x_6 \ge 800\} \ge \delta_2 \\
2x_1 + x_2 + x_3 + x_4 + x_5 + x_6 \le 260 \\
x_1 + x_2 + x_3 + x_4 + x_5 + x_6 \ge 150 \\
7x_1 + 4x_2 + 3x_3 + 5x_4 + 2x_5 + 6x_6 \le 1100 \\
2x_1 + 7x_2 + 3x_3 + 5x_4 + x_5 + 2x_6 \ge 450 \\
x_j \ge 10, j = 1,2,3,4,5,6
\end{cases}
\end{cases}
\tag{5.160}
$$

In order to solve them, we use the fuzzy simulation-based PSA to deal with it. After running, we get the following solutions:

$$(LAM) \quad x^* = (10, 10, 20.7, 54.9, 10, 16.8), \delta^* = (0.46, 0.68),$$

$$(UAM) \quad x^* = (10, 10, 27.7, 66.3, 10, 18.2), \delta^* = (0.56, 0.73).$$

The Matlab® file is presented in A.5.

5.6 The Earth-Rock Work Scheme for Xiluodu Project

In this section we discuss the earth-rock work allocation problem under a fuzzy environment for the Xiluodu project, which is one of the four largest giant hydropower stations in the Jinsha River downstream. The earth-rock work allocation system mainly includes the material balance subsystem and the road transportation subsystem.

5.6.1 Background Statement

There are five main factors in the earth-rock allocation system:

1. *Excavation project.* In an excavation project, such as a dam base, a hydropower plant base, river division holes, and so on, some excavated materials are usable, and some are useless and are discarded, which is decided by the physical property of the materials and the excavation schedule.

2. *Filling project.* In a filling project, such as a rock-fill dam body, an upstream cofferdam and downstream cofferdam, the materials used should satisfy certain physical property requirements.

3. *Transfer station.* Transfer station denotes the site where excavated materials are stored as standby materials for future use. Each transfer station has a limited capability.

4. Dumping site. Dumping site is the site where useless excavated materials are discarded. The reasons include:

(a) Excavated materials do not meet the physical properties of the filling project.

(b) Though physical properties are congruous, there are no feasible roads.

(c) Though physical properties are congruous, the schedule is incongruous, and there are no available transfer stations at that time. Each dump site has a limited capability.

5. *Borrow area.* The borrow area is the special site where materials are mined when materials that come from excavation project and transfer station cannot satisfy the filling demand in quantity or in physical property. Each borrow area has a limited reserve.

The earth-rock allocation system mainly includes the material balance system and the road transportation system. The optimization goal of the earth-rock work allocation problem is to minimize the cost of the system and meanwhile guarantee the construction schedule, to eventually realize a speedy and economical construction.

Earth-rock work balance system. In order to complete a water conservancy and hydropower construction engineering, earth-rock work needs to be transported among some sites in the earth-rock work allocation system. There are five types of earth-rock work transportation relations in the system: A, B, C, D, and E (see Figure 5.8).

In this system, an important concept should be introduced, that is, the matching relation, which reflects the comprehensive cost. The matching relation is usually

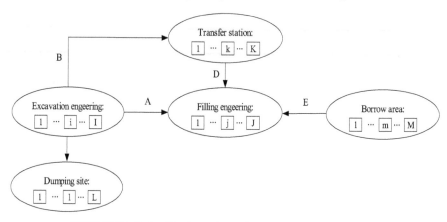

FIGURE 5.8 Earth-rock work transportation relations.

evaluated by considering the following factors:

1. The earth-rock work exploitation cost and the storage cost, which can be apportioned to the number of units of materials.

2. Transport costs related to the distance, road situation, and transport equipment.

3. If the earth-rock work satisfies the filling project and there exist some connected road sections, then we can obtain the corresponding matching data. If one of the two conditions is not satisfied, the matching data is infinite.

Transportation system. In the earth-rock work allocation system, the traffic situation is the key guaranteeing the construction schedule. In particular, traffic intensities of the road sections of some important road sections are the main factors in judging whether an allocation scheme is appropriate or not, since high traffic intensity may usually lead to a delay in the schedule.

We use a number h to denote an elementary road section without any branch. If h section will be passed when the earth-rock work is transported from the excavation project i to the filling project $j, R(i, j, h) = 1$, otherwise, $R(i, j, h) = 0$, then we can use a matrix to denote a transportation scheme. For example, there are two excavation projects i_1 and i_2, two filling projects j_1 and j_2, and the road sections between them are presented in Figure 5.9. We can obtain the following transportation scheme in the Table 5.3 according to Figure 5.9.

TABLE 5.3 Transportation scheme for the example of road sections

Excavation project	Filling project	$R(i, j, h)$				
		S1	S2	S3	S4	S5
i_1	j_1	1	0	1	1	0
i_1	j_2	1	0	1	0	1
i_2	j_1	0	1	1	1	0
i_2	j_2	0	1	1	0	1

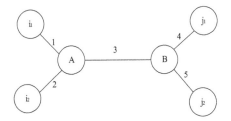

FIGURE 5.9 Example of road sections.

5.6.2 Modeling

The biobjective model for the earth-rock balance and transportation problem is proposed as follows.

5.6.2.1 Notation

The notation we used in this study is listed as follows.

Indices

i excavation project, $i = 1, 2, \cdots, I$
j filling project, $j = 1, 2, \cdots, J$
k transfer station, $k = 1, 2, \cdots, K$
l dumping site, $l = 1, 2, \cdots, L$
m borrow area, $m = 1, 2, \cdots, M$

Fuzzy parameters

$\tilde{C}_{ij}$ the relation of excavation engineering and filling engineering
$\tilde{C}_{ik}$ the relation of excavation engineering and transfer station
$\tilde{C}_{il}$ the relation of excavation engineering and dumping site
$\tilde{C}_{kj}$ the relation of transfer station and filling engineering
$\tilde{C}_{mj}$ the relation of borrow area and filling engineering
$\tilde{A}_i^E$ the quantity of excavation engineering
$\tilde{A}_j^F$ the quantity of filling engineering
$\tilde{V}_m^B$ the storage of borrow area
$\tilde{V}_k^T$ the capacity of transfer station
$\tilde{V}_l^D$ the capacity of dumping site

Certain parameters

$R(a, b, h)$ the relation of transport road sections; it denotes whether h section will be passed from a to b, $R(a, b, h) = 1$ denotes that the road section h will be passed, and $R(a, b, h) = 0$ denotes that the road section h will not be passed

Decision variables

x_{ij}^{EF} transport quantity from excavation engineering to filling engineering
x_{ik}^{ET} transport quantity from excavation engineering to transfer station
x_{il}^{ED} transport quantity from excavation engineering to dumping site
x_{kj}^{TF} transport quantity from transfer station to filling engineering

x_{mj}^{BF} transport quantity from borrow area to filling engineering

5.6.2.2 Objective Functions

The objective for the earth-rock work allocation should be considered is the total costs, which consists of five parts as the following equation:

$$
\begin{aligned}
F = &\sum_{i=1}^{I} \sum_{j=1}^{J} \tilde{C}_{ij} x_{ij}^{EF} + \sum_{i=1}^{I} \sum_{k=1}^{K} \tilde{C}_{ik} x_{ik}^{ET} + \sum_{i=1}^{I} \sum_{l=1}^{L} \tilde{C}_{il} x_{il}^{ED} + \sum_{m=1}^{M} \sum_{j=1}^{J} \tilde{C}_{mj} x_{mj}^{BF} \\
&+ \sum_{k=1}^{K} \sum_{j=1}^{J} \tilde{C}_{kj} x_{kj}^{TF}
\end{aligned}
\tag{5.161}
$$

In order to guarantee the feasibility of the allocation scheme, we need to consider another objective, the peak intensity of road transport, as follows:

$$
\begin{aligned}
Q = \max_{h=1,2,\cdots,H} \Bigg\{ &\sum_{i=1}^{I} \sum_{j=1}^{J} x_{ij}^{EF} R(i,j,h) + \sum_{i=1}^{I} \sum_{k=1}^{K} x_{ik}^{ET} R(i,k,h) + \sum_{i=1}^{I} \sum_{l=1}^{L} x_{il}^{ED} R(i,l,h) \\
&+ \sum_{m=1}^{M} \sum_{j=1}^{J} x_{mj}^{RF} R(m,j,h) + \sum_{k=1}^{K} \sum_{j=1}^{J} x_{kj}^{TF} R(k,j,h) \Bigg\}
\end{aligned}
\tag{5.162}
$$

The objectives of the model is to minimize the total costs and the peak intensity of road transport simultaneously, that is,

$$
\min\{F, Q\}
\tag{5.163}
$$

5.6.2.3 Constraints

According to the earth-rock work allocation problem, the objective functions should subject to some constraints as follows:

1. *Excavation quantity balance constraint.* A certain scheduled quantity of each excavation project in a period should be completed, and all of the earth-rock work in each excavation project should be transported to filling projects, transfer stations, or dumping sites.

$$
\sum_{j=1}^{J} x_{ij}^{EF} + \sum_{k=1}^{K} x_{ik}^{ET} + \sum_{l=1}^{L} x_{il}^{ED} \leq \tilde{A}_{i}^{E}, \quad i = 1, 2, \cdots, I
\tag{5.164}
$$

2. *Filling quantity balance constraint.* A certain scheduled quantity of each filling project in a period should be completed, and the total earth-rock work from excavation projects, transfer stations, and borrow areas to each filling project should be equal to the total filling quantity.

$$
\sum_{i=1}^{I} x_{ij}^{EF} + \sum_{k=1}^{K} x_{kj}^{TF} + \sum_{m=1}^{M} x_{mj}^{BF} \geq \tilde{A}_{j}^{F}, \quad j = 1, 2, \cdots, J
\tag{5.165}
$$

3. *Transfer station capacity constraint.* The difference in earth-rock work quantity between the input from excavation projects and the output to filling projects should be no more than the capacity of the transfer station.

$$\sum_{i=1}^{I} x_{ik}^{ET} - \sum_{j=1}^{J} x_{kj}^{TF} \leq \tilde{V}_k^T, \quad k = 1, 2, \cdots, K \tag{5.166}$$

4. *Exploitation quantity constraint.* In the whole construction process, the total material amount from a borrow area to the filling project and the transfer station should be less than the capacity of this borrow area.

$$\sum_{j=1}^{J} x_{mj}^{BF} \leq \tilde{V}_m^B, \quad m = 1, 2, \cdots, M \tag{5.167}$$

5. *Dumping site capacity constraint.* In the whole construction period, the amount of material from all excavation projects to the dumping site should be less than the capacity of the dumping site.

$$\sum_{i=1}^{I} x_{il}^{ED} \leq \tilde{V}_l^D, \quad l = 1, 2, \cdots, L \tag{5.168}$$

6. *Logical constraint.* The transport quantity should be no less than 0.

$$x_{ij}^{EF} \geq 0, x_{ik}^{ET} \geq 0, x_{il}^{ED} \geq 0, x_{kj}^{TF} \geq 0, x_{mj}^{BF} \geq 0 \tag{5.169}$$

5.6.2.4 Model Formulation

According to (5.40), we can propose the following earth-rock work allocation model (5.170):

$$
\begin{cases}
\min E[F] \\
\min Q \\
\text{s.t.}
\begin{cases}
Me\{ \sum\limits_{j=1}^{J} x_{ij}^{EF} + \sum\limits_{k=1}^{K} x_{ik}^{ET} + \sum\limits_{l=1}^{L} x_{il}^{ED} \leq \tilde{A}_i^E \} \geq \delta_1, \ i = 1, 2, \cdots, I \\
Me\{ \sum\limits_{i=1}^{I} x_{ij}^{EF} + \sum\limits_{k=1}^{K} x_{kj}^{TF} + \sum\limits_{m=1}^{M} x_{mj}^{BF} \geq \tilde{A}_j^F \} \geq \delta_2, \ j = 1, 2, \cdots, J \\
Me\{ \sum\limits_{i=1}^{I} x_{ik}^{ET} - \sum\limits_{j=1}^{J} x_{kj}^{TF} \leq \tilde{V}_k^T \} \geq \delta_3, \ k = 1, 2, \cdots, K \\
Me\{ \sum\limits_{j=1}^{J} x_{mj}^{BF} \leq \tilde{V}_m^B \} \geq \delta_4, \ m = 1, 2, \cdots, M \\
Me\{ \sum\limits_{i=1}^{I} x_{il}^{ED} \leq \tilde{V}_l^D \} \geq \delta_5, \ l = 1, 2, \cdots, L \\
x_{ij}^{EF} \geq 0, x_{ik}^{ET} \geq 0, x_{il}^{ED} \geq 0, x_{kj}^{TF} \geq 0, x_{mj}^{BF} \geq 0
\end{cases}
\end{cases}
\tag{5.170}
$$

By applying rough techniques, we can obtain the LAM and UAM for the earth-rock work allocation problem. Then we transform the LAM and UAM for the earth-rock work allocation problem into their corresponding crisp equivalent models as follows:

$$
\begin{cases}
\min E[F] = \sum_{i=1}^{I} \sum_{j=1}^{J} \left(\frac{(1-\lambda)}{2}(C_{ij} - \alpha_{ij}^{C}) + \frac{C_{ij}}{2} + \frac{\lambda}{2}(C_{ij} + \beta_{ij}^{C}) \right) x_{ij}^{EF} \\
\qquad + \sum_{i=1}^{I} \sum_{k=1}^{K} \left(\frac{(1-\lambda)}{2}(C_{ik} - \alpha_{ik}^{C}) + \frac{C_{ik}}{2} + \frac{\lambda}{2}(C_{ik} + \beta_{ik}^{C}) \right) x_{ik}^{ET} \\
\qquad + \sum_{i=1}^{I} \sum_{l=1}^{L} \left(\frac{(1-\lambda)}{2}(C_{il} - \alpha_{il}^{C}) + \frac{C_{il}}{2} + \frac{\lambda}{2}(C_{il} + \beta_{il}^{C}) \right) x_{il}^{ED} \\
\qquad + \sum_{m=1}^{M} \sum_{j=1}^{J} \left(\frac{(1-\lambda)}{2}(C_{mj} - \alpha_{mj}^{C}) + \frac{C_{mj}}{2} + \frac{\lambda}{2}(C_{mj} + \beta_{mj}^{C}) \right) x_{mj}^{BF} \\
\qquad + \sum_{k=1}^{K} \sum_{j=1}^{J} \left(\frac{(1-\lambda)}{2}(C_{kj} - \alpha_{kj}^{C}) + \frac{C_{kj}}{2} + \frac{\lambda}{2}(C_{kj} + \beta_{kj}^{C}) \right) x_{kj}^{TF} \\
\text{s.t.}
\begin{cases}
Q \le Q_0 \\
\sum_{j=1}^{J} x_{ij}^{EF} + \sum_{k=1}^{K} x_{ik}^{ET} + \sum_{l=1}^{L} x_{il}^{ED} \le A_i^E + (1-\delta_1)\beta_i^A, \ i = 1,2,\cdots,I \\
\sum_{i=1}^{I} x_{ij}^{EF} + \sum_{k=1}^{K} x_{kj}^{TF} + \sum_{m=1}^{M} x_{mj}^{BF} \ge A_j^F - (1-\delta_2)\alpha_j^A, \ j = 1,2,\cdots,J \\
\sum_{i=1}^{I} x_{ik}^{ET} - \sum_{j=1}^{J} x_{kj}^{TF} \le V_k^T + (1-\delta_3)\beta_k^V, \ k = 1,2,\cdots,K \\
\sum_{j=1}^{J} x_{mj}^{BF} \le V_m^B + (1-\delta_4)\beta_m^V, \ m = 1,2,\cdots,M \\
\sum_{i=1}^{I} x_{il}^{ED} \le V_l^D + (1-\delta_5)\beta_l^V, \ l = 1,2,\cdots,L \\
x_{ij}^{EF} \ge 0, x_{ik}^{ET} \ge 0, x_{il}^{ED} \ge 0, x_{kj}^{TF} \ge 0, x_{mj}^{BF} \ge 0
\end{cases}
\end{cases}
\tag{5.171}
$$

and

$$
\begin{cases}
\min E[F] \\
\text{s.t.}
\begin{cases}
Q \le Q_0, \\
\sum_{j=1}^{J} x_{ij}^{EF} + \sum_{k=1}^{K} x_{ik}^{ET} + \sum_{l=1}^{L} x_{il}^{ED} \le A_i^E - \delta_1 \alpha_i^A, \ i = 1,2,\cdots,I \\
\sum_{i=1}^{I} x_{ij}^{EF} + \sum_{k=1}^{K} x_{kj}^{TF} + \sum_{m=1}^{M} x_{mj}^{BF} \ge A_j^F - (1-\delta_2)\alpha_j^A, \ j = 1,2,\cdots,J \\
\sum_{i=1}^{I} x_{ik}^{ET} - \sum_{j=1}^{J} x_{kj}^{TF} \le V_k^T - \delta_3 \alpha_k^V, \ k = 1,2,\cdots,K \\
\sum_{j=1}^{J} x_{mj}^{BF} \le V_m^B - \delta_4 \alpha_m^V, \ m = 1,2,\cdots,M \\
\sum_{i=1}^{I} x_{il}^{ED} \le V_l^D - \delta_5 \alpha_l^V, \ l = 1,2,\cdots,L \\
x_{ij}^{EF} \ge 0, x_{ik}^{ET} \ge 0, x_{il}^{ED} \ge 0, x_{kj}^{TF} \ge 0, x_{mj}^{BF} \ge 0
\end{cases}
\end{cases}
\tag{5.172}
$$

where Q_0 is the maximal traffic intensity of the road.

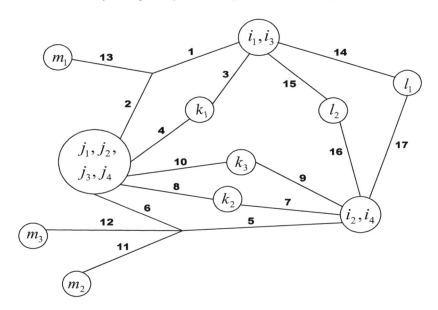

FIGURE 5.10   Road sections.

5.6.3   Solving Process and Comparison Analysis

Let us consider the earth-rock work allocation problem in this water conservancy and hydropower construction engineering. There are $i = 4$ excavation projects, $j = 4$ filling projects, $m = 3$ borrow areas, $k = 3$ transfer stations, and $l = 2$ dumping sites. The road sections of this construction area are presented in Figure 5.10.

Table 5.4 expresses the scheduled quantities of each excavation project and filling project. In a water conservancy and hydropower construction engineering, it is usual to divide the whole construction process into several periods, and so, there are certain scheduled quantities for every excavation project and filling project in each period.

TABLE 5.4   Quantities of EP and FP $(\times 10^4 m^3)$

	EP1	EP2	EP3	EP4	FP1	FP2	FP3	FP4
Quantity	1.2	2	1.5	1.3	1.6	0.9	1.6	1.2

Note: EP stands for excavation project; FP stands for filling project.

Table 5.5 includes the capacities of the borrow areas (BA), the transfer stations (TA) and the dumping sites (DS). Since in the earth-rock work allocation problem the materials are extremely special, and the virtual-actual degree of the earth–rock work is changing all the time, it is very difficult for engineers to give a certain crisp value to the volume of the earth-rock work, so they usually use fuzzy variables to describe it. Since triangular fuzzy numbers are very easily obtained by extending the mean value by adding to the left and right spreads according to their experiences, it is natural to adopt familiar triangular fuzzy numbers. Hence, in Table 5.5 and Table 5.6, there are triangular fuzzy numbers (m, α, β), where m is the mean value with the membership 1, α is the left spread, the membership of $m - \alpha$ is 0 and β is the right spread, and the membership of $m + \beta$ is 0. The values of m, α and β are provided by the engineers. It is noted that, when we give (m, α, β), and the membership function is implicit and easily obtained, the meaning is the same as the form $((m - \alpha, m, m + \beta)$.

TABLE 5.5   Capacities of BA, TS, and DS $(\times 10^4 m^3)$

	1	2	3
BA	(1.1,0.4,0.4)	(0.95,0.2,0.2)	(1,0.4,0.4)
TS	(1.1,0.4,0.4)	(0.5,0.2,0.2)	(0.8,0.4,0.4)
DS	(5,0.2,0.2)	(8,0.3,0.3)	

TABLE 5.6   Match relation

	FP1	FP2	FP3	FP4
EP1	(15,1.8,1.6)	∞	(10,1.2,0.8)	∞
EP2	∞	(5.8,0.8,0.8)	∞	(9.7,1.2,1)
EP3	(13,1.9,1.2)	∞	(5,1.2,0.6)	∞
EP4	∞	(7.9,1,0.5)	∞	(8.2,1.4,1.6)
TS1	(20,3.2,3.5)	∞	(14,1,3)	∞
TS2	∞	(2.3,0.5,0.6)	∞	(3.5,0.6,0.3)
TS3	∞	(4.6,0.8,0.2) ∞ (5.5,0.5,0.5)		
BA1	(10.6,2,1.2)	∞	(21.3,4,5)	∞
BA2	∞	(15.8,2.5,1.5)	(17.6,3,2.5)	
BA3	(13.6,4,5)	∞	(30.5,3,5)	∞

	TS1	TS2	TS3	DS1	DS2
EP1	(20,0.05,0.1)	∞	∞	(12,0.04,0.05)	(10.5,0.03,0.05)
EP2	∞	(2.3,0.03,0.04)	(4.6,0.06,0.03)	(3.5,0.07,0.03)	(2.8,0.04,0.03)
EP3	(14,0.03,0.02)	∞	∞	(11,0.01,0.03)	(3.6,0.05,0.04)
EP4	∞	(3.5,0.05,0.05)	(5.5,0.04,0.04)	(2,0.01,0.01)	(9.4,0.03,0.01)
TS1	∞	∞	∞	∞	∞
TS2	∞	∞	∞	∞	∞
TS3	∞	∞	∞	∞	∞
BA1	∞	∞	∞	∞	∞
BA2	∞	∞	∞	∞	∞
BA3	∞	∞	∞	∞	∞

In Table 5.6 we provide the matching relation among each element in the earth-rock work allocation system. Note that the values of the matching relation are not crisp certain numbers; the values are actually fuzzy variables according to our practical investigation. Because in the construction process of a new water conservancy and hydropower engineering, there is no statistical data of the matching relation, the engineer will evaluate the above corresponding factors that influence the comprehensive cost, and give some fuzzy variables instead of crisp numbers. We also use the triangular fuzzy numbers to describe the matching relation.

According to the matching relationship in Table 5.6, there are some infinite value, and hence, we can remove some unfeasible variables, and can get a simplified model (5.173) for this numerical example.

$$
\left\{
\begin{aligned}
&\min \ [C_{11}x_{11}^{EF} + C_{13}x_{13}^{EF} + C_{22}x_{22}^{EF} + C_{24}x_{24}^{EF} + C_{31}x_{31}^{EF} + C_{33}x_{33}^{EF} + C_{42}x_{42}^{EF} + C_{44}x_{44}^{EF}] \\
&\quad + [C_{11}x_{11}^{TF} + C_{13}x_{13}^{TF} + C_{22}x_{22}^{TF} + C_{24}x_{24}^{TF} + C_{32}x_{32}^{TF} + C_{34}x_{34}^{TF}] \\
&\quad + [C_{11}x_{11}^{BF} + C_{13}x_{13}^{BF} + C_{22}x_{22}^{BF} + C_{24}x_{24}^{BF} + C_{31}x_{31}^{BF} + C_{33}x_{33}^{BF}] \\
&\quad + [C_{11}x_{11}^{ET} + C_{22}x_{22}^{ET} + C_{23}x_{23}^{ET} + C_{31}x_{31}^{ET} + C_{42}x_{42}^{ET} + C_{43}x_{43}^{ET}] \\
&\quad + [C_{11}x_{11}^{ED} + C_{12}x_{12}^{ED} + C_{21}x_{21}^{ED} + C_{22}x_{22}^{ED} + C_{31}x_{31}^{ED} + C_{32}x_{32}^{ED} + C_{41}x_{41}^{ED} + C_{42}x_{42}^{ED}] \\[4pt]
&\text{s.t.}
\begin{cases}
(1)\begin{cases}
Me\{[x_{11}^{EF} + x_{13}^{EF}] + [x_{11}^{ET}] + [x_{11}^{ED} + x_{12}^{ED}] \le \tilde{A}_1^E\} \ge \delta_1, \\
Me\{[x_{22}^{EF} + x_{24}^{EF}] + [x_{22}^{ET} + x_{23}^{ET}] + [x_{21}^{ED} + x_{22}^{ED}] \le \tilde{A}_2^E\} \ge \delta_1, \\
Me\{[x_{31}^{EF} + x_{33}^{EF}] + [x_{31}^{ET}] + [x_{31}^{ED} + x_{32}^{ED}] \le \tilde{A}_3^E\} \ge \delta_1, \\
Me\{[x_{42}^{EF} + x_{44}^{EF}] + [x_{42}^{ET} + x_{43}^{ET}] + [x_{41}^{ED} + x_{42}^{ED}] \le \tilde{A}_4^E\} \ge \delta_1,
\end{cases} \\
(2)\begin{cases}
Me\{[x_{11}^{EF} + x_{31}^{EF}] + [x_{11}^{TF}] + [x_{11}^{BF} + x_{31}^{BF}] \ge \tilde{A}_1^F\} \ge \delta_2 \\
Me\{[x_{22}^{EF} + x_{42}^{EF}] + [x_{22}^{TF} + x_{32}^{TF}] + [x_{22}^{BF}] \ge \tilde{A}_2^F\} \ge \delta_2 \\
Me\{[x_{13}^{EF} + x_{33}^{EF}] + [x_{13}^{TF}] + [x_{13}^{BF} + x_{33}^{BF}] \ge \tilde{A}_3^F\} \ge \delta_2 \\
Me\{[x_{24}^{EF} + x_{44}^{EF}] + [x_{24}^{TF} + x_{34}^{TF}] + [x_{24}^{BF}] \ge \tilde{A}_4^F\} \ge \delta_2
\end{cases} \\
(3)\begin{cases}
Me\{[x_{11}^{ET} + x_{31}^{ET}] - [x_{11}^{TF} + x_{13}^{TF}] \le \tilde{V}_1^T\} \ge \delta_3 \\
Me\{[x_{22}^{ET} + x_{42}^{ET}] - [x_{22}^{TF} + x_{24}^{TF}] \le \tilde{V}_2^T\} \ge \delta_3, \\
Me\{[x_{23}^{ET} + x_{43}^{ET}] - [x_{32}^{TF} + x_{34}^{TF}] \le \tilde{V}_3^T\} \ge \delta_3
\end{cases} \\
(4)\begin{cases}
Me\{x_{11}^{BF} + x_{13}^{BF} \le \tilde{V}_1^B\} \ge \delta_4 \\
Me\{x_{22}^{BF} + x_{24}^{BF} \le \tilde{V}_2^B\} \ge \delta_4 \\
Me\{x_{31}^{BF} + x_{33}^{BF} \le \tilde{V}_3^B\} \ge \delta_4
\end{cases} \\
(5)\begin{cases}
Me\{x_{11}^{ED} + x_{21}^{ED} + x_{31}^{ED} + x_{41}^{ED} \le \tilde{V}_1^D\} \ge \delta_5 \\
Me\{x_{12}^{ED} + x_{22}^{ED} + x_{32}^{ED} + x_{42}^{ED} \le \tilde{V}_2^D\} \ge \delta_5
\end{cases} \\
x_{ij}^{EF} \ge 0, x_{ik}^{ET} \ge 0, x_{il}^{ED} \ge 0, x_{kj}^{TF} \ge 0, x_{mj}^{BF} \ge 0
\end{cases}
\end{aligned}
\right.
$$

$$(5.173)$$

First, we consider the single objective model without considering the traffic intensity constraints, that is, the above single objective model (5.173). According to the LLAM, LUAM, and equivalent transformation theorems, we could get two corresponding models for the earth-rock work allocation problem. Since they are clear, we omit them here.

After solving the single objective lower approximation model and the single objective upper approximation model for the earth-rock work allocation problem, we get the results in Figure 5.11 and Table 5.7.

We explain the above results as follows: Because the feasible region of the UAM is lager than the feasible region of the LAM, we can find a better optimal solution, theoretically. In this example, the optimal objective value of the UAM for the earth-rock work allocation problem is 44.15, which is smaller than 44.64 of the optimal objective value of the LAM.

Since the cost objective is considered, the smaller the objective value the better

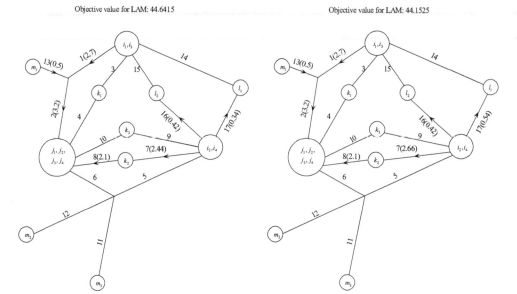

FIGURE 5.11   Allocation scheme without consideration of traffic intensity.

TABLE 5.7   Results of the single objective LAM and UAM

	x_{11}^{EF}	x_{13}^{EF}	x_{33}^{EF}	x_{22}^{TF}	x_{24}^{TF}	x_{11}^{BF}	x_{22}^{ET}	x_{42}^{ET}	x_{22}^{ED}	x_{41}^{ED}	others
LAM	1.1	0.1	1.5	0.9	1.2	0.5	1.58	0.86	0.42	0.34	0
UAM	1.1	0.1	1.5	0.9	1.2	0.5	2	0.66	0	0.54	0

that is, the objective value of the UAM is better than that of the the LAM. We can conclude that: when the feasible region expands, we can usually find better solutions. So, when we use both the LAM and UAM to deal with a model under the fuzzy environment, we can describe the solution as an interval [44.15,44.64]. Because the model is for decision making under fuzzy environments, a crisp certain solution is a little limited, and it is more appropriate to use the interval value to describe the solution or the decision in the fuzzy environment.

Second, let us consider the traffic intensity objective. We can move this objective to the constraints as follows. We determine some tolerance traffic intensities and consider the traffic intensity constraint of each road section. Thus, we can turn model (5.171) and model (5.172) into linear models. The road sections 2 and 6 are the important ones since the earth-rock work from several directions will pass through these roads to the filling projects. These two road sections are the main reasons for the possible infeasibility of the allocation scheme because of the congested traffic situation. So, here we consider the traffic intensity of road sections 2 and 6 only, and we suppose that the tolerance traffic intensities of the other road sections are big enough.

Objective values: [80.015, 83.741]

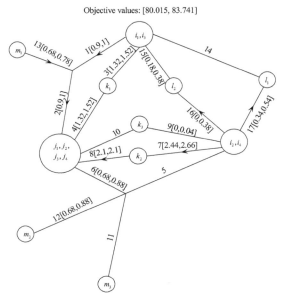

FIGURE 5.12 Allocation scheme with consideration of traffic intensity.

In order to make the allocation scheme feasible, we give two traffic intensity limits and add the following two traffic intensity constraints (5.174) to Model (5.173) as follows:

$$\begin{cases} [x_{11}^{EF} + x_{13}^{EF} + x_{31}^{EF} + x_{33}^{EF}] + [x_{11}^{BF} + x_{13}^{BF}] \le Q_0^2 \\ [x_{22}^{EF} + x_{24}^{EF} + x_{42}^{EF} + x_{44}^{EF}] + [x_{22}^{BF} + x_{24}^{BF}] + [x_{22}^{BF} + x_{24}^{BF}] \le Q_0^6 \end{cases} \quad (5.174)$$

The traffic intensity tolerance of road sections 2 and 6 are $Q_0^2 = 1, Q_0^6 = 1.5$. Then, after solving the LAM and UAM for the earth-rock work allocation problem with the consideration of the traffic intensities, we find that the total cost goes up from [44.15,44.64] to [67.93,83.51], and the allocation scheme is shown in Figure 5.12.

So, after adding the traffic intensity constraints, we can see from Figure 5.12 that the other road sections share some quantities of earth-rock work to alleviate road sections 2 and 6, which leads to a total cost increase. However, the sacrifice of the cost will make the feasibility of the allocation scheme more guaranteed in return.

It is necessary to mention that the above results are obtained based on the following parameters: $\lambda = 0.5, \delta_1 = \delta_2 = \delta_3 = \delta_4 = 0.8$.

Since there are some parameters that include the pessimistic–optimistic index λ and the predetermined confidence levels δ_r in the decision-making model for the earth-rock work allocation, the decision makers can adjust these parameters to obtain different solutions under different levels of the parameters. So, at last, let us do the following sensitivity analysis based on the parameters. The solutions reflect different optimistic–pessimistic attitudes and different confidence levels for the chance constraints. Here we set $\delta_r, r = 1, 2, 3, 4$ as the same δ, and we change the parameters λ and δ and use Table 5.8 to show the different optimal objective values of LAM and UAM with different parameters.

TABLE 5.8   Sensitivity analysis

δ \ λ	0	0.2	0.4	0.6	0.8	1.0
0.5	[70.68,74.27]	[73.62,77.21]	[76.55,80.16]	[79.46,83.08]	[82.33,85.96]	[85.20,88.84]
0.6	[71.34,74.94]	[74.28,77.89]	[77.22,80.84]	[80.13,83.76]	[83.00,86.64]	[85.87,89.52]
0.7	[72.00,75.64]	[74.94,78.59]	[77.88,81.54]	[80.80,84.46]	[83.67,87.34]	[86.54,90.23]
0.8	[72.66,76.36]	[75.60,79.31]	[78.54,82.27]	[81.46,85.20]	[84.34,88.09]	[87.21,90.98]
0.9	[73.32,77.09]	[76.26,80.04]	[79.21,82.99]	[82.13,85.93]	[85.00,88.84]	[87.88,91.74]
1.0	[73.98,77.82]	[76.93,80.77]	[79.87,83.72]	[82.79,86.67]	[85.67,89.59]	[88.55,92.51]

From Table 5.8 we can see that, under the same confidence levels of the chance constraints, when the parameter λ gets bigger, the value of the objective function increases, and vice versa. It is noted that the pessimistic–pessimistic parameter should be understood according to a real-life problem. For the cost objective, λ is actually a pessimistic parameter; and for the profit objective, λ is actually an optimistic parameter. We can conclude like this: Under the same confidence levels of the chance constraints, the result will be more optimistic (pessimistic) when the optimistic (pessimistic) parameter goes up. The decision maker will get the certain optimal result according to his attitude, and we believe it is more appropriate in the real-life uncertain decision making problem.

From Table 5.8 we can also see that, with the same pessimistic–pessimistic parameter, the optimal objective values of the LAM will grow bigger when the confidence level increases, and the optimal objective values of the UAM will also become bigger when the confidence level increases. We can also explain it again as follows. With the same pessimistic–pessimistic attitude, the results will be better for both of the LAM and the UAM when the confidence levels decrease. That is because, when the confidence levels decrease, the feasible region will expand and we can find a better solution in a larger feasible region. In contrast, when the confidence levels increase, the feasible region will shrink, and the solution we can find in a smaller feasible region will be worse.

To sum up, in a maximization (minimization) decision-making problem under fuzzy environment, the decision maker can choose the parameters accordingly for the decision-making model. If he or she holds an optimistic attitude about the objective, he or she can choose the bigger(smaller) optimistic-pessimistic parameter λ, and if he is cautious about the constraints, he can choose larger confidence levels δ. After deciding the parameters, the decision maker will get the solution that meets his or her requirements by using the proposed models and approaches.

Chapter 6

Methodological System for RMODM

Rough multiple objective decision making (RMODM) considers the application of the rough set theory to multiple objective decision making. In this book the rough set theory is used to deal with a series of multiple objective decision-making problems:

- Classic multiple objective decision-making problems

- Bilevel multiple objective decision-making problems

- Random multiple objective decision-making problems

- Fuzzy multiple objective decision-making problems

The applications of the rough set theory in this book mainly refers to the rough approximation technique. Rough approximation techniques here not only refer to the approximation for coefficients, that is, rough intervals, but also to the approximation for various kinds of models themselves.

Multiple objective decision making (MODM) models with uncertain variables, including rough intervals, random variables and fuzzy variables, are meaningless mathematically, and we have to adopt certain philosophies to deal with them according to the decision maker's preferences and habits. Thus the following six kinds of models are proposed:

1. Expected value model (EVM);
2. Chance-constrained model (CCM);
3. Dependent-chance model (DCM);
4. Expectation model with chance constraints (ECM);
5. Chance-constrained model with expectation constraints (CEM);
6. Dependent-chance model with expectation constraints (DEM).

For the random MODM and fuzzy MODM, the integration of rough approximation techniques and above models will make the decision process more flexible and practical. Then we will obtain the corresponding rough decision models.

In some special situations, we can use mathematical tools to transform the models mentioned above into their crisp equivalent models. However, it is difficult to obtain the crisp equivalent models for other cases, and then we can design hybrid intelligent algorithm to get the approximate solutions.

RMODM has been applied to optimization problems such as solid transportation problems [285], reverse logistics problems [286], supply chain planning problems,

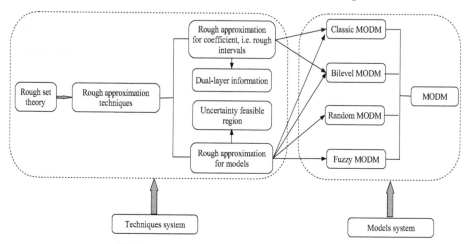

FIGURE 6.1   Development of RMODM.

resource-constrained project scheduling problems and allocation problems [318], inventory problems [272], and so on. It is expected that a rough model will be applied to more and more practical problems with further research.

6.1   Motivation for Researching RMODM

Why should we research RMODM? Let us recall the two foundations: multiple objective decision making and the rough set theory (see Figure 6.1).

Optimization deals with the problem of seeking solutions over a set of possible choices to optimize certain criteria. If there is only one criterion to consider, it becomes a single objective optimization problem, a type studied extensively for the past 50 years. If there is more than one criterion and they must be treated simultaneously, we have a multiple objective optimization problem, which is also called multiple objective decision-making problem. Multiple objective decision making arise in the design, modeling, and planning of many complex real systems in the areas of industrial production, urban transportation, capital budgeting, forest management, reservoir management, layout and landscaping of new cities, and energy distribution, for instance. Almost every important real-world decision problem involves multiple and conflicting objectives that need to be tackled while respecting various constraints, leading to an overwhelming problem complexity. The research of certain MODM can be traced back in to the 19th century. V. Pareto first introduced multiple objective decision-making models from the mathematical point of view in 1896. Later, K. Arrow [10] proposed the concept of efficient points in 1953. MODM has gradually become widespread concern and has developed continually.

Generally speaking, there are five elements in a MODM problem:

1. Decision variable: $\mathbf{x} = (x_1, x_2, \cdots, x_n)^T$.

2. Objective function: $f(\mathbf{x}) = [f_1(\mathbf{x}), f_2(\mathbf{x}), \cdots, f_m(\mathbf{x})], (m \geq 2)$.

3. Feasible solution set: $X = \{\mathbf{x} \in \mathfrak{R}^n | g_i(\mathbf{x}) \leq 0, i = 1, 2, \cdots, p, h_r(\mathbf{x}) = 0, r = 1, 2, \cdots, q\}$.

4. Preference relation: In the image set $f(X) = \{f(\mathbf{x}) | \mathbf{x} \in X\}$, there is a certain binary relation that could reflect the preference of the decision maker.

5. Definition of the solution: Define the optimal solution of f in X based on the known preference relation.

So, an MODM problem can be described as follows:

$$\begin{cases} \max[f_1(\mathbf{x}), f_2(\mathbf{x}), \cdots, f_m(\mathbf{x})] \\ \text{s.t.} \begin{cases} g_i(\mathbf{x}) \leq 0, \ i = 1, 2, \cdots, p \\ h_r(\mathbf{x}) = 0, r = i = 1, 2, \cdots, q \end{cases} \end{cases}$$

The concept of rough set, which was presented by Z. Pawlak in [238, 241], is an important concept since it is applicable in many fields such as artificial intelligence, expert systems, civil engineering, medical data analysis, data mining, pattern recognition, and decision theory. Rough set theory is an effective tool in dealing with insufficient information. For classic MODM, the feasible region may not be precise, so we can apply rough approximation techniques to handle it, and then we have the following model:

$$\begin{cases} \min[f_1(\mathbf{x}), f_2(\mathbf{x}), \cdots, f_m(\mathbf{x})] \\ \text{s.t.} \begin{cases} \mu_X^R(\mathbf{x}) \geq \alpha, \alpha \in (0, 1] \\ X = \{\mathbf{x} \in \mathfrak{R}^n | g_j(\mathbf{x}) \leq 0, j = 1, 2, \cdots, p\} \end{cases} \end{cases}$$

where $\mu_X^R(\mathbf{x})$ is the rough membership function of x belonging to X, $\alpha(\in (0, 1])$ is predetermined rough membership by DM, representing the DM's required accuracy. In addition, the parameters in classic MODM can also be assumed as rough intervals due to dual-layer information.

Actually, in order to make a satisfactory decision in practice, an important problem is to determine the type and accuracy of the information. If complete information is required in the decision-making process, it will mean the expenditure of some extra time and money. If incomplete information is used to make a decision quickly, then it is possible to take nonoptimal action. In fact, we cannot have complete accuracy in both information and decision because the total cost is the sum of the cost spent for running the target system and the cost spent for getting decision information. Since we have to balance the advantage of making better decisions against the disadvantages of getting more accurate information, incomplete information will almost surely be used in the real-life decision process, and uncertain decision making is an important tool in dealing with decision making with imperfect information. The MODM model with uncertain parameters can be formulated by

$$\begin{cases} \max[f_1(\mathbf{x}, \boldsymbol{\xi}), f_2(\mathbf{x}, \boldsymbol{\xi}), \cdots, f_m(\mathbf{x}, \boldsymbol{\xi})] \\ \text{s.t. } g_r(\mathbf{x}, \boldsymbol{\xi}) \leq 0, r = 1, 2, \cdots, p \end{cases} \tag{6.1}$$

where $\xi = (\xi_1, \xi_2, \cdots, \xi_n)$ is a random vector or fuzzy vector. Since the above model is not clear in the mathematical sense, the operator, such as the expected value operator, chance operator are used to make them meaningful mathematically. However, information may be lost in the process of using these operators for random and fuzzy cases. The application of rough approximation techniques makes the decision-making process more flexible and practical.

Several researchers have discussed the combination of rough set theory and all kinds of MODM, including classic MODM, bilevel MODM, random MODM, and fuzzy MODM [272, 309, 311–315, 317–319]. But how to establish the general framework of RMODM and apply RMODM to more complex real problems? This is the strong motivation to study RMODM.

6.2   Physics-Based Model System

Rough multiple objective decision-making refers to application rough set theory to multiple objective decision making. The "rough " referred to here includes not only the rough approximation to feasible sets but also the rough interval parameters. In this book we use four typical problems: site layout planning problems, bilevel allocation problems, project scheduling problems, and allocation problems to illustrate multiple objective rough decision making (MORDM), bilevel multiple objective rough decision making (BL-MORDM), random multiple objective rough decision making (Ra-MORDM), and fuzzy multiple objective rough decision making (Fu-MORDM). Among the various typical problems, we choose the construction site layout planning problem, hierarchical water resource management problem, resource-constrained project scheduling problem, and earth-rock work allocation problem to clarify corresponding rough multiple objective decision making in detail (see Figure.6.2).

In the dynamic construction site layout planning problem, some parameters, such as cost, are highly imprecise as it has to be estimated in terms of a decision maker's subjective experiences and objective historical data. Obviously, the strategy capable of handling the problem under both normal and special conditions is preferred. Based on such considerations, decision makers identified that this kind of variable ranges from a units to b units in most cases and, in some special periods, the range may vary from c units to d units, where $c \leq a \leq b \leq d$. Hence rough intervals $([a,b],[c,d])$ can be used to deal with these kind of uncertain parameters. So it is reasonable for people to believe that the rough reverse logistic problem is more realistic. Thus we develop the multiple objective rough decision-making model for the dynamic construction site layout planning problem.

In the bilevel water resource allocation problem, the presence of uncertainty further amplifies the complexity of the problem. If the decision maker classifies the decision environment into two categories–a normal case and a special case–the deci-

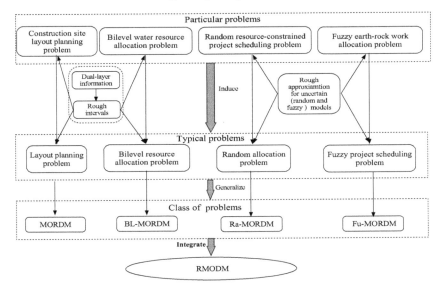

FIGURE 6.2 RMODM problems.

sion parameters are assumed as intervals in each case. For this type of information, the existing methods can neither reflect its dual-layer feature nor entirely pass it to the resulting decisions. Two challenges thus emerge: one is to find an effective expression that could reflect dual-layer information (i.e., not only the parameter's most possible value but also its most reliable value), and the other is to use an appropriate method to generate decisions with dual-layer information directly corresponding to the most possible and reliable conditions of the system. Rough intervals can be a suitable concept to express such information. Obviously, the essence of regional water resources allocation based on water rights under market mechanisms is a two-story structure game decision-making problem.

In the resource-constrained project scheduling problem, when developing the most appropriate schedule for a project, uncertainty is unavoidable for the decision manager. Elements such as processing time, resource quantities, due date of the project, and maximum limited resources need to be considered. The literature above proves the existence of randomness in resource-constrained project scheduling problems with multimode (rc-PSP/mM). However, no one has considered the roughness in rc-PSP/mM which also exists. For instance, let us consider the activity-drilled drain hole (including flexible drain) belonging to our example project drilling grouting, because the dates collected from a practical system often contains noise, which means the dates are imperfect and incomplete. In order to deal with the dates appropriately, we use the rough approximation technique to handle the phenomenon. Thus the random multiple objective rough decision-making model should be built for the resource-constrained project scheduling problem.

In the earth-rock work allocation problem, some parameters are fuzzy variables because of insufficient information. As we apply the rough approximation technique to deal with the fuzzy model, the decision will be more flexible, so it is reasonable to build a fuzzy multiple objective rough decision-making model.

It is noted that the problems introduced in this book are just some example problems, and readers can absolutely extend the application areas.

6.3 Mathematical Model System

First of all, we consider the classic MODM model as

$$\begin{cases} \max[f_1(\mathbf{x}), f_2(\mathbf{x}), \cdots, f_m(\mathbf{x})] \\ \text{s.t. } g_r(\mathbf{x}) \leq 0, r = 1, 2, \cdots, p \end{cases}$$

If the DM can construct approximation for the feasible region of the classic MODM model, we can obtain

$$\begin{cases} \min[f_1(\mathbf{x}), f_2(\mathbf{x}), \cdots, f_m(\mathbf{x})] \\ \text{s.t.} \begin{cases} \mu_X^R(\mathbf{x}) \geq \alpha, \alpha \in (0, 1] \\ X = \{\mathbf{x} \in \Re^n | g_j(\mathbf{x}) \leq 0, j = 1, 2, \cdots, p\} \end{cases} \end{cases}$$

where $\mu_X^R(x)$ is the rough membership function of x belonging to X, and $\alpha(\in (0, 1])$ is the predetermined rough membership by the DM, representing the DM's required accuracy. As the parameters in MODM are treated by the rough approximation technique, we have

$$\begin{cases} \max[f_1(\mathbf{x}, \boldsymbol{\xi}), f_2(\mathbf{x}, \boldsymbol{\xi}), \cdots, f_m(\mathbf{x}, \boldsymbol{\xi})] \\ \text{s.t. } g_r(\mathbf{x}, \boldsymbol{\xi}) \leq 0, r = 1, 2, \cdots, p \end{cases}$$

where $\boldsymbol{\xi} = (\xi_1, \xi_2, \cdots, \xi_n)$ is a rough interval vector.

Meanwhile, the bilevel model with rough intervals vectors can be formulated by

$$\begin{cases} \max\limits_{\mathbf{x}}[F_1(\mathbf{x}, \mathbf{y}, \boldsymbol{\xi}), F_2(\mathbf{x}, \mathbf{y}, \boldsymbol{\xi}), \cdots, F_M(\mathbf{x}, \mathbf{y}, \boldsymbol{\xi})] \\ \text{s.t. } G_r(\mathbf{x}, \mathbf{y}, \boldsymbol{\xi}) \leq 0, r = 1, 2, \cdots, P \\ \text{where } \mathbf{y} \text{ solves} \\ \quad \begin{cases} \max\limits_{\mathbf{y}}[f_1(\mathbf{x}, \mathbf{y}, \boldsymbol{\xi}), f_2(\mathbf{x}, \mathbf{y}, \boldsymbol{\xi}), \cdots, f_m(\mathbf{x}, \mathbf{y}, \boldsymbol{\xi})] \\ \text{s.t. } g_r(\mathbf{x}, \mathbf{y}, \boldsymbol{\xi}) \leq 0, r = 1, 2, \cdots, p \end{cases} \end{cases}$$

where $\boldsymbol{\xi} = (\xi_1, \xi_2, \cdots, \xi_n)$ is a rough interval vector.

Actually, if $\boldsymbol{\xi}$ denotes random and fuzzy vectors, we can obtain random MODM model and fuzzy MODM model.

It is necessary for us to know that the above models are conceptual models rather than mathematical models because we cannot maximize an uncertain quantity. A natural ordership does not exist in an uncertain world. Since uncertain variables exist, the above models are ambiguous. The meaning of maximizing $f_1(\mathbf{x}, \mathbf{y}, \boldsymbol{\xi}), f_2(\mathbf{x}, \mathbf{y}, \boldsymbol{\xi})$,

$\cdots, f_n(\mathbf{x}, \mathbf{y}, \boldsymbol{\xi})$ is unclear, and the constraints $g_r(\mathbf{x}, \mathbf{y}, \boldsymbol{\xi}) \leq 0$, $r = 1, 2, \cdots, p$ do not define a deterministic feasible set. So we need to adopt some philosophies to deal with and make the above model solvable. Philosophy 1–5 will be used to deal with the rough decision-making models.

First, let us consider the objective functions formulated as

$$\max \ [f_1(\mathbf{x}, \boldsymbol{\xi}), f_2(\mathbf{x}, \boldsymbol{\xi}), \cdots, f_m(\mathbf{x}, \boldsymbol{\xi})]$$

where $\boldsymbol{\xi}$ is the rough interval vector (random vector, fuzzy vector).

There are three types of philosophies to handle the objectives.

Philosophy 1. Making the decision with optimizing the expected value of the objectives. That is, maximizing the expected values of the objective functions for the Max problem, or minimizing the expected values of the objective functions for the Min problem.

$$\max \ E[f_1(\mathbf{x}, \boldsymbol{\xi}), f_2(\mathbf{x}, \boldsymbol{\xi}), \cdots, f_m(\mathbf{x}, \boldsymbol{\xi})]$$

or

$$\min \ E[f_1(\mathbf{x}, \boldsymbol{\xi}), f_2(\mathbf{x}, \boldsymbol{\xi}), \cdots, f_m(\mathbf{x}, \boldsymbol{\xi})].$$

Philosophy 2. Making the decision that provides the best optimal objective values with a given confidence level. That is, maximizing the referenced objective values $\bar{f}_i$ subjects to $f_i(\mathbf{x}, \boldsymbol{\xi}) \geq \bar{f}_i$ with a confidence level α_i, or minimizing the referenced objective values $\bar{f}_i$ subjects to $f_i(\mathbf{x}, \boldsymbol{\xi}) \leq \bar{f}_i$ with a confidence level α_i.

$$\begin{cases} \max \ [\bar{f}_1, \bar{f}_2, \cdots, \bar{f}_n] \\ \text{s.t. } Ch\{f_i(\mathbf{x}, \boldsymbol{\xi}) \geq \bar{f}_i\} \geq \alpha_i, \ i = 1, 2, \cdots, n \end{cases}$$

where α_i should be predetermined, $\bar{f}_1, \bar{f}_2, \cdots, \bar{f}_n$ are called critical values, Ch means $Appr, Pr$ or Me.

Philosophy 3. Making the decision with maximizing the chance of the events. That is, maximizing the chance of the events $f_i(\mathbf{x}, \boldsymbol{\xi}) \geq \bar{f}_i$ or $f_i(\mathbf{x}, \boldsymbol{\xi}) \leq \bar{f}_i$.

$$\max[Ch\{f_1(\mathbf{x}, \boldsymbol{\xi}) \geq \bar{f}_1\}, Ch\{f_2(x, \boldsymbol{\xi}) \geq \bar{f}_2\}, \cdots Ch\{f_m(x, \boldsymbol{\xi}) \geq \bar{f}_m\}]$$

where $\bar{f}_i$ should be predetermined, and Ch means $Appr, Pr$, or Me. Then, let us consider the constraints

$$\begin{cases} g_r(\mathbf{x}, \boldsymbol{\xi}) \leq 0 \ r = 1, 2, \cdots, p \\ \mathbf{x} \in X \end{cases}$$

where $\boldsymbol{\xi}$ is the rough intervals vector, random vector, or fuzzy vector.

There are two types of philosophies to handle the constraints.

Philosophy 4. Making the optimal decision subject to the expected constraints. That is,

$$E[g_r(\mathbf{x}, \boldsymbol{\xi})] \leq 0, \ r = 1, 2, \cdots, p.$$

Philosophy 5. Making the optimal decision subjects to the chance constraints.

$$Ch\{g_r(\mathbf{x}, \boldsymbol{\xi}) \leq 0\} \geq \beta_r, \ r = 1, 2, \cdots, p.$$

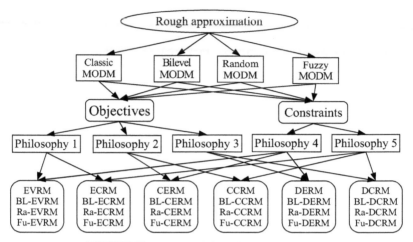

FIGURE 6.3 Decision philosophies system.

By combining the three philosophies for the objective functions and two philosophies for the constraints, we can get several types of models that can deal with the initial multiple objective decision-making models with uncertain parameters. In particular, for random and fuzzy cases, the information may be lost in the transformation process. So the application of rough approximation techniques makes the decision more flexible and practical (see Figure 6.3). The general frameworks for EVM, CCM, and DCM are illustrated as

$$(\textbf{EVM}) \begin{cases} \max E[f_1(\mathbf{x},\boldsymbol{\xi}), f_2(\mathbf{x},\boldsymbol{\xi}), \cdots, f_n(\mathbf{x},\boldsymbol{\xi})] \\ \text{s.t.} \begin{cases} E[g_r(\mathbf{x},\boldsymbol{\xi})] \leq 0, \ r = 1,2,\cdots,p \\ \mathbf{x} \in X \end{cases} \end{cases}$$

and

$$(\textbf{CCM}) \begin{cases} \max \left[\bar{f}_1, \bar{f}_2, \cdots, \bar{f}_n\right] \\ \text{s.t.} \begin{cases} Ch\{f_i(\mathbf{x},\boldsymbol{\xi}) \geq \bar{f}_i\} \geq \alpha_i, \ i = 1,2,\cdots,n \\ Ch\{g_r(\mathbf{x},\boldsymbol{\xi}) \leq 0\} \geq \beta_r, \ r = 1,2,\cdots,p \\ \mathbf{x} \in X \end{cases} \end{cases}$$

where $\alpha_i (i = 1,2,\cdots,m), \beta_r (r = 1,2,\cdots,p)$ are the predetermined confidence levels.

$$(\textbf{DCM}) \begin{cases} \max[Ch\{f_1(\mathbf{x},\boldsymbol{\xi}) \geq \bar{f}_1\}, Ch\{f_2(\mathbf{x},\boldsymbol{\xi}) \geq \bar{f}_2\}, \cdots, Ch\{f_m(\mathbf{x},\boldsymbol{\xi}) \geq \bar{f}_m\}] \\ \text{s.t.} \begin{cases} Ch\{g_r(\mathbf{x},\boldsymbol{\xi}) \leq 0\} \geq \beta_r, \ r = 1,2,\cdots,p \\ \mathbf{x} \in X \end{cases} \end{cases}$$

where $\bar{f}_i (i = 1,2,\cdots,m), \beta_r (r = 1,2,\cdots,p)$ are the predetermined referenced objective values and confidence levels.

For random EVM, if we apply the rough approximation technique to deal with it, we have

$$(\textbf{Ra-EVRM}) \begin{cases} \max\left[E[f_1(\mathbf{x},\boldsymbol{\xi})], E[f_2(\mathbf{x},\boldsymbol{\xi})], \cdots, E[f_m(\mathbf{x},\boldsymbol{\xi})]\right] \\ \text{s.t.} \begin{cases} \alpha(I_r) \geq \theta_r \\ \mathbf{x} \in \bar{I}_r \\ r = 1,2,\cdots,p \end{cases} \end{cases}$$

where $\alpha(I_r)$ expresses the accuracy of the approximation, $I_r = \{\mathbf{x} \in X | g_r(\mathbf{x},\boldsymbol{\xi}) \leq 0\}$, and the similarity relationship is defined as R_{h_r} as follows:

$$\mathbf{x}R_{h_r}\mathbf{y} \Leftrightarrow E[|g_r(\mathbf{x},\boldsymbol{\xi}) - g_r(\mathbf{y},\boldsymbol{\xi})|] \leq h_r,$$

where $\mathbf{x}, \mathbf{y} \in X$, h_r is the deviation which the decision maker permits.

For random CCM, if we apply the rough approximation technique to deal with it, we have

$$(\textbf{Ra-CCRM}) \begin{cases} \max\left[\bar{f}_1, \bar{f}_2, \cdots, \bar{f}_m\right] \\ \text{s.t.} \begin{cases} \alpha(I_r) \geq \theta_r \\ \mathbf{x} \in \bar{I}_r \\ r = 1,2,\cdots,p \end{cases} \end{cases}$$

where $\alpha(I_r)$ expresses the accuracy of the approximation, $I_r = \{\mathbf{x} \in X | g_r(\mathbf{x},\boldsymbol{\xi}) \leq 0, Pr\{f_i(\mathbf{x},\boldsymbol{\xi}) \geq \bar{f}_i\} \geq \alpha_i, i = 1,2,\cdots,m\}$, and the similarity relationship is defined as $R_{h_r}^{\eta_r}$ as

$$\mathbf{x}R_{h_r}^{\eta_r}\mathbf{y} \Leftrightarrow Pr[|g_r(\mathbf{x},\boldsymbol{\xi}) - g_r(\mathbf{y},\boldsymbol{\xi})| \leq h_r] \geq \eta_r.$$

For random DCM, if we apply the rough approximation technique to deal with it, we have

$$(\textbf{Ra-DCRM}) \begin{cases} \max[\delta_1, \delta_2, \cdots, \delta_m] \\ \text{s.t.} \begin{cases} Pr\{f_i(\mathbf{x},\boldsymbol{\xi}) \geq \bar{f}_i\} \geq \delta_i, i = 1,2,\cdots,m \\ \alpha(I_r) \geq \theta_r \\ \mathbf{x} \in \bar{I}_r \\ r = 1,2,\cdots,p \end{cases} \end{cases}$$

where $\alpha(I_r)$ expresses the accuracy of the approximation, $I_r = \{\mathbf{x} \in X | g_r(\mathbf{x},\boldsymbol{\xi}) \leq 0, Pr\{f_i(\mathbf{x},\boldsymbol{\xi}) \geq \bar{f}_i\} \geq \alpha_i, i = 1,2,\cdots,m\}$, and the similarity relationship is defined as $R_{h_r}^{\eta_r}$ as follows:

$$\mathbf{x}R_{h_r}^{\eta_r}\mathbf{y} \Leftrightarrow Pr[|g_r(\mathbf{x},\boldsymbol{\xi}) - g_r(\mathbf{y},\boldsymbol{\xi})| \leq h_r] \geq \eta_r.$$

For fuzzy EVM, if we apply the rough approximation technique to deal with it, we have

$$(\textbf{Fu-EVRM}) \begin{cases} (LAM) \begin{cases} \max[E^{Me}[f_1(\mathbf{x},\boldsymbol{\xi})], E^{Me}[f_2(\mathbf{x},\boldsymbol{\xi})], \cdots, E^{Me}[f_m(\mathbf{x},\boldsymbol{\xi})]] \\ \text{s.t.} \begin{cases} E^{Pos}[g_r(\mathbf{x},\boldsymbol{\xi})] \leq 0, r = 1,2,\cdots,p \\ \mathbf{x} \in X \end{cases} \end{cases} \\ (UAM) \begin{cases} \max[E^{Me}[f_1(\mathbf{x},\boldsymbol{\xi})], E^{Me}[f_2(\mathbf{x},\boldsymbol{\xi})], \cdots, E^{Me}[f_m(\mathbf{x},\boldsymbol{\xi})]] \\ \text{s.t.} \begin{cases} E^{Pos}[g_r(\mathbf{x},\boldsymbol{\xi})] \leq 0, r = 1,2,\cdots,p \\ \mathbf{x} \in X \end{cases} \end{cases} \end{cases}$$

For fuzzy CCM, if we apply the rough approximation technique to deal with it, we have

$$(\textbf{Fu-CCRM}) \begin{cases} (LAM) \begin{cases} \max[\bar{f}_1, \bar{f}_2, \cdots, \bar{f}_m] \\ \text{s.t.} \begin{cases} Pos\{f_i(\mathbf{x}, \boldsymbol{\xi}) \geq \bar{f}_i\} \geq \delta_i, i = 1, 2, \cdots, m \\ Pos\{g_r(\mathbf{x}, \boldsymbol{\xi}) \leq 0\} \geq \theta_r, \ r = 1, 2, \cdots, p \\ \mathbf{x} \in X \end{cases} \end{cases} \\ (UAM) \begin{cases} \max[\bar{f}_1, \bar{f}_2, \cdots, \bar{f}_m] \\ \text{s.t.} \begin{cases} Nec\{f_i(\mathbf{x}, \boldsymbol{\xi}) \geq \bar{f}_i\} \geq \delta_i, i = 1, 2, \cdots, m \\ Nec\{g_r(\mathbf{x}, \boldsymbol{\xi}) \leq 0\} \geq \theta_r, \ r = 1, 2, \cdots, p \\ \mathbf{x} \in X \end{cases} \end{cases} \end{cases}$$

For fuzzy DCM, if we apply the rough approximation technique to deal with it, we have

$$(\textbf{Fu-DCRM}) \begin{cases} (LAM) \begin{cases} \max[\delta_1, \delta_2, \cdots, \delta_m] \\ \text{s.t.} \begin{cases} Pos\{f_i(\mathbf{x}, \boldsymbol{\xi}) \geq \bar{f}_i\} \geq \delta_i, i = 1, 2, \cdots, m \\ Pos\{g_r(\mathbf{x}, \boldsymbol{\xi}) \leq 0\} \geq \theta_r, \ r = 1, 2, \cdots, p \\ x \in X \end{cases} \end{cases} \\ (UAM) \begin{cases} \max[\delta_1, \delta_2, \cdots, \delta_m] \\ \text{s.t.} \begin{cases} Nec\{f_i(\mathbf{x}, \boldsymbol{\xi}) \geq \bar{f}_i\} \geq \delta_i, i = 1, 2, \cdots, m \\ Nec\{g_r(\mathbf{x}, \boldsymbol{\xi}) \leq 0\} \geq \theta_r, \ r = 1, 2, \cdots, p \\ \mathbf{x} \in X \end{cases} \end{cases} \end{cases}$$

In this book, we mainly discuss the first three models, and the techniques are all incorporated when we deal with the expected value model (EVM), the chance-constrained model (CCM), and the dependent-chance model(DCM). The rest of the models, the expectation model with chance constraints (ECM), the chance-constrained model with expectation constraints (CEM), and dependent-chance model with expectation constraints (DEM) can be handled in the same way. The reader can use the model when they use different philosophies, and it is possible to use every model.

6.4   Model Analysis System

For the linear uncertain multiple objective decision-making models,

$$\begin{cases} \max \left[\sum_{j=1}^{n} \tilde{c}_{1j}x_j, \sum_{j=1}^{n} \tilde{c}_{2j}x_j, \cdots, \sum_{j=1}^{m} \tilde{c}_{nj}x_j \right] \\ \text{s.t.} \begin{cases} \tilde{a}_{rj}x_j \leq \tilde{b}_r, \ r = 1, 2, \cdots, p \\ \mathbf{x} \in X \end{cases} \end{cases}$$

where $\tilde{c}_{ij}, \tilde{a}_{rj}, \tilde{b}_r, (i = 1, 2, \cdots, n; r = 1, 2, \cdots, p)$ are uncertain coefficients, that is, the rough interval, random variables and fuzzy variables. We discussed how to transform the three types of objective functions and two types of constraints into their crisp equivalent formulas in detail. In this book, we introduced the equivalent models for EVM, CCM, and DCM in detail and we simplify them to EEVM, ECCM, and EDCM. In addition, we consider the following linear bilevel MODM:

$$
\begin{cases}
\max \left[\tilde{\mathbf{C}}_1^T \mathbf{x} + \tilde{\mathbf{D}}_1^T \mathbf{y}, \tilde{\mathbf{C}}_2^T \mathbf{x} + \tilde{\mathbf{D}}_2^T \mathbf{y}, \cdots, \tilde{\mathbf{C}}_{m_1}^T \mathbf{x} + \tilde{\mathbf{D}}_{m_1}^T \mathbf{y} \right] \\
\text{s.t. } \tilde{\mathbf{A}}_r^T \mathbf{x} + \tilde{\mathbf{B}}_r^T \mathbf{y} \le \tilde{E}_r, r = 1, 2, \cdots, p_1 \\
\text{where } \mathbf{y} \text{ solves} \\
\quad \begin{cases}
\max \left[\tilde{\mathbf{c}}_1^T \mathbf{x} + \tilde{\mathbf{d}}_1^T \mathbf{y}, \tilde{\mathbf{c}}_2^T \mathbf{y} + \tilde{\mathbf{d}}_2^T \mathbf{y}, \cdots, \tilde{\mathbf{c}}_{m_2}^T \mathbf{x} + \tilde{\mathbf{d}}_{m_2}^T \mathbf{y} \right] \\
\text{s.t. } \begin{cases} \tilde{\mathbf{a}}_r^T \mathbf{x} + \tilde{\mathbf{b}}_r^T \mathbf{y} \le \tilde{e}_r, r = 1, 2, \cdots, p_2 \\ \mathbf{x} \in \Re^{n_1}, \mathbf{y} \in \Re^{n_2} \end{cases}
\end{cases}
\end{cases}
$$

where $\tilde{\mathbf{C}}_i = (\tilde{C}_{i1}, \tilde{C}_{i2}, \cdots, \tilde{C}_{in_1})^T, \tilde{\mathbf{D}}_i = (\tilde{D}_{i1}, \tilde{D}_{i2}, \cdots, \tilde{D}_{in_1})^T, \tilde{\mathbf{A}}_r = (\tilde{A}_{r1}, \tilde{A}_{r2}, \cdots, \tilde{A}_{rn_1})^T,$ $\tilde{\mathbf{B}}_r = (\tilde{B}_{r1}, \tilde{B}_{r2}, \cdots, \tilde{B}_{rn_1})^T$, are rough interval vectors, and $\tilde{E}_r$ are rough intervals, $i = 1, 2, \cdots, m_1, r = 1, 2, \cdots, p_1$. At the same time, $\tilde{\mathbf{c}}_i = (\tilde{c}_{i1}, \tilde{c}_{i2}, \cdots, \tilde{c}_{in_2})^T, \tilde{\mathbf{d}}_i = (\tilde{d}_{i1}, \tilde{d}_{i2}, \cdots, \tilde{d}_{in_2})^T, \tilde{\mathbf{a}}_r = (\tilde{a}_{r1}, \tilde{a}_{r2}, \cdots, \tilde{a}_{rn_1})^T, \tilde{\mathbf{b}}_r = (\tilde{b}_{r1}, \tilde{b}_{r2}, \cdots, \tilde{b}_{rn_1})^T$ are rough interval vectors, and $\tilde{e}_r$ are rough intervals, $i = 1, 2, \cdots, m_2, r = 1, 2, \cdots, p_2$.

For the multiple objective decision making with rough intervals parameters, there are 4 basic theorems for handling the objective functions and the constraints: Theorem 2.8, Theorem 2.10, Theorem 2.11, and Theorem 2.14. According to these four theorems, we can get the crisp equivalent models for EVRM, CCRM, and DCRM.

EEVRM

$$
\begin{cases}
\max [\frac{1}{2} \sum\limits_{j=1}^{n} \eta((c_{1j1} + c_{1j2}) + (1 - \eta)(c_{1j3} + c_{1j4}))x_j, \frac{1}{2} \sum\limits_{j=1}^{n} (\eta(c_{2j1} + c_{2j2}) \\
\quad + (1 - \eta)(c_{2j3} + c_{2j4}))x_j, \cdots, \frac{1}{2} \sum\limits_{j=1}^{n} (\eta(c_{mj1} + c_{mj2}) + (1 - \eta)(c_{mj3} + c_{mj4}))x_j] \\
\text{s.t,} \begin{cases} \sum\limits_{j=1}^{n} (a_{rj1} + a_{rj2} + a_{rj3} + a_{rj4})x_j \le b_{r1} + b_{r2} + b_{r3} + b_{r4}, r = 1, 2, \cdots, p \\ x_j \ge 0, j = 1, 2, \cdots, n \end{cases}
\end{cases}
$$

ECCRM

$$
\begin{cases}
\max [\bar{f}_1, \bar{f}_2, \cdots, \bar{f}_m] \\
\text{s.t.} \begin{cases}
\Phi, & \text{if } c_{i4} \le \bar{f}_i \\
\bar{f}_i \le c_{i4} - \frac{(c_{i4} - c_{i3})\beta_i}{1 - \eta}, & \text{if } c_{i2} \le \bar{f}_i \le c_{i4} \\
\bar{f}_i \le \frac{\eta(c_{i4} - c_{i3})c_{i2} + (1 - \eta)(c_{i2} - c_{i1})c_{i4} - \beta_i(c_{i2} - c_{i1})(c_{i4} - c_{i3})}{\eta(c_{i4} - c_{i3}) + (1 - \eta)(c_{i1} - c_{i2})}, & \text{if } c_{i1} \le \bar{f}_i \le c_{i2} \\
\bar{f}_i \le \beta_i c_{i1} + (1 - \beta_i)c_{i4}, & \text{if } c_{i3} \le \bar{f}_i \le c_{i4} \\
X, & \text{if } \bar{f}_i \le c_i
\end{cases}
\end{cases}
$$

EDCRM

$$
\begin{cases}
\max\left[(1-\eta)\frac{c_{14}-\bar{f}_1}{(c_{14}-c_{13})},(1-\eta)\frac{c_{24}-\bar{f}_2}{(c_{24}-c_{23})},\cdots,(1-\eta)\frac{c_{m4}-\bar{f}_m}{(c_{m4}-c_{m3})}\right] \\
\text{s.t.}\begin{cases}
\alpha_r(d-c)+(1-\eta)c\leq 0, & \text{if } c\leq 0\leq a \\
\alpha_r(b-a)(d-c)+\eta a(d-c)+(1-\eta)c(b-a)\leq 0, & \text{if } a\leq 0\leq b \\
\alpha_r(d-c)-(1-\eta)d\leq 0, & \text{if } b\leq 0\leq d \\
\mathbf{x}\geq\mathbf{0}
\end{cases}
\end{cases}
$$

For bilevel multiple objective decision-making with rough interval parameters, there are four basic theorems: Theorem 3.1, Theorem 3.3, Theorem 3.4, and Theorem 3.5. According to these four theorems, we can get the crisp equivalent models for BL-EVRM, BL-CCRM and BL-DCRM.

BL-EEVRM

$$
\begin{cases}
\max[\frac{1}{2}\sum_{j=1}^{n_1}(\eta(C_{1j1}+C_{1j2})+(1-\eta)(C_{1j3}+C_{1j4}))x_j+\sum_{j=1}^{n_2}(\eta(D_{1j1}+D_{1j2}) \\
+(1-\eta)(D_{1j3}+D_{1j4}))y_j,\frac{1}{2}\sum_{j=1}^{n_1}(\eta(C_{2j1}+C_{2j2})+(1-\eta)(C_{2j3}+C_{2j4}))x_j+ \\
\sum_{j=1}^{n_2}(\eta(D_{2j1}+D_{2j2})+(1-\eta)(D_{2j3}+D_{2j4}))y_j,\cdots,\frac{1}{2}\sum_{j=1}^{n_1}(\eta(C_{m_1j1}+C_{m_1j2})+ \\
(1-\eta)(C_{m_1j3}+C_{m_1j4}))x_j+\sum_{j=1}^{n_2}((\eta(D_{m_1j1}+D_{m_1j2})+(1-\eta)(D_{m_1j3}+D_{m_1j4} \\
))y_j] \\
\text{s.t. }\sum_{j=1}^{n_1}\eta(A_{rj1}+A_{rj2})+(1-\eta)(A_{rj3}+A_{rj4})+\sum_{j=1}^{n_2}\eta(B_{rj1}+B_{rj2})+(1-\eta) \\
(B_{rj3}+B_{rj4})\leq\eta(E_{r1}+E_{r2})+(1-\eta)(E_{r3}+E_{r4}),r=1,2,\cdots,p_1 \\
\text{where } y \text{ solves} \\
\begin{cases}
\max[[\frac{1}{2}\sum_{j=1}^{n_2}(\eta(c_{1j1}+c_{1j2})+(1-\eta)(c_{1j3}+c_{1j4}))x_j+\sum_{j=1}^{n_2}(\eta(d_{1j1}+d_{1j2}) \\
+(1-\eta)(d_{1j3}+d_{1j4}))y_j,\frac{1}{2}\sum_{j=1}^{n_1}(\eta(c_{2j1}+c_{2j2})+(1-\eta)(c_{2j3}+c_{2j4}))x_j+ \\
\sum_{j=1}^{n_2}(\eta(d_{2j1}+d_{2j2})+(1-\eta)(d_{2j3}+d_{2j4}))y_j,\cdots,\frac{1}{2}\sum_{j=1}^{n_2}(\eta(c_{m_2j1}+c_{m_2j2})+ \\
(1-\eta)(c_{m_2j3}+c_{m_2j4}))x_j+\sum_{j=1}^{n_2}(\eta(d_{m_1j1}+d_{m_1j2})+(1-\eta)(d_{m_2j3}+d_{m_2j4} \\
))] \\
\text{s.t. }\begin{cases}
\sum_{j=1}^{n_2}\eta(a_{rj1}+a_{rj2})+(1-\eta)(a_{rj3}+a_{rj4})+\sum_{j=1}^{n_2}\eta(b_{rj1}+b_{rj2}) \\
+(1-\eta)(b_{rj3}+b_{rj4})\leq\eta(e_{r1}+e_{r2})+(1-\eta)(e_{r3}+e_{r4}) \\
r=1,2,\cdots,p_2 \\
\mathbf{x}\in\Re^{n_1},y\in\Re^{n_2}
\end{cases}
\end{cases}
\end{cases}
$$

For **BL-ECCRM**, only the upper-level model is presented since the lower-level model has the same form.

$$
\begin{cases}
\max [\bar{F}_1, \bar{F}_2, \cdots, \bar{F}_{m_1}] \\
\text{s.t.} \begin{cases}
\Phi, & \text{if } 0 \le c \\
\beta_r(d-c) + (1-\eta)c \le 0, & \text{if } c \le 0 \le a \\
\beta_r(b-a)(d-c) + \eta a(d-c) + (1-\eta)c(b-a) \le 0, & \text{if } a \le 0 \le b \\
\beta_r(d-c) - (1-\eta)d \le 0, & \text{if } b \le 0 \le d \\
X, & \text{if } d \le 0
\end{cases}
\end{cases}
$$

Similarly, for **BL-EDCRM**, only the upper-level model are presented since the lower-level model has the same form.

BL-DCRM

$$
\begin{cases}
\max \left[(1-\eta)\frac{c_{14}-\bar{f}_1}{(c_{14}-c_{13})}, (1-\eta)\frac{c_{24}-\bar{f}_2}{(c_{24}-c_{23})}, \cdots, (1-\eta)\frac{c_{m4}-\bar{f}_{m_1}}{(c_{m_14}-c_{m_13})} \right] \\
\text{s.t.} \begin{cases}
\alpha_r(d-c) + (1-\eta)c \le 0, & \text{if } c \le 0 \le a \\
\alpha_r(b-a)(d-c) + \eta a(d-c) + (1-\eta)c(b-a) \le 0, & \text{if } a \le 0 \le b \\
\alpha_r(d-c) - (1-\eta)d \le 0, & \text{if } b \le 0 \le d \\
\mathbf{x} \ge \mathbf{0}
\end{cases}
\end{cases}
$$

$$
\begin{cases}
\max \left[\eta\frac{c_{14}-\bar{f}_1}{c_{14}-c_{13}} + (1-\eta)\frac{c_{12}-\bar{f}_1}{c_{12}-c_{11}}, \eta\frac{c_{24}-\bar{f}_2}{c_{24}-c_{23}} + (1-\eta)\frac{c_{22}-\bar{f}_2}{c_{22}-c_{21}}, \cdots, \eta\frac{c_{m_14}-\bar{f}_{m_1}}{c_{m_14}-c_{m_13}} + \right. \\
\left. (1-\eta)\frac{c_{m_12}-\bar{f}_{m_1}}{c_{m_12}-c_{m_11}} \right] \\
\text{s.t.} \begin{cases}
\alpha_r(d-c) + (1-\eta)c \le 0, & \text{if } c \le 0 \le a \\
\alpha_r(b-a)(d-c) + \eta a(d-c) + (1-\eta)c(b-a) \le 0, & \text{if } a \le 0 \le b \\
\alpha_r(d-c) - (1-\eta)d \le 0, & \text{if } b \le 0 \le d \\
\mathbf{x} \ge \mathbf{0}
\end{cases}
\end{cases}
$$

$$
\begin{cases}
\max \left[(1-\eta)\frac{c_{14}-\bar{f}_1}{c_{14}-c_{13}} + \eta\frac{c_{12}-\bar{f}_1}{c_{12}-c_{11}}, (1-\eta)\frac{c_{24}-\bar{f}_2}{c_{24}-c_{23}} + \eta\frac{c_{22}-\bar{f}_2}{c_{22}-c_{21}}, \cdots, (1-\eta)\frac{c_{m_14}-\bar{f}_{m_1}}{c_{m_14}-c_{m_13}} \right. \\
\left. +\eta\frac{c_{m_12}-\bar{f}_{m_1}}{c_{m_12}-c_{m_11}} \right] \\
\text{s.t.} \begin{cases}
\alpha_r(d-c) + (1-\eta)c \le 0, & c \le 0 \le a \\
\alpha_r(b-a)(d-c) + \eta a(d-c) + (1-\eta)c(b-a) \le 0, & a \le 0 \le b \\
\alpha_r(d-c) - (1-\eta)d < 0, & b \le 0 \le d \\
\mathbf{x} \ge \mathbf{0}
\end{cases}
\end{cases}
$$

For random multiple objective decision making, there are three basic theorems for handling the objective functions and the constraints: Theorem 4.1, Theorem 4.4, and Theorem 4.5. According to these three theorems, we can get the crisp equivalent models for Ra-EVRM, Ra-CCRM, and Ra-DCRM.

Ra-EEVRM

$$
\begin{cases}
\max \left[E[\tilde{\mathbf{c}}_1^T \mathbf{x}], E[\tilde{\mathbf{c}}_2^T \mathbf{x}], \cdots, E[\tilde{\mathbf{c}}_m^T \mathbf{x}] \right] \\
\text{s.t.} \begin{cases}
E[\tilde{\mathbf{a}}_r^T \mathbf{x}] \le E[\bar{b}_r] + h_r, r = 1, 2, \cdots, p \\
\mathbf{x} \in X
\end{cases}
\end{cases}
$$

Ra-ECCRM

$$\begin{cases} \max[\bar{f}_1, \bar{f}_2, \cdots, \bar{f}_m] \\ \text{s.t.} \begin{cases} \Phi^{-1}(1-\alpha_i)\sqrt{\mathbf{x}^T V_i^c \mathbf{x}} + \mu_i^{cT}\mathbf{x} \geq \bar{f}_i \\ \mathbf{x} \in \bar{I}_r, r = 1,2,\cdots,p \end{cases} \end{cases}$$

Ra-EDCRM

$$\begin{cases} \max\left[\Phi\left(\dfrac{\bar{f}_1 - \mu_1^{cT}\mathbf{x}}{\sqrt{\mathbf{x}^T V_1^c \mathbf{x}}}\right), \Phi\left(\dfrac{\bar{f}_2 - \mu_2^{cT}\mathbf{x}}{\sqrt{\mathbf{x}^T V_2^c \mathbf{x}}}\right), \cdots, \Phi\left(\dfrac{\bar{f}_m - \mu_m^{cT}\mathbf{x}}{\sqrt{\mathbf{x}^T V_m^c \mathbf{x}}}\right)\right] \\ \text{s.t.} \begin{cases} \alpha(I_r) \geq \theta_r \\ \mathbf{x} \in \bar{I}_r \\ r = 1,2,\cdots,p \end{cases} \end{cases}$$

For fuzzy multiple objective decision making, there are nine basic theorems for handling the objective functions and the constraints: Theorem 5.6, Theorem 5.7, Theorem 5.8, Theorem 5.10, Theorem 5.11, Theorem 5.12, Theorem 5.13, Theorem 5.14 and Theorem 5.15. According to these ine theorems, we can get the crisp equivalent models for Fu-EVRM, Fu-CCRM, and Fu-DCRM.

Fu-EEVRM

$$(E\text{-}LLAM)_1 \quad \begin{cases} \max[f_1(\mathbf{x}), f_2(\mathbf{x}), \cdots, f_m(\mathbf{x})] \\ \text{s.t.} \begin{cases} \frac{r_1+r_2}{2} \leq 0 \\ r_3 \leq 0 \\ x_j \geq 0, j = 1,2,\cdots,n \end{cases} \end{cases}$$

$$(E\text{-}LUAM)_1 \quad \begin{cases} \max[f_1(\mathbf{x}), f_2(\mathbf{x}), \cdots, f_m(\mathbf{x})] \\ \text{s.t.} \begin{cases} \frac{r_2+r_3}{2} \leq 0 \\ r_3 \leq 0 \\ x_j \geq 0, j = 1,2,\cdots,n \end{cases} \end{cases}$$

$$(E\text{-}LLAM)_2 \quad \begin{cases} \max[f_1(\mathbf{x}), f_2(\mathbf{x}), \cdots, f_m(\mathbf{x})] \\ \text{s.t.} \begin{cases} \frac{r_1+r_2}{2} + \frac{r_3^2}{2(r_3-r_2)} \leq 0 \\ r_2 \leq 0 \leq r_3 \\ x_j \geq 0, j = 1,2,\cdots,n \end{cases} \end{cases}$$

$$(E\text{-}LUAM)_2 \quad \begin{cases} \max[f_1(\mathbf{x}), f_2(\mathbf{x}), \cdots, f_m(\mathbf{x})] \\ \text{s.t.} \begin{cases} -\frac{r_2^2}{2(r_3-r_2)} \leq 0 \\ r_2 \leq 0 \leq r_3 \\ x_j \geq 0, j = 1,2,\cdots,n \end{cases} \end{cases}$$

$$(E\text{-}LLAM)_3 \quad \begin{cases} \max[f_1(\mathbf{x}), f_2(\mathbf{x}), \cdots, f_m(\mathbf{x})] \\ \text{s.t.} \begin{cases} \frac{r_2+r_3}{2} - \frac{r_1^2}{2(r_2-r_1)} \leq 0 \\ r_1 \leq 0 \leq r_2 \\ x_j \geq 0, j = 1,2,\cdots,n \end{cases} \end{cases}$$

$$(E\text{-}LUAM)_3 \begin{cases} \max[f_1(\mathbf{x}), f_2(\mathbf{x}), \cdots, f_m(\mathbf{x})] \\ \text{s.t.} \begin{cases} \frac{r_2^2}{2(r_2-r_1)} \le 0 \\ r_1 \le 0 \le r_2 \\ x_j \ge 0, j = 1, 2, \cdots, n \end{cases} \end{cases}$$

$$(E\text{-}LLAM)_4 \begin{cases} \max[f_1(\mathbf{x}), f_2(\mathbf{x}), \cdots, f_m(\mathbf{x})] \\ \text{s.t.} \begin{cases} \frac{r_2+r_3}{2} \le 0 \\ 0 \le r_1 \\ x_j \ge 0, j = 1, 2, \cdots, n \end{cases} \end{cases}$$

$$(E\text{-}LUAM)_4 \begin{cases} \max[f_1(\mathbf{x}), f_2(\mathbf{x}), \cdots, f_m(\mathbf{x})] \\ \text{s.t.} \begin{cases} \frac{r_1+r_2}{2} \le 0 \\ 0 \le r_1 \\ x_j \ge 0, j = 1, 2, \cdots, n \end{cases} \end{cases}$$

where $f_i(x) = \sum_{j=1}^{n} \left(\frac{(1-\lambda)}{2}(c_{ij} - \alpha_{ij}^c) + \frac{c_{ij}}{2} + \frac{\lambda}{2}(c_{ij} + \beta_{ij}^c) \right) x_j, r_1 = \sum_{j=1}^{n} (a_{rj} - \alpha_{rj}^a)x_j - b_r - \beta_r^b, r_2 = \sum_{j=1}^{n} a_{rj}x_j - b_r, r_3 = \sum_{j=1}^{n} (a_{rj} + \beta_{rj}^a)x_j - b_r + \alpha_r^b.$

Fu-ECCRM

$$\begin{cases} (LAM) \begin{cases} \max[\bar{f}_1, \bar{f}_2, \cdots, \bar{f}_m] \\ \text{s.t.} \begin{cases} f_i \le c_i^T x - L^{-1}(1 - \delta_i)\alpha_i^{cT}x, \quad i = 1, 2, \cdots, m, \\ b_r - L^{-1}(1 - \theta_r)\alpha_r^b - a_r^T x - R^{-1}(\theta_r)\beta_r^{aT}x \ge 0, r = 1, 2, \cdots, p \\ \mathbf{x} \ge 0 \end{cases} \end{cases} \\ (UAM) \begin{cases} \max[\bar{f}_1, \bar{f}_2, \cdots, \bar{f}_m] \\ \text{s.t.} \begin{cases} \bar{f}_i \le c_i^T x + R^{-1}(\delta_i)\beta_i^{cT}x, \quad i = 1, 2, \cdots, m, \\ b_r + R^{-1}(\theta_r)\beta_r^b - a_r^T x + L^{-1}(\theta_r)\alpha_r^{aT}x \ge 0, r = 1, 2, \cdots, p \\ \mathbf{x} \ge 0 \end{cases} \end{cases} \end{cases}$$

Fu-EDCRM

$$\begin{cases} (LAM) \begin{cases} \max \left[\frac{c_1^T x - f_1}{\alpha_1^{cT}x}, \frac{c_2^T x - f_2}{\alpha_2^{cT}x}, \cdots, \frac{c_m^T x - f_m}{\alpha_m^{cT}x} \right] \\ \text{s.t.} \begin{cases} b_r + R^{-1}(\theta_r)\beta_r^b - a_r^T x + L^{-1}(\theta_r)\alpha_r^{aT}\mathbf{x} \ge 0, r = 1, 2, \cdots, p \\ \mathbf{x} \ge 0 \end{cases} \end{cases} \\ (UAM) \begin{cases} \max \left[\frac{c_1^T x - f_1}{\beta_1^{cT}x}, \frac{c_2^T x - f_2}{\beta_2^{cT}x}, \cdots, \frac{c_m^T x - f_m}{\beta_m^{cT}x} \right] \\ \text{s.t.} \begin{cases} b_r + R^{-1}(\theta_r)\beta_r^b - a_r^T x + L^{-1}(\theta_r)\alpha_r^{aT}\mathbf{x} \ge 0, r = 1, 2, \cdots, p \\ \mathbf{x} \ge 0 \end{cases} \end{cases} \end{cases}$$

Figure 6.4 illustrates the model analysis process.

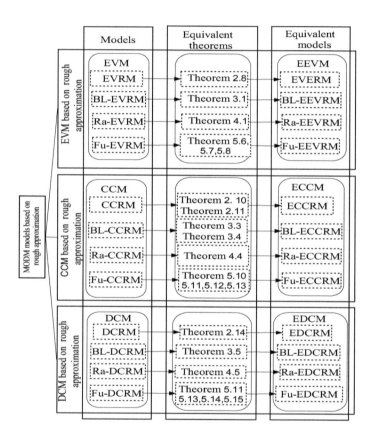

FIGURE 6.4   Model analysis process for linear models.

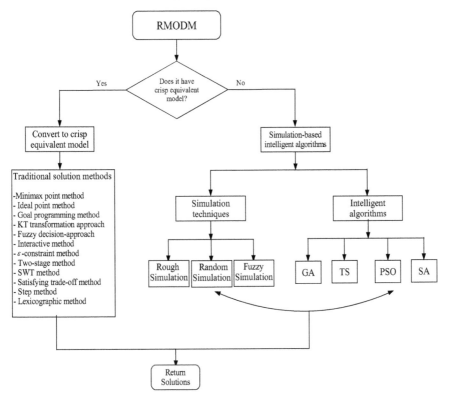

FIGURE 6.5 General flowchart of the algorithms system.

6.5 Algorithm System

For RMODM, two types of algorithm systems are considered. One uses traditional MODM solution techniques as RMODM can be converted into crisp equivalents. The other uses the hybrid intelligent algorithms focusing on those decision-making problem that cannot be converted into crisp equivalent models. The general flowchart of the two algorithm systems is shown in Figure 6.5.

After we get the crisp equivalent models, we could employ the basic solution methods to derive the solution. There are several solution methods introduced in detail the book, which includes:

- Minimax point method

- Ideal point method

- Goal programming method

- KT transformation approach

- Fuzzy decision-approach

- Interactive Method

- ε-constraint method

- Two-stage method

- Satisfying trade-off method

- Step method

- Lexicographic method

The above solution methods are the most popular ones for multiple objective decision making, including bilevel multiple objective decision making, so the decision maker can choose different methods when he or she receives different request or under different conditions.

For the nonlinear rough multiple objective decision-making models, it is very difficult to transform the three types of objective functions and two types of constraints into their crisp equivalents. So, we propose several simulation technique to simulate the objective functions and constraints. There are three kinds of simulations for each kind of uncertainty. In Section 2.3.3.1, Section 2.4.3.1, and Section 2.5.3.1, we propose rough simulation 1 for expected value, rough simulation 2 for critical value and rough simulation 3 for *Appr* measure, respectively. In Section 4.3.3.1, Section 4.4.3.1, and Section 4.5.3.1, we propose random simulation 1 for expected value, random simulation 2 for critical value and random simulation 3 for probability, respectively. In Section 5.3.3.1, Section 5.4.3.1, and Section 5.5.3.1, we propose fuzzy simulation 1 for expected value, fuzzy simulation 2 for critical value, and fuzzy simulation 3 for *Me* measure, respectively.

By combining the simulations techniques and intelligent algorithms, we can create some hybrid algorithms, and we can obtain several kinds of hybrid algorithms (see Figure 6.6).

These simulations will embed into four types of basic intelligent algorithms, which include:

- Genetic algorithm (GA)

- Tabu search algorithm (TS)

- Particle swarm optimization algorithm (PSO)

- Simulated annealing algorithm (SA)

So for the general RMODM, we present the following ideas to design the algorithm. For the linear rough multiple decision making model , we can transform them into some crisp equivalent models and use the above traditional solution methods

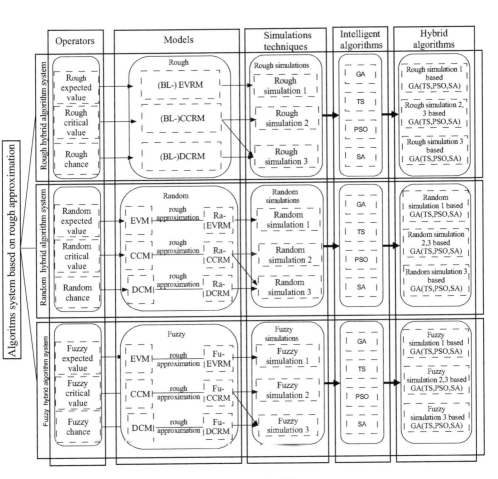

FIGURE 6.6   Hybrid intelligent systems.

to solve them directly. For the other rough multiple decision making models, especially the nonlinear model, we embed the corresponding simulation techniques into the intelligent algorithm to find the solutions.

Application domains for each intelligent algorithm is as follows. Some example areas of the application of GA are

- Scheduling

- Chemistry, Chemical Engineering

- Medicine

- Data Mining and Data Analysis

- Geometry and Physics

- Economics and Finance

- Networking and Communication

- Electrical Engineering and Circuit Design

- Image Processing

- Combinatorial Optimization

Some example areas of the application of TS are

- Combinatorial Optimization

- Machine Learning

- Biochemistry

- Operations Research

- Networking and Communication

Some example areas of the application of PSO are

- Machine Learning

- Function Optimization

- Geometry and Physics

- Operations Research

- Chemistry, Chemical Engineering

- Electrical Engineering and Circuit Design

Some example areas of the application of SA are

- Combinatorial Optimization

- Function Optimization

- Chemistry, Chemical Engineering

- Image Processing

- Economics and Finance;

- Electrical Engineering and Circuit Design

- Machine Learning

- Geometry and Physics

- Networking and Communication

Although we used these four algorithms in the book, there are some other excellent intelligent algorithms, such as the ant colony optimization algorithm (ACO), artificial neural network (ANN), immune algorithms (IA) etc. We expect more advanced intelligent algorithms to be found, and we are willing to use them if they are appropriate in our future research.

6.6   Research Ideas and Paradigm: 5MRP

In the field of decision making, excellent research should integrate the background of the problem, a mathematical model, and an effective solution method with a significant application. However, doing all these together is very difficult. How do we know a problem is meaningful? How to describe this problem through scientific language? How to design an efficient algorithm to solve a practiced problem? Finally how to apply this integrated method to the engineering field? All these questions must be answered under a new paradigm following a certain methodology. This new paradigm will enable researchers to draw scientific results and conclusions under the guidance of science, and will play a significant guiding role in conducting scientific research.

The research idea of 5MRP expresses the initial relationship among Research, Model and Problem. *R* stands for a research system that includes research specifics, research background, research base, research reality, research framework, and applied research; M refers to a model system that includes concept models, physical models, physical and mathematical models, mathematical and physical models, designed model for algorithms, and describes the specific model. P represents a problem system that includes a particular problem, a class of problems, abstract problems, problem restoration, problem solution, and problem settlement. Next, we take RMODM as an example to present how to use 5MPR to start the research.

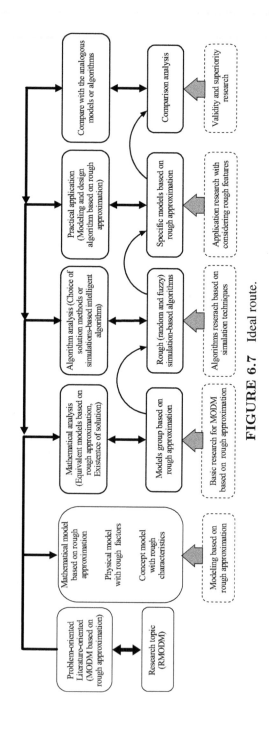

FIGURE 6.7 Ideal route.

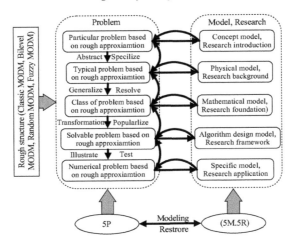

FIGURE 6.8   Problem system.

To summarize the research ideas and the framework of the research work, see Figure 6.7. When research is started, we usually proceed to study a particular problem, which has research value and can be described as a concept model. It is the introduction of the research. After studying the particular problem and the problem with the same essence of the particular problem, we can obtain the typical problem that has universality and can be abstracted to a physical model. It is the background of the research. Then we generalize the typical problem to a class of problems that can be abstracted to common mathematical problems, and we can then propose the mathematical model. It is the foundation of the research. Then we design the algorithm and obtain the model for the procedure of the algorithm. It is the framework of the research. Finally, we should apply the above models to a practical problem and establish a numerical model for the specific problem and employ the algorithm to get the solution to illustrate efficiency and validity. It is the application of the research.

Then we use the following Figures 6.8–6.10 illustrations to describe the relationship between the problem system, model system, and research system.

Figure 6.8 emphasizes the problem system and presents a train thought of to deal with the problem. In real life, many decision making problems must face objective uncertainty or subjective uncertainty. Although probability theory and possibility theory provide alternative techniques, neither of them can depict uncertainty completely. This confusion drives us to start a new research. Then a typical problem based on rough approximation is formed in the brain, and we generalize it to a class-problem based on rough approximation, that is, it represents the whole features of a class of problems based on rough approximation. How should we solve it? This question reminds us that we should convert it to a solvable problem based on rough approximation. Finally, a numerical problem should be presented to confirm that these problems can be solved.

Figure 6.9 emphasizes a model system and presents the series of models that are

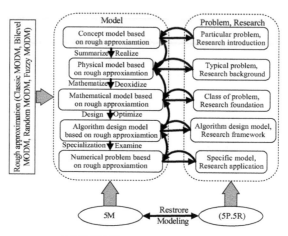

FIGURE 6.9 Model system.

used to deal with the corresponding problems. When we face those particular problems that make us confused, a concept based on rough approximation is formed in our brain. Aiming at the typical problem, a physical model with rough approximation should be constructed to present the real-life structure of those problems. To help decision makers make the decision, a mathematical model based on rough approximation should be constructed to quantitatively analyze those problems. Naturally, after constructing a mathematical model, we should design an algorithm model to solve it. Hence, it is called the algorithm design model based on rough approximation. Finally, a specific model based on rough approximation according to a particular problem is constructed to show the rationality of the proposed model and the efficiency of the proposed algorithm design model. Conversely, a specific model can be used to examine the efficiency and convergence of the designed algorithm model. If so, the algorithm can be applied to optimize the mathematical model. We further deoxidize the mathematical model to a physical model to describe the realistic problem and solve the particular problem.

Figure 6.10 emphasizes the research system and presents the technological process when we face a problem and do our research work. The research process is also rooted in the problem and model. First, a basic research introduction should be done when we face a particular problem based on rough approximation. It means that we should make a basic description and construct a concept model of the particular problem. Second, the research background should be clarified when it is abstracted as a typical problem based on rough approximation. In this process, the essence of the occurrence of the problem should be investigated and then a physical model is given. Third, we should prepare the basic research foundation when the physical model based on rough approximation has abstracted a mathematical model based on rough approximation. The research foundation basically includes the following parts: (i) Which class of problems should these typical problems divide into? (ii)

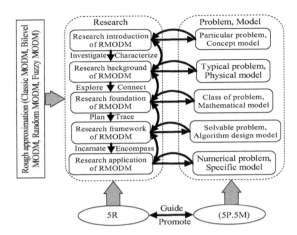

FIGURE 6.10 Research system.

What research has been done about this class of problems? (iii) What literatures are useful for our research about problems, models, algorithms? Fourth, a research framework should be given to help us do the research. Finally, a research application is given to show the whole research process. Conversely, the application should include the whole research framework and reflect the research method. The research framework can also trace the research foundation and further connect the research background. In the end, the research goes back to the research introduction to the particular problem and an optimal strategy is obtained.

Let us propose each step of 5MRP. When we start a research, we usually proceed to study a particular problem that has research value and can be described as a concept model. It is the introduction of the research. After studying the particular problem and the problem with the same essence as the particular problem, then we can obtain the typical problem that has universality and that can be abstracted to a physical model. It is the background of the research. Then we generalize the typical problem ulteriorly to a class of problems that can be abstracted to a common mathematical problem, then we can propose the mathematical model. It is the foundation of the research. Then we design the algorithm and obtain the model for the procedure of the algorithm. It is the framework of the research; At last, we should apply the above models to a practical problem and establish the numerical model for the specific problem and employ the algorithm to get the solution to illustrate the efficiency and validity. It is the application of the research.

We employ Figure 6.11 to present how we use the 5MPR research ideal to do our research: MORDM, BL-MORDM, Ra-MORDM, and Fu-MORDM.

In conclusion, 5RMP is an effective paradigm that can be widely used in various fields of scientific research and can contribute to research in all areas in a standardized and efficient manner. In the area of decision making, 5RMP will be well reflected because of its rigorous logical and effective applicability, and it will play an

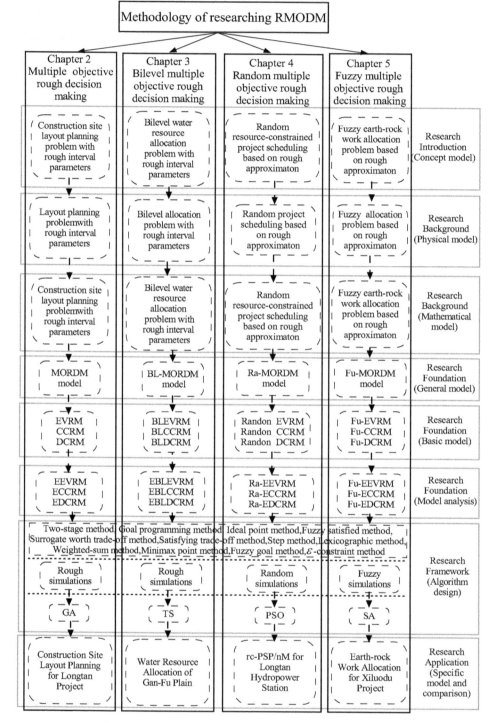

FIGURE 6.11 Methodological system for RMODM.

outstanding guiding role in the practice side of research.

6.7 Future Research

MODM is an effective method to handle decision-making problems. Since the information obtained in real problems is imprecise, the rough set theory provides a good method to deal with it. This book attempts to integrate MODM and rough set theory. As shown by this book, rough set theory and multiple objective decision making (including classic multiple objective decision making, bilevel multiple objective decision making,random multiple objective decision making and fuzzy multiple objective decision making) construct the basis of RMODM. We aim at the fusion of rough set theory and multiple objective decision making theory and both have seen much development. We hope to amalgamate their merits and obtain some good results to allow for further application. Although our attempt is preliminary, it is expected that RMODM will have further improvements and developments with further research in MODM and rough set theory and application to practical problems. Next, we look ahead at future research with models, algorithms, and applications for RMODM.

From the viewpoint of the models, we apply the rough approximation technique to all kinds of MODM (Classic MODM, Bilevel MODM, Random MODM, and Fuzzy MODM). However, rough set theory is a rapidly growing subject, so how to apply the new results in the rough set theory to MODM model may be an interesting area for future research. In our book, we only consider linear models as special cases and discuss their properties. In future research, more models, such as quadratic models and fractional models, will be addressed. In addition, comparison among these models may be a meaningful issue. Moreover, there are many formulation for random MODM and fuzzy MODM, so how to develop other random MODM models and fuzzy MODM models based on rough approximation should be considered.

From the perspective of the algorithms, our book provides quite a few of them to treat RMODM. For some special linear cases, we used traditional algorithms to deal with their crisp equivalents. For the other models that cannot be converted into crisp equivalents, we use the hybrid intelligent algorithms, that is, simulation techniques + intelligent algorithms to obtain the solution. However, the algorithms provided in this book do not include all the traditional algorithms and intelligent algorithms. How to select the algorithms and compare their advantages and disadvantages are the research questions not yet answered. In addition, convergence and convergent speed need further research. Moreover, the improvement and transformation of these algorithms are worthy of discussion.

From the application aspect, RMODM may be applied to four decision-making problems based on rough approximation in this book, that is, the construction site layout planning problem, the bilevel water resource allocation problem, the project scheduling problem, and the allocation problem, all of which occur in engineering. However, the application discussed in this book form only a small part of the actual problems. In the field of management, finance, and manufacturing, there are several problems in which the DM must face multiple objective and the information can be dealt with by rough set theory, and we can apply RMODM to model them. How to depict these objectives and information is worthy of deeper research.

In conclusion, systematic research about RMODM has been provided by this book, and a scientific research area can also be found for future research from these aspects: models, algorithms, and applications.

References

[1] E. Aarts, J. Korst. *Simulated annealing and Boltzmann machines*. New York, NY; John Wiley and Sons, 1988.

[2] M. Abo-Sinna. A bi-level non-linear multi-objective decision making under fuzziness. *OPSEARCH-NEW DELHI*, 2001, **38**(5): 484–495.

[3] B. Abolpour, M. Javan, M. Karamouz. Water allocation improvement in river basin using Adaptive Neural Fuzzy Reinforcement Learning approach. *Applied Soft Computing*, 2007, **7**(1): 265–285.

[4] R. Ahuja, T. Magnanti, J. Orlin. Network flows. *Operations Research Center Working Paper; OR 185-88*, 1988.

[5] A. Al Radif. Integrated water resources management (IWRM): An approach to face the challenges of the next century and to avert future crises. *Desalination*, 1999, **124**(1–3): 145–153.

[6] J. Alander. *An indexed bibliography of genetic algorithms: Years 1957-1993*. Citeseer, 1994.

[7] A. Anagnostopoulos, L. Michel, P. Hentenryck, Y. Vergados. A simulated annealing approach to the traveling tournament problem. *Journal of Scheduling*, 2006, **9**(2): 177–193.

[8] G. Anandalingam. Multi-level programming and conflict resolution. *European Journal of Operational Research*, 1991, **51**(2): 233–247.

[9] P. Angeline. Evolutionary optimization versus particle swarm optimization: Philosophy and performance differences. **In:** *Evolutionary Programming VII*, Springer, Berlin, 1998, 601–610.

[10] K. Arrow, E. Barankin, D. Blackwell. Admissible points of convex sets. *Contributions to the Theory of Games*, 1953, **2**: 87–91.

[11] H. Aytug, M. Lawley, K. McKay, S. Mohan, R. Uzsoy. Executing production schedules in the face of uncertainties: A review and some future directions. *European Journal of Operational Research*, 2005, **161**(1): 86–110.

[12] N. Azizi, S. Zolfaghari. Adaptive temperature control for simulated annealing: A comparative study. *Computers & Operations Research*, 2004, **31**(14): 2439–2451.

[13] T. Bäck, U. Hammel, H. Schwefel. Evolutionary computation: Comments on

the history and current state. *IEEE Transactions on Evolutionary Computation*, 1997, **1**(1): 3–17.

[14] P. Banerjee, M. Jones, J. Sargent. Parallel simulated annealing algorithms for cell placement on hypercube multiprocessors. *IEEE Transactions on Parallel and Distributed Systems*, 1990, 91–106.

[15] J. Bard. An algorithm for solving the general bilevel programming problem. *Mathematics of Operations Research*, 1983, **8**(2): 260–272.

[16] J. Bard. Coordination of a multidivisional organization through two levels of management. *Omega*, 1983, **11**(5): 457–468.

[17] J. Bard. *Practical bilevel optimization: algorithms and applications*. Kluwer Academic, Dordrecht, Netherlands, 1998.

[18] J. Bard, J. Falk. An explicit solution to the multi-level programming problem. *Computers & operations research*, 1982, **9**(1): 77–100.

[19] J. Bard, J. Moore. A branch and bound algorithm for the bilevel programming problem. *SIAM Journal on Scientific and Statistical Computing*, 1990, **11**: 281.

[20] V. Barichard, M. Ehrgott, X. Gandibleux, V. T'Kindt. Multiobjective Programming and Goal Programming: Theoretical Results and Practical Applications. *Lecture Notes in Economics and Mathematical Systems*, 2009, 298.

[21] M. Bastos, C. Ribeiro. Reactive tabu search with path-relinking for the Steiner problem in graphs. **In:** *Proceedings of the third Metaheuristics International Conference*, Pennsylvania, 1999, 31–36.

[22] R. Battiti, G. Tecchiolli, et al. The reactive tabu search. *ORSA journal on computing*, 1994, **6**: 126–126.

[23] A. Baykasoglu, T. Dereli, I. Sabuncu. An ant colony algorithm for solving budget constrained and unconstrained dynamic facility layout problems. *Omega*, 2006, **34**(4): 385–396.

[24] A. Baykasolu, N. Gindy. A simulated annealing algorithm for dynamic layout problem. *Computers & Operations Research*, 2001, **28**(14): 1403–1426.

[25] E. Beale. On minimizing a convex function subject to linear inequalities. *Journal of the Royal Statistical Society. Series B (Methodological)*, 1955, **17**(2): 173–184.

[26] R. Bellman, L. Zadeh. Decision-making in a fuzzy environment. *Management science*, 1970, **17**(4).

[27] R. Benayoun, J. De Montgolfier, J. Tergny, O. Laritchev. Linear programming with multiple objective functions: Step method (STEM). *Mathematical programming*, 1971, **1**(1): 366–375.

[28] A. Bevilacqua. A methodological approach to parallel simulated annealing on an SMP system. *Journal of Parallel and Distributed Computing*, 2002, **62**(10): 1548–1570.

[29] W. Bialas, M. Karwan. Two-level linear programming. *Management Science*, 1984, 1004–1020.

[30] J. Bidot, T. Vidal, P. Laborie, J. Beck. A theoretic and practical framework for scheduling in a stochastic environment. *Journal of Scheduling*, 2009, **12**(3): 315–344.

[31] J. Birge, I. I. for Applied Systems Analysis. *A standard input format for multi-period stochastic linear programs*. International Institute for Applied Systems Analysis, Laxenburg, 1987.

[32] J. Birge, F. Louveaux. *Introduction to stochastic programming*. Springer Verlag, Berlin, 1997.

[33] H. Bjornlund, J. McKay. Factors affecting water prices in a rural water market: A South Australian experience. *Water Resources Research*, 1998, **34**(6): 1563–1570.

[34] J. Blazewicz, et al. Scheduling subject to resource constraints: Classification and complexity. *Discrete Applied Mathematics*, 1983, **5**(1): 11–24.

[35] I. Blöchliger, N. Zufferey. A graph coloring heuristic using partial solutions and a reactive tabu scheme. *Computers & Operations Research*, 2008, **35**(3): 960–975.

[36] F. Boctor. Heuristics for scheduling projects with resource restrictions and several resource-duration modes. *International Journal of Production Research*, 1993, **31**(11): 2547–2558.

[37] F. Boctor. A new and efficient heuristic for scheduling projects with resource restrictions and multiple execution modes. *European Journal of Operational Research*, 1996, **90**(2): 349–361.

[38] E. Bonabeau, M. Dorigo, G. Theraulaz. *Swarm intelligence: From natural to artificial systems*. Oxford University Press, New York, 1999.

[39] A. Bortfeldt. A genetic algorithm for the two-dimensional strip packing problem with rectangular pieces. *European Journal of Operational Research*, 2006, **172**(3): 814–837.

[40] P. Brucker, A. Drexl, R. Möhring, K. Neumann, E. Pesch. Resource-constrained project scheduling: Notation, classification, models, and methods. *European Journal of Operational Research*, 1999, **112**(1): 3–41.

[41] F. Buffa, W. Jackson. A goal programming model for purchase planning. *Journal of Purchasing and Materials Management*, 1983, **19**(3): 27–34.

[42] H. Calvete, C. Galé, P. Mateo. A genetic algorithm for solving linear fractional bilevel problems. *Annals of Operations Research*, 2009, **166**(1): 39–56.

[43] H. Calvete, C. Gale, M. Oliveros, B. Sánchez-Valverde. A goal programming approach to vehicle routing problems with soft time windows. *European Journal of Operational Research*, 2007, **177**(3): 1720–1733.

[44] L. Campos, J. Verdegay. Linear programming problems and ranking of fuzzy numbers. *Fuzzy Sets and Systems*, 1989, **32**(1): 1–11.

[45] W. Candler, R. Townsley. Linear Two-Level Programming Problem. *Computers & Operations Research*, 1982, **9**(1): 59–76.

[46] U. Castellani, A. Fusiello, R. Gherardi, V. Murino. Automatic selection of MRF control parameters by reactive tabu search. *Image and Vision Computing*, 2007, **25**(11): 1824–1832.

[47] V. Černý. Thermodynamical approach to the traveling salesman problem: An efficient simulation algorithm. *Journal of Optimization Theory and Applications*, 1985, **45**(1): 41–51.

[48] A. Cervantes, I. Galvan, P. Isasi. AMPSO: A new particle swarm method for nearest neighborhood classification. *IEEE Transactions on Systems, Man, and Cybernetics, Part B: Cybernetics*, 2009, **39**(5): 1082–1091.

[49] S. Chanas, D. Kuchta. A concept of the optimal solution of the transportation problem with fuzzy cost coefficients. *Fuzzy Sets and Systems*, 1996, **82**(3): 299–305.

[50] V. Chankong, Y. Haimes. *Multiobjective decision making: Theory and methodology*. North-Holland New York, 1983.

[51] A. Charnes, W. Cooper. Management models and industrial applications of linear programming. *Management Science*, 1957, **4**(1): 38–91.

[52] A. Charnes, W. Cooper. Chance-constrained programming. *Management Science*, 1959, **6**(1): 73–79.

[53] A. Charnes, W. Cooper. Deterministic equivalents for optimizing and satisficing under chance constraints. *Operations Research*, 1963, **11**(1): 18–39.

[54] A. Charnes, W. Cooper, R. Ferguson. Optimal estimation of executive compensation by linear programming. *Management Science*, 1955, **1**(2): 138–151.

[55] A. Charnes, W. Cooper, G. Symonds. Cost horizons and certainty equivalents: An approach to stochastic programming of heating oil. *Management Science*, 1958, **4**(3): 235–263.

[56] R. Chelouah, P. Siarry. Tabu search applied to global optimization. *European Journal of Operational Research*, 2000, **123**(2): 256–270.

[57] H. Chen, N. Flann, D. Watson. Parallel genetic simulated annealing: A mas-

sively parallel SIMD algorithm. *IEEE Transactions on Parallel and Distributed Systems*, 1998, **9**(2): 126–136.

[58] K. Chen, T. Li, T. Cao. Tribe-PSO: A novel global optimization algorithm and its application in molecular docking. *Chemometrics and Intelligent Laboratory Systems*, 2006, **82**(1-2): 248–259.

[59] S. Chen, B. Luk, Y. Liu. Application of adaptive simulated annealing to blind channel identification with HOC fitting. *Electronics Letters*, 2002, **34**(3): 234–235.

[60] R. Cheng, M. Gen. An adaptive superplane approach for multiple objective optimization problems. *Tech. Rep.*, Technical Report, Ashikaga Institute of Technology, Ashikaga, 1998.

[61] W. Chiang, R. Russell. A reactive tabu search metaheuristic for the vehicle routing problem with time windows. *INFORMS Journal on computing*, 1997, **9**(4): 417.

[62] S. Chu, M. Zhu, L. Zhu. Goal programming models and DSS for manpower planning of airport baggage service. *Multiple Criteria Decision Making for Sustainable Energy and Transportation Systems*, 2010, 189–199.

[63] M. Clerc. TRIBES-un exemple d'optimisation par essaim particulaire sans parametres de contrôle. *Optimisation par Essaim Particulaire (OEP 2003)*, Paris, France, 2003.

[64] M. Clerc, J. Kennedy. The particle swarm-explosion, stability, and convergence in a multidimensional complex space. *IEEE transactions on Evolutionary Computation*, 2002, **6**(1): 58–73.

[65] C. Coello, G. Pulido, M. Lechuga. Handling multiple objectives with particle swarm optimization. *IEEE Transactions on Evolutionary Computation*, 2004, **8**(3): 256–279.

[66] D. Coit, A. Smith. Reliability optimization of series-parallel systems using a genetic algorithm. *IEEE Transactions on Reliability*, 2005, **45**(2): 254–260.

[67] B. Colson, P. Marcotte, G. Savard. Bilevel programming: A survey. *4OR: A Quarterly Journal of Operations Research*, 2005, **3**(2): 87–107.

[68] B. Colson, P. Marcotte, G. Savard. An overview of bilevel optimization. *Annals of Operations Research*, 2007, **153**(1): 235–256.

[69] D. Cvijovic, J. Klinowski. Taboo search: An approach to the multiple minima problem. *Science(Washington)*, 1995, **267**(5198): 664–664.

[70] Z. Czech, P. Czarnas. Parallel simulated annealing for the vehicle routing problem with time windows. **In:** *10th Euromicro Workshop on Parallel, Distributed and Network-based Processing*, IEEE, 2002, 376–383.

[71] G. Dantzig. Linear programming under uncertainty. *Management Science*, 1955, **1**(3): 197–206.

[72] I. Das, J. Dennis. Normal-boundary intersection: A new method for generating the Pareto surface in nonlinear multicriteria optimization problems. *SIAM Journal on Optimization*, 1998, **8**: 631.

[73] K. Deb, A. Pratap, S. Agarwal, T. Meyarivan. A fast and elitist multiobjective genetic algorithm: NSGA-II. *IEEE transactions on evolutionary computation*, 2002, **6**(2): 182–197.

[74] E. Demeulemeester, W. Herroelen. *Project Scheduling: A Research Handbook*. Kluwer Academic, Dordrecht, Netherland, 2002.

[75] S. Dempe. *Foundations of Bilevel Programming*. Kluwer Academic, Dordrecht, Netherlands, 2002.

[76] S. Dempe. Annotated bibliography on bilevel programming and mathematical programs with equilibrium constraints. *Optimization*, 2003, **52**(3): 333–359.

[77] M. Dempster. *Stochastic Programming: Based on Proceedings*. Academic Press, San Diego, CA, 1980.

[78] R. Diekmann, R. Luling, J. Simon. A general purpose distributed implementation of simulated annealing. **In:** *Fourth IEEE Symposium on Parallel and Distributed Processing*, IEEE, 2002, 94–101.

[79] W. Domschke, G. Krispin. Location and layout planning. *OR Spectrum*, 1997, **19**(3): 181–194.

[80] L. Dos Santos Coelho, P. Alotto. Tribes Optimization Algorithm Applied to the Loney's Solenoid. *IEEE Transactions on Magnetics*, 2009, **45**(3): 1526–1529.

[81] K. Dowsland. Some experiments with simulated annealing techniques for packing problems. *European Journal of Operational Research*, 1993, **68**: 389–399.

[82] K. Dowsland, E. Soubeiga, E. Burke. A simulated annealing based hyper-heuristic for determining shipper sizes for storage and transportation. *European Journal of Operational Research*, 2007, **179**(3): 759–774.

[83] A. Drexl, J. Gruenewald. Nonpreemptive multi-mode resource-constrained project scheduling. *IIE transactions*, 1993, **25**(5): 74–81.

[84] D. Dubois, H. Prade. Operations on fuzzy numbers. *International Journal of Systems Science*, 1978, **9**(6): 613–626.

[85] D. Dubois, H. Prade. *Fuzzy Sets and Systems: Theory and Applications*. Academic Press, San Dieg, 1980.

[86] D. Dubois, H. Prade. The mean value of a fuzzy number. *Fuzzy sets and*

systems, 1987, **24**(3): 279–300.

[87] D. Dubois, H. Prade, H. Farreny, R. Martin-Clouaire, C. Testemale, E. Harding. *Possibility Theory*. Plenum Press, New York, 1988.

[88] L. Duczmal, R. Assuncao. A simulated annealing strategy for the detection of arbitrarily shaped spatial clusters. *Computational Statistics & Data Analysis*, 2004, **45**(2): 269–286.

[89] J. Dupačová. Portfolio optimization via stochastic programming: Methods of output analysis. *Mathematical methods of operations research*, 1999, **50**(2): 245–270.

[90] J. Dupačová. Output analysis for approximated stochastic programs. *Stochastic Optimization: Algorithms and Applications*, 2001, 1–29.

[91] J. Dupačová, G. Consigli, S. Wallace. Scenarios for multistage stochastic programs. *Annals of Operations Research*, 2000, **100**(1): 25–53.

[92] I. Dupanloup, S. Schneider, L. Excoffier. A simulated annealing approach to define the genetic structure of populations. *Molecular Ecology*, 2002, **11**(12): 2571–2581.

[93] J. Dupaová. Applications of stochastic programming: Achievements and questions. *European Journal of Operational Research*, 2002, **140**(2): 281–290.

[94] F. Dweiri, F. Meier. Application of fuzzy decision-making in facilities layout planning. *International Journal of Production Research*, 1996, **34**(11): 3207–3225.

[95] J. Dyer. Interactive goal programming. *Management Science*, 1972, **19**(1): 62–70.

[96] S. Easa, K. Hossain, et al. New mathematical optimization model for construction site layout. *Journal of Construction Engineering and Management*, 2008, **134**: 653.

[97] K. Easter, R. Hearne. Water markets and decentralized water resources management: international problems and opportunities. *Water Resources Bulletin*, 1995, **31**(1): 9–20.

[98] N. Edirisinghe, E. Patterson, N. Saadouli. Capacity planning model for a multipurpose water reservoir with target-priority operation. *Annals of Operations Research*, 2000, **100**(1): 273–303.

[99] A. El-Bouri, N. Azizi, S. Zolfaghari. A comparative study of a new heuristic based on adaptive memory programming and simulated annealing: The case of job shop scheduling. *European Journal of Operational Research*, 2007, **177**(3): 1894–1910.

[100] E. Elbeltagi, T. Hegazy, A. Eldosouky. Dynamic layout of construction temporary facilities considering safety. *Journal of Construction Engineering and Management*, 2004, **130**(4): 534–541.

[101] E. Elbeltagi, T. Hegazy, A. Hosny, A. Eldosouky. Schedule-dependent evolution of site layout planning. *Construction Management and Economics*, 2001, **19**(7): 689–697.

[102] S. Elmaghraby. *Activity Networks: Project Planning and Control by Network Models*. John Wiley & Sons, New York, 1977.

[103] Y. Ermoliev. Methods of stochastic programming. *Monographs in Optimization and Operations Research*, Nauka, Moskwa, 1976.

[104] U. ESCAP. *Principles and Practices of Water Allocation among Water-use Sectors*. UN, 2000.

[105] Y. Feng, B. Zheng, Z. Li. Exploratory study of sorting particle swarm optimizer for multiobjective design optimization. *Mathematical and Computer Modelling*, 2010, **52**(11-12): 1966–1975.

[106] A. Ferguson, G. Dantzig. The allocation of aircraft to routes—an example of linear programming under uncertain demand. *Management Science*, 1956, **3**(1): 45–73.

[107] P. Fetterolf, G. Anandalingam. A Lagrangian relaxation technique for optimizing interconnection of local area networks. *Operations Research*, 1992, **40**(4): 678–688.

[108] J. Fliege. Gap-free computation of Pareto-points by quadratic scalarizations. *Mathematical Methods of Operations Research*, 2004, **59**(1): 69–89.

[109] N. Freed, F. Glover. Simple but powerful goal programming models for discriminant problems. *European Journal of Operational Research*, 1981, **7**(1): 44–60.

[110] A. Garrido. A mathematical programming model applied to the study of water markets within the Spanish agricultural sector. *Annals of Operations Research*, 2000, **94**(1): 105–123.

[111] S. Gass, P. Roy. The compromise hypersphere for multiobjective linear programming. *European Journal of Operational Research*, 2003, **144**(3): 459–479.

[112] S. Geman, D. Geman, K. Abend, T. Harley, L. Kanal. Stochastic relaxation, Gibbs distributions and the Bayesian restoration of images. *Journal of Applied Statistics*, 1993, **20**(5): 25–62.

[113] M. Gen, R. Cheng. *Genetic Algorithms and Engineering Design*. Wiley, New York, 1997.

[114] M. Gen, R. Cheng. *Genetic Algorithms and Engineering Optimization.* Wiley, New York, 2000.

[115] M. Gen, Y. Li, K. Ida. Solving multi-objective transportation problem by spanning tree-based genetic algorithm. *IEICE TRANSACTIONS on Fundamentals of Electronics, Communications and Computer Sciences*, 1999, **82**(12): 2802–2810.

[116] M. Gill, Y. Kaheil, A. Khalil, M. McKee, L. Bastidas. Multiobjective particle swarm optimization for parameter estimation in hydrology. *Water Resources Research*, 2006, **42**(7): W07417.

[117] R. Glauber. Time-dependent statistics of the ising model. *Journal of Mathematical Physics*, 1963, **4**: 294.

[118] F. Glover. Future paths for integer programming and links to artificial intelligence. *Computers & Operations Research*, 1986, **13**(5): 533–549.

[119] F. Glover. A template for scatter search and path relinking. **In:** *Artificial Evolution*, Springer, Berlin, 1998, 13.

[120] F. Glover. Parametric tabu-search for mixed integer programs. *Computers & Operations Research*, 2006, **33**(9): 2449–2494.

[121] A. González. A study of the ranking function approach through mean values. *Fuzzy Sets and Systems*, 1990, **35**(1): 29–41.

[122] S. Greco, B. Matarazzo, R. Slowinski. Rough sets theory for multicriteria decision analysis. *European Journal of Operational Research*, 2001, **129**(1): 1–47.

[123] D. Greening. Parallel simulated annealing techniques. *Physica D: Nonlinear Phenomena*, 1990, **42**(1-3): 293–306.

[124] J. Grzymala-Busse. A new version of the rule induction system LERS. *Fundamenta Informaticae*, 1997, **31**(1): 39.

[125] Y. Haimes, W. Hall, H. Freedman. *MultiobjectiveOoptimization in Water Resources Systems: The Surrogate Worth Trade-Off Method.* Elsevier Science & Technology, Amsterdam, 1975.

[126] P. Hansen. The steepest ascent mildest descent heuristic for combinatorial programming. **In:** *Congress on Numerical Methods in Combinatorial Optimization*, Capri, Italy, 1986, 70–145.

[127] P. Hansen, B. Jaumard, G. Savard. New branch-and-bound rules for linear bilevel programming. *SIAM Journal on Scientific and Statistical Computing*, 1992, **13**: 1194.

[128] H. Harmanani, A. Hajar. Genetic algorithm for solving site layout problem with unequal-size and constrained facilities. *Journal of computing in civil*

engineering, 2002, **16**: 143.

[129] L. He, G. Huang, G. Zeng, H. Lu. Identifying optimal regional solid waste management strategies through an inexact integer programming model containing infinite objectives and constraints. *Waste Management*, 2009, **29**(1): 21–31.

[130] S. Hejazi, A. Memariani, G. Jahanshahloo, M. Sepehri. Linear bilevel programming solution by genetic algorithm. *Computers & Operations Research*, 2002, **29**(13): 1913–1925.

[131] W. Herroelen, R. Leus. Project scheduling under uncertainty: Survey and research potentials. *European Journal of Operational Research*, 2005, **165**(2): 289–306.

[132] J. Higle, S. Sen. *Stochastic Decomposition: A Statistical Method for Large Scale Stochastic Linear Programming*. Kluwer Academic, Dordrecht Netherland, 1996.

[133] S. Ho, S. Yang, G. Ni, E. Lo, H. Wong. A particle swarm optimization-based method for multiobjective design optimizations. *IEEE Transactions on Magnetics*, 2005, **41**(5).

[134] S. Ho, S. Yang, G. Ni, H. Wong. A particle swarm optimization method with enhanced global search ability for design optimizations of electromagnetic devices. *IEEE Transactions on Magnetics*, 2006, **42**(4): 1107–1110.

[135] J. Holland. *Adaptation in Natural and Artificial Systems*. MIT Press Cambridge, MA, USA, 1992.

[136] R. Horst, P. Pardalos, N. Van Thoai. *Introduction to global optimization*. Springer, Berlin, 2000.

[137] J. Hu, E. Goodman, K. Seo, M. Pei. Adaptive hierarchical fair competition (ahfc) model for parallel evolutionary algorithms. **In:** *Proceedings of the Genetic and Evolutionary Computation Conference*, Morgan Kaufmann, San Francisco, 2002, 772–779.

[138] X. Hu, R. Eberhart. Multiobjective optimization using dynamic neighborhood particle swarm optimization. *Proceedings of the Evolutionary Computation*, 2002, 1677–1681.

[139] G. Huang, L. He, G. Zeng, H. Lu. Identification of optimal urban solid waste flow schemes under impacts of energy prices. *Environmental Engineering Science*, 2008, **25**(5): 685–696.

[140] V. Huang, P. Suganthan, J. Liang. Comprehensive learning particle swarm optimizer for solving multiobjective optimization problems. *International Journal of Intelligent Systems*, 2006, **21**(2): 209–226.

[141] J. Ignizio. *Goal Programming and Extensions*. Lexington Books Lexington,

MA, 1976.

[142] Y. Ijiri. *Management Goals and Accounting for Control.* North-Holland Amsterdam, 1965.

[143] G. Infanger. Planning under uncertainty solving large-scale stochastic linear programs. *Tech. Rep.*, Stanford Univ., CA (United States). Systems Optimization Lab., 1992.

[144] L. Ingber. Very fast simulated re-annealing. *Mathematical and Computer Modelling*, 1989, **12**(8): 967–973.

[145] L. Ingber. Simulated annealing: Practice versus theory. *Mathematical and computer modelling*, 1993, **18**(11): 29–57.

[146] L. Ingber. Adaptive simulated annealing (ASA): Lessons learned. *Arxiv preprint cs/0001018*, 2000.

[147] N. Isendahl, A. Dewulf, C. Pahl-Wostl. Making framing of uncertainty in water management practice explicit by using a participant-structured approach. *Journal of Environmental Management*, 2010, **91**(4): 844–851.

[148] H. Ishibuchi, T. Murata. A multi-objective genetic local search algorithm and its application to flowshop scheduling. *IEEE Transactions on Systems, Man, and Cybernetics, Part C: Applications and Reviews*, 1998, **28**(3).

[149] H. Ishibuchi, H. Tanaka. Multiobjective programming in optimization of the interval objective function. *European Journal of Operational Research*, 1990, **48**(2): 219–225.

[150] Y. Ishizuka, E. Aiyoshi. Double penalty method for bilevel optimization problems. *Annals of Operations Research*, 1992, **34**(1): 73–88.

[151] K. Iwamura, M. Horiike. λ Credibility. **In:** *Proceedings of the Fifth International Conference on Information and Management Sciences*, Toyko, 2006.

[152] S. Jacobsen. On marginal allocation in single constraint min-max problems. *Management Science*, 1971, **17**(11): 780–783.

[153] A. Jakeman, R. Letcher, J. Norton. Ten iterative steps in development and evaluation of environmental models. *Environmental Modelling & Software*, 2006, **21**(5): 602–614.

[154] P. Kall. *Stochastic Linear Programming.* Springer-Verlag, Berlin, 1976.

[155] P. Kall, S. Wallace. *Stochastic programming.* Wiley & Son, New York, 1994.

[156] S. Kaplan. Application of programs with maximin objective functions to problems of optimal resource allocation. *Operations Research*, 1974, 802–807.

[157] F. Karray, E. Zaneldin, T. Hegazy, A. Shabeeb, E. Elbeltagi. Tools of soft computing as applied to the problem of facilities layout planning. *IEEE Trans-*

actions on Fuzzy Systems, 2002, **8**(4): 367–379.

[158] E. Karsak, S. Sozer, S. Alptekin. Product planning in quality function deployment using a combined analytic network process and goal programming approach. *Computers & Industrial Engineering*, 2003, **44**(1): 171–190.

[159] A. Kashan, B. Karimi. A discrete particle swarm optimization algorithm for scheduling parallel machines. *Computers & Industrial Engineering*, 2009, **56**(1): 216–223.

[160] A. Kaufmann. *Introduction to the Theory of Fuzzy Subsets*. Academic Press, San Diego, CA, 1975.

[161] K. Kendall, S. Lee. Formulating blood rotation policies with multiple objectives. *Management Science*, 1980, **26**(11): 1145–1157.

[162] J. Kennedy, R. Eberhart. Particle swarm optimization. **In:** *IEEE International Conference on Neural Networks*, vol. 4, IEEE, Piscataway, NJ, 1995, 1942–1948.

[163] J. Kim, M. Gen, K. Ida. Bicriteria network design using a spanning tree-based genetic algorithm. *Artificial Life and Robotics*, 1999, **3**(2): 65–72.

[164] K. Kim, K. Moon. Berth scheduling by simulated annealing. *Transportation Research Part B: Methodological*, 2003, **37**(6): 541–560.

[165] S. Kirkpatrick, C. Gelatt, M. Vecchi. Optimization by simulated annealing. *Science*, 1983, **220**(4598): 671.

[166] R. Kolisch, A. Drexl. Local search for nonpreemptive multi-mode resource-constrained project scheduling. *IIE Transactions*, 1997, **29**(11): 987–999.

[167] R. Kolisch, R. Padman. An integrated survey of deterministic project scheduling. *Omega*, 2001, **29**(3): 249–272.

[168] S. Komlósi. Quasiconvex first-order approximations and Kuhn-Tucker type optimality conditions. *European Journal of Operational Research*, 1993, **65**(3): 327–335.

[169] M. Kondo. On the structure of generalized rough sets. *Information Sciences*, 2006, **176**(5): 589–600.

[170] J. Kortelainen. On relationship between modified sets, topological spaces and rough sets. *Fuzzy Sets and Systems*, 1994, **61**(1): 91–95.

[171] J. Koza. *Genetic Pprogramming: On the Programming of Computers by Means of Natural Selection*. MIT press, Cambridge, MA, 1992.

[172] K. Krickeberg. *Probability Theory*. Addison-Wesley, Boston, MA, 1965.

[173] K. Lam, X. Ning, T. Ng. The application of the ant colony optimization algorithm to the construction site layout planning problem. *Construction Man-*

agement and Economics, 2007, **25**(4): 359–374.

[174] E. Lee, H. Shih. *Fuzzy and Multi-Level Decision Making: And Interactive Computational Approach*. Springer-Verlag, New York, 2000.

[175] J. Lee, S. Kim. Using analytic network process and goal programming for interdependent information system project selection. *Computers & Operations Research*, 2000, **27**(4): 367–382.

[176] S. Lee. *Goal Programming for Decision Analysis*. Auerbach, Philadelphia, PA, 1972.

[177] S. Lee, K. Lee. Synchronous and asynchronous parallel simulated annealing with multiple Markov chains. *IEEE Transactions on Parallel and Distributed Systems*, 2002, **7**(10): 993–1008.

[178] Z. Lee, S. Su, C. Chuang, K. Liu. Genetic algorithm with ant colony optimization (GA-ACO) for multiple sequence alignment. *Applied Soft Computing*, 2008, **8**(1): 55–78.

[179] W. Lei, K. Qi, X. Hui, W. Qidi. A modified adaptive particle swarm optimization algorithm.

[180] J. Leite, B. Topping. Parallel simulated annealing for structural optimization. *Computers & Structures*, 1999, **73**(1-5): 545–564.

[181] D. Li, L. Gao, J. Zhang, Y. Li. Power system reactive power optimization based on adaptive particle swarm optimization algorithm. **In:** *Intelligent Control and Automation, 2006. WCICA 2006. The Sixth World Congress on*, vol. 2, IEEE, 2006, 7572–7576.

[182] H. Li, P. Love. Site-level facilities layout using genetic algorithms. *Journal of Computing in Civil Engineering*, 1998, **12**: 227.

[183] H. Li, P. Love. Genetic search for solving construction site-level unequal-area facility layout problems. *Automation in Construction*, 2000, **9**(2): 217–226.

[184] J. Li, J. Xu. A novel portfolio selection model in a hybrid uncertain environment. *Omega*, 2009, **37**(2): 439–449.

[185] J. Li, J. Xu, M. Gen. A class of multiobjective linear programming model with fuzzy random coefficients. *Mathematical and Computer Modelling*, 2006, **44**(11-12): 1097–1113.

[186] N. Li, J. Cha, Y. Lu. A parallel simulated annealing algorithm based on functional feature tree modeling for 3D engineering layout design. *Applied Soft Computing*, 2010, **10**(2): 592–601.

[187] R. Li. *Multiple Objective Decision Making in a Fuzzy Environment*. Ph.D. Thesis, Kansas State University, KA, 1990.

[188] W. Li, C. McMahon. A simulated annealing-based optimization approach for

integrated process planning and scheduling. *International Journal of Computer Integrated Manufacturing*, 2007, **20**(1): 80–95.

[189] X. Li. A non-dominated sorting particle swarm optimizer for multiobjective optimization. **In:** *Genetic and Evolutionary Computation (GECCO 2003)*, Springer, New York, 2003, 198–198.

[190] Y. Li, G. Huang, Y. Huang, H. Zhou. A multistage fuzzy-stochastic programming model for supporting sustainable water-resources allocation and management. *Environmental Modelling & Software*, 2009, **24**(7): 786–797.

[191] Y. Li, G. Huang, S. Nie. A robust interval-based minimax-regret analysis approach for the identification of optimal water-resources-allocation strategies under uncertainty. *Resources, Conservation and Recycling*, 2009, **54**(2): 86–96.

[192] Y. Li, G. Huang, S. Nie, L. Liu. Inexact multistage stochastic integer programming for water resources management under uncertainty. *Journal of Environmental Management*, 2008, **88**(1): 93–107.

[193] C. Lin, M. Gen. Multi-criteria human resource allocation for solving multistage combinatorial optimization problems using multiobjective hybrid genetic algorithm. *Expert Systems with Applications*, 2008, **34**(4): 2480–2490.

[194] S. Ling, H. Iu, K. Chan, H. Lam, B. Yeung, F. Leung. Hybrid particle swarm optimization with wavelet mutation and its industrial applications. *IEEE Transactions on Systems, Man, and Cybernetics, Part B: Cybernetics*, 2008, **38**(3): 743–763.

[195] B. Liu, K. Iwamura. Modelling stochastic decision systems using dependent-chance programming. *European Journal of Operational Research*, 1997, **101**(1): 193–203.

[196] B. Liu, Y. Liu. Expected value of fuzzy variable and fuzzy expected value models. *Fuzzy Systems, IEEE Transactions on*, 2002, **10**(4): 445–450.

[197] L. Liu, Y. Li, L. Yang. The maximum fuzzy weighted matching models and hybrid genetic algorithm. *Applied Mathematics and Computation*, 2006, **181**(1): 662–674.

[198] Q. Liu, J. Xu. A study on facility location–allocation problem in mixed environment of randomness and fuzziness. *Journal of Intelligent Manufacturing*, 2009, 1–10.

[199] D. Loucks, J. Stedinger, D. Haith. *Water Resource Systems Planning and Analysis*. Prentice-Hall, Englewood Cliff, NJ, 1981.

[200] H. Lu, G. Huang, L. He. Inexact rough-interval two-stage stochastic programming for conjunctive water allocation problems. *Journal of Environmental Management*, 2009.

[201] H. Lu, G. Huang, Y. Lin, L. He. A two-step infinite α-cuts fuzzy linear programming method in determination of optimal allocation strategies in agricultural irrigation systems. *Water Resources Management*, 2009, **23**(11): 2249–2269.

[202] H. Lu, G. Huang, Z. Liu, L. He. Greenhouse gas mitigation-induced rough-interval programming for municipal solid waste management. *Journal of the Air & Waste Management Association*, 2008, **58**(12): 1546–1559.

[203] M. Lu. On crisp equivalents and solutions of fuzzy programming with different chance measures. *INFORMATION-YAMAGUCHI-*, 2003, **6**(2): 125–134.

[204] R. Luce. Semiorders and a theory of utility discrimination. *Econometrica, Journal of the Econometric Society*, 1956, **24**(2): 178–191.

[205] H. Luss. Minimax resource allocation problems: Optimization and parametric analysis. *European Journal of Operational Research*, 1992, **60**(1): 76–86.

[206] S. Mac Lane. *Saunders Mac Lane: A mathematical autobiography*. AK Peters Ltd, London, 2005.

[207] O. Mangasarian. Computable numerical bounds for lagrange multipliers of stationary points of non-convex differentiable non-linear programs. *Operations Research Letters*, 1985, **4**(2): 47–48.

[208] I. Maqsood, G. Huang, J. Scott Yeomans. An interval-parameter fuzzy two-stage stochastic program for water resources management under uncertainty. *European Journal of Operational Research*, 2005, **167**(1): 208–225.

[209] M. Mawdesley, S. Al-Jibouri. Proposed genetic algorithms for construction site layout. *Engineering Applications of Artificial Intelligence*, 2003, **16**(5-6): 501–509.

[210] J. Mayer. *Stochastic Linear Programming Algorithms: A comparison Based on a Model Management System*. CRC, Boca Raton, FL, 1998.

[211] C. Meise. On the convergence of parallel simulated annealing. *Stochastic Processes and their Applications*, 1998, **76**(1): 99–115.

[212] R. Meller, K. Gau. The facility layout problem: Recent and emerging trends and perspectives. *Journal of manufacturing systems*, 1996, **15**(5): 351–366.

[213] N. Metropolis, A. Rosenbluth, M. Rosenbluth, A. Teller, E. Teller, et al. Equation of state calculations by fast computing machines. *The journal of Chemical Physics*, 1953, **21**(6): 1087.

[214] Z. Michalewicz. *Genetic Algorithms + Data Structures*. Springer, New York, 1996.

[215] S. Mishra, A. Ghosh. Interactive fuzzy programming approach to Bi-level quadratic fractional programming problems. *Annals of Operations Research*,

2006, **143**(1): 251–263.

[216] H. Modares, M. Alfiand. Parameter estimation of bilinear systems based on an adaptive particle swarm optimization. *Engineering Applications of Artificial Intelligence*, 2010.

[217] R. Moore. *Interval Analysis*, vol. 60. Prentice-Hall Englewood Cliffs, New Jersey, 1966.

[218] D. Morgan, J. Eheart, A. Valocchi. Aquifer remediation design under uncertainty using a new chance constrained programming technique. *Water Resources Research*, 1993, **29**(3): 551–561.

[219] M. Mori, C. Tseng. A genetic algorithm for multi-mode resource constrained project scheduling problem. *European Journal of Operational Research*, 1997, **100**(1): 134–141.

[220] A. M'silti, P. Tolla. An interactive multiobjective nonlinear programming procedure. *European Journal of Operational Research*, 1993, **64**(1): 115–125.

[221] S. Nahmias. Fuzzy variables. *Fuzzy Sets and Systems*, 1978, **1**(2): 97–110.

[222] W. Nanry, J. Wesley Barnes. Solving the pickup and delivery problem with time windows using reactive tabu search. *Transportation Research Part B: Methodological*, 2000, **34**(2): 107–121.

[223] R. Narasimhan. Goal programming in a fuzzy environment. *Decision Sciences*, 1980, **11**(2): 325–336.

[224] A. Neumaier. *Interval Methods for Systems of Equations*. Cambridge University Press, Cambridge, U.K., 1990.

[225] J. Nieminen. Rough tolerance equality. *Fundamenta Informaticae*, 1988, **11**(3): 289–296.

[226] P. Nijkamp, J. Spronk. *Interactive Multiple Goal Programming*. Vrije Universiteit, Economische Fakulteit, Amsterdam, Netherlands, 1978.

[227] X. Ning, K. Lam, M. Lam. Dynamic construction site layout planning using max-min ant system. *Automation in Construction*, 2010, **19**(1): 55–65.

[228] N. Nudtasomboon, S. Randhawa. Resource-constrained project scheduling with renewable and non-renewable resources and time-resource tradeoffs. *Computers & Industrial Engineering*, 1997, **32**(1): 227–242.

[229] G. Onwubolu. Optimization of milling operations for the selection of cutting conditions using Tribes. *Proceedings of the Institution of Mechanical Engineers, Part B: Journal of Engineering Manufacture*, 2005, **219**(10): 761–771.

[230] G. Onwubolu, B. Babu. TRIBES application to the flow shop scheduling problem. *New Optimization Techniques in Engineering*, **141**: 517–536.

[231] K. ORourke, T. Bailey, R. Hill, W. Carlton. Dynamic routing of unmanned aerial vehicles using reactive tabu search. *Military Operations Research Journal*, 2000, **6**: 33–42.

[232] L. Ozdamar. A genetic algorithm approach to a general category project scheduling problem. *IEEE Transactions on Systems, Man, and Cybernetics, Part C: Applications and Reviews*, 1999, **29**(1): 44–59.

[233] P. Pai, W. Hong. Support vector machines with simulated annealing algorithms in electricity load forecasting. *Energy Conversion and Management*, 2005, **46**(17): 2669–2688.

[234] P. Pai, W. Hong. Software reliability forecasting by support vector machines with simulated annealing algorithms. *Journal of Systems and Software*, 2006, **79**(6): 747–755.

[235] Q. Pan, M. Tasgetiren, Y. Liang. A discrete differential evolution algorithm for the permutation flowshop scheduling problem. *Computers & Industrial Engineering*, 2008, **55**(4): 795–816.

[236] B. Panigrahi, V. Ravikumar Pandi, S. Das. Adaptive particle swarm optimization approach for static and dynamic economic load dispatch. *Energy Conversion and Management*, 2008, **49**(6): 1407–1415.

[237] K. Parsopoulos, M. Vrahatis. Particle swarm optimization method in multiobjective problems. **In:** *Proceedings of the 2002 ACM Symposium on Applied Computing*, ACM, Madrid, 2002, 603–607.

[238] Z. Pawlak. Rough sets, rough functions and rough calculus. *Rough Fuzzy Hybridization: A New Trend in and Decision Making*, 99–109.

[239] Z. Pawlak. Information systems theoretical foundations. *Information Systems*, 1981, **6**(3): 205–218.

[240] Z. Pawlak. Rough sets. *International Journal of Parallel Programming*, 1982, **11**(5): 341–356.

[241] Z. Pawlak. *Rough Sets: Theoretical Aspects of Reasoning About Aata.* Springer, 1991.

[242] Z. Pawlak, A. Skowron. Rough Membership Function: A Tool for Reasoning with Uncertainty. Algebraic Methods in Logic and Computer Science. *Banach Center Publications, Institute of Mathematics, Polish Academy of Sciences, Warsaw*, 1993, **28**: 135–150.

[243] Z. Pawlak, A. Skowron. Rough membership functions. **In:** *Advances in the Dempster–Shafer Theory of Evidence*, John Wiley & Sons, New York, 1994, 271.

[244] L. Polkowski. 8. Mathematical Morphology of Rough Sets. *Bulletin of the Polish Academy of Sciences-Mathematics*, 1993, **41**(3): 241.

[245] E. Porteus, J. Yormark. More on min-max allocation. *Management Science*, 1972, **18**(9): 502–507,

[246] R. Prasad. Characterization of Quasieoncave Function, 1991.

[247] A. Prékopa. *Studies in Applied Stochastic Programming*. SI, Provo, 1978.

[248] A. Prékopa. *Stochastic Programming*. Springer, New York, 1995.

[249] G. Pugh. Fuzzy allocation of manufacturing resources. *Computers & Industrial Engineering*, 1997, **33**(1-2): 101–104.

[250] K. Qin, J. Yang, Z. Pei. Generalized rough sets based on reflexive and transitive relations. *Information Sciences*, 2008, **178**(21): 4138–4141.

[251] M. Rebolledo. Rough intervals–enhancing intervals for qualitative modeling of technical systems. *Artificial Intelligence*, 2006, **170**(8-9): 667–685.

[252] I. Rechenberg. Evolutionsstrategie, Optimierung technischer Systeme nach Prinzipien der biologischen Evolution, Volume 15 von Reihe Problemata. F, 1973.

[253] C. ReVelle. *Optimizing Reservoir Resources: Including a New Model for Reservoir Reliability*. John Wiley & Sons, New York, 1999.

[254] M. Reyes-Sierra, C. Coello. Multi-objective particle swarm optimizers: A survey of the state-of-the-art. *International Journal of Computational Intelligence Research*, 2006, **2**(3): 287–308.

[255] M. Rijckaert, E. Walraven. Geometric programming: Estimation of Lagrange multipliers. *Operations research*, 1985, **33**(1): 85–93.

[256] J. Robinson, S. Sinton, Y. Rahmat-Samii. Particle swarm, genetic algorithm, and their hybrids: optimization of a profiled corrugated horn antenna. **In:** *Antennas and Propagation Society International Symposium*, vol. 1, IEEE, San Francisco, 2002, 314–317.

[257] A. Ruszczyski. An augmented Lagrangian decomposition method for block diagonal linear programming problems. *Operations Research Letters*, 1989, **8**(5): 287–294.

[258] A. Sadagopan, et al. Interactive algorithms for multiple criteria nonlinear programming problems. *European Journal of Operational Research*, 1986, **25**(2): 247–257.

[259] M. Sakawa. Interactive fuzzy goal programming for multiobjective nonlinear problems and its application to water quality management. *Control and Cybernetics*, 1984, **13**(2): 217–228.

[260] M. Sakawa. *Fuzzy sets and Interactive Multiobjective Optimization*. Plenum Press, New York, 1993.

[261] M. Sakawa, I. Nishizaki, Y. Uemura, K. Kubota. Interactive fuzzy programming for multilevel linear programming problems with fuzzy parameters. *Electronics and Communications in Japan (Part III: Fundamental Electronic Science)*, 2000, **83**(6): 1–9.

[262] E. Sawacha, S. Naoum, D. Fong. Factors affecting safety performance on construction sites. *International Journal of Project Management*, 1999, **17**(5): 309–315.

[263] Y. Sawaragi, H. Nakayama, T. Tanino. *Theory of Multiobjective Optimization*. Academic Press, San Diego, 1985.

[264] B. Schandl, K. Klamroth, M. Wiecek. Norm-based approximation in bicriteria programming. *Computational Optimization and Applications*, 2001, **20**(1): 23–42.

[265] M. Schlueter, N. Rüger. Application of a GIS-based simulation tool to illustrate implications of uncertainties for water management in the Amudarya river delta. *Environmental Modelling & Software*, 2007, **22**(2): 158–166.

[266] M. Schniederjans. *Linear Goal Programming*. Petrocelli Books, New York, 1984.

[267] H. Schwefel. *Evolution and Optimum Seeking. Sixth-Generation Computer Technology Series*. Wiley, New York, 1995.

[268] A. Sethi, S. Sethi. Flexibility in manufacturing: a survey. *International Journal of Flexible Manufacturing Systems*, 1990, **2**(4): 289–328.

[269] D. Sha, C. Hsu. A hybrid particle swarm optimization for job shop scheduling problem. *Computers & Industrial Engineering*, 2006, **51**(4): 791–808.

[270] Z. Shangguan, M. Shao, R. Horton, T. Lei, L. Qin, J. Ma. A model for regional optimal allocation of irrigation water resources under deficit irrigation and its applications. *Agricultural Water Management*, 2002, **52**(2): 139–154.

[271] X. Shi, H. Xia. Interactive bilevel multi-objective decision making. *Journal of the Operational Research Society*, 1997, **48**(9): 943–949.

[272] Y. Shi, L. Yao, J. Xu. A probability maximization model based on rough approximation and its application to the inventory problem. *International Journal of Approximate Reasoning*, 2011, (52): 261–280.

[273] H. Shih, Y. Lai, E. Stanley Lee. Fuzzy approach for multi-level programming problems. *Computers & Operations Research*, 1996, **23**(1): 73–91.

[274] P. Siarry, G. Berthiau. Fitting of tabu search to optimize functions of continuous variables. *International Journal for Numerical Methods in Engineering*, 1997, **40**(13): 2449–2457.

[275] R. Slowinski, B. Soniewicki, et al. DSS for multiobjective project scheduling.

European Journal of Operational Research, 1994, **79**(2): 220–229.

[276] R. Slowinski, D. Vanderpooten. Similarity relation as a basis for rough approximations. *ICS Research Report*, 1995, **53**: 95.

[277] R. Slowinski, D. Vanderpooten. A generalized definition of rough approximations based on similarity. *IEEE Transactions on Knowledge and Data Engineering*, 2000, **12**(2): 331–336.

[278] A. Soltani, T. Fernando. A fuzzy based multi-objective path planning of construction sites. *Automation in Construction*, 2004, **13**(6): 717–734.

[279] A. Sprecher, A. Drexl. Multi-mode resource-constrained project scheduling by a simple, general and powerful sequencing algorithm1. *European Journal of Operational Research*, 1998, **107**(2): 431–450.

[280] N. Srinivas, K. Deb. Muiltiobjective optimization using nondominated sorting in genetic algorithms. *Evolutionary computation*, 1994, **2**(3): 221–248.

[281] R. Steuer, E. Choo. An interactive weighted Tchebycheff procedure for multiple objective programming. *Mathematical Programming*, 1983, **26**(3): 326–344.

[282] A. Syarif, Y. Yun, M. Gen. Study on multi-stage logistic chain network: a spanning tree-based genetic algorithm approach. *Computers & Industrial Engineering*, 2002, **43**(1-2): 299–314.

[283] F. Szidarovszky, M. Gershon, L. Duckstein. *Techniques for Multiobjective Decision Making in Systems Management*. Elsevier Science, Amsterdam, 1986.

[284] C. Tang. A max-min allocation problem: Its solutions and applications. *Operations Research*, 1988, **36**(2): 359–367.

[285] Z. Tao, J. Xu. A class of rough multiple objective programming and its application to solid transportation problem. *Sichuan University Research Report*, 2010.

[286] Z. Tao, J. Xu. A goal programming approach for multiple objective programming model with rough interval parameters and its application to reverse logistics problem. *Sichuan University Research Report*, 2010.

[287] P. Toint. Non-monotone trust-region algorithms for nonlinear optimization subject to convex constraints. *Mathematical Programming*, 1997, **77**(3): 69–94.

[288] S. Tsumoto, H. Tanaka. PRIMEROSE: Probabilistic rule induction method based on rough sets and resampling methods. *Computational Intelligence*, 2007, **11**(2): 389–405.

[289] H. Tuy, A. Migdalas, N. Hoai-Phuong. A novel approach to bilevel nonlinear programming. *Journal of Global Optimization*, 2007, **38**(4): 527–554.

[290] A. Tversky. Features of similarity. *Psychological review*, 1977, **84**(4): 327–352.

[291] S. Van de Vonder, E. Demeulemeester, W. Herroelen, R. Leus. The use of buffers in project management: The trade-off between stability and makespan. *International Journal of Production Economics*, 2005, **97**(2): 227–240.

[292] L. Vicente, P. Calamai. Bilevel and multilevel programming: A bibliography review. *Journal of Global Optimization*, 1994, **5**(3): 291–306.

[293] G. Wang, X. Wang, Z. Wan. A fuzzy interactive decision making algorithm for bilevel multi-followers programming with partial shared variables among followers. *Expert Systems with Applications*, 2009, **36**(7): 10471–10474.

[294] L. Wang, L. Fang, K. Hipel. Water resources allocation: A cooperative game theoretic approach. *Journal of Environmental Informatics*, 2003, **2**(2): 11–22.

[295] L. Wang, L. Fang, K. Hipel. Basin-wide cooperative water resources allocation. *European Journal of Operational Research*, 2008, **190**(3): 798–817.

[296] Y. Wang, B. Li, T. Weise, J. Wang, B. Yuan, Q. Tian. Self-adaptive learning based particle swarm optimization. *Information Sciences*, 2010, doi:10.1016/j.ins.2010.07.013.

[297] T. Weir, B. Mond. Sufficient Fritz John optimality conditions and duality for nonlinear programming problems. *Opsearch*, 1986, **23**(3): 129–141.

[298] R. Wets. Stochastic Programming in Handbooks in Operations Research and Management Science, Vol. 1, 1989.

[299] R. Wets, J. Ermoliev. *Numerical Techniques for Stochastic Optimization*. Springer-Verlag, Berlin, 1988.

[300] R. Wets, W. Ziemba. Stochastic programming: State of the art. *Annal of Operations Research*, 1999.

[301] D. White, G. Anandalingam. A penalty function approach for solving bi-level linear programs. *Journal of Global Optimization*, 1993, **3**(4): 397–419.

[302] A. Wierzbicki. Reference Point Approaches Multicriteria Decision Making. *Advances in MCDM Models, Algorithms, and Application*, 1999, 9–13.

[303] S. Wong, W. Ziarko. Comparison of the probabilistic approximate classification and the fuzzy set model. *Fuzzy Sets and Systems*, 1987, **21**(3): 357–362.

[304] H. Wu. The Karush–Kuhn–Tucker optimality conditions in multiobjective programming problems with interval-valued objective functions. *European Journal of Operational Research*, 2009, **196**(1): 49–60.

[305] T. Wu, C. Chang, S. Chung. A simulated annealing algorithm for manufacturing cell formation problems. *Expert Systems with Applications*, 2008, **34**(3): 1609–1617.

[306] W. Xia, Z. Wu. An effective hybrid optimization approach for multi-objective flexible job-shop scheduling problems. *Computers & Industrial Engineering*, 2005, **48**(2): 409–425.

[307] Z. Xiao, M. Hong, J. Li. Intelligent particle swarm optimization in multiobjective optimization. **In:** *The 2005 IEEE Congress on Evolutionary Computation*, Edinburgh, U.K., 2005, 714–719.

[308] J. Xu, Z. Li. Multi-Objective Dynamic Construction Site Layout Planning under Fuzzy Random Environment. *Sichuan University Research Report*, 2010.

[309] J. Xu, Q. Liu, R. Wang. A class of multi-objective supply chain networks optimal model under random fuzzy environment and its application to the industry of Chinese liquor. *Information Sciences*, 2008, **178**(8): 2022–2043.

[310] J. Xu, Y. Tu. A bi-level expected value programming model for resource allocation under fuzzy random environment and its application to regional water planning. *Sichuan University Research Report*, 2010.

[311] J. Xu, L. Yao. A class of expected value multi-objective programming problems with random rough coefficients. *Mathematical and Computer Modelling*, 2009, **50**(1-2): 141–158.

[312] J. Xu, L. Yao. A class of multiobjective linear programming models with random rough coefficients. *Mathematical and Computer Modelling*, 2009, **49**(1-2): 189–206.

[313] J. Xu, L. Yao. A class of bi-level expected value programming with random rough coefficients and its application to production-inventory problem. *Sichuan University Research Report*, 2010.

[314] J. Xu, L. Yao. Random rough variable and random rough programming. *Sichuan University Research Report*, 2010.

[315] J. Xu, L. Yao. *Random-Like Multiple Objective Decision Making*. Springer, Berlin, 2011.

[316] J. Xu, Z. Zhang. A fuzzy random resource-constrained scheduling model with multiple projects and its application to a working procedure in a large-scale water conservancy and hydropower construction project. *Journal of Scheduling*, doi:10.1007/s10951–010–0173–1.

[317] J. Xu, L. Zhao. A class of fuzzy rough expected value multi-objective decision making model and its application to inventory problems. *Computers & Mathematics with Applications*, 2008, **56**(8): 2107–2119.

[318] J. Xu, X. Zhou. A class of fuzzy expectation multi-objective model with chance constraints based on rough approximation and its application in allocation problem. *Sichuan University Research Report*, 2010.

[319] J. Xu, X. Zhou. *Fuzzy-like multiple objective decision making*. Springer, 2010.

[320] J. Xu, X. Zhou, D. Wu. Portfolio selection using λ mean and hybrid entropy. *Annals of Operations Research*, doi:10.1007/s10479–009–0550–3.

[321] R. Yager. A procedure for ordering fuzzy subsets of the unit interval. *Information Sciences*, 1981, **24**(2): 143–161.

[322] R. Yager. On the evaluation of uncertain courses of action. *Fuzzy Optimization and Decision Making*, 2002, **1**(1): 13–41.

[323] J. Yagi, E. Arai, K. Shirase, S. Matsumoto. Action-based union of the temporal opposites in scheduling: Non-deterministic approach. *Automation in Construction*, 2003, **12**(3): 321–329.

[324] L. Yang, L. Liu. Fuzzy fixed charge solid transportation problem and algorithm. *Applied Soft Computing*, 2007, **7**(3): 879–889.

[325] Y. Yao. Two views of the theory of rough sets in finite universes. *International Journal of Approximate Reasoning*, 1996, **15**(4): 291–318.

[326] Y. Yao. Constructive and algebraic methods of the theory of rough sets. *Information Sciences: An International Journal*, 1998, **109**(1-4): 21–47.

[327] Y. Yao. Relational interpretations of neighborhood operators and rough set approximation operators. *Information Sciences*, 1998, **111**(1-4): 239–259.

[328] Y. Yao. Probabilistic approaches to rough sets. *Expert Systems*, 2003, **20**(5): 287–297.

[329] Y. Yao, Y. Chen. Subsystem based generalizations of rough set approximations. *Foundations of Intelligent Systems*, 210–218.

[330] I. Yeh. Construction-site layout using annealed neural network. *Journal of Computing in Civil Engineering*, 1995, **9**: 201.

[331] J. Yeh, J. Fu. Parallel adaptive simulated annealing for computer-aided measurement in functional MRI analysis. *Expert Systems with Applications*, 2007, **33**(3): 706–715.

[332] J. Yisu, J. Knowles, L. Hongmei, L. Yizeng, D. Kell. The landscape adaptive particle swarm optimizer. *Applied Soft Computing*, 2008, **8**(1): 295–304.

[333] E. Youness. Characterizing solutions of rough programming problems. *European Journal of Operational Research*, 2006, **168**(3): 1019–1029.

[334] K. Youssef, H. Yousef, O. Sebakhy, M. Wahba. Adaptive fuzzy APSO based inverse tracking-controller with an application to DC motors. *Expert Systems with Applications*, 2009, **36**(2): 3454–3458.

[335] L. Zadeh. Fuzzy sets. *Information and Control*, 1965, **8**(3): 338–353.

[336] L. Zadeh. The concept of a linguistic variable and its application to approximate reasoning. *Information Sciences*, 1975, **8**(3): 199–249.

[337] L. Zadeh. Fuzzy sets as a basis for a theory of possibility. *Fuzzy Sets and Systems*, 1978, **1**(1): 3–28.

[338] L. Zadeh, C. Desoer. *Linear system theory*. McGraw-Hill New York, 1963.

[339] M. Zeleny. *Multiple Criteria Decision Making*. McGraw-Hill, New York, 1982.

[340] S. Zenios, M. Holmer, R. McKendall, C. Vassiadou-Zeniou. Dynamic models for fixed-income portfolio management under uncertainty. *Journal of Economic Dynamics and Control*, 1998, **22**(10): 1517–1541.

[341] G. Zhang, J. Lu, T. Dillon. Decentralized multi-objective bilevel decision making with fuzzy demands. *Knowledge-Based Systems*, 2007, **20**(5): 495–507.

[342] H. Zhang, C. Tam, H. Li. Multimode project scheduling based on particle swarm optimization. *Computer-Aided Civil and Infrastructure Engineering*, 2006, **21**(2): 93–103.

[343] H. Zhang, J. Wang. Particle Swarm Optimization for Construction Site Unequal-Area Layout. *Journal of Construction Engineering and Management*, 2008, **134**: 739.

[344] W. Zhu. Generalized rough sets based on relations. *Information Sciences*, 2007, **177**(22): 4997–5011.

[345] H. Zimmermann. Description and optimization of fuzzy syatems. *International Journal of General Systems*, 1975, **2**(1): 209–215.

[346] H. Zimmermann. Fuzzy programming and linear programming with several objective functions. *Fuzzy Sets and Systems*, 1978, **1**(1): 45–55.

[347] H. Zimmermann. *Fuzzy Sets, Decision Making, and Expert Systems*. Springer, Berlin, 1987.

[348] P. Zouein, I. Tommelein. Dynamic layout planning using a hybrid incremental solution method. *Journal of Construction Engineering and Management*, 1999, **125**(6): 400–408.

[349] P. Zouein, I. Tommelein. Dynamic layout planning using a hybrid incremental solution method. *Journal of Construction Engineering and Management*, 1999, **125**(6): 400–408.

Appendix A

MATLAB Files

A.1   MATLAB® file for Example 2.9

```
N=10000;
L=0;
for i=1:N
    x1=unifrnd(-1,1);
    x2=unifrnd(-2,2);
    L=L+(1+x1)/(1+x1^2)+(1+x2)/(1+x2^2);
end
L/(2*N)

tic
N=3000;
v=1;
alpha=0.8;
for i=1:N
    x11(i)=unifrnd(0,1);
    x21(i)=unifrnd(1,2);
    x31(i)=unifrnd(2,3);
    x12(i)=unifrnd(-1,3);
    x22(i)=unifrnd(0,3);
    x32(i)=unifrnd(1,5);
    g1(i)=x11(i)+x21(i)^2+x31(i)^3;
    g2(i)=x12(i)+x22(i)^2+x32(i)^3;
end
for j=1:N
    N1=0;
    N2=0;
    for i=1:N
        if g1(i)>=v
            N1=N1+1;
        end
        if g2(i)>=v
            N2=N2+1;
        end
```

```
        end
    if (N1+N2)/(2*N)>=alpha
            v=v+0.005
    end
    plot (j,v);
    hold on;
end
v
toc

tic
 N=10000;
 M=0;
 K=0;
 N1=0;
 N2=0;
 for i=1:N
    x11=unifrnd(1,2);
    x21=unifrnd(2,3);
    x12=unifrnd(0,5);
    x22=unifrnd(1,4);

    if x11^2+x21^2<=18
        M=M+1;
    end
    Tr1=M/N;

    if x12^2+x22^2<=18
        K=K+1;
    end
    Tr2=K/N;
end
Tr=(Tr1+Tr2)/2
toc

tic
e=0;
N=200;
for i=1:N
    rho1=normrnd(0,1);
    rho2=normrnd(1,2);
    L=0;
    for j=1:N
    x11=unifrnd(rho1,rho1+1);
```

```
        x12=unifrnd(rho1-1,rho1+2);
        x21=unifrnd(rho2,rho2+1);
        x22=unifrnd(rho2-1,rho2+2);
    end
    e=e+L/(2*N);
end
e/N
toc

alpha=0.9;
beta=0.9;
N=100;
for i=1:N
    rho1(i)=normrnd(0,1);
    rho2(i)=normrnd(1,2);
end

for i=1:N
    v(i)=0;
    for j=1:N
        x11(j)=unifrnd(rho1(i), rho1(i)+1);
        x12(j)=unifrnd(rho1(i)-1, rho1(i)+2);
        x21(j)=unifrnd(rho2(i), rho2(i)+1);
        x22(j)=unifrnd(rho2(i)-1, rho2(i)+2);
        f1(j)=x11(j)^2+x21(j)^2;
        f2(j)=x12(j)^2+x22(j)^2;
    end
    for k=1:N
        N1(k)=0;
        N2(k)=0;
        for l=1:N
            if f1(l)>=v(i)
            N1(k)=N1(k)+1;
            end
            if f2(l)>=v(i)
            N2(k)=N2(k)+1;
            end
        end
        if (N1(k)+N2(k))/(2*N)>=beta
            v(i)=v(i)+0.05;
        end
    end
end

for j=1:N
```

```
    for k=j+1:N
        if v(j)<v(k)
            temp=v(k);
            v(k)=v(j);
            v(j)=temp;
        end
    end
end

v(alpha*N)

tic
alpha=0.8;
N=500;
for i=1:N
    rho1(i)=normrnd(0,1);
    rho2(i)=normrnd(1,2);
end

for i=1:N
    N1(i)=0;
    N2(i)=0;
    for j=1:N
        x11=unifrnd(rho1(i),rho1(i)+1);
        x12=unifrnd(rho1(i)-1,rho1(i)+2);
        x21=unifrnd(rho2(i),rho2(i)+1);
        x22=unifrnd(rho2(i)-1,rho2(i)+2);
        if x11+x21>=0
            N1(i)=N1(i)+1;
        end
        if x21+x22>=0
            N2(i)=N2(i)+1;
        end
    end
    beta(i)=(N1(i)+N2(i))/(2*N);
end

for i=1:N
    for j=i+1:N
        if beta(i)<beta(j)
            temp=beta(j);
            beta(j)=beta(i);
            beta(i)=temp;
        end
    end
end
```

```
end
beta(alpha*N)
toc


tic
e=0;
Pos=0;
Nec=0;
N=1000;
for i=1:N
    rho1(i)=unifrnd(1,3);
    if 1<rho1(i)&&rho1(i)<2
        mu1=rho1(i)-1;
    end
    if 2<=rho1(i)&&rho1(i)<3
        mu1=3-rho1(i);
    end
    rho2(i)=unifrnd(2,4);
    if 2<rho2(i)&&rho2(i)<3
        mu2=rho2(i)-2;
    end
    if 3<=rho2(i)&&rho2(i)<4
        mu2=4-rho2(i);
    end
    rho3(i)=unifrnd(3,5);
    if 3<rho3(i)&&rho3(i)<4
        mu3=rho3(i)-4;
    end
    if 4<=rho3(i)&&rho3(i)<5
        mu3=5-rho3(i);
    end
    A=[mu1,mu2,mu3];
    mu(i)=min(A);
end

for i=1:N
    E1(i)=0;
    E2(i)=0;
    for j=1:N
        x11=unifrnd(rho1(i),rho1(i)+1);
        x12=unifrnd(rho1(i)-1,rho1(i)+2);
        x21=unifrnd(rho2(i),rho2(i)+1);
        x22=unifrnd(rho2(i)-1,rho2(i)+2);
        x31=unifrnd(rho3(i),rho3(i)+1);
```

```
            x32=unifrnd(rho3(i)-1,rho3(i)+2);
            E1(i)=E1(i)+x11*x21*x31;
            E2(i)=E2(i)+x12*x22*x32;
    end
    E(i)=(E1(i)+E2(i))/(2*N);
end

MIN=E(1);
for i=2:N
    if MIN>E(i)
        MIN=E(i);
    end
end

MAX=E(1); for i=2:N
    if MAX<E(i)
        MAX=E(i);
    end
end

for i=1:N
    r=unifrnd(MIN,MAX);
    if r>=0
        for i=1:N
            if Pos<mu(i)&&E(i)>=r
                Pos=mu(i);
            end
            if Nec>1-mu(i)&&E(i)<r
                Nec=1-mu(i);
            end
        end
        Cr=1/2*(Pos+Nec);
        e=e+Cr;
    end

    if r<0
        for i=1:N
            if Pos<mu(i)&&E(i)<=r
                Pos=mu(i);
            end
            if Nec>1-mu(i)&&(i)>r
                Nec=1-mu(i);
            end
        end
        Cr=1/2*(Pos+Nec);
```

```
        e=e-Cr;
    end


end
E=a+b+e*(MAX-MIN)/N
a=max(MIN,0);
b=min(MAX,0);
fid=fopen('data.txt', 'w');
fprintf(fid, '%6.2f\n', E);
fclose(fid);
toc

tic
alpha=0.9;
beta=0.9;
N=400;
for i=1:N
    rho1(i)=unifrnd(1,3);
    if 1<rho1(i)&&rho1(i)<2
        mu1=rho1(i)-1;
    end
    if 2<=rho1(i)&&rho1(i)<3
        mu1=3-rho1(i);
    end
    rho2(i)=unifrnd(2,4);
    if 2<rho2(i)&&rho2(i)<3
        mu2=rho2(i)-2;
    end
    if 3<=rho2(i)&&rho2(i)<4
        mu2=4-rho2(i);
    end
    rho3(i)=unifrnd(3,5);
    if 3<rho3(i)&&rho3(i)<4
        mu3=rho3(i)-3;
    end
    if 4<=rho3(i)&&rho3(i)<5
        mu3=5-rho3(i);
    end
    A=[mu1,mu2,mu3];
    mu(i)=min(A);
end

for i=1:N
    v(i)=6;
```

```
for j=1:N
    x11(j)=unifrnd(rho1(i),rho1(i)+1);
    x12(j)=unifrnd(rho1(i)-1,rho1(i)+2);
    x21(j)=unifrnd(rho2(i),rho2(i)+1);
    x22(j)=unifrnd(rho2(i)-1,rho2(i)+2);
    x31(j)=unifrnd(rho3(i),rho3(i)+1);
    x32(j)=unifrnd(rho3(i)-1,rho3(i)+2);
    f1(j)=x11(j)^2+x21(j)^2+x31(j)^2;
    f2(j)=x12(j)^2+x22(j)^2+x32(j)^2;
end

for j=1:N
    N1(j)=0;
    N2(j)=0;
    for k=1:N
        if f1(k)>=v(i)
            N1(j)=N1(j)+1;
        end
        if f2(k)>=v(i)
            N2(j)=N2(j)+1;
        end
    end
    if (N1(j)+N2(j))/(2*N)>=alpha
        v(i)=v(i)+0.08;
    end
end
end

r=15;
for i=1:N
MAX=0;
MIN=1;
  for j=1:N
    if MAX<mu(j)&&v(j)>=r
        MAX=mu(j);
    end
    if MIN>1-mu(j)&&v(j)<r
        MIN=1-mu(j);
    end
end
L=(MAX+MIN)/2;
    if L>=alpha;
        r=r+0.1;
    end
    plot(i,r);
```

```
        hold on;
end
toc

alpha=0.9;
N=200;
for i=1:N
    rho1(i)=unifrnd(1,3);
    if 1<rho1(i)&&rho1(i)<2
        mu1=rho1(i)-1;
    end
    if 2<=rho1(i)&&rho1(i)<3
        mu1=3-rho1(i);
    end
    rho2(i)=unifrnd(2,4);
    if 2<rho2(i)&&rho2(i)<3
        mu2=rho2(i)-2;
    end
    if 3<=rho2(i)&&rho2(i)<4
        mu2=4-rho2(i);
    end
    rho3(i)=unifrnd(3,5);
    if 3<rho3(i)&&rho3(i)<4
        mu3=rho3(i)-3;
    end
    if 4<=rho3(i)&&rho3(i)<5
        mu3=5-rho3(i);
    end
    A=[mu1,mu2,mu3];
    mu(i)=min(A);
end

for i=1:N
    Tr1(i)=0;
    Tr2(i)=0;
    for j=1:N
        x11(j)=unifrnd(rho1(i),rho1(i)+1);
        x12(j)=unifrnd(rho1(i)-1,rho1(i)+2);
        x21(j)=unifrnd(rho2(i),rho2(i)+1);
        x22(j)=unifrnd(rho2(i)-1,rho2(i)+2);
        x31(j)=unifrnd(rho3(i),rho3(i)+1);
        x32(j)=unifrnd(rho3(i)-1,rho3(i)+2);
        if x11(j)^2+x21(j)^2+x31(j)^2>=16
            Tr1(i)=Tr1(i)+1;
        end
```

```
        if x12(j)^2+x22(j)^2+x32(j)^2>=16
            Tr2(i)=Tr2(i)+1;
        end
    end
     Tr(i)=(Tr1(i)+Tr2(i))/(2*N);

end

r=0.1;
for i=1:N
    MAX=0;
    MIN=1;
    for j=1:N
        if MAX<mu(j)&&Tr(j)>=r
            MAX=mu(j);
        end
        if MIN>1-mu(j)&&Tr(j)<r
            MIN=1-mu(j);
        end
    end
    L=(MAX+MIN)/2;
    if L>=alpha;
        r=r+0.01;
    end
    plot(i,r);
    hold on;
end
r


N=200; L=0; for i=1:N
    rho11=unifrnd(1,2);
    rho12=unifrnd(0,3);
    rho21=unifrnd(2,3);
    rho22=unifrnd(1,4);
    E11(i)=0;
    E12(i)=0;
    for j=1:N
        x11(j)=unifrnd(rho11-1,rho11+1);
        x12(j)=unifrnd(rho11-2,rho11+2);
        x21(j)=unifrnd(rho21-1,rho21+1);
        x22(j)=unifrnd(rho21-2,rho21+2);
        E11(i)=E11(i)+x11(j)*x21(j);
        E12(i)=E12(i)+x12(j)*x22(j);
    end
```

```
    E1(i)=(E11(i)+E12(i))/(2*N);
    E21(i)=0;
    E22(i)=0;
    for j=1:N
        x11(j)=unifrnd(rho12-1,rho12+1);
        x12(j)=unifrnd(rho12-2,rho12+2);
        x21(j)=unifrnd(rho22-1,rho22+1);
        x22(j)=unifrnd(rho22-2,rho22+2);
        E21(i)=E21(i)+x11(j)*x21(j);
        E22(i)=E12(i)+x12(j)*x22(j);
    end
    E2(i)=(E21(i)+E22(i))/(2*N);
    L=L+E1(i)+E2(i);
end
 L/(2*N)

N=200; L=0; for i=1:N
    rho11=unifrnd(1,2);
    rho12=unifrnd(0,3);
    rho21=unifrnd(2,3);
    rho22=unifrnd(1,4);
    E11(i)=0;
    E12(i)=0;
    for j=1:N
        x11(j)=unifrnd(rho11-1,rho11+1);
        x12(j)=unifrnd(rho11-2,rho11+2);
        x21(j)=unifrnd(rho21-1,rho21+1);
        x22(j)=unifrnd(rho21-2,rho21+2);
        E11(i)=E11(i)+x11(j)*x21(j);
        E12(i)=E12(i)+x12(j)*x22(j);
    end
    E1(i)=(E11(i)+E12(i))/(2*N);
    E21(i)=0;
    E22(1)=0;
    for j=1:N
        x11(j)=unifrnd(rho12-1,rho12+1);
        x12(j)=unifrnd(rho12-2,rho12+2);
        x21(j)=unifrnd(rho22-1,rho22+1);
        x22(j)=unifrnd(rho22-2,rho22+2);
        E21(i)=E21(i)+x11(j)*x21(j);
        E22(i)=E12(i)+x12(j)*x22(j);
    end
    E2(i)=(E21(i)+E22(i))/(2*N);
    L=L+E1(i)+E2(i);
end
```

```
L/(2*N)
```

A.2   MATLAB® file for Example 2.13

```
clc;
clear all;
clf;
figure(1);
N=3;
GEN=100;
POP_SIZE=10;
P_MUTATION=0.2;
P_CROSSOVER=0.3;
 Chr=[];
 Obj=[];
 Ob1=[];
 Ob2=[];
w_1=0.5;
w_2=0.5;

tic
j=1;
while(j<=POP_SIZE)
    x0=unifrnd(0,10);
    y0=unifrnd(0,10);
    z0=unifrnd(0,10);
    t=constraint_check(x0,y0,z0);
    if(t==1)
        Chr=[Chr,[x0,y0,z0]'];
        j=j+1;
    end
end
for gen=1:GEN
    INDX=[];
    z1Max=0;
    z2Max=0;
    rTemp=[];
    eval=[];
    for j=1:POP_SIZE
        t1=Obfunction1(Chr(1,j),Chr(2,j),Chr(3,j));
        if(t1>z1Max)
```

```
        z1Max=t1;
    end
    t2=Obfunction2(Chr(1,j),Chr(2,j),Chr(3,j));
    if(t2>z2Max)
        z2Max=t2;
    end
    u=w_1*t1+w_2*t2;
    Obj=[Obj,u];
    Ob1=[Ob1,t1];
    Ob2=[Ob2,t2];
end
[Obj,INDX]=sort(Obj);
for i=1:POP_SIZE
    t=sqrt(w_1^2*(Ob1(j)-z1Max)^2+w_2^2
    *(Ob2(j)-z2Max)^2);
    rTemp=[rTemp,t];
end
rMax=max(rTemp);
rMin=min(rTemp);
rr=unifrnd(0,1);
for i=1:POP_SIZE
    r_x(i)=sqrt(w_1^2*(Obfunction1(Chr(1,j),
    Chr(2,j),Chr(3,j))
    -z1Max)^2+w_2^2*(Obfunction2(Chr(1,j),
    Chr(2,j),Chr(3,j))-z2Max)^2);
    t=(rMax-r_x(i)+rr)/(rMax-rMin+rr);
    eval=[eval,t];
end
temp=[];
qTemp=[];
qTemp=[qTemp,0];
q(1)=eval(1);
qTemp=[qTemp,q(1)];
for i=2:POP_SIZE
    q(i)=q(i-1)+eval(i);
    qTemp=[qTemp,q(i)];
end
for i=1:POP_SIZE
    r=unifrnd(0,q(POP_SIZE));
    for j=1:POP_SIZE
        if(r>=qTemp(j) && r<qTemp(j+1))
            temp=[temp,[Chr(1,j),Chr(2,j),
            Chr(3,j)]'];
        break;
    end
```

```
        end
    end

    for i=1:POP_SIZE
        for k=1:N
            Chr(k,i)=temp(k,i)
        end
    end
    pop=POP_SIZE/2;
    for i=1:pop
        if (unifrnd(0,1)>P_CROSSOVER) continue;
        end
        j=floor(unifrnd(1,POP_SIZE));
        jj=floor(unifrnd(1,POP_SIZE));
        r=unifrnd(0,1);
        for k=1:N
            x(k)=r*Chr(k,j)+(1-r)*Chr(k,jj);
            y(k)=r*Chr(k,jj)+(1-r)*Chr(k,j);
        end
        if(constraint_check(x(1),x(2),x(3))==1)
            for k=1:N
                Chr(k,j)=x(k)
            end
        end
        if(constraint_check(y(1),y(2),y(3))==1)
            for k=1:N
                Chr(k,jj)=y(k)
            end
        end
    end
    INFTY=10;
    precision=0.0001;
    for i=1:POP_SIZE
        if (unifrnd(0,1)>P_MUTATION) continue;
        end
        for k=1:N
            x(k)=Chr(k,i);
        end
        for k=1:N
            if(unifrnd(0,1)<0.5)
            direction(k)=unifrnd(-1,1);
            else direction(k)=0;
            end
        end
        infty=unifrnd(0,INFTY);
```

```
        while(infty>precision)
            for j=1:N
                y(j)=x(j)+infty*direction(j)
            end
            if(constraint_check(y(1),y(2),y(3))==1)
                for k=1:N
                    Chr(k,i)=y(k);
                end
                break;
            end
            infty=unifrnd(0,infty);
        end
    end
scatter(Obj(POP_SIZE),gen,8,'r*');
axis([0 GEN 0 50]);
title('Search
process');
legend('Best so far');
hold on; pause(0.005);
hold off;
toc end

function [t] = constraint_check(x1,x2,x3)
t=0;
rfs=0;
if((x1>=0)
&(x2>=0) & (x3>=0))
    if((x1+x2+x3<=10) & (3*x1+5*x2+3*x3>=4))
        t=1;
    end
end

function [Obfunction1] = Obfunction1(X,Y,Z)
N=20;
M=20;
Obfunction1=0;
v=[];
E=[];
for i=1:N
    T=0;
    x1=unifrnd(5,7);
    if 5<x1&&x1<6
        mu1=x1-5;
    end
    if 6<=x1&&x1<7
```

```
        mu1=7-x1;
    end

    x2=unifrnd(6.5,10);
    if 6.5<=x2&&x2<8
        mu2=2*(x2-6.5)/3;
    end
    if 8<=x2&&x2<=10
        mu2=0.5*(10-x2);
    end
    for j=1:M
        k1=normrnd(x1,2);
        k2=normrnd(x2,1);
        T=T+(3*k1^2*X-2*k1*k2*Y+1.3*k2^2*Z);
    end
    E=[E,T/M];
    v=[v,min(mu1,mu2)];
end
MIN=min(E);
MAX=max(E);
for k=1:N
    r=unifrnd(MIN,MAX);
    b1=0;
    b2=0;
    if r>=0
        for i=1:N
            if E(i)>=r&&b1<=v(i)
                b1=v(i);
            end
            if E(i)<r&&b2<=v(i)
                b2=v(i);
            end
        end
         Obfunction1=Obfunction1+(b1+1-b2)/2;
    else
        for i=1:N
            if E(i)<=r&&b1<v(i)
                b1=v(i);
            end
            if E(i)>r&&b2<v(i)
                b2=v(i);
            end
        end
        Obfunction1= Obfunction1-(b1+1-b2)/2;
    end
```

```
end if MIN<=0
    a=0;
else
    a=MIN;
end if MAX>=0
    b=0;
else
    b=MAX;
end
Obfunction1=Obfunction1*(MAX-MIN)/N+a+b;


function [Obfunction2] = Obfunction2(t1,t2,t3)
N=20;
M=20;
Obfunction2=0;
v=[];
E=[];
for i=1:N
    T=0;
    x3=unifrnd(4,6);
    if 4<=x3&&x3<5
        mu3=x3-4;
    end
    if 5<=x3&&x3<=6
        mu3=6-x3;
    end

    x4=unifrnd(5,8);
    if 5<=x4&&x4<7
        mu4=0.5*(x4-5);
    end
    if 7<=x4&&x4<=8
        mu4=8-x4;
    end
    for j=1:M
        T=T+2.5*(normrnd(x3,1.5))^2*t1+
        3*normrnd(x3,1.5)*normrnd(x4,2)*t2
        +5*(normrnd(x4,2))^2*t3;
    end
    E=[E,T/M];
    v=[v,min(mu3,mu4)];
end
MIN=min(E);
 MAX=max(E);
```

```
 for k=1:N
    r=unifrnd(MIN,MAX);
    b1=0;
    b2=0;
    if r>=0
        for i=1:N
            if E(i)>=r&&b1<=v(i)
                b1=v(i);
            end
            if E(i)<r&&b2<=v(i)
                b2=v(i);
            end
        end
         Obfunction2=Obfunction2+(b1+1-b2)/2;
    else
        for i=1:N
            if E(i)<=r&&b1<v(i)
                b1=v(i);
            end
            if E(i)>r&&b2<v(i)
                b2=v(i);
            end
        end
        Obfunction2= Obfunction2-(b1+1-b2)/2;
    end
end if MIN<=0
    a=0;
else
    a=MIN;
end if MAX>=0
    b=0;
else
    b=MAX;
end Obfunction2=Obfunction2*(MAX-MIN)/N+a+b;
```

A.3   MATLAB® file for Example 3.4

```
N_jinji=1000;
N_ycss=1000;
j=0;
while(j<1)
```

```
    x0=unifrnd(0,5);
    y0=unifrnd(0,5);
    t=constraint_check(x0,y0);
    if(t==1)
        j=j+1;
    end
end

Tlist_x=[x0];
Tlist_y=[y0];
for i=1:N_jinji
    j=0;
    h=5;
    z0=0;
    z1=0;
    z_ciyou=0;
    x_ciyou=0;
    y_ciyou=0;


    z0=x0^2+y0^2;
    for i=1:N_ycss
        r=unifrnd(-1,1);
        R=unifrnd(-1,1);
        x1=x0+r*h;
        if(x1<0)
            continue
        end
        y1=y0+R*h;
        if(y1<0)
            continue
        end
        if(x1+y1>5)
            continue
        end
        if(x1==find(Tlist_x) & y1==find(Tlist_y))
            continue;
        end
        z1=x1^2+y1^2;
        if(z1<z0)
            if(z1>z_ciyou)
                z_ciyou=z1;
                x_ciyou=x1;
              y_ciyou=y1;
            end
```

```
            continue
        elseif(z1==z0)
            Tlist_x=[Tlist_x,x1];
            Tlist_y=[Tlist_x,y1];
            continue
        else
            j=j+1;
            break;
        end
    end
    if(j==0)
        t=constraint_check(x_ciyou,y_ciyou);
        if(t==1)
            x0=x_ciyou;
            y0=y_ciyou;
        end
    end
    if(j==1)
        t=constraint_check(x1,y1);
        if(t==1)
            x0=x1;
            y0=y1;
            z0=z1;
            Tlist_x=[x0];
            Tlist_y=[y0];
        end
    end
    scatter(x_ciyou,y_ciyou,8,'bo')
    scatter(Tlist_x,Tlist_y,8,'r*');
    axis([0 5 0 5]);
    title('Search process');
    legend('Best so far', 'Sub-optimal so far');
    hold on;
    pause(0.005);
end
hold off;
Tlist_x
Tlist_y
z=x0^2+y0^2


function[t]=constraint_check(x1,x2)
t=0;
rrs=0;
if((x1>=0) & x2>=0))
```

```
    if(x1+x2<=5)
      rrs=rrsimulation(x1,x2);
         if(rrs<=7)
             t=1;
         end
    end
end

function [rrsimulation]=rrsimulation(x1,x2)
Mcount=100;
Racount=100;
E_1=0;
E_2=0;
 lambda_1=0;
  lambda_2=0;
  rrsimulation=0;

for i=1:Mcount
    a=0;
    b=0;
    lambda_1=unifrnd(1,2);
    for i=1:Racount
        a=a+((x1-normrnd(lambda_1,1))^2
        +(x2-normrnd(lambda_1,1))^2)^0.5;
    end
    E_1=E_1+a/Racount;

    lambda_2=unifrnd(0,3);
    for i=1:Racount
        b=b+((x1-normrnd(lambda_2,1))^2
        +(x2-normrnd(lambda_2,1))^2)^0.5;
    end
    E_2=E_2+b/Racount;
end
rrsimulation=(E_1+E_2)/(2*Mcount);
```

A.4   MATLAB® file for Example 4.4

```
clc;
clear all;
clf;
```

```
figure(1);
c1=1.4962;
c2=1.4962;
w=0.7298;
MaxDT=100;
N=10;
P=[];
v=[];
pbest=[];
Gmax=0;
p=[];
con=0;

tic
j=1;
while(j<=N)
    x0=unifrnd(0,5);
    y0=unifrnd(0,5);
    t=constraint_check(x0,y0);
    if(t==1)
        P=[P,[x0,y0]'];
        u=unifrnd(0,5);
        v=[v,u];
        j=j+1;
    end
end for Gen=1:MaxDT
    pbest=P;
    for i=1:N
        g=fitness(P(1,i),P(2,i));
        p=[p,g];
        if(p(i)>=Gmax)
            Gmax=p(i);
            gbest_1=P(1,i);
            gbest_2=P(2,i);
        end
        v(i)=w*v(i)+c1*unifrnd(0,1)*sqrt((pbest(1,i)-
        P(1,i))^2+(pbest(2,i)-P(2,i))^2)+c2*unifrnd(0,1)
        *sqrt((gbest_1-P(1,i))^2+(gbest_2-P(2,i))^2);
        t1=P(1,i)+v(i);
        t2=P(2,i)+v(i);
        con=constraint_check(t1,t2);
        if(con==0)
            continue;
        end
        P(1,i)=t1;
```

```
        P(2,i)=t2;
        t=fitness(P(1,i),P(2,i));
        if (t>p(i))
            pbest(1,i)=P(1,i);
            pbest(2,i)=P(2,i);
            p(i)=t;
        end
        if (t>Gmax)
            Gmax=t;
            gbest_1=P(1,i);
            gbest_2=P(2,i);
            T_1=Ob1(gbest_1,gbest_2);
            T_2=Ob2(gbest_1,gbest_2);
        end
    plot(Gen,t,'--b.');
    axis([0 MaxDT 0 8]);
    title('Searching process');
    legend('Best so far');
    hold on;
    pause(0.005);
    grid on;
    end
    plot(Gen,Gmax,'--r.');
    axis([0 MaxDT 0 8]);
    title('Searching process');
    legend('Fitness value','Best so far');
    hold on;
    pause(0.005);
    grid on;
    plot(Gen,T_1,'-b.');
    plot(Gen,T_2,'--r.');
    axis([0 MaxDT 0 10]);
    title('Searching process');
    legend('H2','H1');
    hold on;
    pause(0.005);
    grid on;
end
gbest_1
gbest_2
T_1
T_2
Gmax
toc
```

```
function [t] = constraint_check(x1,x2)
t=0;
 rrs=0;
 if((x1>=0) &(x2>=0))
    if(x1+x2<=5)
            t=1;
    end
end


function fitness=fitness(x1,x2);
w1=0.5;
w2=0.5;
fitness=w1*Ob1(x1,x2)+w2*Ob2(x1,x2);

function Ob1=Ob1(x1,x2)
N=50;
M=50;
EXP=0;
for i=1:N
    T=0;
    t1=normrnd(3,1);
    t2=normrnd(2,0.5);
    for i=1:M
        T=T+sqrt((normrnd(t1,1)-x1)^2
        +(normrnd(t2,1)-x2)^2);
    end
    EXP=EXP+T/M;
end Ob1=EXP/N;

function Ob2=Ob2(x1,x2)
N=20;
M=20;
EXP=0;
for i=1:N
    T=0;
    t1=normrnd(3,1);
    t2=normrnd(2,0.5);
    for i=1:M
        T=T+sqrt((normrnd(t1,1)+x1)^2
        +(normrnd(t2,1)+x2)^2);
    end
    EXP=EXP+T/M;
end
Ob2=EXP/N;
```

A.5 MATLAB® file for Example 5.13

```
tic
initial_temperature=20;
finish_temperature=1;
cooling_rate=0.95;
M=100;
k=1;
x=unifrnd(0,2*pi);
f0=sin(x);
T=initial_temperature;
MIN=0;

while T>=finish_temperature
    n=1;
    while n<M
        x=unifrnd(0,8*pi);
        f=sin(x);
        diff=f-f0;
        if diff<0
            f0=f;
        else
            if exp(-diff/T)>rand(1)
                f0=f;
            end
        end
        if MIN>f0
            MIN=f0;
        end
        n=n+1;
        plot(n,MIN);
        hold on;
    end
    fprintf(fp,'%3.6f ',f0);
    k=k+1;
    T=T*cooling_rate;
end
MIN
toc
```

Index

Milton Keynes UK
Ingram Content Group UK Ltd.
UKHW031139141024
449569UK00024B/1222